Nonlinear Stochastic Mechanics

International Union of Theoretical and Applied Mechanics

N. Bellomo · F. Casciati (Eds.)

Nonlinear Stochastic Mechanics

IUTAM Symposium, Turin, 1991

Springer-Verlag
Berlin Heidelberg New York
London Paris Tokyo
Hong Kong Barcelona Budapest

Prof. Nicola Bellomo

Dept. of Mathematics
Politecnico di Torino
Corso Duca degli Abruzzi 24
I-10129 Torino
Italy

Prof. Fabio Casciati

Dept. of Structural Mechanics
University of Pavia
Via Abbiategrasso 211
I-27100 Pavia
Italy

ISBN-13: 978-3-642-84791-2 e-ISBN-13: 978-3-642-84789-9
DOI: 10.1007/978-3-642-84789-9

Softcover reprint of the hardcover 1st edition 1992

Typesetting: Camera ready by authors

61/3020-5 4 3 2 1 0 – Printed on acid-free paper

Scientific Committee

S.T.Ariaratnam, Canada
N.Bellomo, Italy (Co-Chairman)
P.Bernard, France
V.V. Bolotin, USSR
F.Casciati, Italy (Co-Chairman)
M.Hazewinkel, The Netherlands
R.N.Iyengar, India
F.Kozin +, USA
Y.K.Lin, USA
S.Nakagiri, Japan
J.B.Roberts, U.K.
W.Schiehlen, Germany
K.Sobczyk, Poland
F.Ziegler, Austria

Local Organizing Committee

I.Bonzani, Polytechnic of Torino
P.Colombi, University of Pavia
G.Comini, University of Pavia
M.Pandolfi-Bianchi, Polytechnic of Torino
A.Repaci, Polytechnic of Torino

Sponsors

- International Union of Theoretical & Applied Mechanics (IUTAM);
- Italian Association of Theoretical and Applied Mechanics (AIMETA): Group of Stochastic Mechanics;
- National Council for the Research (CNR): Mathematical Committee, Technological Committee, National Group of Earthquake Mitigation;
- Cassa Risparmio Torino (CRT);
- Minister for University and Scientific and Technological Research (MURST);
- Polytechnic of of Torino, Department of Mathematics;
- University of Pavia, Department of Structural Mechanics;

Editor's Preface

The Symposium, held in Torino (ISI, Villa Gualino) July 1-5, 1991 is the sixth of a series of IUTAM-Symposia on the application of stochastic analysis to continuum and discrete mechanics. The previous one, held in Innsbruck (1987), was mainly concentrated on qualitative and quantitative analysis of stochastic dynamical systems as well as on bifurcation and transition to chaos of deterministic systems.

This Symposium concentrated on fundamental aspects (stochastic analysis and mathematical methods), on specific applications in various branches of mechanics, engineering and applied sciences as well as on related fields as analysis of large systems, system identification, earthquake prediction. Numerical methods suitable to provide quantitative results, say stochastic finite elements, approximation of probability distribution and direct integration of differential equations have also been the object of interesting presentations.

Specific topics of the sessions have been: Engineering Applications, Equivalent Linearization of Discrete Stochastic Systems, Fatigue and Life Estimation, Fluid Dynamics, Numerical Methods, Random Vibration, Reliability Analysis, Stochastic Differential Equations, System Identification, Stochastic Control.

We are indebted to the IUTAM Bureau for having promoted and sponsored this Symposium and the Scientific Committee for having collaborated to the selection of participants and lecturers as well as to a prompt reviewing of the papers submitted for publication into these proceedings. A special thank is due to Frank Kozin: the organization of this meeting was for him very important; he missed the meeting but his organizer ability was present.

In addition, we need expressing our gratitude to the participants who have contributed to enlight the Symposium with excellent scientific contributions and with stimulating discussions during the variuos sessions. These alive and qualified participation certainly contributed to make the meeting a successful one.

N.Bellomo F.Casciati

Address of the Secretary-General of IUTAM

Mr. Chairman,
Ladies and gentlemen,

On behalf of the International Union of Theoretical and Applied Mechanics I would like to welcome you at the Symposium on Nonlinear Stochastic Mechanics.

The proposal to arrange this Symposium was accepted in August 1988 by the General Assembly of IUTAM. Since that time both chairmen, Professor Bellomo and Professor Casciati have done an excellent job preparing such an outstanding scientific programme. The chairmen have been advised by the Scientific Committee in selecting participants and papers.

The International Union of Theoretical and Applied Mechanics appoints the members of the Scientific Committee with great care. And, therefore, I will especially call Professors

S. T. Ariaratnam from Canada,
P. Bernard from France,
V. V. Bolotin from USSR,
M. Hazewinkel from The Netherlands,
R. N. Iyengar from India,
Y. K. Lin from USA,
S. Nakagiri from Japan
J. B. Roberts from UK,
K. Sobczyk from Poland,
F. Ziegler from Austria.

However, one member of the Scientific Committee is no longer with us.

Professor Frank Kozin from the United States of America died in spring 1990. He served from the early beginning in 1988 as a very active member of the Scientific Committee and participated in the first Committee meeting during December 1989 in person, too. Thus, his ideas have still great influence on this Symposium. You have already honoured him earlier this morning, I thank you very much for this remembrance of a great scientist.

Italy is a member of the International Union of Theoretical and Applied Mechanics since 1949. The first IUTAM Symposium in this country took place as early as 1950 followed by three others. The following topics were treated:

1950 in Pallanza, Lago Maggiore:
General Mechanics.

1960 in Varenna:
Cosmical Gas Dynamics IV

1982 in Torino:
Modern Developments in Analytical Mechanics

1985 in Udine:
Dynamics of Multibody Systems

Thus, we are opening today the second IUTAM Symposium in Torino, in field of Solid mechanics, too.

The science of stochastic mechanics has attracted the consideration of IUTAM since more than one decade. The first symposium in the field was held in

1976 in Southampton, England

Then, it followed

1982 the Symposium in Frankfurt/Oder

and

1987 the Innsbruck Symposium under
the chairmenship of the Professors
Schueller and Ziegler.

(Two other closely related symposia were held in 1972 in Coventry, England on Stability of Stochastic Dynamical Systems and 1984 in Stockholm, Sweden on Probabilistic Methods in Mechanics of Solids and Structures).

And, I am sure, additional symposia will be organized by IUTAM in the years to come.

In addition, it has to be mentioned that in the field of Nonlinear Dynamics stochastic methods are getting more important, too.

Another trend is the strong impact of computers on mechanics, both theoretical and applied. The development of computational methods should, however, not be understood as a new branch of science. There is a real large amount of knowledge on mechanical material and mechanical systems available that has to be included in computational methods. Therefore, the Bureau of IUTAM feels that computers should support mechanical sciences and not vice versa.

Ladies and gentlemen, enjoy the honour and pleasure attending a IUTAM Symposium, which means a first class symposium.

Thank you.

Werner Schiehlen

Final Address at the Closing Ceremony

Ladies and gentlemen,

As a member of the Bureau let me at first say a few words about the philosophy of the International Union of Theoretical and Applied Mechanics:

The roofing society was founded in the twenties to enhance the exchange of scientific ideas and to promote the personal understandings among the international group of scientists. The IUTAM-symposia are an outcome of that philosophy and, in addition, are designed to promote a more narrow field of mechanics and to define the current state-of-the-art. Participation is by invitation only. What about our special Stochastics Conference, the 6th in a row? Well, that field is not narrow in the measures of mechanics even if you exclude linear problems.

What has been deterministic is the planning: the program, including 15 sessions had not to be changed. All speakers, counted in mathematical terms, ie exempt a very few one, showed up. The session titles give a good indication of the flavor of this symposium: Identification, associated with inverse problems, Equivalent Linearization, Reliability, Fatigue, Stochastic Differential Equations and, a completely new invention that to my charging was especially fruitful namely Engineering Applications, a little excursion into Fluid Dynamics, Control, Random Vibrations (in a more explicit form) and Numerical Methods.

It is too early to examine the quality of achievements of this symposium, I would like to leave this to the acceptance of the proceedings. Formally I can say, the symposium was a success, all goals of IUTAM have been met, lively discussions, a mixture of high ranking "older" scientists who feel young and beginners in our field who feel already mature.

What was random in the program: the very first beginning, the start of the morning sessions and the patience of the session chairmen. But thanks to the speakers we could close all the sessions in time to reach the splendid social program.

Some work remains to the chairmen. A report of the Symposium has to be prepared for the Annual Report of IUTAM. And the collected manuscripts will be edited as proceedings of the Symposium to be published by Springer-Verlag. I am convinced that Professor Bellomo and Professor Casciati will accomplish this work also in the same outstanding manner.

Before I close with the thanks of the IUTAM-bureau to the chairmen Professor Bellomo and Professor Casciati and to their staff as well as to all the contributors and participants let me put an invitation to you, to continue this kind of symposia with a good random flavor in 3 to 5 years from now. What would be needed is a carefully filed program proposal on a form that can be received from the secretary-general Professor Schiehlen, Stuttgart. Presumptive chairmen should contact collegues of high reputation from several countries to suggest a scientific committee. I can tell from personal experience it is a pleasure to work for such a symposium in the field of random processes.

Have a good journey back home and we thank you all again for your participation.

Franz Ziegler
Member of the IUTAM-Bureau

Table of Contents

Statistical Linearization and Large Excitation of Nonlinear Stochastic Mechanical Systems

Nguyen Dong Anh, Ralf Krause and Werner Schiehlen

Institute B of Mechanics, University of Stuttgart, Pfaffenwaldring 9,
7000 Stuttgart 80, FRG

Summary

In mechanical systems nonlinear effects due to cubic stiffness and Coulomb friction are often observed. The behavior of such systems is analysed for different colored noise excitations, particularly the softening Duffing oscillator. The statistical linearization is used for obtaining mean square responses and then the mean square jump phenomenon is discussed. It is shown that the jumps can occur in the Duffing oscillator with softening stiffness. It is also shown that the softening oscillator does not exhibit stationary response for some range of the excitation bandwidth. Moreover, in this range the softening system may exhibit a nonstationary response increasing to infinity with the time. In the case of stationary responses the agreement between simulation and statistical linearization results is very good. The response of a single-degree-of-freedom spring-mass system with viscous and Coulomb friction with colored noise excitation using the technique of statistical linearization is also discussed. Further a good agreement between the simulation and analytical results is observed.

1 Introduction

A linearly damped system with cubic nonlinear stiffness known as Duffing oscillator is often used to describe mechanical structures undergoing large deflections, Kauderer /1/. Such nonlinear systems may exhibit jump phenomena that can not occur in linear systems. The existence of mean square jump phenomena in Duffing oscillator under narrow-band random excitation has been first observed by Lyon et. al. /2/.

In some applications a restoring term is found to be of the softening kind. Examples occur in the equations of motion of a pendulum undergoing large amplitude oscillations or a ship executing large amplitude rolling motions. A softening Duffing oscillator,

N. Bellomo · F. Casciati (Eds.)
Nonlinear Stochastic Mechanics
IUTAM Symposium Turin

excited by white noise with nonzero mean has been considered in Roberts, Spanos /3/. It was shown that for a given mean value of the excitation, positive solutions for mean square response obtained by statistical linearization exist only if the nonlinearity parameter is greater than a critical negative value. It was mentioned, however that, a strictly stationary solution may not exist. Thus, it is important to obtain a sufficient condition of existence of a stationary response of the softening Duffing oscillator.

The application of statistical linerization, Lin /4/, Müller, Popp /5/ and Anh /6/, to a system with softening characteristic requires that it is always stable. First of all, the statistical linearization can only yield a stationary solution, provided that the 'equivalent' linear natural frequency is positive.

The influence of Coulomb friction on oscillating systems is also a question of interest in many fields of engineering. Coulomb friction arises, e.g., when a solid slides on a dry surface. The behavior of such a system driven by colored noise of order two will be shown, too.

2 Duffing Oscillator

The Duffing oscillator describes a linearly damped, single–degree–of–freedom system with cubic nonlinear stiffness. The resulting forced vibrations are found from the following equation

$$\ddot{x} + 2\xi\omega_n\dot{x} + \omega_n^2\,(1 + \gamma\, x^2\,)x = f(t) \tag{2.1}$$

where $f(t)$ is an external excitation of colored noise, ξ is the damping ratio and ω_n is the undamped natural frequency of the corresponding linear system $(\gamma = 0)$. The nonlinearity parmeter γ can be either positive (hardening Duffing oscillator) or negative (softening Duffing oscillator), respectively. A softening Duffing oscillator excited by white noise with non–zero mean has been considered by Roberts /3/. The application of the statistical linearization to the Duffing system with softening characteristic requires that it is always stable. The plot of the restoring force $g(x) = \omega_n^2\,(1 + \gamma\, x^2\,)x$ and the potential energy $U(x) = \omega_n^2\,(\frac{x^2}{2} + \frac{\gamma}{4}x^4\,)$ are shown in Fig. 1. There is a 'potential well' centered at the position of the stable, static equilibrum, $x = 0$. As x increases from zero, $(x > 0)$ the restoring force reaches an absolute maximum value at $\bar{x} = \frac{1}{\sqrt{-3\gamma}}$. When the oscillator is randomly excited the response will eventually 'escape' from the potential well. However, escaping is not possible if x does not exceeds $\bar{x}$, since the restoring force progressively increases with displacement x. If x exceed $\bar{x}$, or falls below $-\bar{x}$, escape is possible, since the restoring force is decreasing with the displacement.

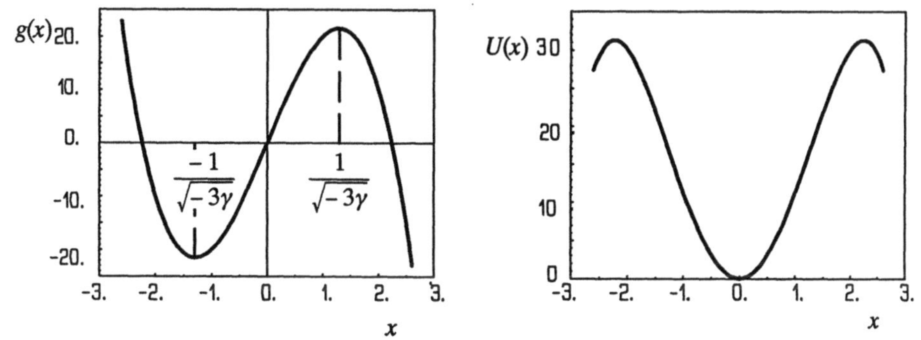

Fig. 1: Restoring force g(x) and potential energy U(x) for negative value of γ.

Then, in order to get a stationary response it should be required that the probability

$$P\left\{ \frac{-1}{\sqrt{-3\gamma}} < x < \frac{1}{\sqrt{-3\gamma}} \right\} \rightarrow 1 \tag{2.2}$$

or $$P\{ 1 + 3\gamma x^2 > 0 \} \rightarrow 1 \tag{2.3}$$

where $P(x)$ is the probability of the event x. If x is assumed to be a normal distributed process as by the statistical linearization method one gets

$$P\{ 1 + 3\gamma x^2 > 0 \} = \operatorname{erf}\left[\frac{1}{\sqrt{-6\gamma}\,\sigma_x} \right] , \tag{2.4}$$

where σ_x is the standard deviation of process x and erf( ) is the error function. The condition (2.3) is sufficient to keep the considered softening system stable. In particular, the condition (2.3) results in

$$E\{ 1 + 3\gamma x^2 \} > 0 \tag{2.5}$$

or $$1 + 3\gamma\sigma_x^2 > 0 . \tag{2.6}$$

It will be shown later that condition (2.6) guarantes a positive 'equivalent' linear natural frequency. In fact, using the statistical linearization method, eq. (2.1) is approximated by the linear equation

$$\ddot{x}_o + \beta_e \dot{x}_o + \omega_e^2 x_o + b = f(t) \tag{2.7}$$

where $$x(t) = x_o(t) + m \quad , E\{x_o(t)\} = 0 . \tag{2.8}$$

The mean response m may be different from zero, due to the softening characteristic of the system. But by Anh et. al /7/ it is shown that $m = 0$. The parameters β_e, ω_e^2 and b are determined by minimizing the difference function

$$e = E\{ \omega^2 (x_o + \gamma x_o^3) + 2\zeta\omega_n \dot{x}_o - \beta_e \dot{x}_o - \omega_e^2 x_o - b \} . \tag{2.9}$$

Then one gets

$$\ddot{x} + 2\xi\omega_n \dot{x} + \omega_n^2 (1 + 3\gamma\sigma_x^2) x = f(t) . \tag{2.10}$$

Now, two cases of the external excitation f(t) will be considered: either a first order colored noise or a second order colored noise.

3 First Order Colored Noise Response

Let the excitation f(t) in (2.10) be a first order colored noise q(t). Then, one obtains

$$\begin{aligned} &\ddot{x} + 2\zeta\omega_n \dot{x} + \omega_n^2 (1 + \gamma x^2) x = q(t) \\ &\dot{q} + a\, q = \sigma_1 \sqrt{2\, a}\ \xi(t) . \end{aligned} \tag{3.1}$$

The equivalent linear equation (2.10) reads as

$$\ddot{x} + 2\zeta\omega_n \dot{x} + \omega_n^2 (1 + 3\gamma\, \sigma_x^2) x = q(t) . \tag{3.2}$$

The mean square σ_x^2 can be calculated by an integration of the power spectral density of (3.2) and the excitation q(t) about the frequency domain, Müller and Schiehlen /9/. One gets

$$\sigma_1^2 (a + 2\zeta\omega_n) - 2\zeta\omega_n^3\sigma_x^2(1 + 3\gamma\sigma_x^2)(a^2 + 2\zeta\omega_n a + \omega_n^2(1 + 3\gamma\sigma_x^2)) = 0 . \tag{3.3}$$

For the given parameters ζ, ω_n, γ, σ_1 and a the mean square value σ_x^2 can be found as a root of the third order equation (3.3) which may have one or three real solutions.

3.1 Softening Duffing Oscillator

For softening Duffing oscillator where the nonlinearity parameter γ is negative, the mean square value σ_x^2 can be found from (3.3), however, the additional condition (2.6) has to be considered. Then one gets for the critical value $\sigma_x^2 = \dfrac{1}{-3\gamma}$ and for $\sigma_x^2 \to \infty$ from (3.3)

$$f\left(\frac{1}{-3\gamma}\right) = -\sigma_1^2\,(a + 2\zeta\omega_n) < 0\ ,$$

$$f(\infty) \sim 18\zeta\omega_n^5\,\gamma^2\,(\infty)^3 > 0\ , \tag{3.4}$$

i. e. there is a positive value σ_{x3}^2: $\frac{-1}{3\gamma} < \sigma_{x3}^2 < \infty$ satisfying $f(\sigma_{x3}^2) = 0$.

Noting that $\gamma < 0$ it follows

$$3\gamma\sigma_{x3}^2 + 1 < 0\ . \tag{3.5}$$

Therefore, the solution σ_{x3}^2 does not satisfy condition (2.6) and must be rejected. The equation (3.3) can be rewritten in the form

$$\sigma_x^6 + A\sigma_x^4 + B\sigma_x^2 + C = 0 \tag{3.6}$$

where

$$A = \frac{(a^2 + 2\zeta\omega_n a + 2\omega_n^2)}{3\gamma\omega_n^2}\ ,$$

$$B = \frac{(a^2 + 2\zeta\omega_n a + 2\omega_n^2)}{9\gamma^2\omega_n^2}\ ,$$

$$C = \frac{-\sigma_1^2\,(a + 2\zeta\omega_n)}{18\zeta\gamma^2\omega_n^5}\ . \tag{3.7}$$

Eq. (3.6) has three real positive roots if $\gamma < 0$ and $D(a, \sigma_1, \zeta, \omega_n, \gamma) < 0$ where

$$D = \frac{q^2}{4} + \frac{p^3}{27}\ ,\quad q = \frac{2}{27}A^3 - \frac{1}{3}AB + C\ ,\quad p = \frac{-A^2}{3} + B\ . \tag{3.8}$$

From the three positive roots of eq. (3.6), the lowest one will be taken as the mean square of response, because of the probability (2.4) calculated for this solution will have a maximum value (i.e. it will be closer to one). Fig. 2 shows plots of σ_x^2 versus the bandwidth parameter a for $\zeta = 0.1$, $\sigma_1 = 1.5$, $\omega_n = 2$ and $\gamma = -0.2$. There are three values of σ_x^2 for $0 < a < a_1 = 0.38$ and $a_2 = 2.7 < a$. For $a_1 < a < a_2$ there is only one solution σ_{x3}^2, which does not fit (2.6). Then, the jump phenomenon of the softening Duffing oscillator under a first order colored noise excitation is as follows: The observation is started at a value of $a > a_2$ on the lower branch. As the value of a is reduced σ_x^2 slowly increases until $a = a_2$. As a decreases further, the stationary response of the softening system does not exist for

$a_1 < a < a_2$, however, the softening system exhibits a non–stationary response which increases to infinity when $t \rightarrow \infty$.

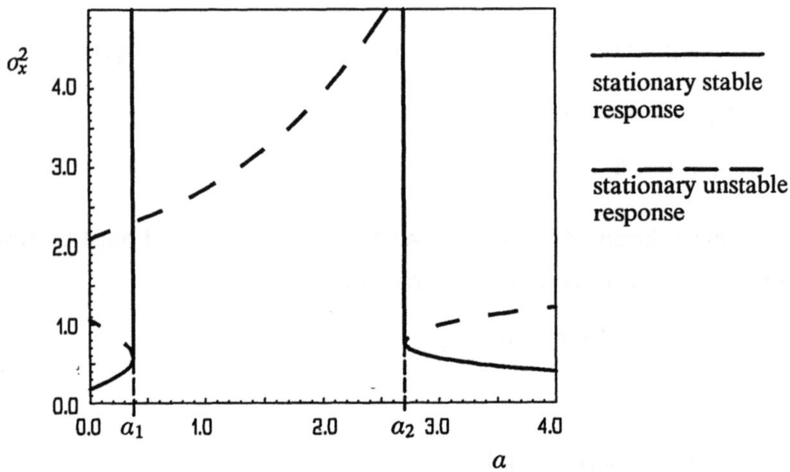

Fig. 2: Mean square σ_x^2 versus bandwidth parameter a .

Thus, the range $a_1 < a < a_2$ of the bandwidth parameter is in this sense 'dangereous' and the softening system has an infinite jump, while the harmonically excited softening Duffing system exhibits only a finite jump, see Lyon et. al. /2/. At the value $a = a_1$ the stationary response occurs again and as a is reduced from a_1 to zero, σ_x^2 slowly decreases on the lower branch. It should be noted, that the probabilistic condition (2.3) must be checked for every value of mean square response on the curve in Fig. 2.

In Fig. 3 is shown a comparison of analytical $\sigma^2_{St.L}$ and numerical results σ^2_{Sim} for $\omega_n = 5$, $\zeta = 0.1$, $\gamma = -0.2$. There is a very good agreement between both curves for low $\sigma_1 = 3.09$ and high excitation $\sigma_2 = 5.74$ intensity. For higher excitation becomes the numerical solution unstable for $2 < a < 5$. It can be shown with (2.3) that the probability P_{σ_2} for $\sigma_2 = 5.74$ is less than one, see also Fig. 3.

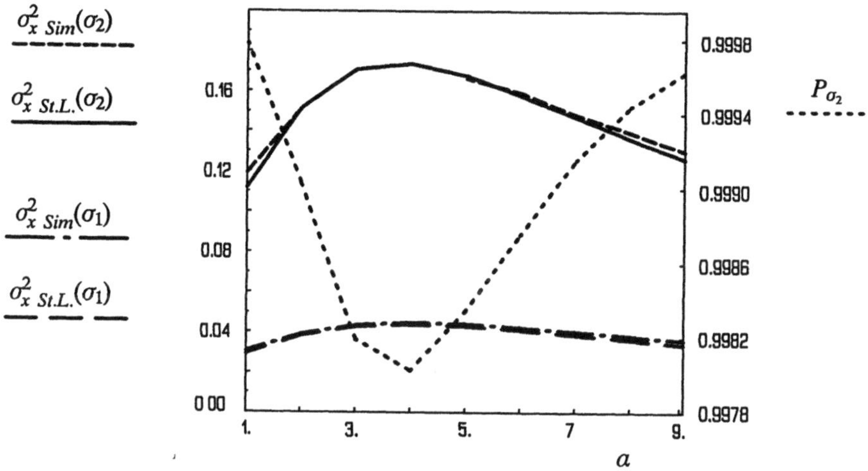

Fig. 3: Mean square $\sigma^2_{x\,\mathrm{Sim}}$ for simulation and $\sigma^2_{x\,\mathrm{St.L.}}$ for statistical linerization versus bandwidth parameter a for different excitation.

3.2 Second Order Colored Noise

Let f(t) be now a second order colored noise, resulting in the following system:

$$\ddot{x} + 2\zeta\omega_n\dot{x} + \omega_n^2\,(1 + \gamma\,x^2)x = p(t)\ , \tag{3.9}$$

$$\ddot{p} + 2h\dot{p} + (h^2 + \Omega^2)\,p = 2\sigma_2\sqrt{h(h^2 + \Omega^2)}\ \ \xi(t)\ . \tag{3.10}$$

The equivalent linear system takes the form

$$\ddot{x} + 2\zeta\omega_n\dot{x} + \omega_n^2\,(1 + 3\gamma\,\sigma_x^2)x = p(t)\ . \tag{3.11}$$

The investigation here extends the results present by Richard et. al. /8/, since the random excitation is different.

From (3.11) for the mean square value response one gets the following equation

$$\begin{aligned}\sigma_x^2 = \{\ &\sigma_2^2\,(h(h^2 + \Omega^2) + \zeta\omega_n^3(1 + 3\ \gamma\sigma_x^2) + 4h\zeta\omega_n(h + \zeta\omega_n))\ \}\\ &\{\zeta\omega_n^3(1 + 3\gamma\sigma_x^2)\ [\ (h^2 + \Omega^2 - \omega_n^2(1 + 3\gamma\sigma_x^2))^2 +\\ &+ 4(h + \zeta\omega_n)\,(h\omega_n^2(1 + 3\gamma\sigma_x^2) + \zeta\omega_n(h^2 + \Omega^2))\ ]\}^{-1}\ .\end{aligned} \tag{3.12}$$

Eq.(3.12) is a fourth order equation in σ_x^2. An examination of the signs of the terms of this equation shows for a hardening system ($\gamma > 0$) that out of the four roots, one must be real and negative, one must be real and positive and two may be real or complex conjugate. For a softening system ($\gamma < 0$) two or all four roots may be real and positive. By Anh et. al. /7/ it is shown that for a hardening Duffing oscillator jump phenomena can only occur in the case of narrow bandwidth excitation and not for wide bandwidth excitation. It is noted again that only the lowest positive root of eq. (3.12) will be taken for mean square of response.

The jump phenomena of the softening Duffing oscillator subject to second order colored noise excitation is shown in Fig. 4: The observation starts at a frequency $\Omega < \Omega_1$. As the value of Ω is reduced, σ_x^2 slowly increases until $\Omega = \Omega_1$. As Ω decreases further a jump at $\Omega = \Omega_1$ takes place with an related decrease in σ_x^2, after which σ_x^2 decreases further with decreasing Ω. As in the case of first order colored noise excitation, the probabilistic condition (2.3) must be checked for these results.

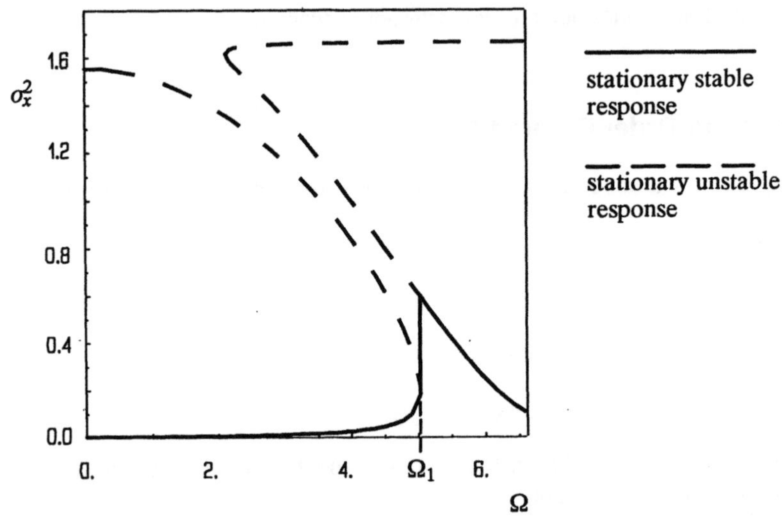

Fig. 4: Mean square σ_x^2 versus Ω^2 for softening Duffing oscillator.

For the softening Duffing oscillator with excitation of second order (3.10) the comparision does not show big differences between analytical investigations $\sigma_{St.L.}^2$ and numerical results σ_{Sim}^2 for $\zeta = 0.1$, $\omega_n = 5$, $\gamma = -0.2$ and $h = 0.1$, see Fig. 5. But again it is important to check the condition (2.3). For low excitation $\sigma_1 = 1.47$ is a stable response expected because the probability $P\{1 + 3\gamma x^2 > 0\} \rightarrow 1$. The numerical results show such stable behavior. For the large excitation $\sigma_2 = 4.58$ the probabilistic condition (2.3) for P decreases for $3 \leq \Omega, = 7$. Therefore the system becomes unstable and no numerical solution exists, Anh et. al. /7/.

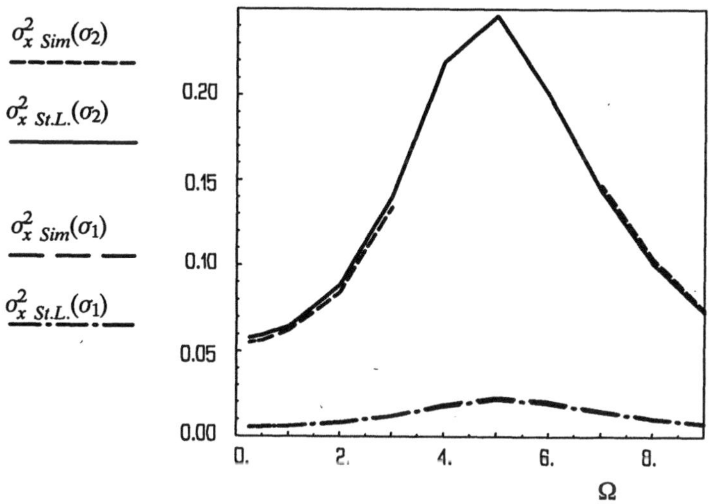

Fig. 5: Mean square $\sigma^2_{x\,Sim}$ for simulation and $\sigma^2_{x\,St.L.}$ for statistical linerization versus bandwidth parameter a for different excitation.

4 Coulomb Friction

Consider a system which consists of a random force excited mass with Coulomb friction, Levitan /10/. The equation of motion for the system has the form

$$\ddot{x} + 2h\dot{x} + \frac{R}{m}\text{sign}(\dot{x}) + \omega_n^2 x = p(t) \tag{4.1}$$

with the linear damping parameter d, the friction force R, natural frequency ω_n, the mass m and the excitation $p(t)$. It is assumed that the random excitation is sufficiently large and the mass will move continously and in the limit case $t \to \infty$ the sytem will exhibit a stationary response. In this case, application of the statistical linearization method to (4.1) is possible replacing it with the linear equation

$$\ddot{x} + 2\left\{h + \sqrt{\frac{1}{2\pi}\frac{R}{m\sigma_{\dot{x}}^2}}\right\}\dot{x} + \omega_n^2 x = p(t) \ . \tag{4.2}$$

4.1 Second Order Colored Noise Excitation

Consider the case where the random excitation in (4.1) is a second order colored noise p(t) as (3.10). In this case the corresponding equation (4.2) results in the following equations for $\sigma_{\dot{x}}^2$ and σ_x^2:

$$1 = \frac{\sigma_2^2\,(\overline{\alpha}^2 + \Omega^2)[\mu g + (\overline{\alpha} + h)\sigma_{\dot{x}}\,]}{(\mu g + h\sigma_{\dot{x}})\;B(\sigma_{\dot{x}})}\;, \tag{4.3}$$

$$\sigma_x^2 = \frac{\sigma_2^2\,\sigma_{\dot{x}}}{[\mu g + h\sigma_{\dot{x}}]\omega_n^2}\,\frac{A(\sigma_{\dot{x}})}{B(\sigma_{\dot{x}})} \tag{4.4}$$

where

$$\mu = \sqrt{\frac{1}{2\pi}\frac{R}{m}}\;,$$

$$\omega_n^2 = h^2 + \Omega^2\;,$$

$$\overline{\alpha} = \pi\frac{h\Omega^2}{h^2 + \Omega^2}\;,$$

$$\begin{aligned} A(\sigma_{\dot{x}}) &= \overline{\alpha}(\alpha^2 + \Omega^2)\,\sigma_{\dot{x}}^2 + \omega_n^2(\mu g + h\sigma_{\dot{x}})\sigma_{\dot{x}} + 4\overline{\alpha}(\mu g + h\sigma_{\dot{x}}) \\ &\quad (\mu g + (\overline{\alpha} + h)\sigma_{\dot{x}})\;, \\ B(\sigma_{\dot{x}}) &= (\overline{\alpha}^2 + \Omega^2 - \omega_n^2)^2\,\sigma_{\dot{x}}^2 + 4[\mu g + (\overline{\alpha} + h)\sigma_{\dot{x}}] \\ &\quad [\mu g(\overline{\alpha}^2 + \Omega^2) + (\overline{\alpha}\omega_n^2 + h(\overline{\alpha}^2\Omega^2))\sigma_{\dot{x}}]\;. \end{aligned} \tag{4.5}$$

The investigation of Anh et. al. /7/ shows that eq. (4.3) has one positive solution if

$$\sigma_2^2 > 4\mu^2 = \frac{2}{\pi}\left(\frac{R}{m}\right)^2\;. \tag{4.6}$$

In Fig. 6 is given a solution for $\sigma_2 = 3$, $h = 0.1$, $R/m = 0.2$ and the bandwidth parameter $\overline{\alpha} = 0.1$ or $\overline{\alpha} = 4.0$. It becomes clear, that the resonant effect of narrow band excitation is significant for small parameter $\overline{\alpha}$. Other investigations are made for different values of the parameters of the excitation and the system, the solutions of (4.3) and (4.4) are shown in Tab 1. There is also a comparison between the simulation $\sigma_{x\,Sim}^2$, $\sigma_{\dot{x}\,Sim}$ and analytical solutions $\sigma_{x\,St.L.}^2$, $\sigma_{\dot{x}\,St.L.}$ of system (4.1) subject to the second order colored noise excitation (3.10). The meansquares $\sigma_{\dot{x}}^2$ and σ_x^2 show a good agreement.

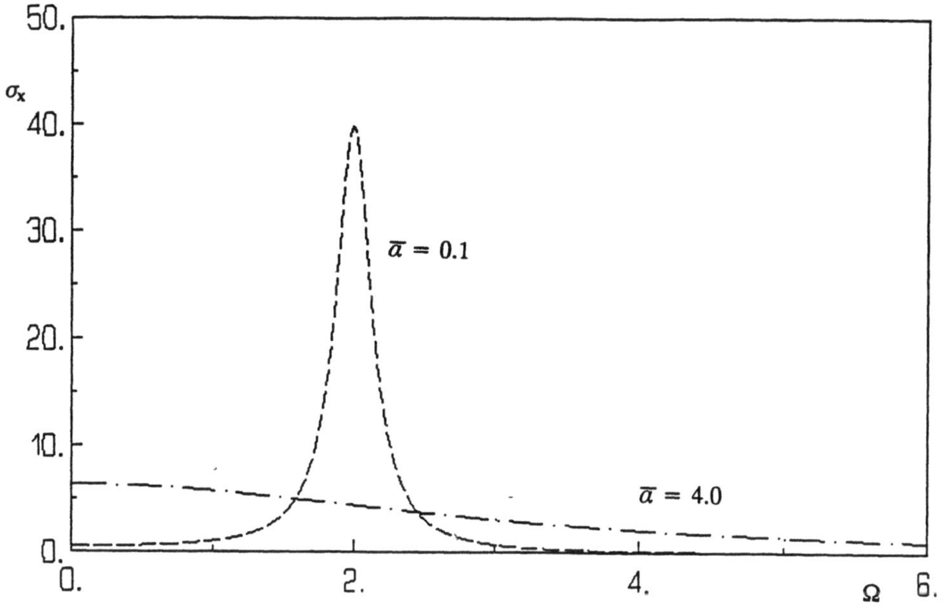

Fig. 6: Deviation σ_x versus excitation frequency Ω for $\overline{a} = 0.1$ and $\overline{a} = 4.0$

N	σ_2^2	$\overline{a}$	Ω	h	ω_n	$\sqrt{\frac{1}{2\pi}\frac{R}{m}}$	$\sigma^2_{\dot{x}\,Sim}$	$\sigma^2_{\dot{x}\,St.L.}$	$\sigma^2_{x\,Sim}$	$\sigma^2_{x\,St.L.}$
1	3.6466	0.5	3	0.5	16	0.1	.02575	.02607	6.E−5	6.1E−5
2	4.6201	0.1	3	0.5	16	0.1	.26568	.02670	7.5E−5	7.6E−5
3	4.6334	0.1	3	0.5	2	0.3	1.5118	1.557	0.3391	0.3950
4	"	0.1	3	0.5	2	0.5	1.4109	1.4504	0.2858	0.3255
5	"	0.1	3	0.5	2	0.8	1.2907	1.3202	0.2321	0.2553
6	"	0.1	3	0.5	2	1.0	1.2198	1.2410	0.2040	0.2192
7	"	0.1	3	0.5	2	1.5	0.9812	1.0297	0.1313	0.1443
8	"	0.1	3	0.5	2	1.86	0.8124	0.830	0.0893	0.0960

Tab. 1: Data of Simulation and Statistical Linearization.

5 Conclusion

The method of statistical linearization is applied for the analysis of the behavior of systems with cubic stiffness or Coulomb friction. For Duffing oscillators the mean square jump phenomenon is found by the analytical investigation. It is shown for the softening Duffing oscillator that in some cases of the parameters the formal application of the statistical linearization yields a mean square value of stationary response which does not exsist in reality. In these cases the presented probabilistic condition is a necessary but not sufficient condition of existence of a stationary response in Duffing oscillators. In cases where the Duffing system has a stationary response the statistical linearization technique gives good results of the mean square value of this stationary response.

In the case of a randomly excited system with Coulomb and viscous damping, the technique of statistical linearization provides an efficient tool for analysis of the stationary response. The comparison of analytical solutions with simulations show that also for a strong influence of the nonlinearity the method is very reliable.

6 References

1 Kauderer, H.: Nichlineare Mechanik. Springer–Verlag, Berlin, 1958.

2 Lyon, R. et al.: Response of a hard–spring oscillator to narrow band excitation. J.Acoust. Soc. Am., 1961, 33, pp.1404 - 1411.

3 Roberts, J.B., Spanos P.D.: Random vibration and statistical linearization. John Wiley & Sons, N. Y., 1990

4 Lin, Y. K.: Probabilistic theory of structural dynamics. Mc Graw– Hill, N. Y., 1967.

5 Müller, P. C., Popp, K., Schiehlen, W.: Berechnungsverfahren stochastischer Fahrzeugschwingungen. Ing.– Arch. 49, 1980, pp. 235–254.

6 Nguyen Dong Anh: Influence of different types of periodic and random perturbations on oscillating nonlinear mechanical system. Doctoral Thesis. Ukranian Acad. Sci., Institute of Math. Kiev , 1986.

7 N. D. Anh, R. Krause, W. Schiehlen: Statistical linerization and large excitation of nonlinear stochastic mechanical systems. Zwischenbericht 54. Institute B of Mechanics, University of Stuttgart, Stuttgart, FRG, 1990.

8 Richard, K.,Anand, G.V.: Nonlinear resonance in strings under narrow–band random excitation. J. Sound. Vib.,1983, 86, pp. 85–98.

9 Müller P. C., Schiehlen, W.: Linear Vibrations. Martinus Nyhoff Publ., 1985.

10 Levitan E. S.: Forced oscillation of a spring–mass system having combined Coulomb and viscous damping. J. of the Acoust. Soc. of America 32, pp 1265–1269 (1960).

On the Localization Phenomenon in Randomly Disordered Engineering Structures

S.T. Ariaratnam
Solid Mechanics Division
Faculty of Engineering, University of Waterloo
Waterloo, Ontario, Canada N2L 3G1
Wei-Chau Xie
CANDU Operations, Atomic Energy of Canada Limited
Mississauga, Ontario, Canada L5K 1B2

Abstract

In this paper, two examples of mono-coupled randomly disordered periodic structures, namely a large shear building and a long beam continuous over several supports are studied. The localization factor is related to the largest Lyapunov exponent of a product of random wave transfer matrices. By using a formula due to Furstenberg for the largest Lyapunov exponent of a product of independent and identically distributed random matrices, the localization factors are calculated by numerical simulation and compared with results obtained by a perturbation method.

1. Introduction

Many engineering structures are designed in the form of identical elements which are assembled end-to-end to form a spatially periodic structure, for example, long space antennae or periodic truss structures. These periodic structures behave like band pass filters. If energy dissipation due to damping is neglected, there is a significant distinction between frequency *pass-bands* where wave propagates without attenuation and frequency *stop-bands* where there is attenuation as the wave propagates (Brillouin, 1946).

However, due to defects in manufacture and assembly, no structure designed as a periodic structure can be perfectly periodic. Disorder can occur in the geometry of configurations and material properties of the structure. In disordered periodic structures, wave amplitudes with frequencies in the pass-bands will also be attenuated; the steady state response of the structure decays exponentially away from the source. It is therefore of importance to study the localization behaviour of disordered periodic structures and evaluate the *localization factors*, the exponential rates at which the amplitudes of waves propagating in the structures decay. As an application, the extent of damage that is spread in a structure due to defect or damage at some location can be estimated. The localization behaviour may also be used to serve as a "damping" mechanism in periodic structures, especially for structures in space where atmospheric damping is negligibly small. The first study of the localization phenomenon in the context of structural dynamics is due to Kissel(1988).

In this paper, the localization factor is related to the largest Lyapunov exponent of the structure. The formulation for evaluating the largest Lyapunov exponent for discrete random dynamical systems is applied to evaluate the localization factors of disordered periodic structures. The localization behaviour of a tall shear building and

N. Bellomo · F. Casciati (Eds.)
Nonlinear Stochastic Mechanics
IUTAM Symposium Turin

that of a long continuous beam over many supports are investigated to illustrate the method developed.

2. Wave transfer matrix

Consider an element numbered n in a periodic linear structure (Figure 1). The element is modelled by the transfer matrix, $\mathbf{T}$, which relates a state vector $\mathbf{x}_{n-1}$ on the left-side of element n to that on the right side, $\mathbf{x}_n$, by the linear transformation $\mathbf{x}_n = \mathbf{T}\mathbf{x}_{n-1}$. In linearly elastic engineering structures, the state vector $\mathbf{x}_n$ usually involves generalized displacements and generalized forces. The transfer matrix can be derived from the dynamic equations of motion of the structure by, possibly, the finite difference or finite element method. Since the dimension of the state vector is even, the transfer matrix is a square matrix of even dimension. For a perfectly periodic structure, each element of the structure is identical; therefore the transfer matrix for each element is the same. The state vector after n elements is related to that at the beginning by $\mathbf{x}_n = \mathbf{T}^n\mathbf{x}_0$.

In this paper, only one dimensional periodic structures are considered, in which the transfer matrices are of dimension 2. Such structures are called mono-coupled structures. According to the nature of the eigenvalues of the transfer matrices, the waves propagating in periodic structures are classified as travelling waves and attenuating waves, which occur in alternating frequency bands known as pass-bands and stop-bands, respectively.

The eigenvalues of the transfer matrix are complex and of the form $e^{\pm ik}$ when $|\mathrm{tr}(\mathbf{T})|<2$; the corresponding frequency is in the pass-band and the waves travel in the form of $e^{\pm ikn}$, where k is the real wave number, the positive sign indicating left-travelling waves and the negative sign right-travelling waves. On the other hand, if $|\mathrm{tr}(\mathbf{T})|>2$, the eigenvalues of the transfer matrix are real and of the form $e^{\pm\alpha}$ or $e^{\pm\alpha+i\pi}$ ($\alpha\in\mathbf{R}$); the corresponding frequency is in the stop-band and the wave amplitudes after traversing n elements are attenuated by the factor $e^{\pm\alpha n}$, in which the real exponent α implies non-travelling or attenuating waves.

When the periodic structure is randomly disordered due to variability in geometry, material and manufacturing conditions, the transfer matrix for each element is not the same, but is a function of the parameters of disorder. In this case, the state vector after n elements is related to the state vector at the beginning by $\mathbf{x}_n = \mathbf{T}_n\mathbf{T}_{n-1}\cdots\mathbf{T}_1\mathbf{x}_0$, where $\mathbf{T}_1, \mathbf{T}_2, \cdots, \mathbf{T}_n$ are *random* matrices.

The wave transfer matrix $\mathbf{W}_n$ is a linear transformation relating the left (or backward) and right (or forward) travelling wave amplitudes, $\overleftarrow{A}$ and $\overrightarrow{B}$, at two neighbouring sections of the element n of the structure (Figure 2):

$$\begin{Bmatrix} \overleftarrow{A}_n \\ \overrightarrow{B}_n \end{Bmatrix} = \mathbf{W}_n \begin{Bmatrix} \overleftarrow{A}_{n-1} \\ \overrightarrow{B}_{n-1} \end{Bmatrix}. \tag{2.1}$$

The matrix $\mathbf{W}_n$ is related to the transfer matrix $\mathbf{T}_n$ by the transformation

$\mathbf{W}_n = \mathbf{X}^{-1}\mathbf{T}_n\mathbf{X}$, the columns of the matrix $\mathbf{X}$ being the eigenvectors of the transfer matrix $\mathbf{T}$ of the perfectly periodic structure.

For a perfectly periodic structure, the wave transfer matrix is simply

$$\mathbf{W}_n = \mathbf{W} = \begin{bmatrix} e^{ik} & 0 \\ 0 & e^{-ik} \end{bmatrix}, \tag{2.2}$$

while for a disordered periodic structure it is of the form

$$\mathbf{W}_n = \begin{bmatrix} 1/t_n & -r_n/t_n \\ -(r_n/t_n)^* & (1/t_n)^* \end{bmatrix}. \tag{2.3}$$

In equation (2.3), t_n is the transmission coefficient, which is the complex amplitude of a wave emerging from the right of the nth element when a wave of amplitude unity is incident at the left; while r_n is the reflection coefficient, which is the complex amplitude of the reflected wave when a wave of amplitude unity is incident from the left.

When the periodic structure is disordered, after n elements the wave amplitudes are related to those at the beginning through a matrix in the form of a product of random matrices: $\mathbf{y}_n = \mathbf{C}_n\mathbf{y}_0$, where $\mathbf{C}_n = \prod_{j=1}^{n}\mathbf{W}_j$, $\mathbf{W}_j$, $j=1,2,\cdots$ being the wave transfer matrices of the individual elements.

In the following section, a formula is developed for evaluating the largest Lyapunov exponent associated with a product of random matrices, which will be seen to be the localization factor of the randomly disordered periodic structure.

3. Formula for evaluating the largest Lyapunov exponent

We define a Lyapunov exponent by

$$\lambda(\mathbf{y}_0) = \lim_{n\to\infty}\frac{1}{n}\log\|\mathbf{y}_n(\mathbf{y}_0)\|.$$

To establish a formula for $\lambda_{\max}$, the maximum Lyapunov exponent, it is necessary to make the assumption that the matrices $\mathbf{W}_n$ are stationary, ergodic and identically distributed. The asymptotic properties of the product of such random matrices were first studied by Furstenberg (1963). The following derivation is due to Khas'minskii (1967) and is applicable to matrices of any dimension.

Let $\mathbf{s}_n = \mathbf{y}_n/\|\mathbf{y}_n\|$. Then $\mathbf{s}_n = \mathbf{W}_n\mathbf{y}_{n-1}/\|\mathbf{W}_n\mathbf{y}_{n-1}\| = \mathbf{W}_n\mathbf{s}_{n-1}/\|\mathbf{W}_n\mathbf{s}_{n-1}\|$, $\|\mathbf{s}_n\| = 1$. Hence $\{\mathbf{s}_n, n=1,2,\cdots\}$ represents a Markov chain on the unit hypersphere $\|\mathbf{s}\| = 1$. Suppose that the chain is ergodic with invariant probability measure $\nu(d\mathbf{s})$. Letting $\rho_n = \log\|\mathbf{y}_n\|$, it is easily seen that $\rho_n = \rho_0 + \sum_{k=1}^{n}\log\|\mathbf{W}_k\mathbf{s}_{k-1}\|$. Since the matrices $\mathbf{W}_1, \mathbf{W}_2, \cdots, \mathbf{W}_n, \cdots$ are independent, the pair $(\mathbf{W}_n, \mathbf{s}_{n-1})$ also forms a Markov chain. Under sufficiently broad conditions (Khas'minskii, 1980) the chain satisfies the strong law of large numbers, and hence

$$\lim_{n\to\infty}\frac{\rho_n}{n} = \lim_{n\to\infty}\sum_{k=1}^{n}\frac{1}{n}\log||\mathbf{W}_k\mathbf{s}_{k-1}|| = E[\log||\mathbf{Ws}||],$$

with probability one (w.p.1), so that

$$\lambda = \lim_{n\to\infty}\frac{1}{n}\log||\mathbf{y}_n|| = \int\int\log||\mathbf{Ws}||P(d\mathbf{W})\nu(d\mathbf{s}), \tag{3.1}$$

where $P(d\mathbf{W})$ is the common probability distribution of the matrices $\mathbf{W}_1$, $\mathbf{W}_2$, $\cdots$ The exponent λ given by the above formula has been shown to correspond to the largest Lyapunov exponent λ_{max} w.p.1 (Oseledec, 1968). In order to obtain the probability measure $\nu(d\mathbf{s})$, the invariance of probability mass on the unit hypersphere $||\mathbf{s}||=1$ is employed. Referring to Figure 3, if the region R_1 is mapped to region R_2 by the transformation $\mathbf{W}$, then equating the probability masses on R_1 and R_2,

$$\int 1_{R_1}(\mathbf{s})\nu(d\mathbf{s}) = \int\int 1_{R_1}\left(\frac{\mathbf{Ws}}{||\mathbf{Ws}||}\right)P(d\mathbf{W})\nu(d\mathbf{s}), \tag{3.2}$$

where $1_R(\cdot)$ is the indicator function which takes the value unity when the argument lies on R and zero if it lies elsewhere. It is usually difficult to solve this integral equation for $\nu(d\mathbf{s})$ analytically even in the two-dimensional case when the hypersphere becomes the unit circle S^1. Approximate methods or numerical simulation techniques have to be employed.

4. Relationship between the largest Lyapunov exponent and the localization factor

In view of the difficulty of solving for the probability measure $\nu(d\mathbf{s})$, the discussion from now on will be restricted to mono-coupled, one-dimensional structures whose transfer matrices are of dimension 2. The matrix $\mathbf{C}_n = \prod_{j=1}^{n}\mathbf{W}_j$ is of the form

$$\mathbf{C}_n = \begin{bmatrix} 1/\tau_n & -\rho_n/\tau_n \\ -(\rho_n/\tau_n)^* & (1/\tau_n)^* \end{bmatrix}, \tag{4.1}$$

where τ_n, ρ_n, $(|\rho_n|<1)$, are the transmission and reflection coefficients, respectively of the n—element disordered structure. The eigenvalues of $\mathbf{C}_n^*\mathbf{C}_n$ are then

$$\sigma_{1,2}^2 = \frac{1}{|\tau_n|^2}(1\pm|\rho_n|)^2. \tag{4.2}$$

Hence, $\sigma_{max}=(1+|\rho_n|)/|\tau_n|$, $\sigma_{min}=(1-|\rho_n|)/|\tau_n|$. Now, by Rayleigh's principle, since $\mathbf{C}_n^*\mathbf{C}_n$ is real, symmetric and positive definite,

$$\sigma_{max}^2 \le \frac{\mathbf{y}_0^*\mathbf{C}_n^*\mathbf{C}_n\mathbf{y}_0}{||\mathbf{y}_0||^2} \le \sigma_{max}^2,$$

so that $\sigma_{min}||\mathbf{y}_0|| \le ||\mathbf{y}_n|| \le \sigma_{max}||\mathbf{y}_0||$. Hence,

$$\log\sigma_{min} + \log||\mathbf{y}_0|| \le \log||\mathbf{y}_n|| \le \log\sigma_{max} + \log||\mathbf{y}_0||.$$

Dividing by n and taking the limit as $n \to \infty$, we get

$$\lambda_{\max} = \lim_{n\to\infty} \frac{1}{n} \log \sigma_{\max} = -\lim_{n\to\infty} \frac{1}{n} \log |\tau_n|. \tag{4.3}$$

Therefore, $|\tau_n| \approx e^{-n\lambda_{\max}}$ as $n \to \infty$ Since $|\tau_n|^2$ is the ratio of transmitted energy to incident energy for the disordered structure, the amplitude of a wave travelling in the structure propagates according to $e^{-n\lambda_{\max}}$ as $n \to \infty$ w.p.1. Hence the largest Lyapunov exponent is in fact the localization factor.

5. Evaluation of localization factor for mono-coupled systems

Let $\boldsymbol{\alpha} = (\alpha_1, \alpha_2, \cdots)^T$ be the random vector parameter of disorder having mean value zero. Then the wave transfer matrix is a function of $\boldsymbol{\alpha}$, $\mathbf{W}(\boldsymbol{\alpha})$. In two-dimensions, a unit vector $\mathbf{s}$ is of the form $\mathbf{s} = \frac{1}{\sqrt{2}}(e^{i\phi}, e^{-i\phi})^T$ where, in view of $\mathbf{s}(\phi+\pi) = -\mathbf{s}(\phi)$, we may take $0 \leq \phi \leq \pi$; Then the formula (3.1) for $\lambda_{\max}$ may be written

$$\lambda_{\max} = \int\int \log\|\mathbf{W}(\alpha)\mathbf{s}(\phi)\| p(\alpha)\mu(\phi) d\alpha d\phi, \tag{5.1}$$

where $p(\boldsymbol{\alpha})$ is the probability density of $\boldsymbol{\alpha}$ and $\nu(d\mathbf{s}) = \mu(\phi)d\phi$. If, as a first-order approximation, $\mu(\phi)$ is taken to be the uniform density $1/\pi$, the formula (5.1) yields a first perturbation approximation to $\lambda_{\max}$ for small disorder (Kissel, 1988, Ariaratnam, 1990), which coincides with the formula

$$\lambda_{\max} \approx \frac{1}{2} \sum_j \sigma_j^2 \left[\frac{\partial^2}{\partial \alpha_j^2} \log \left| \frac{1}{t(\alpha)} \right| \right]_{\alpha_j = 0}, \tag{5.2}$$

where σ_j^2 is the variance of α_j, $t(\alpha)$ is the transmission coefficient of an element.

Kissel (1988), Cai and Lin (1990) have developed numerical schemes to evaluate λ using equation (4.3) by directly multiplying the random matrices. However, for one dimensional disordered periodic structures a more efficient method which utilizes equation (5.1) is as follows:

Suppose the wave transfer matrix given in equation (2.3) can be written as

$$\mathbf{W}_n = \begin{bmatrix} a_1 + ia_2 & b_1 + ib_2 \\ b_1 - ib_2 & a_1 - ia_2 \end{bmatrix}. \tag{5.3}$$

Under the operation of $\mathbf{W}_n$, a given unit vector $\mathbf{s}_n = \frac{1}{\sqrt{2}}\{e^{i\phi_n} \quad e^{-i\phi_n}\}^T$ becomes the vector

$$\hat{\mathbf{s}}_{n+1} = \mathbf{W}_n \mathbf{s}_n = \rho_{n+1} \frac{1}{\sqrt{2}}\{e^{i\phi_{n+1}} \quad e^{-i\phi_{n+1}}\}^T,$$

or, the new unit vector

$$\mathbf{s}_{n+1} = \frac{1}{\sqrt{2}}\{e^{i\phi_{n+1}} \quad e^{-i\phi_{n+1}}\}^T, \qquad \hat{\mathbf{s}}_{n+1} = \rho_{n+1}\mathbf{s}_{n+1}, \tag{5.4}$$

where

$$\rho_{n+1}^2 = [(a_1+b_1)\cos\phi_n - (a_2-b_2)\sin\phi_n]^2 + \\ + [(a_2+b_2)\cos\phi_n + (a_1-b_1)\sin\phi_n]^2 , \tag{5.5}$$

$$\phi_{n+1} = \tan^{-1}\Big[\frac{(a_2+b_2)\cos\phi_n + (a_1-b_1)\sin\phi_n}{(a_1+b_1)\cos\phi_n - (a_2-b_2)\sin\phi_n}\Big] . \tag{5.6}$$

Then, given an initial value ϕ_0, the iteration formula (5.6) can be applied to obtain the probability measure $\nu(ds)$ or the density $\mu(\phi)$ which, after several iterations, will become independent of the initial value ϕ_0 due to the ergodicity of the Markov chain $\{\phi_n, n{=}0,1,,2, \cdots\}$. Equation (3.10) can now be employed to evaluate λ numerically.

6. Examples of localization in mono-coupled engineering structures

Several examples can be found in engineering of nominally periodic structures such as long space antennae, tall shear buildings, long transmission pipes continuous over many supports, and railway tracks. In the following, two examples of periodic structures are studied in some detail.

6.1. Shear buildings

Many engineering structures are formed by connecting relatively stiff beams to flexible columns, some examples being tall shear buildings and long space antennae. These structures can all be modelled by the same mathematical model.

(i) Formulation

Consider a tall shear building as shown in Figure 4. The height of each story is h; the total flexural rigidity of all the columns on each story is ΣEI; the concentrated mass of each floor is m. The stiffness of each floor is considered much larger than the total flexural rigidity of all the columns on each story.

The elements of the shear building are chosen as shown in Figure 4. The transfer matrix for an element relates the displacements v and shear forces s at neighbouring floors

$$\begin{Bmatrix} v_n \\ s_n \end{Bmatrix} = \mathbf{T} \begin{Bmatrix} v_{n-1} \\ s_{n-1} \end{Bmatrix}, \quad \mathbf{T} = \begin{bmatrix} \cos\overline{\omega} & -\sin\overline{\omega}/\overline{\omega}^3 \\ \overline{\omega}^3 \sin\overline{\omega} & \cos\overline{\omega} \end{bmatrix},$$

where $\overline{\omega} = (mh^3/12\Sigma EI)^{1/2}\omega$.

The frequency of the pass-band is determined by the condition $|\mathrm{tr}(\mathbf{T})|<2$, i.e. $|\cos\overline{\omega}|<1$, which is satisfied by all frequencies except $\overline{\omega} = n\pi$, $n = 0,1,2, \cdots$. Hence, the stop-bands are actually "stop-points".

Suppose that the mass m of each floor is randomly disordered so that $m = m_0(1+\alpha)$, where m_0 is the mean mass and α is a zero mean random variable with variance σ^2. Then the transfer matrix of the disordered structure is

$$\mathbf{T}(\gamma) = \begin{bmatrix} \cos(\sqrt{\gamma}\,\overline{\omega}) & -\sin(\sqrt{\gamma}\,\overline{\omega})/(\sqrt{\gamma}\,\overline{\omega})^3 \\ (\sqrt{\gamma}\,\overline{\omega})^3 \sin(\sqrt{\gamma}\,\overline{\omega}) & \cos(\sqrt{\gamma}\,\overline{\omega}) \end{bmatrix}, \tag{6.1}$$

and first-row elements of the wave transfer matrix are

$$\frac{1}{t} = \cos(\sqrt{\gamma}\,\overline{\omega}) + i\,\frac{1}{2}\Big[(\sqrt{\gamma})^3 + \frac{1}{(\sqrt{\gamma})^3}\Big]\sin(\sqrt{\gamma}\,\overline{\omega})\,,$$

$$-\frac{r}{t} = i\,\frac{1}{2}\Big[(\sqrt{\gamma})^3 - \frac{1}{(\sqrt{\gamma})^3}\Big]\sin(\sqrt{\gamma}\,\overline{\omega})\,,$$

where $\gamma = 1 + \alpha$.

A first order approximation of the largest Lyapunov exponent, given by equation (5.2), is

$$\lambda \approx \frac{1}{2}\sigma^2\Big[\frac{d^2}{d\alpha^2}\log\Big|\frac{1}{t(\alpha)}\Big|\Big]_{\alpha=0}$$

$$= \frac{9}{8}\sigma^2\sin^2(\overline{\omega}), \qquad \overline{\omega} \neq n\pi, \;\; n = 0,1,2,\cdots. \tag{6.2}$$

In order to check the validity of this approximation, a computer simulation was performed.

Suppose that the random variable α has a normal distribution, i.e. $\alpha \sim N(0,\sigma^2)$. To obtain the probability density $\mu(\phi)$, the iteration equation (5.6) is employed, where

$$a_1(\alpha) = \cos(\sqrt{\alpha+1}\,\overline{\omega})\,, \qquad a_2(\alpha) = \frac{1}{2}\Big[(\sqrt{\alpha+1})^3 + \frac{1}{(\sqrt{\alpha+1})^3}\Big]\sin(\sqrt{\alpha+1}\,\overline{\omega})\,,$$

$$b_1(\alpha) = 0\,, \qquad b_2(\alpha) = \frac{1}{2}\Big[(\sqrt{\alpha+1})^3 - \frac{1}{(\sqrt{\alpha+1})^3}\Big]\sin(\sqrt{\alpha+1}\,\overline{\omega})\,.$$

Taking $\sigma = 0.1$ and the number of iterations $N = 10^6$, the probability density $\mu(\phi)$ was obtained at 500 discrete points $\mu(\phi_n)$, $\phi_n = n\pi/500$, $n = 0,1,\cdots,499$ in the interval $0 \leq \phi < \pi$. A typical plot of $\mu(\phi)$ as a function of the non-dimensional frequency $\overline{\omega}$ is shown in Figure 5. By varying $\overline{\omega}$, it was found that, except when $\overline{\omega}$ was very small (e.g. $\overline{\omega} = 0.2$), the discrepancy between the actual probability density $\mu(\phi)$ and the uniform distribution $\mu^0(\phi) = 1/\pi$ was very significant especially when $\overline{\omega}$ was near the ends of pass-bands.

Equation (5.1) is now employed to evaluate the largest Lyapunov exponent λ. The numerical results are plotted in Figure 6. For the purpose of comparison, the uniform distribution $\mu^0 = 1/\pi$ is also used in equation (5.1) to find λ. It can be seen clearly that when $\overline{\omega}$ is at the middle of pass-bands, these two results agree well, especially for small $\overline{\omega}$. However, near the ends of pass-bands, the two results differ significantly due to the discrepancy between the actual probability density $\mu(\phi)$ and the uniform distribution μ^0 as noted above.

6.2. Long continuous beam on simple supports

The dynamics and localization behaviour of a long continuous beam on several simple supports have been studied by Lin (1962), Lin and McDaniel (1969), Lin and Yang (1974), Kissel (1988) , and Cai and Lin (1990). Since the procedure of analysis is the same as that in shear buildings, this section will emphasize mainly the differences between the results obtained in the present study and those of the previous investigations.

Consider a long multi-span simply supported Euler-Bernoulli beam as shown in Figure 7. The length of each span is l; the mass per unit length is μ; the flexural rigidity is EI. The transfer matrix for a span relates the angles of rotation ϕ and non-dimensional bending moments at adjacent supports (Yang and Lin, 1975, Lin, 1976):

$$\begin{Bmatrix} \phi_n \\ (\frac{lM}{EI})_n \end{Bmatrix} = \mathbf{T} \begin{Bmatrix} \phi_{n-1} \\ (\frac{lM}{EI})_{n-1} \end{Bmatrix}, \qquad \mathbf{T} = \begin{bmatrix} \cos k & \beta/\sqrt{\bar{\omega}} \\ -\sqrt{\bar{\omega}}\sin^2 k/\beta & \cos k \end{bmatrix}, \tag{6.3}$$

where

$$\cos k = \frac{\sinh\sqrt{\bar{\omega}}\cos\sqrt{\bar{\omega}} - \cosh\sqrt{\bar{\omega}}\sin\sqrt{\bar{\omega}}}{\sinh\sqrt{\bar{\omega}} - \sin\sqrt{\bar{\omega}}}, \qquad \beta = \frac{C_4}{S_3},$$

$$S_3 = \frac{\sinh\sqrt{\bar{\omega}} - \sin\sqrt{\bar{\omega}}}{2}, \qquad C_4 = \frac{\cosh\sqrt{\bar{\omega}}\cos\sqrt{\bar{\omega}} - 1}{2},$$

and the non-dimensional frequency $\bar{\omega} = \omega\left(\mu l^4/EI\right)^{\frac{1}{2}}$. Suppose that the length l of each span is disordered and let $\eta = l/\langle l\rangle$, where $\langle l\rangle = l_0$. The elements of the wave transfer matrix are found to be

$$\frac{1}{t} = \cos(k\eta) + i\left[\frac{\beta(\eta)\sin k}{2\beta} + \frac{\beta\sin^2(k\eta)}{2\beta(\eta)\sin k}\right],$$

$$-\frac{r}{t} = i\left[\frac{\beta\sin^2(k\eta)}{2\beta(\eta)\sin k} - \frac{\beta(\eta)\sin k}{2\beta}\right].$$

In the above expression, whenever the argument η appears, it is implied that the underlying trigonometric and hyperbolic functions should have $\sqrt{\bar{\omega}}$ replaced by $\sqrt{\bar{\omega}}\,\eta$.

The pass-bands are determined by the condition $|\mathrm{tr}(\mathbf{T})| < 0$, i.e.

$$\frac{\sinh\sqrt{\bar{\omega}}\cos\sqrt{\bar{\omega}} - \cosh\sqrt{\bar{\omega}}\sin\sqrt{\bar{\omega}}}{\sinh\sqrt{\bar{\omega}} - \sin\sqrt{\bar{\omega}}} < 1\,. \tag{6.4}$$

Suppose that η is normally distributed with $\langle\eta\rangle = 1$, $\mathrm{var}\eta = \sigma$. We define a new random variable $\alpha = \eta - 1$, so that $\langle\alpha\rangle = 0$, $\langle\alpha^2\rangle = \sigma^2$. Equation (5.6) is used to find the probability density $\mu(\phi)$, where the coefficients are given by

$$a_1(\alpha) = \cos k(\alpha+1), \qquad a_2(\alpha) = \frac{\beta(\alpha+1)\sin k}{2\beta} + \frac{\beta\sin^2 k(\alpha+1)}{2\beta(\alpha+1)\sin k},$$

$$b_1(\alpha) = 0, \qquad b_2(\alpha) = -\frac{\beta(\alpha+1)\sin k}{2\beta} + \frac{\beta\sin^2 k(\alpha+1)}{2\beta(\alpha+1)\sin k},$$

Taking $\sigma = 0.1$ and the number of iterations $N = 10^7$, the probability density $\mu(\phi)$ was obtained at 500 discrete points $\mu(\phi_n)$, $\phi_n = n\pi/500$, $n = 0, 1, \cdots, 499$. A typical result for $\mu(\phi)$ is shown in Figure 8. Equation (5.1) is then applied to determine the largest Lyapunov exponent λ; the numerical results are plotted in Figure 9. The uniform distribution $\mu^0 = 1/\pi$ is also used in equation (5.1) to obtain λ. The result of first approximation is also plotted in Figure 9 for comparison.

Similar observations as those obtained in shear buildings can also be made. When

$\overline{\omega}$ is near the ends of pass-bands, the probability density $\mu(\phi)$ differs from the uniform distribution μ^0 significantly, which causes large discrepancies in the values of the largest Lyapunov exponent.

7. Conclusions

In this paper, the localization phenomenon in randomly disordered mono-coupled engineering structures has been investigated. The localization factor, which is the exponential rate of attenuation of amplitude of waves travelling in the disordered structure, was related to the largest Lyapunov exponent of the structure. Furstenberg's theorem on the asymptotic behaviour of a product of random matrices was employed to study the largest Lyapunov exponent. A perturbation method was applied to obtain a first approximation of the largest Lyapunov exponent. The validity of the approximate result was checked by digital computer simulation. It was found that in the middle of pass-bands, the approximate results agree well with those obtained by computer simulation. However, at the ends of pass-bands, the discrepancy between the two results was very significant, especially for high wave frequencies.

Acknowledgement

This research was supported by the Natural Sciences and Engineering Research Council of Canada through Grant No.A-1815.

References

Ariaratnam, S.T. (1990) Localization in randomly disordered periodic structures. *Proceedings of Workshop on Stochastic Structural Dynamics and Stability* (August 29, 1989, ed. G.I. Schuëller, Institute of Engineering Mechanics, University of Innsbruck, Austria), 41-55.

Brillouin, L. (1946) *Wave Propagation in Periodic Structures: Electric Filters and Crystal Lattices* (McGraw-Hill, New York; reprinted 1953 by Dover Publications, New York).

Cai, G.Q. and Lin, Y.K. (1990) Localization of wave propagation in disordered periodic structures. (Private communication).

Furstenberg, H. (1963) Noncommuting random products. *Transactions of the American Mathematical Society* **108**(3), 377-428.

Khas'minskii, R.Z. (1967) Necessary and sufficient conditions for the asymptotic stability of linear stochastic systems. *Theory of Probability and Its Applications* **12**, 144-147 (English translation).

Khas'minskii, R.Z. (1980) *Stochastic Stability of Differential Equations* (Sijthoff & Noordhoff) (English translation).

Kissel, G.I. (1988) *Localization in Disordered Periodic Structures* (PhD Thesis, Department of Aeronautics and Astronautics, Massachusetts Institute of Technology, USA).

Lin, Y.K. (1962) Free Vibration of continuous beams on elastic supports. *International Journal of Mechanical Sciences* **4**, 409-423.

Lin, Y.K. (1976) Random vibration of periodic and almost periodic structures. *Mechanics Today* **3**, 93-124.

Lin, Y.K. and McDaniel, T.J. (1969) Dynamics of beam-type periodic structures. *Journal of Engineering for Industry, ASME Transactions* **91**, 1133-1141.

Lin, Y.K. and Yang, J.N. (1974) Free vibration of a disordered periodic beam. *Journal of Applied Mechanics* **41**, 383-391.

Oseledec, Y.I. (1968) A multiplicative ergodic theorem. Lyapunov characteristic numbers for dynamical systems. *Transactions of the Moscow Mathematical Society* **19**, 197-231 (English translation).

Yang, J.N. and Lin, Y.K. (1975) Frequency response functions of a disordered periodic beam. *Journal of Sound and Vibration* **38**, 317-340.

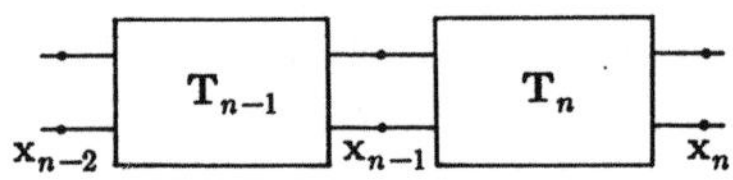

Fig. 1 Transfer matrix

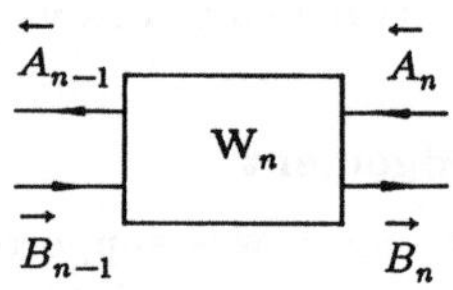

Fig. 2 Wave transfer matrix

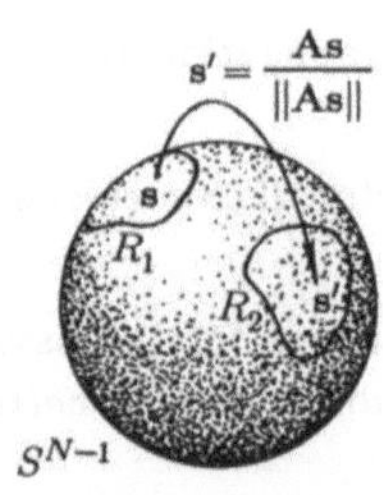

Fig. 3 Markov chain on unit hypersphere S^{N-1}

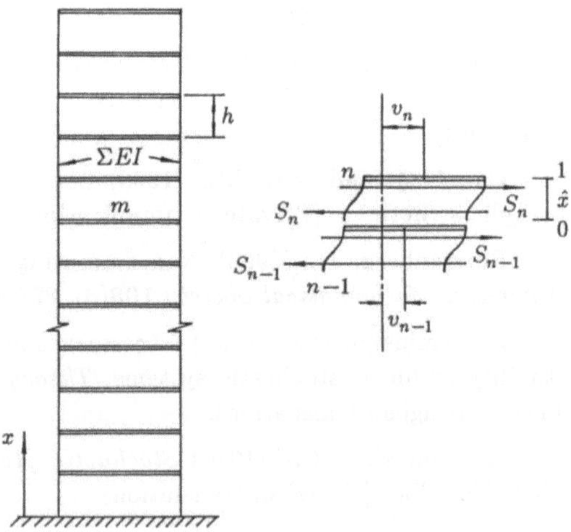

Fig. 4 Shear building

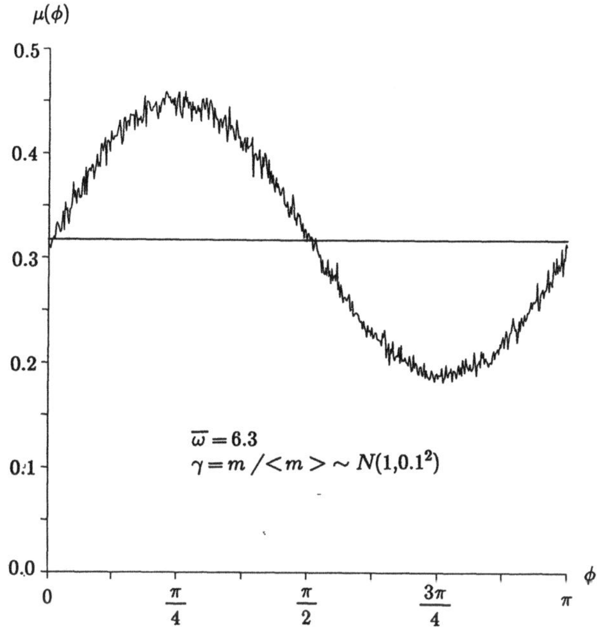

Fig. 5 Invariant probability density $\mu(\phi)$

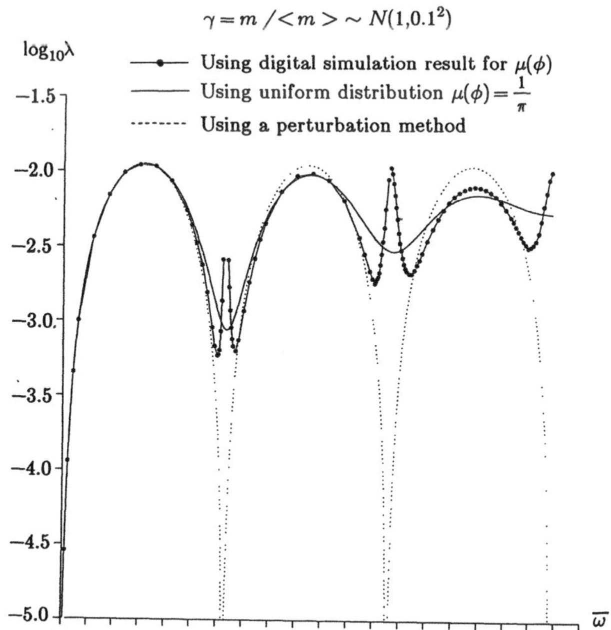

Fig. 6 Localization factor for shear building

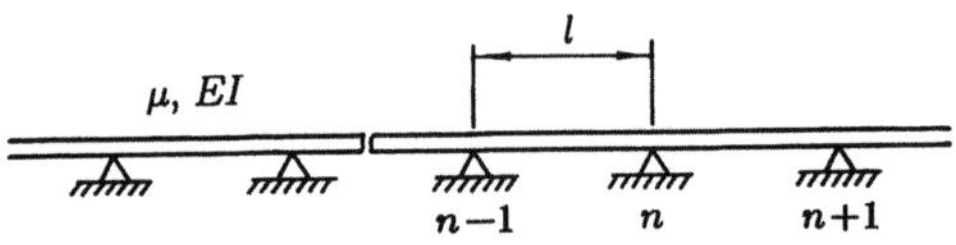

Fig. 7 Long continuous beam on simple supports

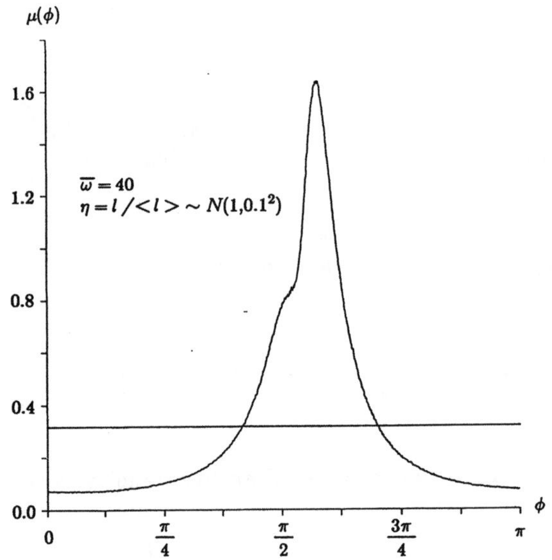

Fig. 8 Invariant probability density $\mu(\phi)$

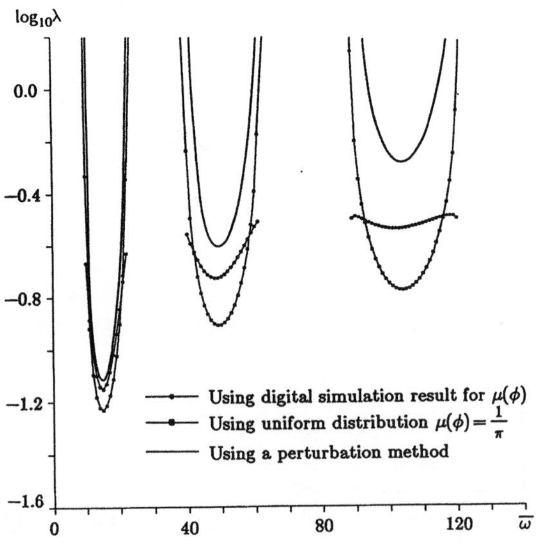

Fig. 9 Localization factor for continuous beam

Nonlinear Dynamic Analysis of Cooling Towers under Stochastic Wind Loading

G. Augusti[a], C. Borri[b], W. Zahlten[c]

[a] Dept. of Structural and Geotechnical Engrg, Università di Roma "La Sapienza", via Eudossiana 18, I-00184 Roma, ITALY
[b] Dept. of Civil Engrg, Università di Firenze, via di S. Marta 3, I-50139 Firenze, ITALY
[c] Inst. für Statik und Dynamik, Ruhr-Universität, Universitätsstr. 150, D-4630 Bochum, GERMANY

Abstract

A numerical model for simulating the nonlinear dynamic response of natural draught cooling towers to wind loading is introduced.

Geometrical and physical nonlinearities are considered in the F.E. modeling of the doubly-curved thin shell structure. An original simulation procedure of the artificial wind loading is used, generating directly the multi-correlated field of the (stochastic) time series of the external pressures. The correlation structure of the fluctuating pressure is imposed on the basis of experimental data obtained from wind tunnel tests.

Some preliminary analyses, carried out for checking the performance and reliability of the model, are finally reported.

1. INTRODUCTION

Wind loading is generally recognized as the most important external action for the design of cooling tower shells. Though by its nature a time-dependent process, it is treated by the engineering codes, e.g. [11], as quasi-static with a suitable amplification factor to take care of dynamic effects.

The coming harmonization of the European codes via the new Eurocodes has led to increased interest in the collapse behaviour of cooling towers. Questions of durability and safety tend to play a greater role in the future, and numerical simulations may help to gain insights into the mechanical behaviour.

The present paper reports on an ongoing research [1,2] that ultimately aims at comparing the quasi-static approach with a stochastic, time-dependent one, allowing a critical evaluation of the dynamic amplification factors.

2. MECHANICAL MODELING AND COMPUTATIONAL ASPECTS

The shell is modeled as a multi-layered continuum whose mechanical variables are related to those of a suitable reference surface, usually the middle surface. Its geometry is described by convective Gaussian coordinates, while a tensorial shell theory of finite rotations [3] is employed to model the kinematic behaviour. The material behaviour, on the other hand, is treated on the level of material points. The prime sources of physical nonlinearity considered here are tension cracking, crushing of concrete in compression, yielding of reinforcement and bond. The individual models are described in detail in [3], special attention is given to the strain-softening of concrete by implementing the plastic-fracturing theory of Bazant/Kim [4].

A tensor-oriented discretization strategy [13] of the principle of virtual work yields the element matrices, leading to the well known tangential equation of motion:

$$\mathbf{M}\cdot\ddot{\overset{+}{\mathbf{V}}}+\mathbf{C}\cdot\dot{\overset{+}{\mathbf{V}}}+\mathbf{K}_t\cdot\overset{+}{\mathbf{V}}=\mathbf{P}(t)-\mathbf{F}_m-\mathbf{F}_c-\mathbf{F}_\sigma \tag{1}$$

where $\mathbf{M}$, $\mathbf{C}$, $\mathbf{K}_t$ $\overset{+}{\mathbf{V}}$, and $\mathbf{P}(t)$ are respectively the mass, damping and tangent stiffness matrices, the incremental displacements and the vector of external loads; $\mathbf{F}_m$, $\mathbf{F}_c$ and

N. Bellomo · F. Casciati (Eds.)
Nonlinear Stochastic Mechanics
IUTAM Symposium Turin

$\mathbf{F}_\sigma$ the internal forces due to inertia, damping and deformation. For the present study, a four node element with bicubic shape functions and 48 DOF has been employed [14].

The damping matrix in (1) is computed as a combination of the mass and linear stiffness matrices (Rayleigh damping) and remains constant during the deformation process. The geometrical and physical nonlinearities are restricted to the tangent stiffness matrix. For the solution of (1) a classic approach has been chosen: time integration by the implicit Newmark algorithm in an incremental formulation [5] with Newton/Raphson iterations during each time step to attain equilibrium states.

3. SIMULATION OF THE WIND LOADING

Wind loading can be considered as the most important external excitation, which acts almost permanently during the life of a cooling tower.

After the sensational collapse of the Ferrybridge towers in 1965, intense studies showed that dynamic effects are comparable, in magnitude, to the static ones. Usually, the engineering methods developed since the beginning of the seventies for predicting the dynamic response to turbulent wind have been based on the approach in frequency domain. More recently, deterministic methods in Time Domain have been employed, paying increasing attention to the time series approach. These allow the investigation of the structural response, even in the case of significantly nonlinear behaviour. The main advantages of this approach can be summarized in a more accurate modeling of certain full-scale data and in relating the results to the theories governing the wind flow around a bluff body [10].

The simulation of the wind action is performed by means of artificial generation of the external pressure field, i. e. of a set of multi-correlated time series of the wind pressure acting on the structure; the auto- and cross-correlation structure must therefore be reproduced as closely as possible to the effective actual one. The adopted procedure can be considered as an extension to the case of shell structures of the techniques developed in [4,5], where the attention was mainly focused on line-like structures.

3.1 Wind action modeling

The wind action at time t in an arbitrary point $P_i = (z_i, \theta_i)$ of the structure is given by the local instantaneous pressure $p_{w,i}(t)$ that can be expressed as

$$p_{w,i}(t) = p_w(z_i,\theta_i,t) = \frac{1}{2} \cdot \rho_a \cdot \overline{C}_p(z_i,\theta_i) \cdot V^2(z_i,\theta_i,t) \tag{2}$$

where $\overline{C}_p(z_i,\theta_i)$ is the effective average pressure coefficient at the relevant point, $V(z_i,\theta_i,t)$ the relative velocity of the incident flow and ρ_a the mass density of the air. The contribution of the so-called "virtual mass" (inertia of the air masses involved in the motion of the structure) is neglected in (2), but can be easily taken into account.

Expressing $V(z_i,\theta_i,t)$ as the sum of a time-independent (or varying quasi-statically) average velocity $V_m(z_i)$ and a fluctuating component $v(z_i,\theta_i,t)$, one can easily derive

$$p_{w,i}(t) = \frac{1}{2} \cdot \rho_a \cdot \overline{C}_p(z_i,\theta_i) \cdot V_m^2(z_i) + \rho_a \cdot \overline{C}_p(z_i,\theta_i) \cdot V_m(z_i) \cdot v(z_i,\theta_i,t) \tag{3}$$

having neglected nonlinear terms in $v^2(z_i,\theta i_i,t)$. Therefore, the actual pressure is expressed as sum of a static component (time-independent) $\overline{p}_i$ plus a fluctuating part $p_i(t)$; the latter one is considered as a Gaussian, stationary random process with zero mean. Indicating by $\overline{q}(z_i)$ the mean kinetic pressure at height z_i

$$\overline{q}(z_i) = \frac{1}{2} \cdot \rho_a \cdot V_m^2(z_i) \tag{4}$$

eq. (3) becomes

$$p_{w,i}(t) = \overline{p}_i + p_i(t) = \overline{C}_p(z_i,\theta_i)\cdot\overline{q}(z_i) \; + \; 2\cdot\overline{C}_p(z_i,\theta_i)\cdot\overline{q}(z_i)\cdot\frac{v(z_i,\theta_i,t)}{V_m(z_i)} \tag{5}$$

3.2 Static action

The static component of the wind pressure, given by

$$\overline{p}_i = \overline{C}_p(z_i,\theta_i)\cdot\overline{q}(z_i) \tag{6}$$

is determined on the basis of the given average wind profile and requires the knowledge of the distribution of the mean pressure coefficients $\overline{C}_p$ for the whole structure. The $\overline{C}_p$ values depend strongly on the surface roughness: their negative negative maximum gets lower as the external roughness increases, as one can derive from the German Code VGB-BTR [11].

The values of $\overline{C}_p$ assumed for the present work have been taken from wind tunnel experiments [8,9].

As for the internal suction, the experimental tests mentioned above suggest a value constant in time and height, given by

$$\overline{p}_{int}(z_i) \;\approx\; -0.45\cdot\overline{q}_H \tag{7}$$

having indicated by $\overline{q}_H$ the kinetic pressure at the top of the tower: eq. (7) thus yields a corrective term which adds (algebraically) to eq. (6).

3.3 Modeling of the fluctuating action

The fluctuating component of the instantaneous pressure value can be expressed as

$$p_i(t) = 2\cdot\overline{C}_p(z_i,\theta_i)\cdot\overline{q}(z_i)\cdot\frac{v(z_i,\theta_i,t)}{V_m(z_i)} = 2\cdot\overline{p}_i\cdot\frac{v(z_i,\theta_i,t)}{V_m(z_i)} \tag{8}$$

The reproduction of the actual time-dependent distribution of the local pressure can be performed through the artificial generation of the time series of the fluctuating pressure at each arbitrary point P_i of the structure.

The basic idea of the proposed procedure is to initialize the simulation starting from the knowledge of the cross-covariance matrices $R^P_{ij}(\tau)$ between the values at point P_i and P_j. These can be written as

$$R^P_{ij}(\tau) \equiv R_p(z_i,\theta_i,z_j,\theta_j,\tau) = \sigma_p(z_i,\,\theta_i)\cdot\sigma_p(z_j,\theta_j)\cdot\rho(z_i,\theta_i,z_j,\theta_j,\tau) \qquad i,j=1,\ldots N \tag{9}$$

where $\sigma_p(z_i,\theta_i)$ and $\sigma_p(z_j,\,\theta_j)$ are respectively the RMS of the mean wind pressure at point i and j and $\rho(z_i,\theta_i,z_j,\theta_j,\tau)$ are the correlation coefficients at time lag τ. The values of the RMS can be easily evaluated as

$$\sigma^P_i \equiv \sigma_p(z_i,\theta_i) = 2\cdot I_u(z_i)\cdot\overline{q}(z_i)\cdot\overline{C}_p(z_i,0)\cdot c_\sigma(\theta_i) = 2\cdot I_u(z_i)\cdot\overline{p}(z_i,0)\cdot c_\sigma(\theta_i) \tag{10}$$

$\overline{C}_p(z_i,0)$ being the average pressure coefficients at the stagnation point, $c_\sigma(\theta_i)$ the RMS value of the normalized pressure, given by

$$c_\sigma(\theta_i) \equiv \frac{\sigma_p(z_i,\theta_i)}{\sigma_p(z_i,0)} \tag{11}$$

and $I_u(z_i)$ the turbulence intensity factor defined as

$$I_u(z_i) \equiv \frac{\sigma_v(z_i)}{V_m(z_i)} \tag{12}$$

The cross-covariance function is therefore determined once the correlation coefficients $\rho_{ij}(\tau)$ and $c_\sigma(\theta_i)$ for the whole structure are known at the time lags as needed by the generation algorithm described below. These data can be derived from direct measurements carried out on existing towers or (as in the case of this work) through wind tunnel tests. Fig. 1 shows the field of the cross-covariance function obtained by elaboration of the data measured by Kasperski [9] and employed for the present analysis: the covariance between two points P_i and P_j along the circumference at one level versus the angles i and k is plotted on the left of Fig. 1, while on the right the cross-covariance $R_{ij}(0)$ between points belonging to two different levels is shown. The $c_\sigma(\theta_i)$ values have been also derived from the above mentioned experimental tests carried out by the Ruhr-University at Bochum.

3.4 Artificial generation of the fluctuating pressure

The knowledge of the effective correlation structure allows the simulation of the fluctuating wind pressure via artificial generation of a set of multi-correlated time series.

Moreover, the direct generation of a pressure field presents two main advantages:

- the transformation of the wind velocity into wind pressure via "traditional" (and roughly approximated) pressure coefficients can be avoided;
- the simulation procedure becomes much more efficient; the integration of the spectra of the wind velocity (necessary to determine the correlation matrices) is avoided, together with all uncertainties due to the use of the spectra (proposed in the literature) which, usually, do not fit perfectly the real ones.

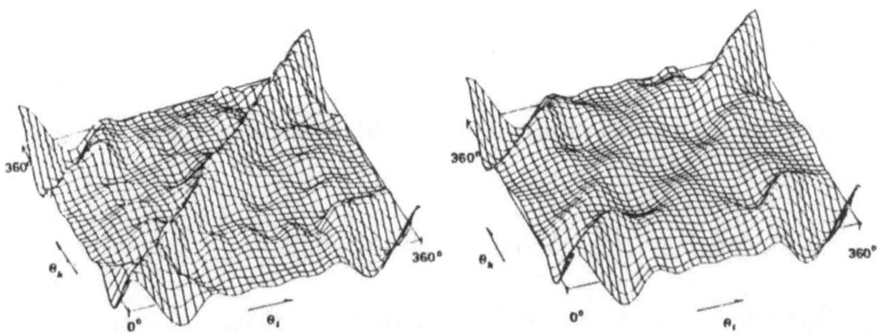

Fig. 1: Cross-covariance functions for one level (left) and two different levels (right)

The reliability of the numerical filtering techniques based on ARIMA (Auto Regressive Integrated Moving Average) models for simulating time processes of wind velocity and pressures has been investigated by Reed and Scanlan [10]. The performed statistical tests (autocorrelation and partial autocorrelation function of the estimated residuals, the 95% confidence intervals for the model parameters, values of the residual variance to provide goodness-of-fit information) demonstrated that AR(1) models are appropriate for velocity time histories with mean V_m less than 10 m/sec and AR(2) are more appropriate for higher values of V_m.

The analyses carried out on the pressure-differences show a dependence of the autoregressive model on the circumferential location (angle): AR(2) models give the best results in the windward and leeward regions, while AR(1) models are superior at about 100° from windward (transition).

The above considerations together with the need of uniformity in the generation procedures have thus led to the choice of AR(2) models for all circumferential regions, even if the simulation by models with different autoregressive order is possible with the adopted algorithm [6,7]. It has been modified for taking into account the effective correlation structure obtained from the above mentioned experimental tests. The procedure can be summarized as follows:

a) indicating with $p_i(t)$ the fluctuating pressure at time t for a given location, one can explicitly write the autoregressive filter as

$$p_i(t) = \sum_{k=1}^{\gamma} \sum_{j=1}^{N} \Phi_{ij}(k)\, p_j(t-k\Delta_t) + N_i(t), \qquad i = 1,..N \tag{13}$$

where γ is the order of the filter ($\gamma = 2$ in this case), $\mathbf{\Phi}(k) = [\Phi_{ij}(k)]$, $k=1,\ldots,\gamma$ are the so-called autoregressive matrices and $\mathbf{N}(t) = \{N_i(t)\}$ is the vector of the random shocks;

b) in a first step of the procedure the matrices $\mathbf{\Phi}(k)$ may be determined from the solution of the linear systems (hypersystem)

$$R^P_{im}(-l\Delta_t) = \sum_{k=1}^{\gamma} \sum_{j=1}^{N} \Phi_{ij}(k)\, R^P_{jm}[(k-l)\Delta_t)], \qquad l=1,\ldots,\gamma; \quad i,m=1,\ldots,N \tag{14}$$

being $\mathbf{R}^P(\tau)$ the known covariance matrix of the $p_i(t)$ processes at time lag τ. In the hypersystem of eq. (14) the autoregressive coefficients $\Phi_{ij}(k)$ are the only unknowns and can therefore be calculated immediately;

c) the second step consists in evaluating the cross-covariance matrix $R^N_{ij}(0)$ of the random shocks; this can be done writing

$$R^P_{il}(0) = \sum_{k=1}^{\gamma} \sum_{j=1}^{N} \Phi_{ij}(k)\, R^P_{jl}(k\Delta_t) + R^N_{il}(0), \qquad i,l=1,\ldots,N \tag{15}$$

and, after substitution of the autoregressive matrices computed from eq. (14) into eq. (15), determining the matrix $\mathbf{R}^N(0)$ for generating the processes $N_i(t)$ as linear combination of a set of uncorrelated random noises $\epsilon_j(t)$. In this way one can write:

$$N_i(t) = L_{ij} \cdot \epsilon_j(t) \tag{16}$$

where L_{ij} can be obtained from the factorization of $\mathbf{R}^N(0)$, so that $\mathbf{L} \times \mathbf{L}^T = \mathbf{R}^N(0)$. This factorization is possible only if the matrix is semi-definite positive: in case of positive definiteness of matrix $\mathbf{R}^N(0)$, $\mathbf{L}$ is its Choleski factor.

4. A SIMPLE NUMERICAL STUDY

An existing cooling tower (Fig. 2), to be found at Gersteinwerk, Germany, has been chosen to test the numerical procedure described in the foregoing. An extensive collapse simulation under quasi-static loading conditions has been performed in [12]. This numerical model, however, had to be simplified considerably in order to keep the computational effort within acceptable limits. The current discretization (Fig. 2) uses 8 elements along the circumference and 4 elements along the meridian, the uppermost very small one modeling the ring stiffener. The discretization has been refined in the windward region for a better consideration of the high variation of the external pressure. Six layers have been defined for the thickness.

Following section 3, the total pressure has been split into a static part p_i and a fluctuating part $p_i(t)$

$$p_{w,i}(t) = \bar{p}_i + p_i(t) \tag{17}$$

At the present moment the available data only cover the correlation structure along the circumference at the top level for time lag 0. The missing data are approximated as follows:

- a perfect correlation ($\rho_{ij}(\tau) = 1$, i,j with $\theta_i = \theta_j$) has been assumed along the meridians with the same correlation structure for each level;
- the correlation at time lag $\neq 0$ is assumed to be given by (γ is the order of the autoregressive filter, here $\gamma = 2$):

$$\rho_{ij}(k\Delta_t) = \frac{\rho_{ij}(0)}{k+1}, \qquad k=0,\ldots,\gamma \tag{18}$$

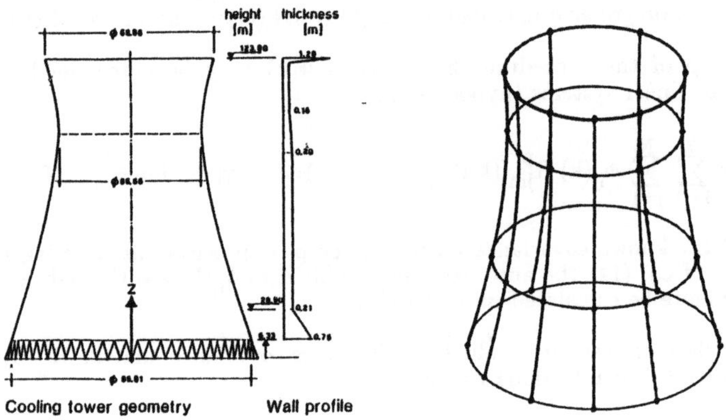

Fig. 2: Cooling tower at Gersteinwerk : structural geometry and F.E. model

The external pressure field is interpolated by bilinear shape functions within each element, ensuring C_0-continuity. Since the nodal values may vary independently in time, it is necessary to re-compute the energetically equivalent vector of nodal forces for each time lag anew.

Fig. 3: Static load-displacement curve of node 91

A nonlinear static analysis under the load history $g + \lambda \cdot w$ (g=dead weight, w=wind load, λ=load factor) has been performed first in order to analyze the collapse mechanism of the structure. A study of the load-displacement curve of node 91 (Fig. 3) reveals the following: the structure behaves linearly until extensive tension cracking sets in. Along the almost horizontal branch (crack plateau) the forces released by the cracked concrete are re-distributed to the reinforcement. After only a small load increment the steel becomes plastic and the structure fails. This behaviour, especially the relatively short phase of nonlinearity between the onset of tension cracking and final failure, is typical for natural draught cooling towers [12]. Fig. 4 shows the computed crack evolution of the inner and outer surfaces during the static loading.

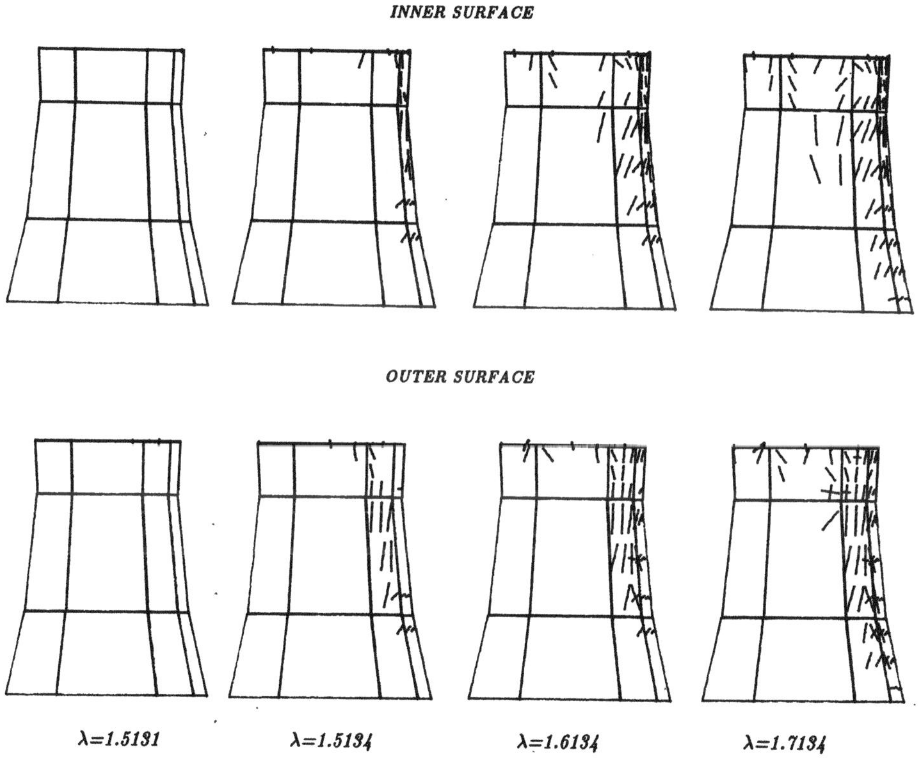

Fig. 4: Crack evolution

For the dynamic analysis eight pressure time series of 60 sec duration have been generated for the upper rim; for instance Fig. 5 shows the pressure history of node 141. The statically deformed configuration under the reference wind load has been calculated; then the turbulent part of the pressure has been added.

To determine the limit load multiplier λ, time-domain analyses have to be performed for several assumed values of λ until a failure condition is found. So far, two such analyses of 20 sec duration have been performed: each took about 40 hours CPU time on a HP-Apollo DN 5500 computer. The displacements calculated for node 91 are shown in Figs. 6a and 6b, which respectively refer to $\lambda = 1.0$ and $\lambda = 1.2$: for comparison's sake both the linear (circles) and nonlinear (triangles) results are plotted.

Under static loads there is only negligible tension cracking (c.f. Fig. 4) so that linear and nonlinear histories start from the same point.

For $\lambda = 1.0$ the structure remains almost wholly elastic; the two plots show only slight differences; the structural response fluctuates about the static mean value since no permanent deformations take place.

For $\lambda = 1.2$, on the contrary, after approx. 10 sec there is a drift-off of the displacement due to extensive tension cracking and the crushing of concrete in some points; the reinforcement, however, remains in the elastic domain.

The current analyses last 20 sec only: a time interval too short for deciding whether the structure will eventually fail or whether a stable oscillation about a new mean deformation is possible. Further numerical studies are needed to answer this open question.

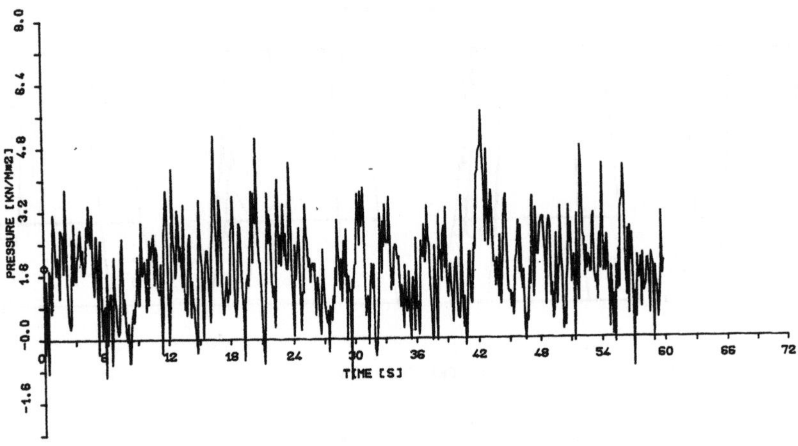

Fig. 5: Example of pressure time history (node 141)

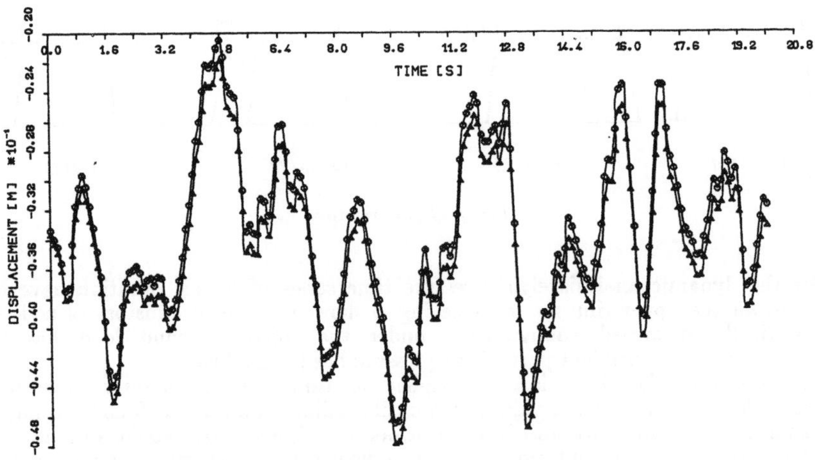

Fig 6a: Simulated displacement histories of node 91 for $\lambda = 1.0$

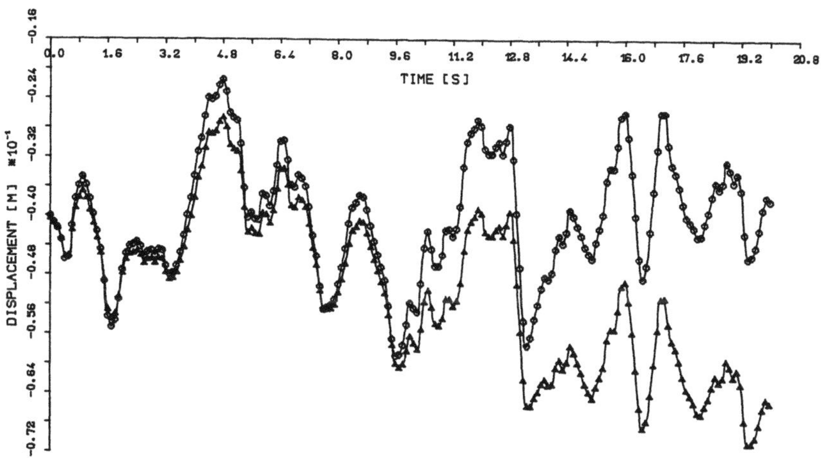

Fig 6b: Simulated displacement histories of node 91 for $\lambda = 1.2$

5. CONCLUDING REMARKS

The first results of an ongoing research have been presented whose final objective is an accurate modeling of the fully nonlinear dynamic response of cooling tower shells subjected to simulated wind loading. The multi-correlated field of monodimensional time processes of the external pressure is directly generated instead of velocity histories. This represents a significant progress in the accuracy of representation of the local turbulence, avoiding the use of the pressure coefficients, which sometimes give just a rough approximation of the aerodynamic phenomena.

The future developments will aim, once available the experimental data needed for overcoming the limitations still imposed on the generation procedure, at studying refined discretized structural models with particular enphasis on the effects of physical nonlinearities.

6. ACKNOWLEDGMENTS

The writers are indebted to Dr.-Ing. M. Kasperski and Prof. Dr.-Ing. H.-J. Niemann for the kind and helpful collaboration in supporting the research with experimental data. This study has been supported by research fonds granted by the Italian Ministry (MURST) through the Universities of Rome and Florence, and by a research convention with CRIS-ENEL (Milan); the A. von Humboldt Foundation (Bonn) is also acknowledged for a research fellowship awarded to W.Z.

7. REFERENCES

1 Borri C., Zahlten W., Nonlinear static and dynamic analysis of cooling tower shells, Proc. 10th (Ital.) Nat. Cong. on Theor. and Appl. Mechanics, A.I.Me.T.A., Pisa, (1990), I 244-250

2 Borri C., Bartoli G., Zahlten W., Dinamica non lineare di torri refrigeranti soggette all'azione del vento, 1st Ital. Nat. Conf. on Wind Engrg., IN-VENTO-90, G. Augusti/C. Borri/P. Spinelli Eds., Firenze, (1990)

3 Basar Y., Zur Struktur konsistenter inkrementeller Theorien für geometrisch nichtlineare Flächentragwerke und deren Operatordarstellung, Ingenieurarchiv, Vol. 56, (1986) 209-220

4 Bazant Z.P., Kim S.-S., Plastic-Fracturing Theory for Concrete, Journ. of Eng. Mech. Div., Proc. ASCE, Vol. 105, (1979) 407-428

5 Zahlten W., A Contribution to the Physically and Geometrically Nonlinear Computer Analysis of General Reinforced Concrete Shells, Techn. Rept. 1990-2, Dept. of Struct. Engrg., Univ. Bochum, (1990)

6 Borri C., Generation procedures of stationary random processes simulating wind time series, Visiting lecture at the Univ. of Bochum, SFB 151, Rept. ST 11/88, Dept. of Civil Engrg., Univ. of Florence, (1988)

7 Borri C., Zahlten W., Fully simulated nonlinear analysis of large structures subjected to turbulent, artificial wind, Mech. of Struct. and Machines, Vol. 19, No. 2 (1991), in print

8 Niemann H.J., Reliability of current design methods for wind-induced stresses, Proc. of the 2nd Int. Symp. on Natural Draught Cooling Towers, Bochum (FRG), (1984), 119-135

9 Kasperski M., Idenfikation bemessungskritischer Lastanteile aus stochastischen Prozessen am Beispel der Naturzugkühltürme unter Windbelastung, Techn. Rept. 1988-6, Dept. of Struct. Engrg., Univ. Bochum, 1988

10 Reed D.A., Scanlan R.H., Time series analysis of cooling tower wind loading, Journ. of Struct. Eng., ASCE, Vol. 109, No. 2, (1983), 538-554

11 Bautechnik bei Kühltürmen, VGB-Richtlinien, Teil II - Bautechnische Richtlinien, Verl. VGB-Kraftwerkstechnik GmbH, Essen (1979)

12 Krätzig W.B., Gruber K., Zahlten W., Numerical collapse analyses for a verification of safety and reliability of large natural draught cooling towers, Techn. Rept., Dept. of Struct. Engrg., Univ. Bochum, in print

13 Harte R., Eckstein U., Derivation of geometrically nonlinear finite shell elements via tensor notation, Int. J. Num. Meth. Eng., Vol. 23, (1986), 367-384

14 Eckstein U., Nichtlineare Stabilitätsberechnung elastischer Schalentragwerke, Techn. Rept. 1983-3, Dept. of Struct. Engrg., Univ. Bochum, (1983)

An Application to the Solution of the Fokker-Planck-Kolmogorov Equation for Non Linear Oscillators under Stochastic Loading

A. BARATTA & G. ZUCCARO

"Dipartimento di Scienza delle Costruzioni"
Faculty of Architecture
University of Naples - Italy

Summary

In a previous paper [1] a procedure for obtaining an approximate solution of the Fokker-Planck-Kolmogokov (FPK) equation, as applied to a single degree of freedom (sdof) oscillator with an unilateral barrier was proposed. Some results were obtained, showing an encouraging agreement with the results obtained via Monte Carlo simulation. In the conclusion of the same paper some ideas for further research and for improving the approximation were outlined.
In particular, the procedure requires a bivariate integration in the probability phase-space. Apart from the fact that numerical integration, as executed in that version, is very time-consuming, it was also identified as a possible source of errors, due to the difficulty in locating the significant domain of the probability functions that are to be integrated, while evolving in time. Therefore the possibility of an analytical closed-form integration is explored here, in order to evaluate also its influence on the accuracy of the results, as a first step towards the validation of the procedure.

Basic Ideas for First Order Solution of the FPK Equation.

The purpose of the study is to find an approximate solution procedure for nonlinear problems relating to free oscillations and/or the dynamical response of a sdof elastic system under the action of a random shot-noise type forcing funtion f(t). The motion of the system is governed by the equation

$$\tilde{f}(t) = \ddot{y} + g(y,\dot{y} \mid \alpha, \omega_0) \quad , \text{ with } \alpha = \zeta\, \omega_0 \tag{1}$$

where ζ is the damping coefficient, ω_0 is the free undamped natural frequency, y is the displacement of the mass, and the superimposed dot denotes a time derivative.
If one considers the particular case where $\tilde{f}(t)$ is a Gaussian non stationary white noise (nswn) with zero mean value, and with intensity factor I(t), such that:

N. Bellomo · F. Casciati (Eds.)
Nonlinear Stochastic Mechanics
IUTAM Symposium Turin

$$E[\tilde{f}(t)\tilde{f}(t+\tau)] = I(t)\delta(\tau) \tag{2}$$

then the displacement y(t) and the relevant velocity $\dot{y}(t)$ are components of a Markov vector, and the instantaneous joint transition probability density $p(y,\dot{y};t)$ obeys the classical FPK equation

$$\frac{\partial p}{\partial t} = \frac{\partial}{\partial \dot{y}}[p(y,\dot{y};t)g(y,\dot{y}|\alpha,\omega_0)] - \dot{y}\frac{\partial p}{\partial \dot{y}} + \frac{1}{2}I(t)\frac{\partial^2 p}{\partial \dot{y}^2} \tag{3}$$

where

$$g(y,\dot{y} \mid \alpha, \omega_0) \tag{4}$$

is a nonlinear restoring force, and

$$p(y,\dot{y};t) = p_0(y,\dot{y}) \tag{5}$$

are the initial conditions.

Let us assume now that no approximate analytical solutions are available to solve the problem governed by (1) and that one is obliged to search for other methods, involving a numerical solution of the differential equation.

The basic idea adopted here, as in [1], is to force the solution of the FPK equation to have the form of a Gaussian joint density function in y and $\dot{y}$.

Let $N(y,\dot{y};t)$ and h(t), respectively, denote a time-dependent Gaussian jdf and a time-dependent normalizing factor, whose role is to eventually allow truncationof the fundamental Gaussian $N(y,\dot{y};t)$. One can then write the solution of the problem in the form:

$$p(y,\dot{y};t) = h(t)\, N(y,\dot{y};t) \tag{6}$$

The variation of $N(y,\dot{y};t)$ with time is expressed by updating the parameters $\mu_y(t)$, $\mu_{\dot{y}}(t)$, $\sigma_y(t)$, $\sigma_{\dot{y}}(t)$, $\rho_{y\dot{y}}(t)$, which are respectively the values of the displacement and the relevant velocity, the standard deviations of the same quantities, and the coefficient of correlation.

$$N(y,\dot{y};t) = N[\, y,\dot{y} \mid \mu_y(t), \mu_{\dot{y}}(t), \sigma_y(t), \sigma_{\dot{y}}(t), \rho_{y\dot{y}}(t)\,] \tag{7}$$

Now let us write the FPK equation in the form,

$$p(y,\dot{y};t+dt) - p(y,\dot{y};t) + \{\frac{\partial}{\partial \dot{y}}[p(y,\dot{y};t)g(y,\dot{y}|\alpha,\omega_0)] - \dot{y}\frac{\partial p}{\partial y} + \frac{1}{2}I(t)\frac{\partial^2 p}{\partial \dot{y}^2}\}dt \quad (8)$$

assuming that

$$\frac{\partial N}{\partial y} = B(y,\dot{y};t)N(y,\dot{y},t) \quad (9)$$

$$\frac{\partial N}{\partial \dot{y}} = A(y,\dot{y};t)N(y,\dot{y},t) \quad (10)$$

$$\frac{\partial^2 N}{\partial \dot{y}^2} = [C(y,\dot{y};t) + A^2(y,\dot{y},t)]N(y,\dot{y},t) \quad (11)$$

on introducing the fundamental distribution $N(y,\dot{y};t)$ into the FPK eqn. one gets:

$$h(t+dt)\, N(y,\dot{y};t+dt) = h(t)\, N(y,\dot{y};t)\, [\, 1 + D(y,\dot{y};t)\, dt\,] \quad (12)$$

where

$$D(y,\dot{y};t) = A(y,\dot{y};t)g(y,\dot{y}|\alpha,\omega_0) + \frac{\partial g(y,\dot{y})}{\partial \dot{y}} - B(y,\dot{y},t)\dot{y}$$

$$+ \frac{1}{2}I(t)[C(y,\dot{y};t) + A^2(y,\dot{y},t)] \quad (13)$$

By multiplying both members of eqn. (12) by $y^r\, \dot{y}^s$ and taking expectations one obtains

$$h(t+dt) \iint y^r\, \dot{y}^s\, N(y,\dot{y};t+dt)\, dy\, d\dot{y} =$$

$$= h(t) \iint y^r\, \dot{y}^s\, N(y,\dot{y};t)\, [1 + D(y,\dot{y};t)\, dt]\, dy\, d\dot{y} \quad (14)$$

where the symbol $\iint$ denotes integration on R^2 between $-\infty$ and $+\infty$ with respect to $(y,\dot{y})$. Hence :

$$E(y^r\, \dot{y}^s \mid t+dt) = [h(t)/h(t+dt)]\ [E(y^r\dot{y}^s|t) + E_{rs}(\, t)\ dt] \quad (15)$$

integrated numerically on the time, where

$$E_{rs}(\, t) = \iint y^r\, \dot{y}^s\, N(y,\dot{y};t)\, D(y,\dot{y};t)\ dy\, d\dot{y} \quad (16)$$

Thus the problem is to calculate integrals in phase-space to obtain the transitory state at the (t+dt) instant of the expected displacement, and velocity and the relevant standard deviations.

Now if one considers the factors $N(y,\dot{y};t)$ in equations (9), (10) and (11) in the form:

$$A(y,\dot{y};t) = a_0(t) + a_1(t)\,y + a_2(t)\,\dot{y} \tag{17}$$

$$B(y,\dot{y};t) = b_0(t) + b_1(t)\,y + b_2(t)\,\dot{y} \tag{18}$$

$$C(y,\dot{y};t) + A^2(y,\dot{y};t) = c_0(t) + a_0^2(t) + a_1^2(t)\,y^2 + a_2^2(t)\,\dot{y}^2 +$$

$$+ 2\,a_0(t)\,a_1(t)\,y + 2\,a_0(t)\,a_2(t)\,\dot{y} + 2\,a_1(t)\,a_2(t)\,y\,\dot{y} \tag{19}$$

where:

$$a_0 = -\frac{\sigma_{\dot{y}}\rho_{y\dot{y}}\mu_y - \sigma_y\mu_{\dot{y}}}{\sigma_{\dot{y}}^2\sigma_y(1-\rho_{y\dot{y}}^2)} \qquad a_1 = +\frac{\rho_{y\dot{y}}}{\sigma_y\sigma_{\dot{y}}(1-\rho_{y\dot{y}}^2)} \qquad a_2 = -\frac{1}{\sigma_{\dot{y}}^2(1-\rho_{y\dot{y}}^2)}$$

$$b_0 = -\frac{\sigma_y\rho_{y\dot{y}}\mu_{\dot{y}} - \sigma_{\dot{y}}\mu_y}{\sigma_y^2\sigma_{\dot{y}}(1-\rho_{y\dot{y}}^2)} \qquad b_1 = -\frac{1}{\sigma_y^2(1-\rho_{y\dot{y}}^2)} \qquad b_2 = +\frac{\rho_{y\dot{y}}}{\sigma_y\sigma_{\dot{y}}(1-\rho_{y\dot{y}}^2)}$$

$$c_0 = -\frac{1}{\sigma_{\dot{y}}^2(1-\rho_{y\dot{y}}^2)}$$

we can write equation (13) in the more general form:

$$D(y,\dot{y};t) = g(y,\dot{y})\,[\,a_0(t) + a_1(t)\,y + a_2(t)\,\dot{y}\,] +$$

$$+ \frac{\partial g(y,\dot{y})}{\partial \dot{y}} -$$

$$- \dot{y}\,[\,b_0(t) + b_1(t)\,y + b_2(t)\,\dot{y}\,] + \frac{1}{2}\,I(t)\,[\,c_0(t) + a_0^2(t) +$$

$$+ a_1^2(t)\,y^2 + a_2^2(t)\,\dot{y}^2 + 2\,a_0(t)\,a_1(t)\,y + 2\,a_0(t)\,a_2(t)\,\dot{y} +$$

$$+ 2\,a_1(t)\,a_2(t)\,y\,\dot{y}\,] \tag{20}$$

and equation (16) becomes

$$E_{rs}(t) = a_0(t) \iint y^r \dot{y}^s \; g(y,\dot{y}) \; N(y,\dot{y};t) \, dy \, d\dot{y} +$$

$$+ a_1(t) \iint y^{r+1} \dot{y}^s \, g(y,\dot{y}) \, N(y,\dot{y};t) \, dy \, d\dot{y} +$$

$$+ a_2(t) \iint y^r \dot{y}^{s+1} \, g(y,\dot{y}) \, N(y,\dot{y};t) \, dy \, d\dot{y} +$$

$$\iint \frac{\partial g(y,\dot{y})}{\partial \dot{y}} \cdot y^r \cdot \dot{y}^s \cdot N(y,\dot{y};t) \cdot dy d\dot{y}$$

$$- b_0(t) \iint y^r \dot{y}^{s+1} \, N(y,\dot{y};t) \, dy \, d\dot{y} - b_1(t) \iint y^{r+1} \dot{y}^{s+1} \, N(y,\dot{y};t) \, dy \, d\dot{y} -$$

$$- b_2(t) \iint y^r \dot{y}^{s+2} \, N(y,\dot{y};t) \, dy \, d\dot{y} +$$

$$+ \frac{1}{2} I(t) \, [\, (c_0(t) + a_0^2(t)) \iint y^r \dot{y}^s \, N(y;\dot{y};t) \, dy \, d\dot{y} +$$

$$+ a_1^2(t) \iint y^{r+2} \dot{y}^s \, N(y,\dot{y};t) \, dy \, d\dot{y} + a_2^2(t) \iint y^r \dot{y}^{s+2} \, N(y,\dot{y};t) \, dy \, d\dot{y} +$$

$$+ 2 \, a_0(t) \, a_1(t) \iint y^{r+1} \dot{y}^s \, N(y,\dot{y};t) \, dy \, d\dot{y} +$$

$$+ 2 \, a_0(t) \, a_2(t) \iint y^r \dot{y}^{s+1} \, N(y,\dot{y};t) \, dy \, d\dot{y} +$$

$$+ 2 \, a_1(t) \, a_2(t) \iint y^{r+1} \dot{y}^{s+1} \, N(y,\dot{y};t) \, dy \, d\dot{y}]. \tag{21}$$

<u>A Particular class of non linear Restoring Forces</u>

As particularly simple case one considers, now, $g(y,y)$, in the form:

$$g(y,\dot{y}) = \sum_{i=0,n} \sum_{j=0,m} g_{ij} \cdot y^i \cdot \dot{y}^j \tag{22}$$

with

$$\frac{\partial g}{\partial \dot{y}} = \sum_{i=0,n} \sum_{j=1,m} j \cdot g_{ij} \cdot y^i \cdot \dot{y}^{j-1} \tag{23}$$

In this case:

$$E_{rs}(t) = a_0(t) \sum g_{ij} \iint y^{r+i} \dot{y}^{s+j} \, N(y,\dot{y};t) \, dy \, d\dot{y} \; +$$

$$+ a_1(t) \Sigma\ g_{ij} \iint y^{r+i+1} \dot{y}^{s+j} N(y,\dot{y};t)\, dy\, d\dot{y} +$$

$$+ a_2(t) \Sigma\ g_{ij} \iint y^{r+i} \dot{y}^{s+j+1} N(y,\dot{y};t)\, dy\, d\dot{y} +$$

$$+ \Sigma_1\ j\, g_{ij} \iint y^{r+i} \dot{y}^{s+j-1} N(y,\dot{y};t)\, dy\, d\dot{y} -$$

$$- b_0(t) \iint y^r \dot{y}^{s+1} N(y,\dot{y};t)\, dy\, d\dot{y} - b_1(t) \iint y^{r+1} \dot{y}^{s+1} N(y,\dot{y};t)\, dy\, d\dot{y} -$$

$$- b_2(t) \iint y^r \dot{y}^{s+2} N(y,\dot{y};t)\, dy\, d\dot{y} +$$

$$+ \frac{1}{2} I(t) [(c_0(t) + a_0^2(t)) \iint y^r \dot{y}^s N(y,\dot{y};t)\, dy\, d\dot{y} +$$

$$+ a_1^2(t) \iint y^{r+2} \dot{y}^s N(y,\dot{y};t)\, dy\, d\dot{y} + a_2^2(t) \iint y^r \dot{y}^{s+2} N(y,\dot{y};t)\, dy\, d\dot{y} +$$

$$+ 2\, a_0(t)\, a_1(t) \iint y^{r+1} \dot{y}^s N(y,\dot{y};t)\, dy\, d\dot{y} +$$

$$+ 2\, a_0(t)\, a_2(t) \iint y^r \dot{y}^{s+1} N(y,\dot{y};t)\, dy\, d\dot{y} +$$

$$+ 2\, a_1(t)\, a_2(t) \iint y^{r+1} \dot{y}^{s+1} N(y,\dot{y};t)\, dy\, d\dot{y}]. \qquad (24)$$

where the symbols Σ and Σ_1 respectively denote

$$\sum_{i=0,n} \sum_{j=0,m} \quad \text{and} \quad \sum_{i=0,n} \sum_{j=1m}$$

or, more explicitly:

$$E_{rs}(t) = a_0(t) \Sigma\, g_{ij}\ E[y^{r+i} \dot{y}^{s+j}] + a_1(t) \Sigma\, g_{ij}\ E[y^{r+i+1} \dot{y}^{s+j}] +$$

$$+ a_2(t) \Sigma\, g_{ij}\ E[y^{r+i} \dot{y}^{s+j+1}] + \Sigma_1\ j\, g_{ij}\ E[y^{r+i} \dot{y}^{s+j-1}] -$$

$$- b_0(t)\ E[y^r \dot{y}^{s+1}] - b_1(t) E[y^{r+1} \dot{y}^{s+1}] - b_2(t) E[y^r \dot{y}^{s+2}] +$$

$$+ \frac{1}{2} I(t) \{ (c_0(t) + a_0^2(t)) E[y^r \dot{y}^s] + a_1^2(t)\ E[y^{r+2} \dot{y}^s] +$$

$$+ a_2^2(t) E[y^r \dot{y}^{s+2}] + 2\, a_0(t)\, a_1(t)\ E[y^{r+1} \dot{y}^s] +$$

$$+ 2\, a_0(t)\, a_2(t) E[y^r \dot{y}^{s+1}] + 2\, a_1(t)\, a_2(t) E[y^{r+1} \dot{y}^{s+1}] \} \qquad (25)$$

The integrals in E_{rs} coincide with the moments of the Gaussian bivariate P.M.F.

Numerical Test: Non-Smooth Non-linear Restoring Force

Let us consider now a sdof elastic system, with one elastic barrier (fig 1). The equation that governs the motion of the system under the action of the random forcing function $\tilde{f}(t)$ is

$$\ddot{y} + 2\alpha\dot{y} + \omega_0^2 y = \tilde{f}(t) - R(y,\dot{y}) \tag{26}$$

where $R(y,\dot{y})$ represent the reactive force of the obstacle at the impact and is of the type

$$R(y,\dot{y}) = (1 + r)(\dot{y}^-)^2 \, \delta(y - y^*) \tag{27}$$

where y^* is the position of the barrier, r is the restitution factor and the superscripts + and - denote, respectively, the values of the function before and after the impact time. For $r=1$ the impact is elastic and no energy is dissipated at impact.

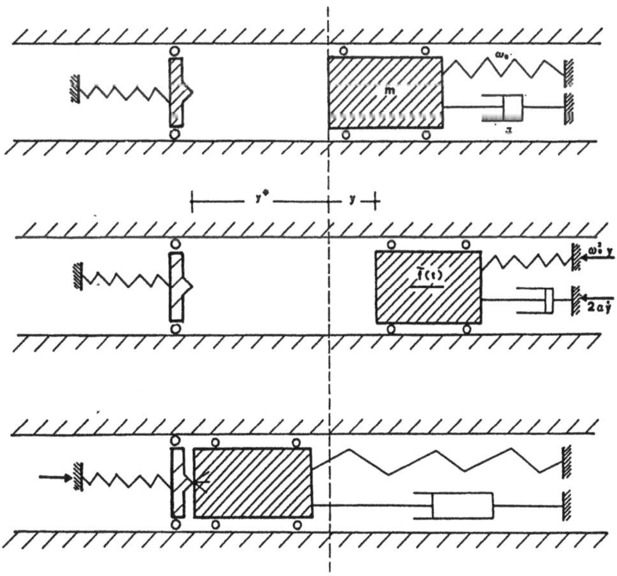

fig.1 The oscillator under consideration.

If $y(0) > y^*$ the solution $y(t)$ to equation (26) obeys the conditions:

$$y(t) > y^* \qquad 0 < t < \infty \tag{28}$$

$$\dot{y}^+ = - r\,\dot{y}^- \tag{29}$$

The function $R(y,\dot{y})$ introduces a strong non-linearity into eqn. (26); therefore the solution is discontinuous in the time derivative. This circumstance causes much difficulty in handling the equation analytically.
Thus the problem can be approached by using the "basic ideas" introduced in the last section, see eqns. (6), (7), and (8), by assuming that the Gaussian jdf in y and $\dot{y}$ is truncated for $y < y^*$.
Now let us approximate the non-smooth restoring force

$$g^*(y,\dot{y}) = 2\alpha\dot{y} + \omega_0^2 y + (1+r)(\dot{y}^-)^2\,\delta(y - y^*) \tag{30}$$

by the more amenable expression

$$g(y,\dot{y}) = 2\alpha\dot{y} + \omega_0^2 y + [1 - H(y - y^*)]\,[K_0\,\delta(y - y^*) + C_0\dot{y}] \tag{31}$$

where H (.) is the Heaviside step function, K_0 is the stiffness of the spring, producing the impulsive term once the impact on the barrier occurs and C_0 represents the value of the damping of the dashpot, simulating the loss of kinetic energy at impact.

For $y > y^*$ $\qquad K_0 = C_0 = 0.$

In the paper a numerical test is reported for the case of a purely elastic impact ($C_0 = 0$) and a rigid ostacle ($K_0 \to \infty$), see [1] for details.

The restoring forces are:

$$g(y,\dot{y}) = -K_0 y^* + (\omega_0^2 + K_0)y + (2\alpha + C_0)\dot{y}$$

$$\frac{\partial g(y,\dot{y})}{\partial \dot{y}} = 2\alpha + C_0 \qquad \text{for } y < y^* \tag{32}$$

and

$$g(y,\dot{y}) = \omega_0^2 y + 2\alpha\dot{y}$$

$$\frac{\partial g(y,\dot{y})}{\partial \dot{y}} = 2\alpha \qquad \text{for } y \geq y^* \tag{33}$$

Remembering equations (9), (10), (11), and the factors of the (17), (18), (19), by substituting the (32) and (33) in the (20) and by integrating both members of the eqn.(12) on R^2 with respect to $(y,\dot{y})$, one obtains:

$$h(t+dt) = h(t)\,[\,1 + E_{00}(t)\;dt\,] \tag{34}$$

By taking expectations of both members of the (12) with respect to y and $\dot{y}$, one gets:

$$\mu_y(t+dt) = \frac{\mu_y(t) + E_{10}(t)dt}{[1 + E_0(t)dt]} \tag{35}$$

$$\mu_{\dot{y}}(t+dt) = \frac{\mu_{\dot{y}}(t) + E_{01}(t)dt}{[1 + E_0(t)dt]} \tag{36}$$

Eventually, by taking second-order moments of both members of the (12) with respect to $(y-\mu_y)^2$, $(\dot{y}-\mu_{\dot{y}})^2$, $(y-\mu_y)(\dot{y}-\mu_{\dot{y}})$ it is possible to calculate updated variances and covariance respectively, as follows:

$$\sigma_y^2(t+dt) = \frac{\sigma_y^2(t) + [E_{20}(t) + \mu_y^2 E_{00}(t) - 2\mu_y E_{10}(t)]dt}{[1 + E_{00}(t)dt]} \tag{37}$$

$$\sigma_{\dot{y}}^2(t+dt) = \frac{\sigma_{\dot{y}}^2(t) + [E_{02}(t) + \mu_{\dot{y}}^2 E_{00}(t) - 2\mu_{\dot{y}} E_{01}(t)]dt}{[1 + E_{00}(t)dt]} \tag{38}$$

$$K_{y\dot{y}}(t+dt) = \frac{K_{y\dot{y}}(t) + [E_{11}(t) - \mu_{\dot{y}} E_{10}(t) - \mu_y E_{01}(t) + \mu_y \mu_{\dot{y}} E_{00}(t)]dt}{[1 + E_{00}(t)dt]} \tag{39}$$

where μ_y and $\mu_{\dot{y}}$ denote $\mu_y(t+dt)$ and $\mu_{\dot{y}}(t+dt)$.

The results reported in this paper are obtained by evaluation of the integrals with respect of y in closed-form, whilst the integration with respect of $\dot{y}$ has been evaluated numerically.

It is worth noting that to compute the (24) one has to split the elements contained equations (32) or (33) according to the position of y respect the barrier.

One observes that for the oscillator without the barrier, the equation (24) can be used in the above mentioned form, in this case the factors of equation (22) are:

$$g_{00} = -K_0 y^*$$
$$g_{10} = \omega_0^2 + K_0 \qquad \text{for } y < y^* \qquad (40)$$
$$g_{01} = 2\alpha + C_0$$

One considers, then, the oscillator of fig.1 with the following numerical characteristics: $\omega_0 = 10\, s^{-2}$, α=0.1, $y^* = -5$ cm, $K_0 = 10.000\, s^{-2}$, $C_0 = 0$.

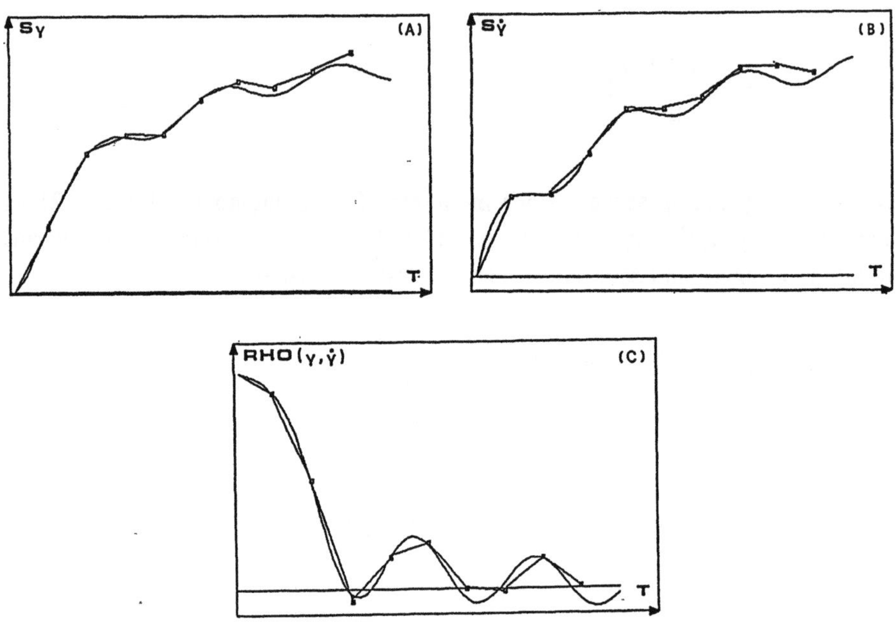

fig.2 Monte Carlo versus calculated statistics: (K_0=0.) (a) standard deviation of displacement; (b) standard deviation of velocity; (c) coefficient of correlation of displacement and velocity

The analysis was concerned with the response of oscillator with homogeneous initial conditions, acted an nswn random forcing function with zero mean value and an intensity function given by

$$I(t) = I_0 \exp[-2(t/T)^2] \qquad (41)$$

where I_0 - F_0^2 ΔT is the initial intensity of the input nswn acceleration.

$F_0 = 10.000$ cm s^{-2} is the assumed initial mean square value of the forcing funtion and $\Delta T = 0.001$ is the time lag used to simulate the nswn, while T is the strong-phase duration of the excitation (5 s).

Once the simulation has been performed over the chosen time interval via the Monte Carlo method [1] one compare the results of the simulation versus the approximate results obtained through the procedure proposed in this paper.

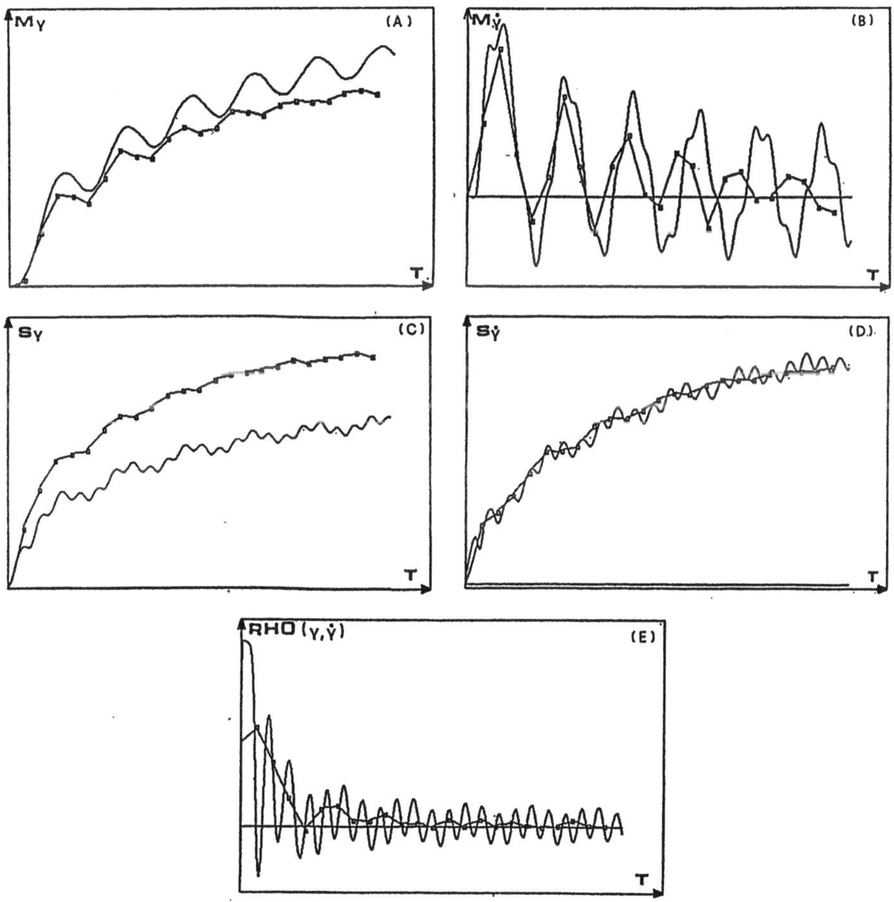

fig.3 Monte Carlo versus calculated statistics: (K_0=10.000) (a) Expected displacement; (b) Expected velocity; (c) standard deviation of displacement; (d) standard deviation of velocity; (e) coefficient of correlation of displacement and velocity

The behaviour of the moments of the response is investigated through the computation of the integrals in eqns. (34)-(39):

$$
\begin{aligned}
m_y(t) &= h(t)\ E_{10} \\
m_{\dot y}(t) &= h(t)\ E_{01} \\
s_y{}^2(t) &= h(t)\ \{E[y^2\ \dot y^0] + m_y{}^2(t)\,E[y^0\ \dot y^0] - 2m_y(t)\,E[y^1\ \dot y^0]\} \qquad (42)\\
s_{\dot y}{}^2(t) &= h(t)\ \{E[y^0\ \dot y^2] + m_{\dot y}{}^2(t)\,E[y^0\ \dot y^0] - 2m_{\dot y}(t)\,E[y^0\ \dot y^1]\} \\
K_{y\dot y}(t) &= h(t)\ \{E[y^1\ \dot y^1] - m_y(t)\,E[y^1\ \dot y^0] - m_{\dot y}(t)\,E[y^0\ \dot y^1] + m_y(t)\,m_{\dot y}(t)\,E[y^0\ \dot y^0]\}
\end{aligned}
$$

In fig 2 the results of the investigation of the transitory response of the system without the barrier ($K_0 = 0$) are shown whilst in fig. 3 the results obtained in presence of a rigid obstacle ($K_0 = 10.000$) are reported.
In both case the simulation values are reported, on the same scale and on the same diagrams as the values obtained through the eqns. (42), as squares joined by lines. As is shown in the diagrams the agreement between the Monte Carlo simulation and the theory is quite good, expecially if one considers that the fitting is very much improved from the results obtained in the previous paper [1], where the integrations were performed only numerically. Therefore the integration R^2 with respect of both of the dimensions $(y,\dot y)$ will certainly improve the results, expecially if one takes into consideration saving in computer time, which might allow one to adopt a smaller interval of integration on (t).

Conclusions

The procedure seems to confirm the promising performance noted in the original paper. It is worthwhile to observe, however, that the procedure can treat any case involving non linearity and of transitory problem.
In judging the results that have been obtained in this paper one should consider that the kind of non linearity (unilaterality) that has been studied is also, by itself, very sharp and difficult to handle, in the deterministic range.
Future work will be concerned with closed form solutions for a wide class of non-linear smooth restoring spring. The computerized procedure will be optimized to solve any kind of smooth non linear problem, as indicated in the second section. The input will be restricted, to the g_{ij} coefficients only (22) (23), and the possibility of building into the procedure the values of the integrals in closed form will lead to a considerable improvement in the time efficiency of the computer performance and the precision of the final results.

References:

1 A. Baratta, Dynamics of a Single-Degree-Of-Freedom System with a Unilateral Obstacle, Euromech 250 Colloquium on Nonlinear Structural System under Random Conditions, Como, Italy, June 19-23, 1989.

2 Y.K.Lin, Probabilistic Theory of Structural Dynamics, McGraw-Hill, NewYork,1976.

3 G. Augusti, A. Baratta and F. Casciati, Probabilistic Methods in Structural Engineering, Chapman & Hall, London, 1984.

4. M.F. Dimentberg and A.I. Menyailow, Response Analysis of vibroimpact systems with random excitation, in: I. Elishakoff and R.H.Lyon (Eds.), Random- Vibration- Status and Recent Developments, Elsevier, Amsterdam, 1986.

5. Y.K.Lin, Application of Markov process theory to nonlinear random vibration problems, in: G.I. Shueller and M. Shinozuka (Eds.), Stochastic Methods in Structural Dynamics, Martinus Nijhoff, Dordrecht, The Netherlands, 1987.

6. A. Baratta, probabilistic treatment of structural problems with unilateral constraints, in: G. Del Piero and F. Maceri (Eds.), Unilateral Problems in Structural Analysis, Vol. 2 (CISM, Courses and Lectures, 304), Springer-Verlag, Vienna, 1987.

7. M.F. Dimentberg and A.I. Menyailow, Statistical dynamics of vibroimpact systems, private comunication, 1987.

8. L. Ascione and C. Bilotti, Il problema di contatto dinamico con attrito di una piastra su fondazine elastica a due parametri, Proc. IX Natl. Conf. of the Italian Association of Theoretical and Applied Mechanics (AIMETA), Bari, Italy, Vol. II, 1988, pp. 485-488.

Robust Numerical Solution of the Transient Fokker-Planck Equation for Nonlinear Dynamical Systems

L.A. Bergman
Dept. of Aero. and Astro. Engineering
University of Illinois
Urbana, Illinois 61801-0236

B.F. Spencer, Jr.
Dept. of Civil Engineering
University of Notre Dame
Notre Dame, Indiana 46556-0767

Summary

The finite element method is applied to the solution of the transient Fokker-Planck equation for several often cited nonlinear stochastic systems accurately giving, for the first time, the joint probability density function of the response for a given initial distribution. The method accommodates nonlinearity in both stiffness and damping as well as both additive and multiplicative excitation, although only the former is considered herein. Several systems are examined, including linear, Duffing, and Van der Pol oscillators, to illustrate the robustness and accuracy of the finite element method.

Introduction

Engineering structures are often subjected to dynamic loads that are most appropriately characterized as random processes. Furthermore, these structures generally exhibit some nonlinearity, whether geometric, material or some combination. As is well known, the determination of response for even the simplest nonlinear stochastic system of engineering interest can be formidable.

For linear systems subjected to white or linearly filtered white (colored) additive noise, the response is known to be Gaussian, and its evolution, for a simple linear oscillator, was first discussed by Wang and Uhlenbeck (1945). These results have been further generalized to N dimensions and, along with response statistics, can now be found in many standard texts (Lin, 1967; Nigam, 1983).

For nonlinear systems subjected to additive white noise excitation, exact solutions for the response process exist only for certain scalar systems (Caughey and Dienes, 1961). However, it is known that the response of nonlinear multi-dimensional systems subjected to white noise excitation forms a vector Markov process, with a transition probability density function satisfying both the forward (Fokker-Planck) and the backward Kolmogorov equations (Caughey, 1963; Caughey, 1971). Unfortunately, these equations remain unsolved except for the linear system of arbitrary dimension and certain scalar nonlinear systems.

The long time, or stationary, response of certain nonlinear systems has been obtained through solution of the steady-state Fokker-Planck equation (Caughey, 1963; Soize, 1988). This, too, is restrictive in scope and has led to the development of a vast body of approximation theory by which the stationary response statistics of more general nonlinear systems can be estimated. Several methods that are now routinely applied are stochastic equivalent linearization (Roberts and Spanos, 1990) and stochastic averaging (Roberts and Spanos, 1986). Both of these can provide reasonable estimates of response statistics but often fail to adequately predict statistics of related extremal distributions, and so forth.

N. Bellomo · F. Casciati (Eds.)
Nonlinear Stochastic Mechanics
IUTAM Symposium Turin

Galerkin based methods have been applied to the solution of the transient Fokker-Planck equation for some nonlinear systems. Bolotin (1967) employed the Galerkin method to investigate the stability of a single degree of freedom model of a thin elastic curved panel. Utilizing global methods with eigenfunctions of the Fokker-Planck equation for the linear system as trial functions, Atkinson (1973) studied the response process of several second order nonlinear oscillators. In other global analyses, using Hermite polynomials as trial functions, Bhandari and Sherrer (1968) examined the response of both one and two degree of freedom oscillators. Wen (1975) extended these results to include transient responses, and later (Wen, 1976) the results for hysteretic systems through implementation of first order filters. More recently, the finite element method has been used to solve the steady-state Fokker-Planck equation of some nonlinear systems by Langley (1985) and Langtangen (1991).

Monte Carlo methods have also been applied to the solution of these problems, as have numerical diffusion techniques (*i.e.* cell mapping methods), both of which were pioneered by Crandall and coworkers (1970, 1971). More recently, Sun and Hsu (1988, 1990) have developed a generalized cell mapping procedure to extend these results to nonlinear second order systems.

Within the past ten years, a Petrov-Galerkin finite element method has been used extensively to solve the backward Kolmogorov equation with absorbing boundaries to obtain the solution of the first passage problem for a variety of two and three dimensional dynamical systems incorporating nonlinearity as well as multiplicative excitation (Bergman and Spencer, 1983; Spencer, 1986; Bergman, 1989). Here the transient equations have been successfully solved, giving the probability functions of first passage time as well as failure statistics.

In this paper, the finite element method is applied to the transient Fokker-Planck equation giving, for the first time, the solution for the probability density function for oscillator problems with large nonlinearities in both stiffness and damping. Initial emphasis is on systems of two dimensions subjected to additive excitation. Several systems are examined, and where possible, results are compared with those in the literature.

Formulation

An appropriate stochastic differential equation describing a time-independent discrete dynamical system subject to additive white noise is given by

$$\frac{d\mathbf{x}}{dt} = \mathbf{g}(\mathbf{x}) + \mathbf{H}\mathbf{w}(t) \tag{1}$$

Here, $\mathbf{x}$ and $\mathbf{g} \in R^n$, $\mathbf{w}(t) \in R^m$ is a vector white noise, and $\mathbf{H}$ is an $n \times m$ dimensional matrix of constant coefficients. The moments of the white noise are

$$E[\mathbf{w}(t)] = 0 \tag{2}$$

$$E[\mathbf{w}(t)\mathbf{w}^T(t+t')] = 2\mathbf{D}\delta(t') \tag{3}$$

where $\delta(\cdot)$ is the Dirac delta function and $\mathbf{D}$ is an $m \times m$ matrix with constant coefficients.

It is well known that Eq. (1)-(3) define a vector Markov process, the behavior of which is completely determined by the transition probability density function $p(\mathbf{x}, t | \mathbf{x}_0)$, where $\mathbf{x}_0 = \mathbf{x}(0)$, which satisfies both the forward Kolmogorov (Fokker-Planck) equation

$$\frac{\partial p}{\partial t} = \mathcal{L}_{\mathbf{x}} p \quad \text{on}\,\Omega \tag{4}$$

and backward Kolmogorov equation

$$\frac{\partial p}{\partial t} = \mathcal{L}^*_{\mathbf{x}_0} p \quad \text{on}\,\Omega \tag{5}$$

subject to the initial condition

$$p(\mathbf{x}, 0|\mathbf{x}_0) = \delta(\mathbf{x}-\mathbf{x}_0) \tag{6}$$

The operators $\mathcal{L}_{\mathbf{x}}$ and $\mathcal{L}^*_{\mathbf{x}_0}$ are, respectively,

$$\mathcal{L}_{\mathbf{x}} u = \sum_{i=1}^{n}\sum_{j=1}^{m}\sum_{k=1}^{m}\sum_{l=1}^{n} D_{jk}\frac{\partial}{\partial x_l}\left[H_{ij}\frac{\partial}{\partial x_l}(H_{lk}u)\right] - \sum_{l=1}^{n}\frac{\partial}{\partial x_l}(g_l u) \tag{7}$$

$$\mathcal{L}^*_{\mathbf{x}_0} v = \sum_{i=1}^{n}\sum_{j=1}^{m}\sum_{k=1}^{m}\sum_{l=1}^{n} D_{jk}H_{ij}\frac{\partial}{\partial x_{0_l}}\left(H_{lk}\frac{\partial v}{\partial x_{0_l}}\right) + \sum_{l=1}^{n} g_l\frac{\partial v}{\partial x_{0_l}} \tag{8}$$

where D_{ij} is the $(i\text{-}j)^{th}$ element of $\mathbf{D}$, and are formal adjoints.

If the domain of interest Ω extends infinitely in all dimensions, boundary conditions need not be prescribed for either Eq. (4) or (5). However, when the domain is finite in one or more dimensions (*e.g.* as occurs in the first passage problem), boundary conditions are required and are often imposed based upon physical understanding of the problem.

When random initial conditions are prescribed, for example in the two dimensional problem as a symmetric binormal density given by

$$f_{\mathbf{X}_0}(\mathbf{x}_0) = f_{\mathbf{X}_0}(x_{0_1}, x_{0_2}) = \phi\left(\frac{x_{0_1}-\mu_{X_{01}}}{\sigma_0}\right)\phi\left(\frac{x_{0_2}-\mu_{X_{02}}}{\sigma_0}\right) \tag{9}$$

where $\phi(\cdot)$ is the standard normal density function, and $\boldsymbol{\mu}_{x_0}$ is the expected value of the initial condition vector $\mathbf{X}_0$, the solution of Eq. (7) is a joint probability density given by

$$f_{\mathbf{X}}(x_1, x_2, t) = \int_{-\infty}^{\infty}\int_{-\infty}^{\infty} p(\mathbf{x}, t|\mathbf{x}_0) f_{\mathbf{X}_0}(\mathbf{x}_0)\, d\mathbf{x}_0 \tag{10}$$

which, as $t \to \infty$, is the stationary probability density function, $f_{\mathbf{X}}(\mathbf{x})$.

Simple post processing leads to the marginal density functions

$$f_{x_1}(x_1, t) = \int_{-\infty}^{\infty} f_{\mathbf{X}}(x_1, x_2, t)\, dx_2 \tag{11}$$

$$f_{x_2}(x_2, t) = \int_{-\infty}^{+\infty} f_{\mathbf{X}}(x_1, x_2, t)\, dx_1 \tag{12}$$

and upcrossing rate of a level a

$$\nu_a^+(a,t) = \int_0^\infty x_2 f_{\mathbf{X}}(a, x_2, t)\, dx_2 \tag{13}$$

The solution of Eq. (7) in terms of $f_{\mathbf{X}}$, for n greater than one and for arbitrary $\mathbf{g}(\mathbf{x})$, generally defies analytical solution. Restricting n to two, or perhaps three, the finite element method provides a convenient and efficient method to obtain approximate solutions to a wide range of problems.

Discretization

Consider the Fokker-Planck equation, Eq. (4), with $n = 2, m = 1$ and $\mathbf{g} = [0\ 1]^T$. We seek to solve this by a straightforward Bubnov-Galerkin finite element method. Thus, within a single element, the joint probability density given in Eq. (10) is interpolated according to

$$f_{\mathbf{X}}(\mathbf{x}, t) = \sum_{r=1}^{n_p} N_r(\mathbf{x}) f_r^e(t) \tag{14}$$

where the $N_r(\mathbf{x}), r = 1, \ldots, n_p$, are the element trial functions of class C^0, n_p is the number of nodes in a single element, and the f_r^e are the nodal values of the density. This expression is substituted into the weak form of Eq. (4) on the element domain, Ω_e, leading to expressions for the element "stiffness" and "mass" matrices

$$k_{rs} = \int_{\Omega_e} \{D \frac{\partial N_r}{\partial x_2} \frac{\partial N_s}{\partial x_2} + N_r \left(g_1 \frac{\partial N_s}{\partial x_1} + g_2 \frac{\partial N_s}{\partial x_2} \right) + N_r N_s \left(\frac{\partial g_1}{\partial x_1} + \frac{\partial g_2}{\partial x_2} \right) \}\, dx_1 dx_2 \tag{15}$$

and

$$m_{rs} = -\int_{\Omega_e} N_r(\mathbf{x}) N_s(\mathbf{x})\, dx_1 dx_2 \tag{16}$$

Transformation of each element matrix into global coordinates and summation over the number of elements in the mesh, N_e, leads to the global matrix evolution equation

$$\mathbf{M}\dot{\mathbf{f}} + \mathbf{K}\mathbf{f} = 0 \tag{17}$$

subject to the initial condition

$$\mathbf{f}(0) = \mathbf{f}_{\mathbf{X}_0} \tag{18}$$

where $\mathbf{f}_{\mathbf{X}_0}$ is a vector of global dimension representing the initial condition given in Eq. (9).

Equation (17) is further discretized in time using the well known Crank-Nicholson method (*cf.* Spencer, 1986). A rectangular mesh comprised of bilinear four node quadrilateral elements is generated over the domain of calculation, which is sufficiently large to minimize the diffusion of probability mass through the boundary.

The Linear Oscillator

Simple verification of the procedure is afforded by the solution for the linear oscillator response. Here,

$$\mathbf{g}(\mathbf{x}) = \begin{bmatrix} x_2 \\ -2\zeta\omega_0 x_2 - \omega_0^2 x_1 \end{bmatrix} \tag{19}$$

and

$$\mathbf{H} = \begin{bmatrix} 0 \\ 1 \end{bmatrix} \tag{20}$$

Oscillator parameters are given by $\zeta = 0.05$ and $\omega_0 = 1$; excitation by $D = 0.1$; and initial conditions by $\mathbf{x}_0 = [5 \;\; 5]^T$ and $\sigma_0^2 = 1/9$. The domain for computational purposes is given by $-10 \leq x_1, x_2 \leq 10$.

The evolution of the solution is presented as a contour plot in Fig. 1 where we have normalized time to be in natural periods (*i.e.* $\tau = \omega t/2\pi$). Thus, for a lightly damped system, one revolution in the phase space would occur approximately at $\tau = 1$. Figure 1 depicts $f_{\mathbf{X}}(x_1, x_2, \tau)$ overlaid at several time steps to show the progression. Figure 2 provides a three dimensional view of Fig. 1. Solutions at time steps 1.0625, 1.1875 and 1.3125 natural periods have been omitted for clarity

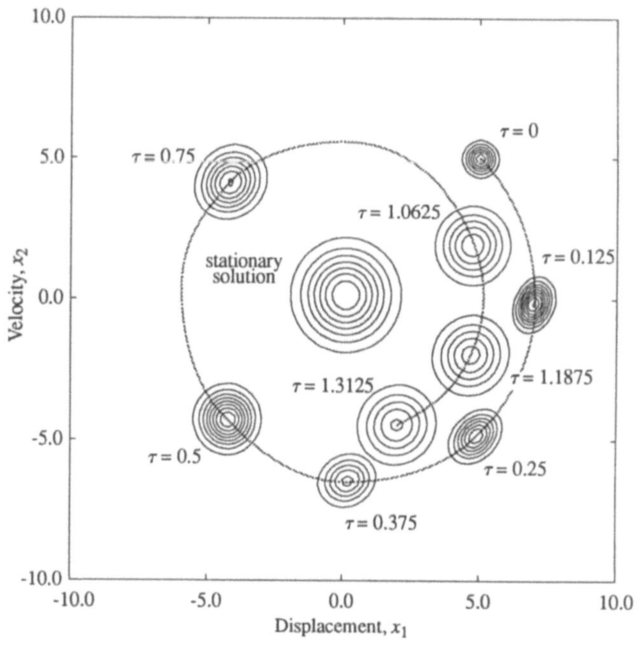

Figure 1. Contour plot of the evolution of the probability density function for the linear oscillator. $\zeta = 0.05$.

of presentation. The probability density at the origin, $f_{\mathbf{X}}(0, 0, \tau)$, vs. time is given in Fig. 3 where it is compared with the exact solution. As is seen here, the finite element solution is indiscernible from the exact solution. Finally, a cross sections of the stationary probability density $f_{\mathbf{X}}(x_1, x_2)$ at $x_2 = 0$ is shown and compared with exact values in Fig. 4. Again, there is excellent agreement between the exact and finite element solutions.

Having validated this numerical approach, we investigate two important nonlinear oscillators in the next sections.

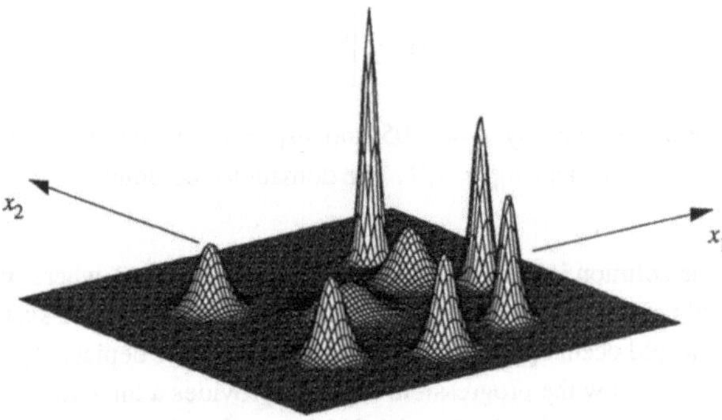

Figure 2. Evolution of the probability density function for the linear oscillator. $\zeta = 0.05$.

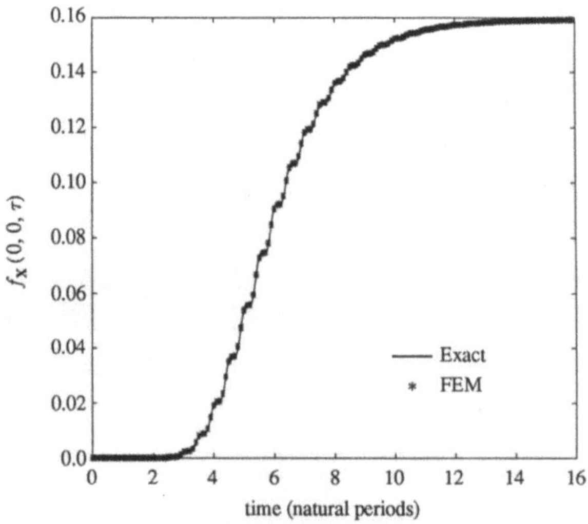

Figure 3. Probability density function for the linear oscillator evaluated at $x_1 = x_2 = 0$. $\zeta = 0.05$.

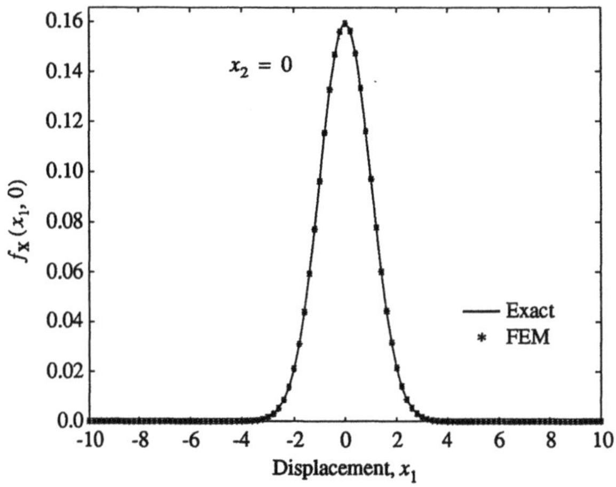

Figure 4. Cross section of the stationary probability density function for the linear oscillator. $\zeta = 0.05$.

The Duffing Oscillator

Here,

$$\mathbf{g}(\mathbf{x}) = \begin{bmatrix} x_2 \\ -2\zeta\omega_0 x_2 - \omega_0^2 x_1 (\gamma + \epsilon x_1^2) \end{bmatrix} \tag{21}$$

and

$$\mathbf{H} = \begin{bmatrix} 0 \\ 1 \end{bmatrix} \tag{22}$$

The second system examined provides a test of the robustness of the method, as the system embodies two stable equilibrium states. Thus, the probability density evolves in time from a unimodal to a bimodal distribution. The oscillator parameters are given by $\zeta = 0.2$, $\omega_0 = 1$, $\gamma = -1$ and $\epsilon = 0.1$; excitation by $D = 0.4$; initial conditions by $\mathbf{x}_0 = [0 \;\; 10]^T$; and $\sigma_0^2 = 1/2$. The domain for computational purposes is given by $-15 \le x_1 \le 15$, $-20 \le x_2 \le 20$.

The evolution of the solution is given in Fig. 5 in the form of a contour plot. The stationary density function is shown and compared with the exact solution in Fig. 6. Again, the finite element results are indiscernible from the exact stationary solution. Figure 7 provides a three-dimensional view of the stationary density function.

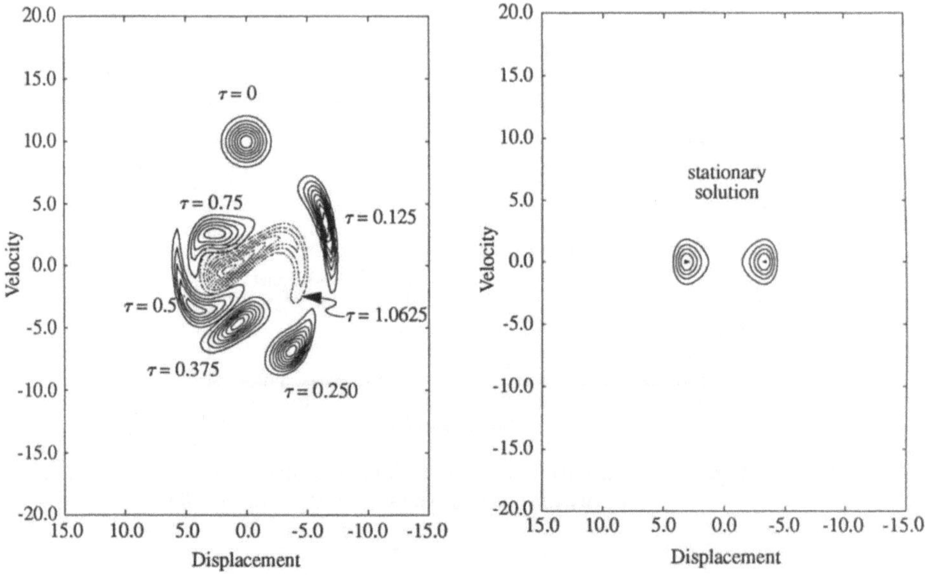

Figure 5. Contour plot of the evolution of the probability density function for the Duffing oscillator. $\zeta = 0.2$.

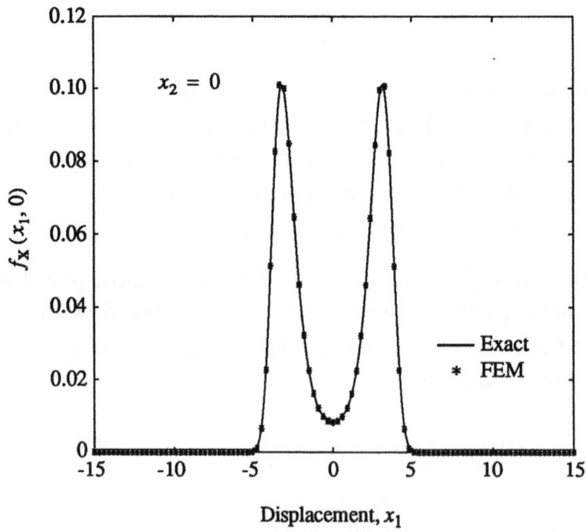

Figure 6. Cross section of the stationary probability density function for the Duffing oscillator. $\zeta = 0.2$.

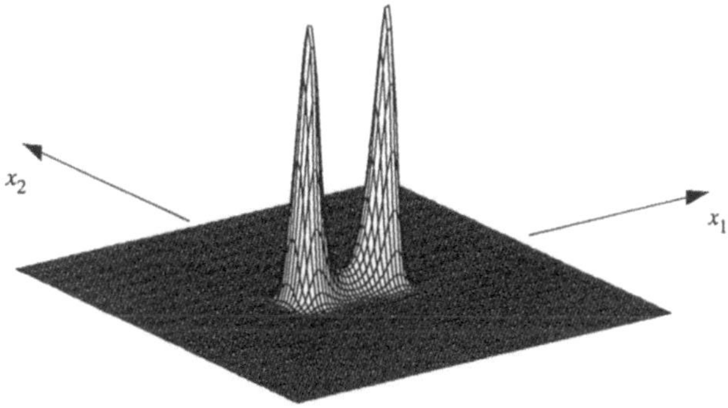

Figure 7. Stationary probability density function for the Duffing oscillator. $\zeta = 0.2$.

The Van der Pol Oscillator

Here,

$$\mathbf{g}(\mathbf{x}) = \begin{bmatrix} x_2 \\ -2\zeta\omega_0 x_2(\epsilon x_1^2 - 1) - \omega_0^2 x_1 \end{bmatrix} \tag{23}$$

and

$$\mathbf{H} = \begin{bmatrix} 0 \\ 1 \end{bmatrix} \tag{24}$$

The oscillator exhibits limit cycle behavior which provides an additional test of the method's robustness. The oscillator parameters are given by $\zeta = 0.05$, $\omega_0 = 1$, and $\epsilon = 1$; excitation by $D = 10$; and initial conditions by $\mathbf{x}_0 = [4 \;\; 4]^T$ and $\sigma_0^2 = 1/2$. The domain for computational purposes is given by $-10 \le x_1, x_2 \le 10$.

The evolution of the probability density function is shown in Fig. 8, and the three-dimensional view of the stationary density function is given in Fig. 9.

Conclusions

The finite element method has been applied to the solution of the Fokker-Planck equation to study the evolution of the transient probability density function of the linear and several nonlinear second order oscillators subjected to additive white noise excitation. Standard Bubnov-Galerkin discretization over a rectangular domain using 4-node quadrilateral elements of class C^0 has been shown to produce excellent results, even for nonlinearities leading to bifurcatory and limit cycle behavior in the system.

While the evolution of the joint probability density function has long been known for linear systems of arbitrary dimension, such results are still largely unknown for nonlinear systems beyond one dimension. The robustness and accuracy of the finite element method clearly demonstrated in this paper make it the method of choice for future study.

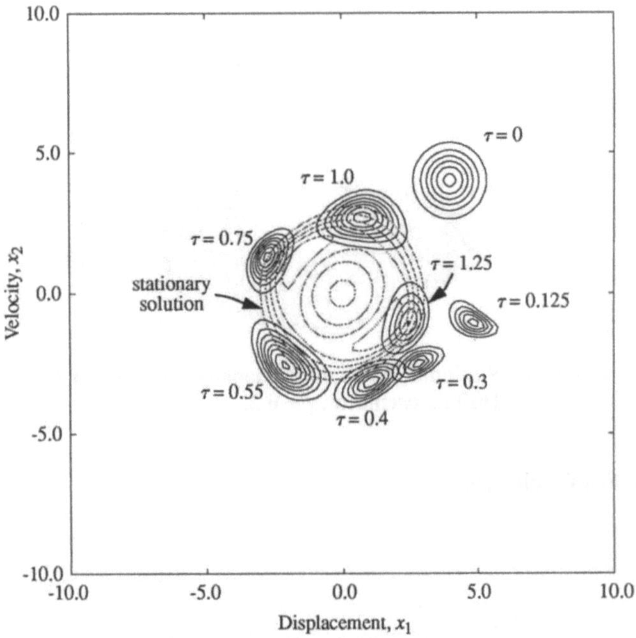

Figure 8. Contour plot of the evolution of the probability density function for the Van der Pol oscillator. $\zeta = 0.05$.

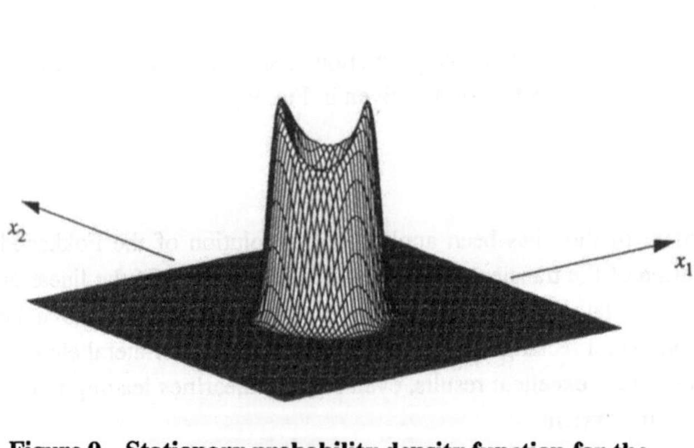

Figure 9. Stationary probability density function for the Van der Pol oscillator. $\zeta = 0.05$.

Acknowledgment

The authors are grateful to the National Center for Supercomputing Applications for a seed grant and access to the Cray Y-MP computer. The second author acknowledges the partial support of the National Science Foundation under Grant # BCS-9006781.

References

1 Atkinson, J.D. Eigenfunction Expansions for Randomly Excited Non-Linear Systems, *Journal of Sound and Vibration*, Vol. 30, No. 2, pp. 153-172, 1973.

2 Bergman, L.A. Numerical Solutions of the First Passage Problem in Stochastic Structural Dynamics, in *Computational Mechanics of Probabilistic and Reliability Analysis*, ElmePress International, Lausanne, Switzerland, pp. 479-508, 1989.

3 Bergman, L.A. and Spencer, B.F., Jr. Solution of the First Passage Problem for Simple Linear and Nonlinear Oscillators by the Finite Element Method, Department of Theoretical and Applied Mechanics, University of Illinois at Urbana-Champaign, *T & AM Report No. 461*, 1983.

4 Bhandari, R.G. and Sherrer, R.E. Random Vibrations in Discrete Nonlinear Dynamic Systems, *Journal of Mechanical Engineering Science*, Vol. 10, No. 2, pp. 168-174, 1968.

5 Bolotin, V.V. Statistical Aspects in the Theory of Structural Stability, in *Dynamic Stability of Structures*, (George Herrmann, Ed.), pp. 67-81, Pergamon Press, New York, 1967.

6 Caughey, T.K. and Dienes, J.K. The Behavior of Linear Systems with White Noise Input, *Journal of Mathematical Physics*, Vol. 32, pp. 2476-2479, 1962.

7 Caughey, T.K. Derivation and Application of the Fokker-Planck Equation to Discrete Linear Dynamic Systems Subjected to White Noise Excitation, *Journal of the Acoustical Society of America*, Vol. 35, No. 11, pp. 1683-1692, 1963.

8 Caughey, T.K. Nonlinear Theory of Random Vibrations, in *Advances in Applied Mechanics* Vol. 11 (Chia-Shun Yih, Ed.), Academic Press, pp. 209-253, 1971.

9 Crandall, S.H., Chandiramani, K.L. and Cook, R.G. Some First-Passage Problems in Random Vibration, *Journal of Applied Mechanics, ASME*, Vol. 33, pp. 532-538, 1966.

10 Crandall, S.H. First Crossing Probabilities of the Linear Oscillator, *Journal of Sound and Vibration*, Vol. 12, No. 3, pp. 285-299, 1970.

11 Langley, R.S. A Finite Element Method for the Statistics of Non-Linear Random Vibration, *Journal of Sound and Vibration*, Vol. 101, No. 1, pp. 41-54, 1985.

12 Langtangen, H.P. A General Numerical Solution Method for Fokker-Planck Equations with Applications to Structural Reliability, *Probabilistic Engineering Mechanics*, Vol. 6, Nol. 1, pp. 33-48, 1991.

13 Lin, Y.K. *Probabilistic Theory of Structural Dynamics*, Mc-Graw Hill, New York, 1967.

14 Nigam, N.C. *Introduction to Random Vibrations*, MIT Press, Cambridge, 1983.

15 Roberts, J.B. and Spanos, P.D. Stochastic Averaging: An Approximate Method of Solving Random Vibration Problems, *International Journal of Non-Linear Mechanics*, Vol. 21, No. 2, pp. 111-134, 1986.

16 Roberts, J.B. and Spanos, P.D. *Random Vibration and Statistical Linearization,* Wiley, Chichester, 1990.

17 Soize, C. Steady-State Solution of Fokker-Planck Equation in Higher Dimension, *Probabilistic Engineering Mechanics,* Vol. 3, No. 4, pp. 196-206, 1988.

18 Spencer, B.F., Jr. *On the Reliability of Nonlinear Hysteretic Structures Subjected to Broadband Random Excitation,* Lecture Notes in Engineering (series editors: C.A. Brebbia and S.A. Orszag), Vol. 21, Springer-Verlag, 1986.

19 Sun, J.-Q. and Hsu, C.S. First-Passage Time Probability of Non-Linear Stochastic Systems by Generalized Cell Mapping Method, *Journal of Sound and Vibration*, Vol. 124, 233-248, 1988.

20 Sun, J.-Q. and Hsu, C.S. The Generalized Cell Mapping Method in Nonlinear Random Vibration Based Upon Short-Time Gaussian Approximation, *ASME Journal of Applied Mechanics,* Vol. 57, pp. 1018-1025, 1990.

21 Wang, M.C. and Uhlenbeck, G. On the Theory of Brownian Motion II, *Reviews of Modern Physics,* Vol. 17, No. 2-3, pp. 323-342, 1945. Reprinted in *Selected Papers on Noise and Stochastic Processes* (N. Wax, Ed.), Dover, New York, 1954.

22 Wen, Y.K. Approximate Method for Nonlinear Random Vibration, *Journal of the Engineering Mechanics Division, ASCE,* Vol. 101, No. EM4, pp. 389-401, 1975.

23 Wen, Y.K. Method for Random Vibration of Hysteretic Systems, *Journal of the Engineering Mechanics Division, ASCE,* Vol. 102, No. EM2, pp. 249-263, 1976.

About Stochastic Linearization

P.BERNARD

Laboratoire de Mathématiques Appliquées.
Université Blaise Pascal. Clermont Ferrand

Summary
This contribution is devoted to some remarks about stochastic linearization and different ways to define it. It is shown that the classical equivalent linearization is not available for power spectral density matching to the dynamics of non-linear oscillators with impacts. Other linear -and also non-linear- approximations are introduced.

1 Introduction

Our goal is to define effective methods for providing information on the dynamics of non-linear oscillators with impacts. The origin of this problem is in the dynamics of heat generators pipes in nuclear power plants. The source of randomness is the turbulent flow of high temperature water circulating in and around the pipes in the heat-exchanger. The impacts are with the supports: for reasons of thermal stresses, gaps must exist between the pipes and their supports. The case of bilateral symmetric gaps is considered here. The impact is modelled by a conservative linear spring, the stiffness of which takes account of three effects:
- the stiffness of the support,
- the stiffness induced by the deformation of the pipe during the impact,
- the stiffness due to neglecting high frequency modes.

The industrial need is to define linear computational algorithms taking into account the most significant among the non-linear effects due to rapidly repeated impacts, with a view to the design of a reliability computational code.

This problem gives us the opportunity to make some remarks about the classical statistical linearization method, and to make some remarks as an introduction to new ones.

2 Some preliminary remarks

Let us consider a non-linear system driven by a white or coloured noise and let us use the classical representation by a stochastic differential system of Itô or Stratonovitch type (from now on, we will use only the Itô formulation). Then consider the Fokker-Planck-Kolmogorov (FPK) partial differential equation for the probability transition density, which is supposed to exist. Note that, from a result of N.Bouleau and F.Hirsch ([4]), it does exist in almost all practical situations. This is a parabolic PDE, which is degenerate, and which is impossible to solve analytically in almost all cases. The complete solution of this equation is the only way to get the law of the response of the system *as a process*. Assuming the existence of an invariant probability measure for the Itô system, it can be obtained as a solution of the stationary FPK equation, which is obtained from the

N. Bellomo · F. Casciati (Eds.)
Nonlinear Stochastic Mechanics
IUTAM Symposium Turin

preceding by setting the time derivative term to zero. Conditions for the existence of this stationary solution are given by R.Z.Hasminskii ([7]). In the case of single degree of freedom (SDOF) oscillators with linear damping, this stationary equation can be solved analytically. This is also the case for some particular multi-degree of freedom (MDOF)oscillators ([6], [12], [13]). However, the stationary solution in this case is just a one-lag probability, that is to say the probability law of the response $X(t)$ at one single time t, and not its law as a process. Hence, this does not give any information on two-lag probabilities, and in particular on the correlation function or the spectral power density. In fact, this does not give any real indication of the dynamics of the oscillator.

The example of the SDOF oscillator with impacts is illustrative of this fact: as can be seen by a glance at formula (4), the stationary law of the velocity of the impacting oscillator is the same as the stationary law of the linear oscillator without impact.

Hence, one cannot expect, from any kind of linearization based upon only the one-lag stationary probability, to do better and find a satisfactory approximation to the dynamic characteristics.

3 Exact formulas for the SDOF oscillator with impacts

3.1 FPK equation

The dynamical equation for this oscillator driven by a white noise, e denoting the value of the gap at rest, and k' the impact stiffness, is the following:

$$m\ddot{q}(t) + c_0\dot{q}(t) + \frac{dU}{dx}(q(t)) = f(t) \tag{1}$$

where

$$U(q) = \frac{1}{2}k_0 q^2 + \frac{1}{2}k'[(q-e)^2 H(q-e) + (q+e)^2 H(-q-e)]$$

The associated Itô stochastic differential equation (SDE), available in the phase space, putting

$$X = \begin{bmatrix} q \\ \dot{q} \end{bmatrix}$$

is the following:

$$\dot{X}(t) = b(X(t))dt + \sigma dW_t \tag{2}$$

where

$$b(x) = \begin{bmatrix} x_2 \\ -m^{-1}\frac{dU}{dx_1} - m^{-1}c_0 x_2 \end{bmatrix}$$

$$x = \begin{bmatrix} x_1 \\ x_2 \end{bmatrix}$$

The associated Fokker-Planck equation for the transition probability is then:

$$\frac{\partial}{\partial t}p(t,y;s,x) = \frac{1}{2}\sigma^2\frac{\partial^2}{\partial y_2^2}p(t,y;s,x) - y_2\frac{\partial}{\partial y_1}p(t,y;s,x)$$

$$-m^{-1}[k_0 y_1 + k'((y_1-e)H(y_1-e) + (y_1+e)H(-y_1-e))\frac{\partial}{\partial y_2}p(t,y;s,x) \tag{3}$$

with the initial condition :

$$lim_{t\downarrow s}p(t,y;s,x)dy = \delta_x(dy)$$

The stationary FKP equation is obtained by setting to zero the time derivative term in (3). It can then be computed analytically, and the stationary solution can be expressed as follows:

$$p(x_1,x_2) = Cexp[-\frac{2c_0}{\sigma^2}U(x_1)]exp[-\frac{1}{2}\frac{2c_0m}{\sigma^2}x_2^2] \tag{4}$$

One can immediately see the two important following facts from this expression :
1) *The position and the velocity are independant at any given time t,*
2) *The stationary law of the velocity of the impacting oscillator is the same as the stationary law of the velocity of the same oscillator without impact.*
(This two facts are not true for MDOF systems)

3.2 Equivalent linearization in the classical sense

Let us recall in a few lines the principle of the method. Consider the case of an oscillator with non-linear stiffness, the motion of which is described by:

$$\ddot{y} + 2\xi\omega_0\dot{y} + h(y) = f(t) \tag{5}$$

The classical way to define a linear oscillator approximating the preceding is to decompose the non-linear term of (5) into a linear part and a residue:

$$\ddot{y} + 2\xi_e\omega_e\dot{y} + \omega_e^2 y + [h(y) - \omega_e^2 y] = f(t) \tag{6}$$

and to choose ω_e so as to minimize the quantity $E[(h(y) - \omega_e^2 y)^2]$. It is noted here that it is not correct to present this as arising from the difference between equation (5) and the linear equation

$$\ddot{y} + 2\xi_e\omega_e\dot{y} + \omega_e^2 y = f(t) \tag{7}$$

since the y are not the same in these two equations.
From the present presentation, it is clear that the y in the residue is the non-linear solution, which is unknown. Hence the idea to replace this y by the solution of the linear equivalent equation. But as this equation depends upon coefficients which are unknown, we get a fixed point problem to solve. In the general case we do not know if there is a solution; there can be several, and we have no theoretical estimation of the error in any sense between the true solution and the equivalent linearized one.

Also, the solution y is Gaussian. Hence, we can use the integration by part formula

$$E(h(q).q^T) = E(\nabla h(q))E(q.q^T) \tag{8}$$

available for any differentiable $h : \mathbb{R}^n \rightarrow \mathbb{R}^n$ and Gaussian q. In the case of non-differentiable h, this formula can be interpreted in the Gaussian distribution calculus sense ([1], [9]).

In the particular case of a SDOF oscillator with impacts, the fixed point problem is the following:

$$k_e = k_0 + 2k'(1 - \Phi(\bar{e}\sqrt{\frac{k_e}{k_0}})) \tag{9}$$

where $\bar{e} = e/\sigma_0$.

This problem has one and only one solution; this can be proved using a fixed point theorem for continuous mappings from a compact interval into itself. Let us remark here that the more usefull algorithm is not the well-known $x_{n+1} = f(x_n)$; because the function f is decreasing, the following algorithm is more efficient:

$$x_{n+1} = f(x_n); \quad x_{n+2} = \frac{1}{2}(x_n + x_{n+1}).$$

3.3 Variance matching equivalent linearization

The point of view is different here. We look for a linear SDE whose solution has the same second moments as the true solution, as discussed by F.Kozin in Innsbruck's IUTAM symposium on nonlinear stochastic dynamic ([8]). We work in the state-space. Let

$$dX_t = g(X_t, t)dt + \sigma dW_t \tag{10}$$

be a non-linear SDE with constant σ. Then, the variance matrix Γ_t satisfies the differential relation:

$$\dot{\Gamma}_t = E[X_t.g^T(X_t) + g(X_t).X^T(t) + \sigma\sigma^T] \tag{11}$$

In the linear case with time-dependent coefficients

$$dX_t = A_t X_t dt + \sigma dW_t \tag{12}$$

and the variance matrix Γ_t of X_t satisfies the following Ricatti differential equation:

$$\dot{\Gamma}_t = \Gamma_t A_t^T + A_t \Gamma_t + \sigma\sigma^T \tag{13}$$

Denoting by A_e the time dependent linear characteristic of the sought equivalent linear equation, the following relation will hold:

$$\Gamma A_e^T + A_e \Gamma = E[X_t.g^T(X_t) + g(X_t).X^T(t)] \tag{14}$$

One solution is:

$$A_e(t) = E(g(X_t)X_t^T).\Gamma_t^{-1} \tag{15}$$

In other words, this coefficients are just the coefficients of the linear regression of $g(X_t)$ on X_t. In the stationary case, they will not depend upon t.

The coefficients of this equivalent linearization can now be computed for the oscillator with impacts. The damping coefficient does not change, and the stiffness is given by:

$$k_e = E(q_t h(q_t))/E(q_t^2) \tag{16}$$

with

$$h(q) = k'[(q - e)H(q - e) + (q + e)H(-q - e)] \tag{17}$$

As we know the true probability law of $q(t)$, we can compute exactly k_e. The following result is obtained:

$$k_e = k_0 n/d \tag{18}$$

where:

$$\begin{aligned} n &= \frac{1}{2} - \Phi(-\bar{e}) + \frac{1}{(1+C)^{1/2}} exp(-\frac{1}{2}\frac{C\bar{e}^2}{1+C})\Phi(-\frac{\bar{e}}{(1+C)^{1/2}}) \\ d &= \frac{1}{2} - \Phi(-\bar{e}) - \frac{\bar{e}}{\sqrt{2\pi}} exp(-\frac{\bar{e}^2}{2}) + [\frac{1+C+(C\bar{e})^2}{(1+C)^{5/2}}\Phi(-\frac{\bar{e}}{(1+C)^{1/2}}) \\ &+ \frac{\bar{e}(1+2C)}{\sqrt{2\pi}(1+C)^2} exp(-\frac{1}{2}\frac{\bar{e}^2}{1+C})] exp(-\frac{1}{2}\frac{C\bar{e}^2}{1+C}) \end{aligned} \tag{19}$$

The constant C is the normalisation constant of (4), and $\bar{e} = \frac{e}{\sigma_0}$ is the reduced gap, a dimensional parameter which is the quotient of the gap by the standard deviation of the response of the linear oscillator without impacts to the same level of white noise excitation.

To conclude this section, let us recall that these coefficients are those which are estimated by the maximum-likeliwhood estimators given by F.Kozin [8].

4 Power spectrum approximation and spectral linearization

By simulation, and statistical estimation, the spectral power density of the stationary response can be obtained (fig.2). One finds that this is not at all the shape of the spectral density of any harmonic oscillator. Hence, this is one more reason to consider that equivalent linearization is not satisfactory to approximate the dynamics of the impacting oscillator.

Several methods can be tried here to approximate this spectral power density. The goal is not to review all possible methods; moreover, only methods available for MDOF systems have been considered.

As a first try Wiener Chaos decomposition was considered. This is a way to obtain an approximation of the response *as a process* and not just as a random variable for a fixed time. The best Gaussian process approximation of a given process is its projection on the first Wiener Chaos ([1],[9]). This can be expressed:

$$Y(t) = E(X(t)) + \int_0^t h(t,s)dW_s \tag{20}$$

where W_t is a Wiener process, the integral an Ito stochastic integral, and h is the first *Wiener kernel*. Unfortunately, it was found to be impossible to obtain an effective algorithm for computing this kernel. The only point worth mentioning here is that this projection *is not* the same as the equivalent linearization, even if some authors thought they had proved the equivalence of these two notions. One way to explain this is the fact that the first chaos projection has no reason to yield a Markov process, whilst equivalent linearization always leads to a Markov process.

4.1 Spectrum approximation via stochastic averaging

This approximation is available for lightly damped oscillators, with wide-band stochastic excitation [11]. The first step is an appropriate change of variables, such that the new quantities are an energy (enveloppe) and a phase, satisfying a SDE with a small parameter. The stochastic averaging method is then applied. This method can be applied to a Duffing oscillator, or a more general oscillator with polynomial non-linear stiffness, which can itself be an approximation by Equivalent non-Linearization (ENL) similar to Caughey's method [5]. Let us also mention a work in progress by R.Bouc on this question [3].

4.2 Spectral Linearization

The problem considered in this section is the following: given a power density spectrum, obtained from a SDOF non-linear system by measure, or by simulation, or by approximate calculus, define a MDOF linear system, the density spectrum of which when driven by a white noise is an acceptable approximation to the given spectrum. A solution to this problem can be proposed, using time dicretization and ARMA processes. The main point in this method is the relation between continuous time and discrete time. More precisely, the following problem has to be solved: given an ARMA process which we know is the time-discretization of the response of an harmonic oscillator excited by a white noise, characterize this oscillator. In fact, an easy answer is obtained from the fact that the poles of the linear discrete time filter defining this ARMA process are

$$exp(-\xi\omega_0 h \pm i\omega_d h) \tag{21}$$

where ξ and ω_0 are the damping and pulsation of the oscillator, $\omega_d = \omega_0\sqrt{1-\xi^2}$, and h is the time discretization step.
Hence, a classical ARMA identification procedure in the spectral field, such as Whittle algorithm, will give a linear system *spectrally equivalent* to the non-linear one. This was applied to the harmonic oscillator with impacts in [14]. Some results are on figure 4.

5 Equivalent non-linearization

The case of non-linear stiffness only is considered here. Among possible non-linearities, some are 'nicer' than others. For example, Duffing oscillators are not too bad, since the stationary probability and some approximation to the correlation function can be computed. It will be the case of some classes of polynomial non-linearities. This is why approximation of the given non-linearity by a polynomial non-linearity of this kind can be tried. The argument is the same as in the case of equivalent linearization: write

$$\begin{aligned} \ddot{z}_t + 2\xi\omega_0\dot{z}_t + \omega_0^2 z_t + k(z_t) &= f(t) \\ k(z_t) &= az_t + bz_t^3 \end{aligned} \tag{22}$$

where a and b are chosen such as to minimize $E(\varepsilon^2)$, with $\varepsilon(z) = h(z) - k(z)$.
Hence:

$$\begin{aligned} aE(z^2) + bE(z^4) = E(zh(z)) \\ aE(z^4) + bE(z^6) = E(z^3h(z)) \end{aligned} \tag{23}$$

This is to be computed when the law of z is the true stationary probability of the oscillator with impacts, which is unknown. Hence, we can use the approximate non-linear solution, but there is again a fixed point problem, and tedious computations. Moreover, in the MDOF cases, this non-linear approximation law is unknown too.

So, the linear equivalent Gaussian probability can be tried. This gives some results (fig.6). But a glance at the results shows that this Gaussian probability does not weight the right displacements. For example, what happens for a displacement near the gap value e is underestimated, while it gives a very important contribution to dynamical effects. It seems better to use the uniform probability on an interval $[-\alpha\sigma_e,\ \alpha\sigma_e]$ with some $\alpha > 0$, $2 < \alpha < 3$ for example.

Results of the computation of exact values for a and b using the variance-matching equivalent linearization are:

$$\begin{aligned} a &= \omega_1^2[2\Phi(-X_e) - \frac{1}{\sqrt{2\pi}}X_e exp(-X_e^2/2)] \\ b &= \frac{1}{\sqrt{2\pi}}\frac{\omega_1^2}{3(\sigma_e^2)}X_e exp(-X_e^2/2) \end{aligned} \tag{24}$$

where

$$X_e = \bar{e}\frac{\sigma_\infty}{\sigma_e}.$$

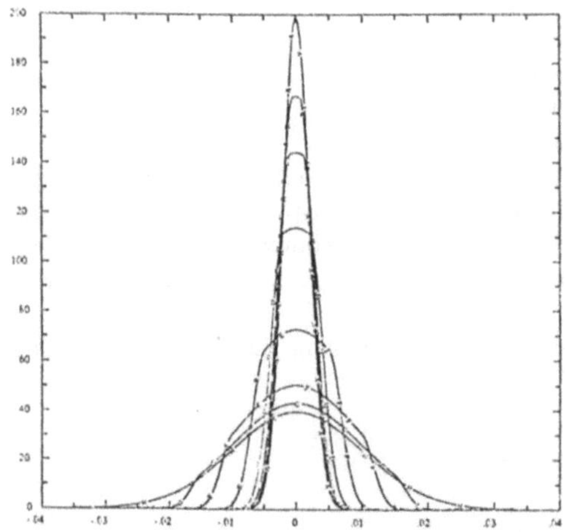

fig.1
Invariant probability density of a SDOF oscillator with impacts for different values of the gap

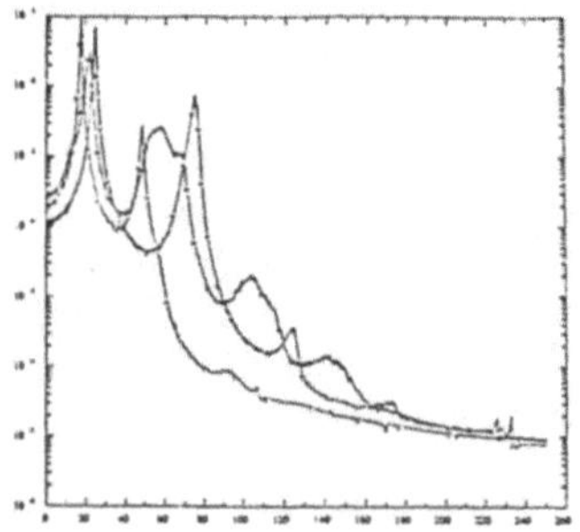

fig.3
Power spectrum of a MDOF oscillator with impact for different values of the gap

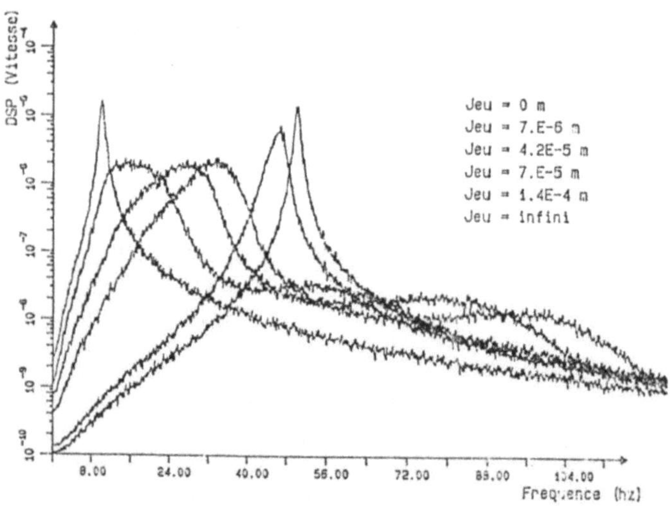

fig.2
Power spectrum of a SDOF oscillator with impacts for different values of the gap

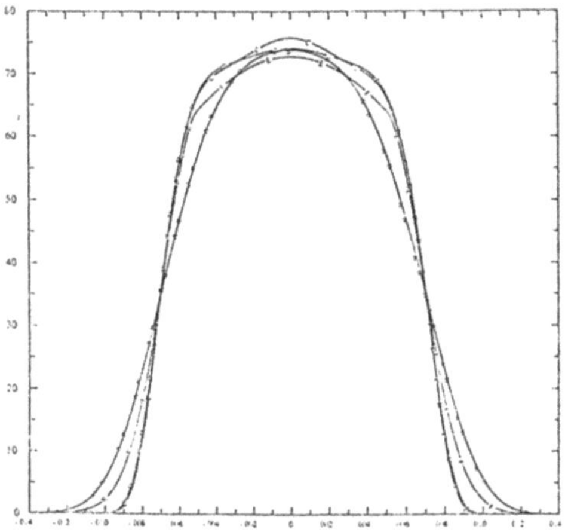

fig.5
Invariant probability density of a SDOF oscillator with impacts and of some equivalent non linearization

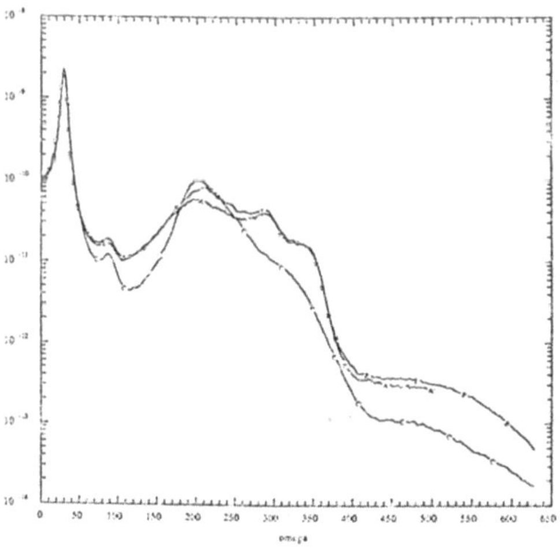

fig.6
Spectrum of a 2DOF oscillator with impacts and of its Duffing equivalent non-linearization

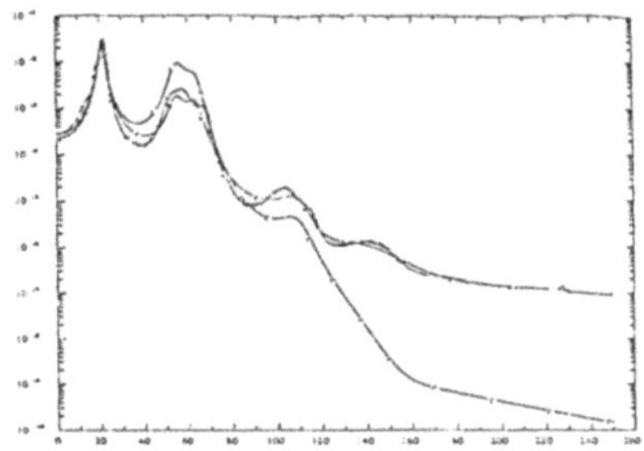

gap:1.E-5m.

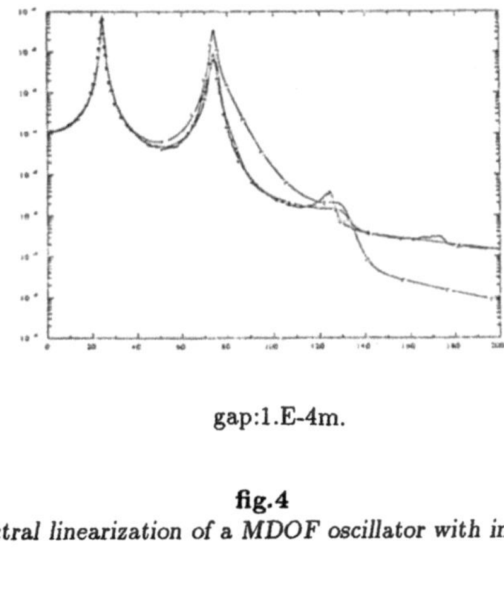

gap:1.E-4m.

fig.4
Spectral linearization of a MDOF oscillator with impacts

Figures are reproduced from References [2] and [13].

References

[1] P.Bernard: Méthodes Mathématiques d'Etude des Vibrations Aléatoires et Analyse sur les Espaces Gaussiens. Thèse de Doctorat d'Etat. Université Blaise Pascal.(1990).

[2] P.Bernard-M.Fogli: Dynamique Stochastique Non Linéaire des Structures - Application à l'Etude des Vibrations Aléatoires de Structures Tubulaires en Présence de Chocs. Rapport de contrat EDF-SEPTEN/Université Blaise Pascal. 1990.

[3] R.Bouc: Power spectral density of a weakly damped strongly non-linear random oscillation, and stochastic averaging. Publication du L.M.A., octobre 1991.

[4] N.Bouleau-F.Hirsch: Propriétés d'absolue continuité dans les espaces de Dirichlet et application aux équations différentielles stochastiques. Séminaire de Probabilités XX. Lecture Notes in Mathematics **1204**, Springer Verlag (1986).

[5] Caughey T.K.: On the Response of Non-Linear Oscillators to Stochastic Excitation. Prob. Eng. Mech. **1** 1 2-4 (1986).

[6] T.K.Caughey-F.Ma: International Journal of Nonlinear Mechanics. **17**, 3, 1983 (137-142).

[7] R.Z.Hasminskii: Stochastic Stability of Differential Equations. Sijthoff & Noordhoff (1980).

[8] F.Kozin: The method of Statistical Linearization for Nonlinear Stochastic Vibrations. IUTAM Symposium, Innsbruck. Springer Verlag, 1988 (45-56).

[9] P.Krée: Methods of Distributions in the Modelling of Random Vibrations. Nonlinear Structural Systems under Random Conditions. Proceedings of the Euromech Colloquium, Como. Edited by F.Casciatti, I.Elishakoff, J.B.Roberts. Structural Safety, special issue, **8** 1-4 , 131-142 (1990)

[10] J.B.Roberts-P.D.Spanos: Random Vibration and Statistical Linearization. J.Wiley & Sons (1990).

[11] J.B.Roberts-P.D.Spanos: Stochastic Averaging: an Approximate Method for Solving Random Vibration Problems. Int. J. Non-linear. Mech. **21** 111-134 (1986).

[12] C.Soize: Steady-state solution of Fokker-Planck equation in higher dimension. Probabilistic Ingineering Mechanics, **3** 4 196-206 (1988).

[13] C.Soize: Exact Stationary Response of Multi-dimensional nonlinear Hamiltonian Dynamical Systems under Parametric and External Stochastic Excitations. To appear in Jal of Sound and Vibrations.

[14] M.Taazount: Etude Dynamique des Structures à Chocs sous Sollicitations Aléatoires. Thèse de l'Université Blaise Pascal. 1991.

Fatigue Life Estimation under Non-Stationary Stochastic Processes

Matěj BÍLÝ, Vladimír KLIMAN
Institute of Materials and Machine Mechanics of the Slovak Academy of Sciences, Bratislava, Czechoslovakia

Summary

The paper presents classification, identification and analysis of non-stationary process properties both in the scope of the correlation theory and counting methods. It is shown how such processes are simulated in real time to obtain the experimental life assessment, or exploited for the computational life estimation using fatigue damage accumulation hypotheses.

Introduction

Despite a long time battle against fatigue, designing to avoid fatigue failures of dynamically loaded components still remains one of the more difficult tasks for an engineer to face. This is mainly due to the fact that many of the factors involved in fatigue processes have only been partially established and are largely empirical. As some experimental results suggest, one of such factors is the non-stationary behaviour of operating loading processes which manifests itself in more than 50% of all practical cases. Taking into account that the process non-stationarities have clearly adverse effects on the resulting fatigue life, it is worth elaborating the problem of fatigue under non-stationary stochastic loads and propose some practically realizable ways of its solution. Three partial fields are to be considered here, covering both the experimental and computational estimations of fatigue lives, and in two directions, generally accepted in fatigue, i.e.

- for the macroblock representation of a non-stationary stochastic process, and
- for a true stochastic non-stationary process.

Classification and identification of non-stationary process properties

Classification of non-stationary stochastic processes may be accomplished in relation to their characteristics or parameters (moment estimators). Thus we can get a non-stationary process with

- a time-dependent first probability density function $f(x,t)$,
- a time-dependent autocorrelation function $B(t,\tau)$,
- a time-dependent power spectral density $S(f,t)$,

N. Bellomo · F. Casciati (Eds.)
Nonlinear Stochastic Mechanics
IUTAM Symposium Turin

- a higher order probability density function $f(x_1, x_2, ..., t)$, etc.

In practice the non-stationary properties are assessed according to their moment estimators, so we get a non-stationary process with

- a time-dependent mean value $\mu(t)$,
- a time-dependent variance $\sigma^2(t)$ [standard deviation $\sigma(t)$],
- both time-dependent mean value $\mu(t)$ and variance $\sigma^2(t)$,
- another time-dependent characteristic parameter, etc.

It is obvious that various combinations can occur here. Moreover, in the literature also other classifications of non-stationary or stationary features of stochastic processes have arisen [1].

A special kind of practically useful processes are so-called *separable non-stationary processes* $y(t)$ which can be expressed as

$$y(t) = A(t)\ x(t) \quad \text{or} \quad y(t) = x(t) + B(t),$$

where $A(t)$ and $B(t)$ are deterministic functions and $x(t)$ is a stationary stochastic process.

Identification of the non-stationary stochastic process properties is a rather awkward and not easily performed procedure. In the most correct mathematical way the non-stationary characteristics should be evaluated as *ensemble averages* from more realizations at discrete time intervals t_i (Fig. 1). Thus for $i = 1,2,...$ we can get a set of

probability density functions (PDFs) $f(x,t)$ and their *statistical moments*

$$m(t_i) = \int_{-\infty}^{\infty} x(t_i)\ f(x,t_i)\ dx,$$

$$s^2(t_i) = \int_{-\infty}^{\infty} [x(t_i) - \mu(t_i)]^2\ f(x,t_i)\ dx,$$

autocorrelation functions (ACFs)

$$R(t_i,\tau) = \int_{-\infty}^{\infty} x(t_i)\ x(t_i+\tau)\ f(x,t_i)\ dx$$

and power spectral densities (PSDs)

$$S(t_i,f) = \int_{-\infty}^{\infty} R(t_i,\tau)\ exp(-i2\pi f\tau)\ d\tau.$$

Unfortunately, in practical applications a sufficient number of process realizations obtained under identical conditions are not usually available and, even worse, these statistical characteristics can hardly be directly used in fatigue. For this reason so-called *parametric evolutionary* statistical characteristics are preferred defined by the following expressions:

parametric evolutionary mean value

$$[m(t)]^T = \frac{1}{T} \int_{t-T}^{t} x(\xi)\ d\xi,$$

parametric evolutionary variance

$$[s^2(t)]^T = \frac{1}{T} \int_{t-T}^{t} \{x(\xi) - [m(t)]^T\}^2 \, d\xi,$$

parametric evolutionary autocorrelation function

$$[R(t,\tau)]^T = \frac{1}{T} \int_{t-T}^{t} x(\xi) \, x(\xi+\tau) \, d\xi, \text{ and}$$

parametric power spectral density

$$[S(t,f)]^T = \int_{-\infty}^{\infty} R(t,\tau) \, exp(-i2\pi f\tau) \, d\tau.$$

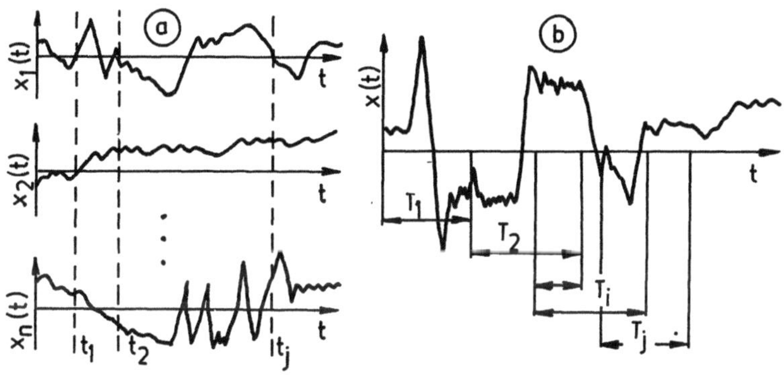

Fig. 1. Scheme of calculation of ensemble averaged (a) and parametric (b) statistical characteristics

Considering that the aim of our analysis is the fatigue life assessment, one can adopt for this purpose an entirely different approach based, for example, on some **counting methods**. Nowadays the most commonly used is *the rain flow method* (RFM) whose principle consists in determining the number of occurrences of partial process ranges with variable mean values and amplitudes which are further collected into a set of blocks forming a macroblock of harmonic cycles (Fig.2). It has a fundamental advantage compared with the correlation theory method as it can be with some caution equally well applied to both stationary and non-stationary processes, and its results can be directly used both for the computational and experimental fatigue life estimation.

Another promising method of the non-stationary process analysis based on **Markov chains** of the first or higher orders. If it is applied to the process ordinates then as a result the *transition probability matrix* between subsequent ordinates i and j is obtained, containing the transition probabilities p_{ij}. If instead of the process ordinates the upper and lower process envelopes are considered, we can obtain the *transition probability matrix*

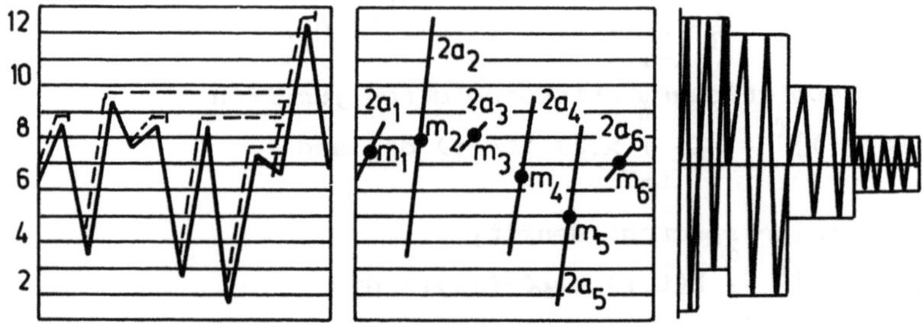

Fig. 2. Principle of rain flow analysis of stochastic process yielding macroblock of sinusoidal cycles

between envelopes, describing both the process and envelope properties, even taking into account the time intervals between transitions. These matrices can then be used for laboratory purposes enabling the non-stationary stochastic process simulation.

Simulation of non-stationary processes

In order to estimate the fatigue life corresponding to the typical use conditions it is indispensable to simulate the operating stochastic processes which are further used as inputs of loading machines or formulae for the fatigue life estimation. Here the progress has reached a more advanced stage compared with the analysis and so practically any non-stationary process can be simulated **in real time** providing its time-dependent statistical characteristics are available [1].

One important problem arising here is the criterion of **quality of the simulation**, or the criterion of the choice of "*the most representative parameters*", in which the simulated and operating processes coincide. The ideal universal method should guarantee the identity of all parameters. This is, however, not always desirable (e.g. at various design stages the corresponding information may not be available), so different opinions on the relative importance of various characteristics of the simulated process must be taken into account.

In accordance with the stochastic process analysis, information about the operating process and the philosophy of the fatigue damage accumulation there are various methods of the non-stationary stochastic process simulation, which could be classified into three groups.

(a) The simplest simulation method is based on a **spectrum of blocks** (macroblock shown in Fig. 2). During the simulation they can be organized in an *ascending-descending* manner, or can be *randomly distributed* to simulate the fatigue damage caused by an operating process. Despite all these disputable facts, the results of experimentally estimated fatigue lives according to this method are reliable, especially if the simulated process represents **strain**, characterizing in the best way the cyclic material behaviour. Sometimes even individual cycles are simulated, forming in the whole the macroblock.

(b) According to our experience, a very promising way of the non-stationary stochastic process simulation seems to be the method exploiting the *transition probability matrix* between the process or upper and lower envelope ordinates, respectively. If two or more adjacent ordinates are taken into account, *Markov chains* of the second or higher orders are simulated (which takes time and from the point of view of fatigue brings hardly estimable advantages).

(c) A more variable and complicated situation occurs when we want to simulate the non-stationary stochastic process with a prescribed *probability density function* $f(x,t)$, and/or *power spectral density* $S(f,t)$. According to the mathematical models developed, two important cases can be considered:

(c1) *Arbitrarily distributed white noise*
White noise is an uncorrelated stochastic process (with a constant power spectral density) with some probability density function of ordinates, which contains certain parameters; if the simulated process is non-stationary then these parameters should be time-dependent deterministic functions (at least one of them). As an example suppose that we want to simulate a three-parameter **Weibull process** with the probability density function

$$f(x,t) = A(t)[x - a(t)]^{m(t)}\exp\{-\frac{A(t)[x - a(t)]^{m(t)+1}}{m(t) + 1}\}.$$

After a linear transformation

$$y(t) = \frac{A(t)}{m(t) + 1}[x(t) - a(t)]^{m(t)+1} \qquad (1)$$

with the Jacobian $J = A(t)\,[x(t) - a(t)]^{m(t)+1}$ we get
$f(y,t) = f(x,t)/J = exp(-y)$.
Thus the resulting process $y(t)$ is stationary and has a normalized exponential PDF. The simulation algorithm is obtained after inverse transformation of eqn (1), i.e.

$$x(t)=a(t) + [\frac{m(t) + 1}{A(t)}\,y(t)]^{1/[m(t)+1]}.$$

In an analogous way one can obtain any other analytical PDF.

It is worth mentioning that if the PDF parameters are constants or stationary stochastic processes then the resulting process is also stationary but with a composed complex PDF in the latter case. Even a more complicated process results when the PDF parameters are non-stationary functions with their time-dependent deterministic parameters.

(c2) *Process with a time-dependent PSD*

Suppose that such a non-stationary stochastic process $y(t)$ is the output of a linear dynamic system excited by some input non-stationary process $x(t)$. In the convolution form it can be expressed as

$$y(t) = \int_{-\infty}^{\infty} h(\xi)\, x(t-\xi)\, d\xi, \qquad (2)$$

where $h(\xi)$ is the unit impulse response function.

Using the Wiener-Khinchin relationship

$$S_y(f) = \int_{-\infty}^{\infty} R(t,\xi)\, \exp(-i2\pi f\tau)$$

and inserting for $R_y(t,\tau)$ from its definition $R_y(t,\tau) = \langle y(t)\, y(t+\tau)\rangle$, where the symbols $\langle\ \rangle$ denote the mean value operator, and for $y(t)$ from eqn (2), then after rather tedious rearrangement [1] we finally get

$$S(t,f) = H(t,f)\, S_x(t,f),$$

where

$$H(t,f) = H(f)\, K(t,f), \quad H(f) = \int_{0}^{\infty} h(\vartheta)\, \exp(-i2\pi f\tau)\, d\tau,$$

$$K(t,f) = \int_{0}^{\infty} k(t,\xi)\, \exp(2\pi f\xi), \quad k(t,\xi) = \psi(t,\xi)\, h(\xi),$$

$$\psi(t,\xi) = R_x(t-\xi,\vartheta)/R(t,\vartheta) \quad \text{for } R_x(t,\vartheta) \neq 0.$$

Providing the PSDs $S_y(t,f)$, $S_x(t,f)$ and function $\psi(t,\xi)$ are known we obtain the unknown unit impulse response function $h(\xi)$ and so the corresponding dynamic system. When this is done, then eqn (2) can be rewritten into the approximate discrete form [1]

$$y_i = \sum_{j=0}^{\infty} h(j\Delta t)\, x(t-j\Delta t)\Delta t = \sum_{j=0}^{\infty} h_j\, x_{i-j}\Delta t \cong \sum_{j=0}^{\infty} c_j\, x_{i-j}.$$

When simulating the pseudo-random numbers representing the process ordinates one should be aware that theoretically they range from minus to plus infinity or from zero to plus infinity. Even if in practice the simulation algorithms have certain limitations, the maximum number which can be generated (although with a very low probability) may still exceed, say, the ultimate tensile strength of a loaded component and cause an immediate failure. This naturally does not correspond to real operating conditions and so to avoid it, a certain maximum limit is to be prescribed. For a Gaussian process it usually makes three mean square values s because the probability of its exceeding is negligible; but for other types of processes such a fixed limit is obviously incorrect as the probability of appearance of higher amplitudes is fairly high, as obvious from Fig. 3. This is why any practical choice of the cut-off level must take into account not only the load process itself but also the material properties. In fatigue this means that exceeding this cut-off level should not bring about a substantial fatigue life change.

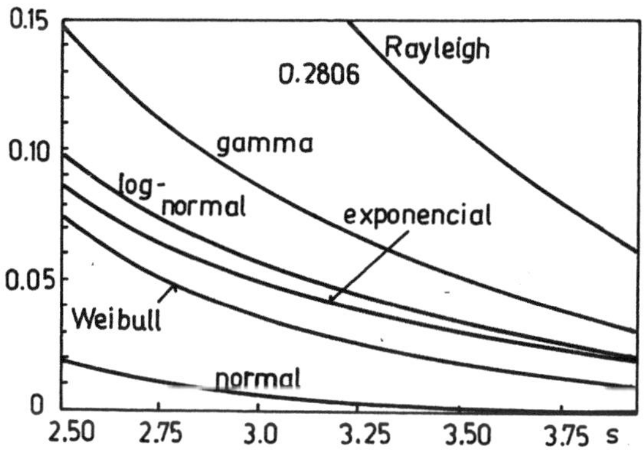

Fig. 3. Tail parts of various probability density functions

Finally, it is worth mentioning that all the simulation algorithms developed are fast enough to be used in **real time** when computing fatigue damage or running a computer-controlled servohydraulic loading system.

Fatigue life assessment

A vast amount of hypotheses, theories, views, as well as personal tastes have yielded a variety of recommended guide lines for the fatigue life assessment. Nevertheless, whatever their origin or philosophy, they all are based on

- material properties under sinusoidal loading (Wöhler curve, Manson-Coffin curve and cyclic stress-strain curve),
- load-time history or better, the local stress-time and/or strain-time history at the most critical location, and
- fatigue damage accumulation hypothesis, relating the material properties to the operating load (strain) history.

As far as the **material properties** are concerned then it is now generally accepted that the most descriptive material characteristic is the *Manson-Coffin curve* representing the relation between the sinusoidal strain amplitude ε_a (plastic strain amplitude ε_{ap}) and the number of cycles to fracture N_f in the form

$$\varepsilon_a = \frac{\sigma'_f}{E}(2N_f)^b + \varepsilon'_f(2N_f)^c,$$

where σ', ε', b, c and E are the material parameters. For this reason the operating process should represent strain and not stress as it still often happens.

A great many **fatigue damage accumulation hypotheses** have been proposed, based both on the macroblock representation ex-pressed by the formula $\sum (N_i/N_{fi}) f(\varepsilon_{ai}, \sigma_{ai}, a_1, a_2, \ldots) = A$, or on the correlation theory characteristics, generally expressed as $T = F[S(f), f(x), b_1, b_2, \ldots]$ [2,5], where N_i and N_{fi} are the number of cycles and the number of cycles to fracture at the *ith* macroblock level, respectively, characterized by its stress σ_{ai} or strain ε_{ai} amplitude, $S(f)$ - power spectral density of the operating process; $f(x)$ - probability density function; $a_1, a_2, \ldots, b_1, b_2, \ldots, A$ - parameters. All these hypotheses have certain advantages and drawbacks and there is probably no good reason for recommending any of them as the best one. Intuitively one perhaps feels that the more experimentally determined parameters a hypothesis contains, the better the approximation of experimental results should be but the more complicated hypotheses require more data and thus its use may be problematic. Our long experience suggests that it is useful to apply a few fatigue damage accumulation hypotheses in parallel and thus get a certain "scatter" and "mean value" of the fatigue life.

Because as usually the operating strain processes are measured at plain locations near a stress-strain concentrator (notch), it is necessary to transform them to the corresponding notch root. This may bring about certain difficulties.

First of all, it is obvious that because of the cyclic stress-strain behaviour, the loaded material does not follow the Hook's law and so the plain strain stochastic process ordinates cannot be simply multiplied by the elastic strain concentration factor to obtain their notch root pairs. One could use some other approaches based on the cyclic stress-strain curve [3,4] but for stochastic processes the results are doubtful and have not been experimentally verified yet. According to our opinion such a transformation should follow step-by-step the cyclic stress-strain curve and continuously recalculate the process ordinates. Such a computer programme has been developed in our institute.

Further, Fig. 4 suggests that the original stationary plain stress or strain process can become after the transformation clearly non-stationary. This can especially happen when the process has a non-zero mean. This unpleasant fact causes profound difficulties and so brings additional reasons why to pay an increased attention to non-stationary processes.

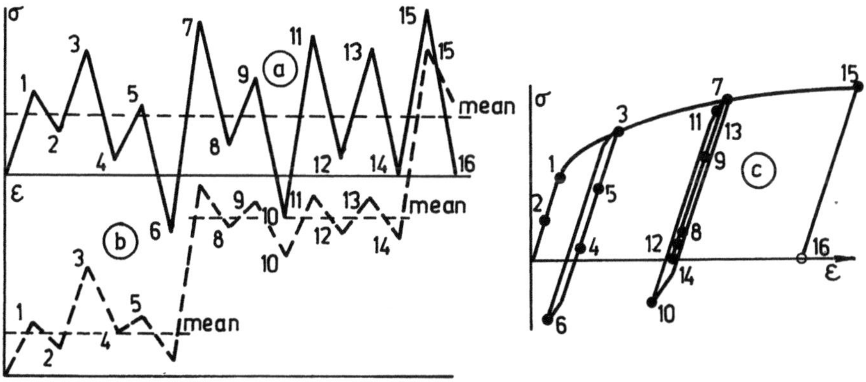

Fig. 4. Stress (strain)-time history at plain location (a) with its strain-time pair in notch root (b) and corresponding stress-strain diagram

What to do then in practice?

Consider a non-stationary stochastic process $\sigma(t)$ or $\varepsilon(t)$ (generally $x(t)$ as illustrated in Fig. 1), which is randomly segmented in k overlapping (as T_i and T_j) or non-overlapping parts (as T_1 and T_2). Using the two parametric *rain flow method*, analyze these k segments $T_1, T_2, \ldots, T_k$ and obtain k macroblocks with variable amplitudes x_{ai} and mean values x_{mi}. The resulting macroblocks are further inserted to any (reasonable) fatigue damage accumulation hypothesis, reflecting the block mean values x_{mi} and as a result the k estimations of fatigue life are obtained. Performing this procedure for all segments, we can design a set of life distribution functions for all selected hypotheses and thus estimate the scatter and mean value of the expected fatigue life λ_{MEAN} or L_{MEAN} (Fig. 5). If all this is performed correctly then the non-stationarity manifests itself in an increased scatter.

It is obvious that the number of segments k and their length T_j should be sufficient for the set of life estimations could be statistically processed. This decision obviously depends on the process length available, on the techniques used (the on-line analysis or analysis of a taped process facilitating segmentation with overlapping), on the knowledge of physical reasons for the non-stationary process behaviour, etc.; without doubts it requires an experienced analyst.

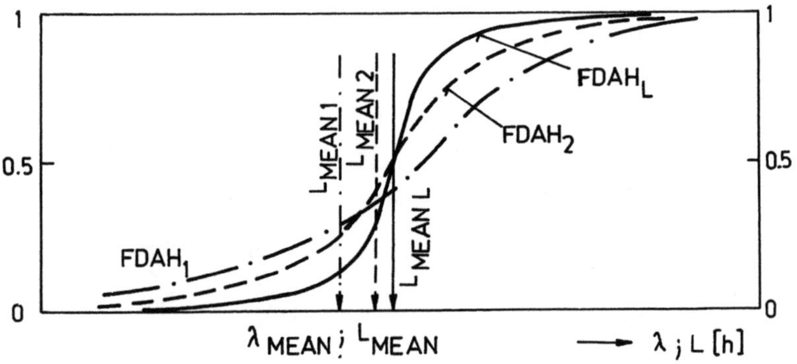

Fig. 5. Distribution functions of fatigue lives for various fatigue damage accumulation hypotheses (FDAH) obtained from one segmented non-stationary process

Analogically one can also proceed when assessing fatigue endurance experimentally under a simulated non-stationary process. Here the macroblocks corresponding to partial segments are reproduced (in the best way in a random manner); the result is again expressed in the probabilistic sense as the mean life, variance, probability of failure, etc.

Conclusions

Although applications of non-stationary stochastic processes in the fatigue field have been quite rare up to now, probably mostly in cases with only research aspects, their actual occurrence and consequent adverse influence on the resulting fatigue life should inevitably raise our interest to them. The research and applied effort should cover

- monitoring and analysis of various typical operating stochastic processes coped with physical explanations of reasons for their eventual non-stationary behaviour,
- elaboration of practical methods of the non-stationary stochastic process analysis in the scope of the correlation theory,
- elaboration of new fatigue damage accumulation hypotheses which could more fully reflect non-stationary process behaviour,
- experimental verification of influences of various kinds of non-stationarities on the resulting fatigue life,
- development of experimentally based recommendations for transformation of stochastic processes into stress-strain concentrators,

- experimental and computational verification of the accuracy of numerical methods used for the simulation of stochastic processes, especially with non-stationary properties.

References

[1] ČAČKO, J.-BÍLÝ, M.-BUKOVECZKY, J.:Random Processes: Measurement, Analysis and Simulation. Elsevier, Amsterdam 1988.

[2] BÍLÝ, M.: Dependability of Mechanical Systems. Elsevier, Amsterdam 1989.

[3] KLESNIL, M. et al.: Cyclic Deformation and Fatigue of Metals. Elsevier, Amsterdam 1991 (in press).

[4] XIULIN, Z.-CHAO, L.-SITAO,Z.:An Approximate Formula for Calculating Local Strain Range. In: Localized Damage, Computer Aided Assessment and Control. Vol. 3. Advanced Computational Methods. Proc. First Int. Conf. Springer Verlag, Berlin 1990, pp. 253-259.

[5] KLIMAN, V.: Fatigue Life Prediction for a Material under Programmable Loading using the Cyclic Stress-strain Properties. Mater. Sci. Engng, 68, 1984, pp. 1-10.

Application of Nonlinear Stochastic Mechanics in Offshore Engineering

Peter Bjerager[1], Ivar Langen[2], Steven R. Winterstein[3], Tom Marthinsen[4], and Daniel Karunakaran[5]

[1] *Veritas Sesam Systems A.S, P.O. Box 300, N-1322 Høvik, Norway*
[2] *Statoil, P.O. Box 300, N-4001 Stavanger, Norway*
[3] *Department of Civil Engineering, Stanford University, Stanford, CA 9430 5-4020*
[4] *Saga Petroleum a.s., P.O. Box 490, N-1301 Sandvika, Norway*
[5] *SINTEF, N-7034 Trondheim, Norway*

ABSTRACT

Nonlinear stochastic mechanics are widely used in the field of offshore engineering. In this paper we survey four applications: nonlinear collapse and fatigue analysis of jacket structures, dynamic analysis of jack-up structures, springing and slow-drift response of a tension leg platform, and hysteretic foundation behavior of a concrete gravity platform.

1. INTRODUCTION

Offshore engineering poses many challenges for structural and soil engineers, oceanographers, and hydrodynamicists. In particular, as current offshore oil reserves become depleted, oil exploration and production is moving into increasing water depths and offshore platforms are installed at water depths of several hundred meters. For example, fixed offshore structures exist at water depth exceeding 300 meters.

Nonlinear phenomena are present in both the loading process, the structural behavior and, for fixed platforms, in the soil mechanics. Nonlinearities in the loading arise from nonlinear waves (non-Gaussian sea surface elevation) and nonlinearities in force mechanisms, as given for example by the Morison equation or second-order diffraction theory. The structure itself may exhibit various nonlinearities. This is the case for example for fixed, dynamically sensitive jack-up and deep-water jacket structures, tethered tension leg platforms, and moored semi-submersible platforms. Depending on soil conditions, nonlinear soil effects may influence both fixed structures - either gravity-based or on piles - and temporarily installed jack-up structures with footings at the sea bed.

Extensive analysis of the behavior and strength of offshore structures takes place throughout the life cycle of the structure, i.e. in concept evaluation, design, construction and operation. In operation, analyses are performed because of damage caused by dropped objects, platform/ship collisions, and explosion damages. Furthermore, a need for substantial re-design of the (deck) structure often arises.

N. Bellomo · F. Casciati (Eds.)
Nonlinear Stochastic Mechanics
IUTAM Symposium Turin

Stochastic methods are today being used for most non-trivial offshore structures. Significant research efforts back these applications. From 1983, the strategic research program "Reliability of Marine Structures" has taken place at A.S Veritas Research, Norway. The program is a joint effort by Det norske Veritas, Saga Petroleum, Statoil and Conoco. At Stanford University, USA, the industry-sponsored "Reliability of Marine Structures Program" is undertaking research in the same area. A co-operation between the two research programs exists. Some of the work reported here has been produced by these programs; other portions have resulted from ongoing collaborations between SINTEF, Statoil and A/S Norske Shell.

The present paper illustrates four uses of nonlinear stochastic methods in offshore engineering. These cases are (1) inspection planning for jacket structures by use of nonlinear collapse and fatigue analysis, (2) dynamic analysis of a jack-up structure, (3) response analysis with respect to slow-drift and springing motions of a tension leg platform, and (4) analysis of the effect of hysteretic soil behavior on a concrete gravity platform.

2. COLLAPSE AND FATIGUE ANALYSIS OF JACKET STRUCTURE

The jacket structure is the most used type of offshore structure, with more than 3000 in operation around the world at present. This represents about 95% of all offshore structures used for oil production. The jacket platform is a truss-work tower consisting of tubular members with a deck on the top and pinned to the sea bed by steel piles, Figure 1. The design originates presumably from Texas as onshore drilling and production towers wandered into the Gulf of Mexico. The deck loads and environmental loads are transferred to the foundation through the legs, of which there may be 4 to 16. The legs are stiffened with bracing which also will carry the resultant horizontal forces. Most of the existing jackets are located in shallow water (less than 50 meters). Such smaller jackets are transported to the field location on a barge and lifted into position by a crane. Larger jackets are either transported lying on a barge and then pushed into the water, or towed to the field location in a self-floating condition. In either case the structure is positioned in an upright condition by a weight-buoyancy system.

For shallow water jackets in the Gulf of Mexico, the dominant in-service problem is corrosion. Furthermore, in this region collapse due to hurricanes is of concern. In the North Sea, an important failure mode for deeper water jacket platforms is fatigue. For this (and other reasons), large resources have been spent every year in inspection of these structures. From the (lack of) findings, the required amount of inspection has been questioned, leading to new methods of inspection planning based on nonlinear collapse analysis with probabilistic fatigue and fracture models. In this way, it has been possible to reduce the costs of inspection in the North Sea.

2.1 Nonlinear collapse analysis

Design of jacket structures are based on linear elastic analysis of the structure with codified requirements to each of the structural element (members, joints, etc.). However, in requalification of existing structures nonlinear quasi-static collapse (or push-over) analysis is being used increasingly (Lloyd and Clawson, 1983; Moan et al. 1985; Stewart et al, 1988; see also Figure 2). Such an analysis determines the so-called reserve and residual strength of the structure in relation to first yielding. Efficient collapse analyses of jacket-type structures

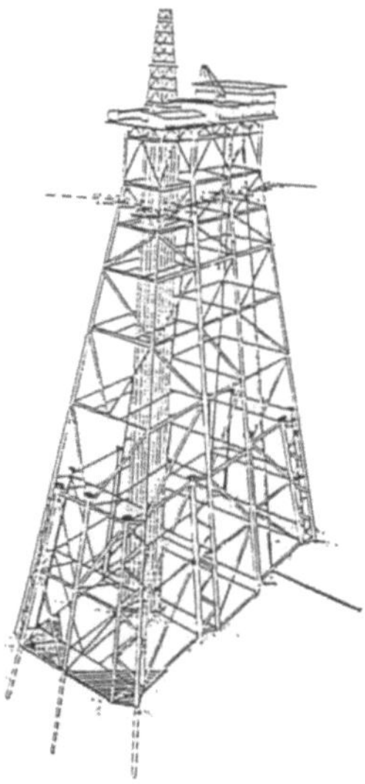

Figure 1: A typical jacket structure.

can be performed by use of the idealized structural unit method utilizing a coarse element mesh in which each beam is modelled as a large-displacement finite element (Ueda and Rashed, 1974; Hellan 1990).

Corresponding to the deterministic collapse analysis, stochastic or probabilistic collapse methods have been developed. The failure path method is based on a member replacement technique in which a member at yielding/failure is replaced by a force (Guenard, 1984; Bjerager et al, 1987, De and Cornell, 1990). Also methods based on plastic limit analysis have been developed (e.g. Holm et al, 1989). Finally, methods based on consistent stochastic finite element formulations are under study.

The primarily outcome of a stochastic collapse analysis is the probability that the structure fails due to an extreme (wave) loading condition (storm). In the analysis, a probabilistic model for the extreme event will be adopted, representing, for example, the worst storm in a year. The collapse analysis can be used to determine the importance of single members in the structure and, as described below, to define the requirement to these as regards fatigue failure (Lotsberg and Kirkemo, 1989).

2.2 Fatigue and fracture analysis

Fatigue failure is an important failure mode for deeper water jackets, such as many of those in the North Sea (Almar-Næss, 1985). Probabilistic methods to analyze this failure mode have been developed over the last decade of so. Models based on the S/N-approach as well as the Paris-Erdogan crack growth law have been established (Wirsching 1984; Madsen et al, 1986; Carr et al, 1986; Madsen, 1987; Kirkemo 1988). Based on a probabilistic modeling of the loading (wave process), as well as on the uncertainties in stress amplitudes and material properties, reliability measures regarding fatigue failure of a structural member can

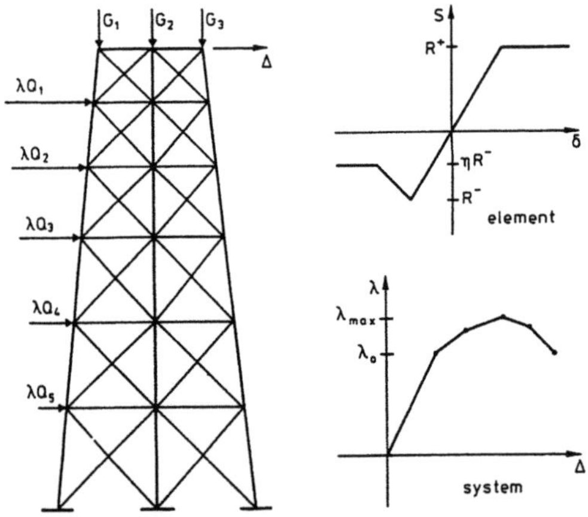

Figure 2: Illustration of static collapse analysis of a jacket structure. The horizontal loading of the structure is increased proportionally beyond failure of first element until the maximum load is reached. System reserve strength is expressed as the ratio between these two load levels.

be established for given reference periods. The methods are used for inspection planning, as outlined in the next section.

Finally it is noted that recently the work has started on formulating system fatigue reliability models for structures where several members may fail in fatigue, or a combination of member failures due to fatigue and overload make take place (Karsan and Kumar, 1990; Karamchandani et al, 1991).

2.3 Inspection planning

The ability of probabilistic methods to account consistently for additional information as provided through inspection and monitoring of offshore structures can be considered the foremost benefit of the methods. With presently available computation methods such as those in (PROBAN, 1989), it is possible to evaluate reliabilities conditional on a number of different inspection findings. These may include both direct observations on the basic variables (material samples) as well as indirect observations (such as "no cracks can be detected"). The uncertainty in the inspection result can be included and the effect of inspection evaluated. The current state of these methods are reviewed in Madsen and Sørensen (1990) and in Skjong and Torhaug (1991).

Typical results from reliability updating through inspection are shown in Figure 3. Each curve in the figure represents a *new state* of information (e.g. after one, two, etc. inspections). Often the reliability measure will be increased due to the new information because the new information may often be "as expected" (no cracks detected) and thereby the uncertainty is reduced, or the inspection finding will show an "un-expected state" (presence of crack) and a repair will take place. Again, this may increase the confidence in the structure.

The results for a single structural member, as shown in Figure 3, are used for inspection planning by introducing a required reliability (or, better, a failure rate) level at which the measure for the considered structural member should stay above. The curves in

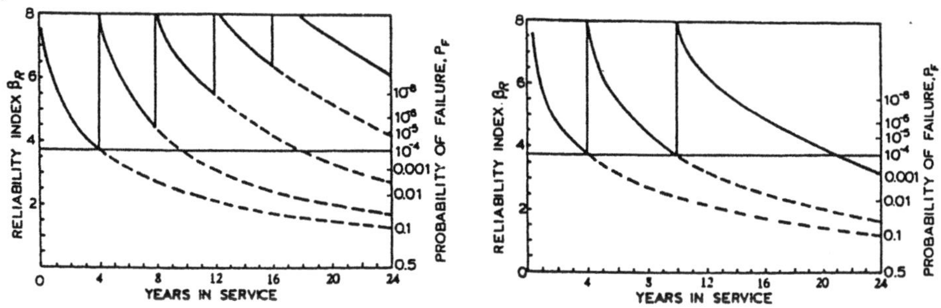

Figure 3: Reliability curves for different levels of information (before and after inspections) for a single structural member (Kirkemo 1988).

the figure can therefore be used to estimate an appropriate time for next inspection (when the current reliability curve down-crosses the threshold value).

The threshold value for a particular member is determined by used of a stochastic nonlinear collapse analysis. Considering the relation

$$P[\text{ collapse } | \text{ member } j \text{ failed }] \cdot P[\text{ member } j \text{ failed }] > P_{prescribed} \tag{1}$$

the right hand side is a prescribed value to be used for the structure under consideration. The first term on the left hand side is determined by a nonlinear collapse analysis and thereby a requirement to the failure probability for member j is established. This value (or a derived regarding failure rate) is used in the inspection planning as regards this member j. By this approach, the importance of the difference members in relation to the reserve strength of the structure is taken into account.

More consistent approaches are under study where a full systems analysis with respect to both fatigue and overload is considered. Furthermore, it is finally mentioned that the present considerations only deals with inspection planning for existing structures. It is possible already at the design stage to take into account expected future costs for inspection and repair and thereby to optimize the design in this respect as well.

3. DYNAMIC ANALYSIS OF JACK-UP PLATFORM

A typical jack-up structure consists of a main platform deck and 3, 4 or more legs that can be lowered to serve as sea bed support, Figure 4. The main platform deck will therefore be lifted above the water on location during drilling or production. However, the deck unit is watertight and it has buoyancy and stability to serve as a transport unit with elevated legs during transit over the ocean. The legs may be pipes perforated with holes or they may be truss-work. The jack-up is mostly used as a drilling platform and only about 1% of the production platforms in the world are jack-up platforms.

Jack-up platforms often show significant dynamic behavior due to their high structural flexibility. Furthermore, the wave loading is of a nonlinear nature due to drag forces

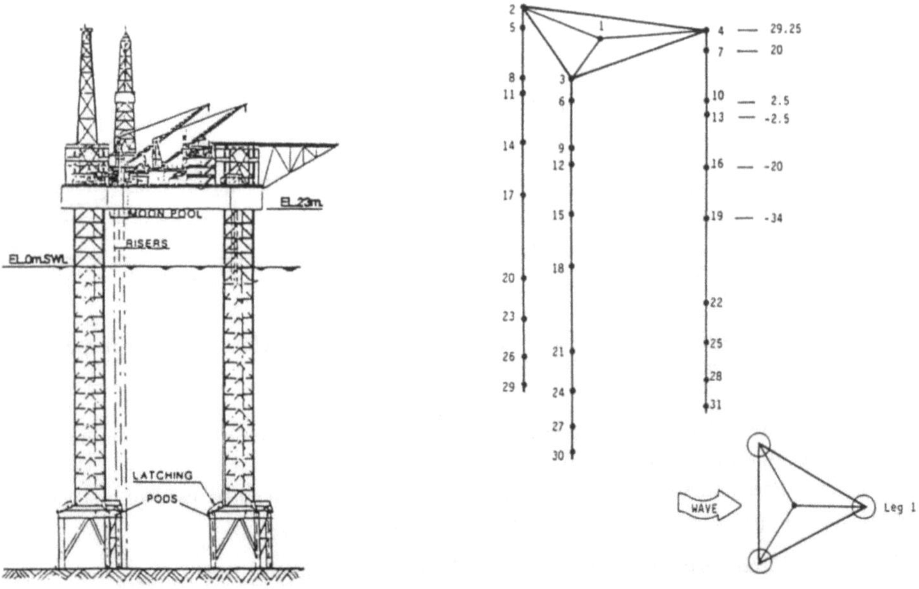

Figure 4: General view and computer model of the jack-up platform.

and varying water level during a load cycle. The need for including these nonlinear effect properly in a jack-up response analysis will be demonstrated in this section. Further details of the analysis are given in Karunakaran (1990) and in Karunakaran et al (1991a).

Generally, jack-up platforms also exhibit nonlinear behavior due to P–Δ effects and hysteretic soil-structure interaction. These effects are not considered in this review, where a linear structural model is used.

3.1 Structural model

Figure 4 shows a jack-up platform designed as a production facility to operate at a water depth of 110 m. The platform is supported by three legs, each of which rests on a pod. The pods are piled in the sea bottom, and each of the legs consists of a triangular truss structure where the typical member diameter range from 0.2 to 0.8 meter. The leg-soil interaction is modelled by linear springs. The springs are varied between a fixed base and a hinged base, resulting in natural periods ranging from 4.4 to 10 sec.

The finite element model of the platform is also shown in Figure 4. The model consists of 31 nodes and 33 beam elements. The top side mass of 16800 tons is lumped to the top four nodes. The structural mass and the added mass of the legs are lumped at appropriate nodal points. A Rayleigh damping model is adopted for the soil and structural damping together, using 2% damping levels in the first and third modes. Hydrodynamic damping is included by using relative velocities in the Morison equation.

3.2 Nonlinear dynamic response analysis

The nonlinear wave loading is modelled by the Morison equation using relative velocity, and by integration of wave forces to the instantaneous water surface with a vertically stretched Airy wave theory (Wheeler, 1970) for the wave kinematics.

The dynamic equation of motion is solved in the time domain using step-by-step integration, assuming that the platform motion is small compared to the length of the exciting waves. The procedure consists of the following steps:

- Simulation of time series of waves and water particle kinematics from a directional wave spectra density using a FFT technique. The sea state is described by a single or double peaked JONSWAP spectrum (Torsethaugen, 1987) with significant wave height H_S and peak period T_P as parameters. In cases of short crested seas the spreading is described by a cosine function with exponent 4.

- Calculation of wave forces time step by time step directly from the simulated water particle kinematics and calculation of dynamic response time series, using the Newmark integration method with constant average acceleration.

- Estimation of extreme response and response transfer functions to be used in long term response analysis. The transfer functions are obtained by dividing the autospectra of the dynamic response by the autospectrum of the wave process.

3.3 Linear analysis

For comparison, a linear analysis of the jack-up platform has been performed. In this case the wave loads are computed from linear wave theory, integrating the forces to the mean water level. The drag term in the Morison equation is linearized using a stochastic linearization technique (Leira, 1987). The equation of motion is solved in the frequency domain.

3.4 Estimation of long term extreme response

The most consistent way to describe the extreme response and fatigue damage is to use long term statistics (Farnes, 1990). The main steps in the procedure are:

- Establish a scatter diagram for the sea state parameters H_S and T_P from a long term joint probability distribution for these parameters. The scatter diagram is divided into areas in which the response function and the Weibull parameters for the response maxima are close to invariant.

- Calculate dynamic response for the sea states which represent the midpoints in the scatter diagram areas and estimate response transfer functions and Weibull shape parameters.

- Establish the long term distribution of the response by accumulating response probability distributions for each sea state in the defined scatter diagram weighted with their probability of occurrence, using the estimated transfer functions and Weibull parameters together with the associated theoretical wave spectral model.

- Estimate extreme response at a specified return period and fatigue life from the established long term distributions.

3.5 Results

The static response function for base shear is compared for the linear case and for different sea states in the nonlinear case. The linear function shows clear cancellation effects for certain frequencies due to phase differences between loads on the different legs. This means that structures with natural periods of 5.3 and 9 sec., coinciding with troughs in the response function, have less resonant response than those with natural periods between 6 and 7 sec. This effect is not as pronounced for short crested waves. The same peaks and troughs are also evident in the nonlinear case for low sea states. For higher sea states, however, the nonlinear response function differs considerably in the higher frequencies. This is due to the super-harmonic load transfer.

A comparison between the long term response estimated from the nonlinear and linear analysis has also been performed. The nonlinear analysis gives significantly higher long term response than the linear analysis for cases where most of the contribution to the long term response comes from higher sea states in the nonlinear analysis. The increase in response is largest for the double peaked wave spectrum model (80% for a natural period of 4 sec.). In cases where the nonlinear long term response is governed by resonant sea states, which is the case for natural periods between 6 and 7 sec., the linear analysis gives slightly larger response. The reason for this is that in the linear analysis the linearization sea state is at $H_S = 10$ meters, while the most contributions come from sea states around $H_S = 6$ meters and thus the loading is overestimated.

From the sample maxima distributions for base shear at various sea states for the case with natural period 5.3 sec., it is seen that the nonlinear response shows nearly Gaussian behavior in the lower sea states, while at higher sea states the behavior is very nonlinear. This explains why the nonlinear long term response is more dominated by extreme sea states. In this study there is no significant influence from the super-harmonics on the long term response.

Comparing fatigue life estimated from the linear and the nonlinear analysis, respectively, it can be concluded that the linear analysis overpredicts the fatigue life by about 20% (Karunakaran et al, 1991b). If the Airy model is used in the nonlinear analysis, the over-prediction is about 40%. Thus, linear dynamic analyses lead to non-conservative fatigue life estimates for the jack-up platform considered.

4. RESPONSE ANALYSIS OF TENSION LEG PLATFORM

A tension leg platform is a floating, semi-submersible offshore structure with buoyancy that exceeds its weight, whereby pretension is caused in the tethers fixed at the bottom of the sea, Figure 5. Thus, the TLP behaves as an inverted pendulum with motions damped by hydrodynamic mass and excited by waves, current and wind. The TLP concept was developed by Conono Inc. (Curtis, 1984), and it is economical for water depths of about 300-600 meters. The first operational TLP was installed at the Hutton field in the British North Sea in the summer of 1984, and the second TLP in operation is the Jolliet structure in the Gulf of Mexico. Currently, the Snorre TLP is being constructed for the Norwegian North Sea, and a fourth TLP is planned to develop the Heidrun field at Haltenbanken,

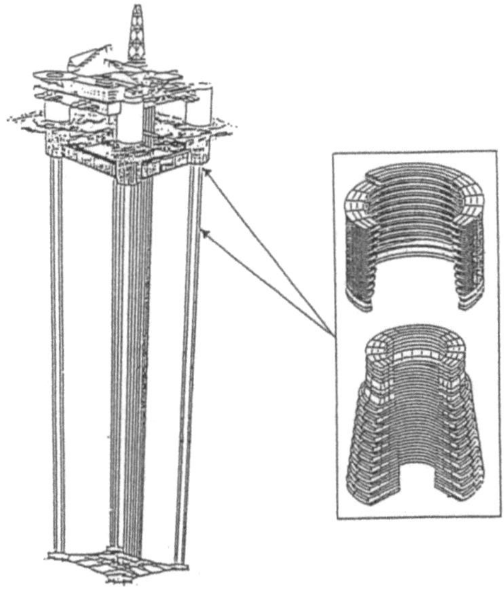

Figure 5: Snorre TLP with the connector between a tether and the hull of the structure.

offshore mid-Norway. While the first three TLPs have steel hulls, the Heidrun TLP will be a concrete structure. Use of the TLP concept is being considered for other deep water fields.

The natural periods of fixed offshore structures, such as jackets and jack-ups, increase roughly linearly with water depth. As the water depth increases, it becomes increasingly difficult to design these structures to avoid significant dynamic response. An alternative design concept for these deeper waters is given by compliant structures, such as the TLP.

Natural periods of the TLP are typically around 1–2 minutes for horizontal ("slow-drift") motions, and about 2–3 seconds for vertical ("springing") modes. These structural frequencies avoid most of the power in the wave elevation spectrum, which is typically concentrated in the 5–20 second range. Nonlinear effects, however, give additional forces whose frequencies are sums and differences of wave elevation frequencies. The sum-frequency terms drive resonant springing response, while horizontal slow-drift motions are controlled by difference-frequency contributions.

We focus here on vertical "springing" motions of the TLP, which contribute to tether fatigue and to platform accelerations. In addition to nonlinear effects, an additional feature of the problem is noteworthy: the resulting force power spectrum is found to oscillate rather rapidly near typical heave, roll, and pitch natural frequencies. This suggests that these frequencies and associated damping ratios should be selected with care, and that uncertainty in these quantities may have significant impact.

4.1 Methodology Overview

A general analysis approach has been developed to include these nonlinear force effects on springing response, and to put these into perspective in view of uncertain structural and sea state properties. Its features are as follows:

Nonlinear response model: To include nonlinear, non-Gaussian effects on TLP forces and responses, these quantities are modelled through second-order Volterra series. The physical system is thus represented by first- and second-order transfer functions between (first-order, Gaussian) wave elevation and TLP forces and responses. These transfer functions are based on wave diffraction theory, appropriate for large-volume TLPs. They have been evaluated numerically for the Snorre platform (Molin and Chen, 1990). The corresponding structural response is found from a linear six-degree-of-freedom model (Marthinsen, 1989; Marthinsen, 1991).

Nonlinear response analysis: The Volterra model of springing response is similar to that used for various studies of slow-drift TLP motion (Neal, 1972; Vinje, 1984; Naess, 1985; Naess, 1986). As in those studies, eigenvalue analysis is used to recast the second-order springing response in terms of independent normal variables (Kac and Siegert, 1947). The present analysis of this model is somewhat simpler, however. Response moments are found directly from these eigenvalues, and extreme responses are estimated from moment-based Hermite transformation models (Winterstein and Marthinsen, 1991a). These models are also currently being applied to predict non-Gaussian effects on TLP tether fatigue. (Winterstein and Marthinsen, 1991b). Such effects are expected to be notable for springing responses, whose nonlinear effects are typically largest for sea states with small-amplitude, high-frequency response cycles.

Uncertainty analysis: Uncertainty is included in sea state and structural parameters through FORM/SORM (first- and second-order reliability methods). These estimate the probability that various extreme response levels will be exceeded, and separate the effect of various uncertainty sources. The foregoing nonlinear response analysis is performed successively (as the limit-state function in a FORM analysis) until the "most likely" failure point is found.

4.2 Numerical Results

Figure 6 shows how various response statistics change with significant wave height H_S. Response statistics for tether tension are shown at left, and for heave acceleration at right. The solid line shows results for a sea state with "typical" period T_P at the median fractile p=0.5 for the given H_S value. The other curves show the effects of somewhat higher (p=0.84) and lower (p=.16) periods T_P. The RMS and kurtosis values are shown, while the associated skewness values are nearly zero and hence omitted. The resulting estimates of extreme responses are found from the Hermite model (Winterstein, 1988; Winterstein and Ness, 1989).

Note that for both tension and accelerations, larger standard deviations and extreme responses are associated with smaller sea state periods. This is because such sea states emphasize second-order nonlinear effects. These effects are also found to be far greater for accelerations than for tether tensions. This can be seen by noting their larger kurtosis values, and is physically due to the amplification of higher (sum) frequencies in these acceleration responses. Note also that high kurtosis values may be found despite the rather light damping present (1%).

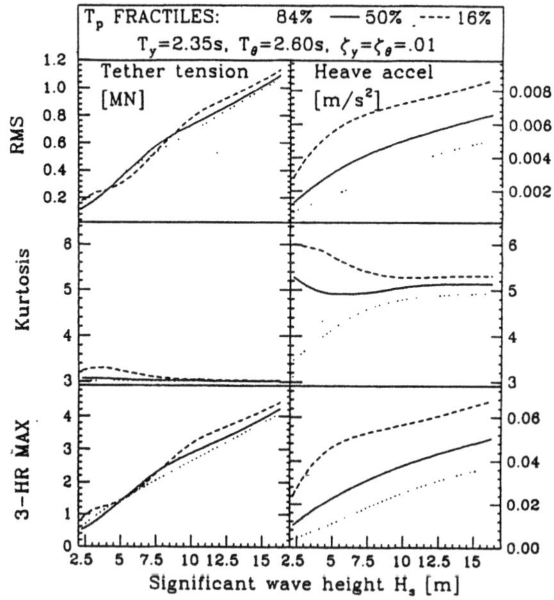

Figure 6: Response for various T_p fractiles.

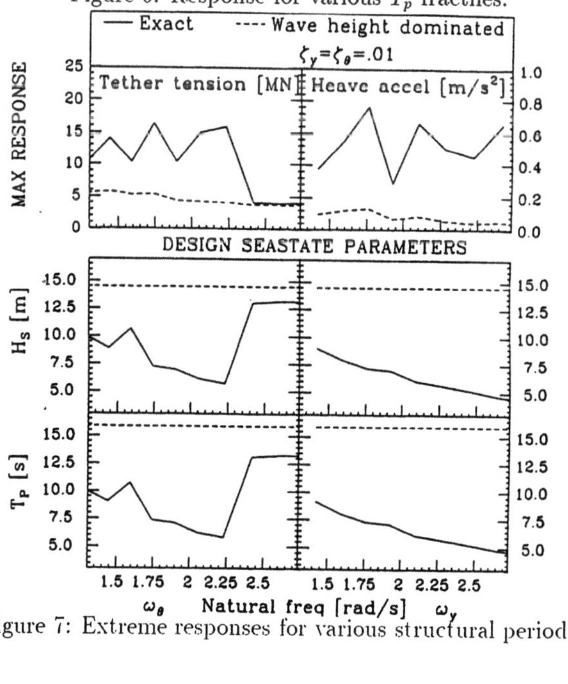

Figure 7: Extreme responses for various structural periods.

4.3 100–year Response Levels

In reliability analysis and design, one often seeks the response level with a specified mean return period, such as 100 years. For fixed offshore structures, it is common to estimate this response from a design sea state with the 100-year H_S value and typical associated T_P. For a Northern North Sea environment typical of the Snorre platform location, this "wave height driven" sea state is (Haver and Nyhus, 1986)

$$H_S = H_{100} = 14.5 \text{ m.}; \quad \text{associated 50\%-fractile } T_P = 15.9 \text{ sec.} \tag{2}$$

Alternatively, to maximize second-order effects one might consider an alternate, "wave period driven" sea state:

$$T_P = 2T_n = 4.7 \text{ sec.}; \quad \text{associated } H_S \text{ from 100-year contour } = 4.6 \text{ m.} \tag{3}$$

in terms of the relevant natural period T_n. Numerical values here apply to the heave acceleration, for which T_n is the vertical natural period T_y=2.35 sec.

Figure 7 shows actual 100–year responses predicted by FORM (solid line), together with the design sea state parameters it gives. Also shown are corresponding estimates based on the wave-height-driven sea state in Eq. 2. Results are shown versus the governing natural frequency—the heave frequency ω_y for heave acceleration and the pitch frequency ω_θ for tether tension—on the 100-year extreme responses.

Note first that over a broad range of ω_y, the period-driven sea state in Eq. 3 well approximates the heave acceleration design sea state. For example, Figure 7 shows design sea state periods between T_P=5.0 sec. and 8.0 sec. for T_y between about 2.5 sec. and 4.0 sec. Note also that while the design sea state parameters H_S and T_P vary smoothly with ω_y, the resulting extreme heave acceleration does not. This is due to similar irregularities in the heave force transfer function (Winterstein and Marthinsen, 1991b).

Results for tether tension depend on ω_θ. The base case ω_θ=2.67, a first estimate for the Snorre TLP, is roughly wave height dominated. The actual 100-year sea state parameters are roughly H_S=13.0 m., T_P=13.1 sec.; however, the sea state in Eq. 2 predicts the maximum response to within 10%. However, when $\omega_\theta < 2.25$ the critical sea states become closer to those of heave acceleration; i.e., more nearly driven by wave period. In this range the extreme tether tension increases, exceeding the wave height dominated result by factors of 2–3. Again, the irregular variation of extreme response with ω_θ mirrors that of the pitch moment transfer function (Winterstein and Marthinsen, 1991b). Note that because of limited transfer function data, second-order effects are not included for frequencies above 2.8 [rad/sec]. Also, the tension results neglect set-down effects, which were found relatively small in some trial calculations.

4.4 Summary: TLP Springing Response

- Tether tensions and platform motions show clear effects of force nonlinearity, both on response variance and higher moments. These effects are most pronounced on acceleration responses, which amplify sum-frequency contributions (Figure 6).

- TLP force transfer functions, which reflect local geometry, oscillate rather rapidly near typical heave, roll, and pitch natural frequencies. Hence, extreme springing responses show similar oscillations with these natural frequencies (Figure 7). The

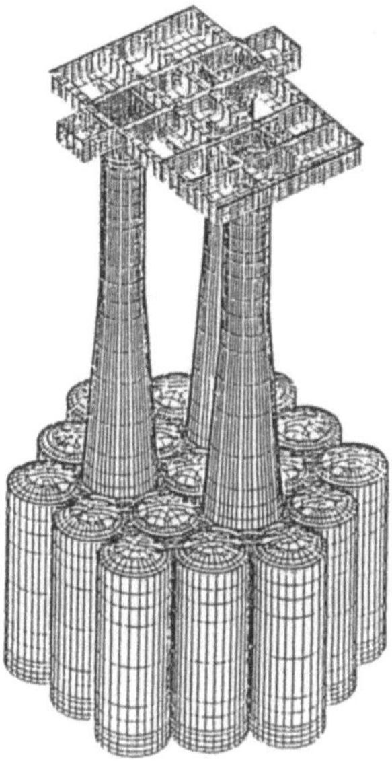

Figure 8: Concrete gravity structure.

inherent uncertainty both in these frequencies and in force transfer functions should also be noted.

- In predicting extreme springing responses, it is found useful to consider several design sea states: one with extreme H_S ("wave height driven," as in Eq. 2), and another with resonant second-order effects ("wave period driven," as in Eq. 3).

5. HYSTERETIC BEHAVIOR OF A GRAVITY PLATFORM

A concrete gravity platform consists normally of a large cellular caisson supporting 3–4 towers with a deck structure on top (Figure 8). The structure withstands external loading by its weight, as it is simply "planted" on the sea bed. The basic concept of the gravity platform is to attain stability in the permanent location without special piling. The substructure (caisson and shafts) is made of concrete, while the deck is a steel structure to reduce the self-weight during tow-out and thereby allow more equipment to be installed before this operation takes place.

A number of gravity platforms are in operation in the British and Norwegian North Sea in water depths up to 217 meters. These platforms have shown a mainly linear behavior with dynamic effects being less important. This may change when moving to larger water depths and softer soil conditions, as will be demonstrated for the gravity platform shown in Figure 9 which is developed by A/S Norske Shell for the Troll field in the North Sea. The water depth is 302 meters and the soil is normally consolidated soft clay, soft at the top with increasing strength by depth. For additional details see Leira et al (1992).

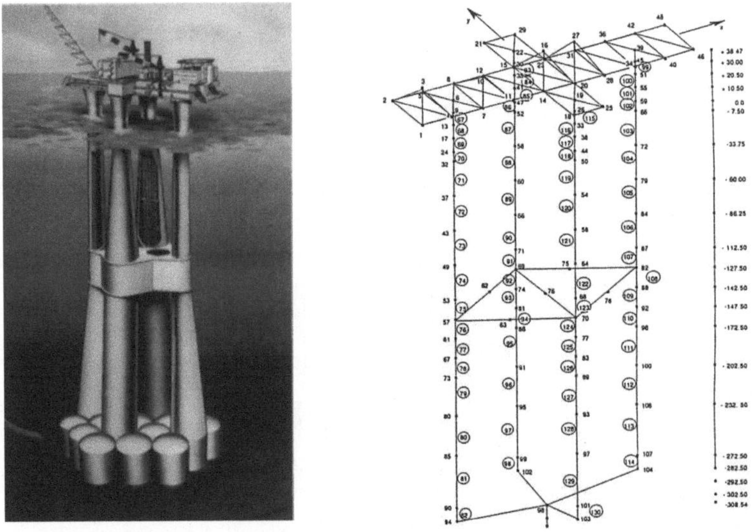

Figure 9: General view and computer model of the gravity platform.

The platform differs from ordinary concrete gravity platforms in that the caisson is very low—only 30 meters high—and the four high slender shafts are connected at midpoint by a horizontal frame. Furthermore, a new foundation solution has been adopted where the walls of the cellular caisson have been extended below the bottom forming cylindrical skirts penetrating 36 meters into the sea bed. The major part of the load is thus transferred to the soil via these skirts. The caisson consists of 19 cells each with diameter 32 meters. The outer diameter of the caisson is 161 meters and the total height of the concrete structure 368 meters.

5.1 Structural model

The model of the platform used for dynamic analysis is shown in Figure 9. It consists of 107 nodes and 132 beam elements. The topside mass of 25600 tons and the structural and hydrodynamic mass are lumped at the appropriate nodal points. The soil structure interaction is represented by a nonlinear hysteretic stiffness model of the kinematic hardening type at a nodal point 6 meters below the sea bed. The structural damping and the hydrodynamic damping together are assumed to be 1.5% modelled as Rayleigh damping at the first and third natural frequencies. No soil damping is added in the nonlinear analysis as it is already included in the hysteresis model.

The wave loading is obtained using the Morison equation with frequency dependent inertia coefficient and linear Airy wave theory for the wave kinematics. Integration to instantaneous water surface is not included since this effect is found not to be important for the response quantities studied in this case.

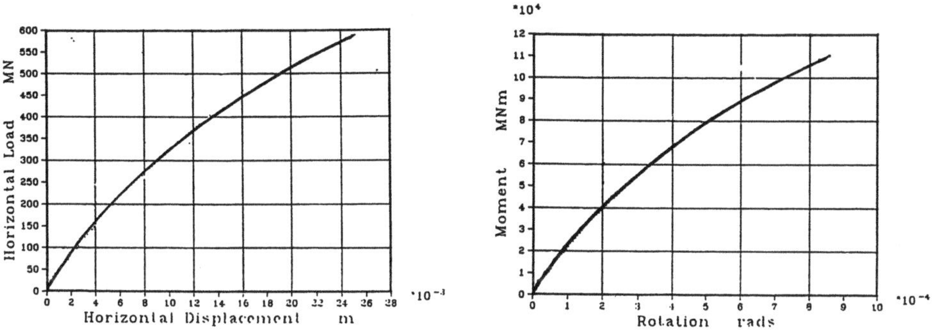

Figure 10: Nonlinear soil stiffness curves.

5.2 Soil model

Nonlinear soil stiffness curves for horizontal displacement and mudline rotation are presented in Figure 10. The curves are obtained from lower limit soil data (25% below mean) via a nonlinear FEM analysis of the foundation (Svanø et al, 1991). These curves represent the backbone curve in the kinematic hardening soil model with nested yield surfaces (Prevost, 1978) which has been used in the present analysis.

In a linear analysis of the platform the nonlinear model is replaced by so-called standard deviation springs and viscous dampers. In this method the linear spring stiffness for a specific sea state is taken as the relation between the standard deviation of the mudline force and the standard deviation of the corresponding displacement obtained from the maxima distributions of these quantities. The maxima distribution of the forces are obtained from the maxima distribution of the total wave loading using an appropriate dynamic amplification factor. The corresponding displacement distributions are determined from the load distribution using the nonlinear backbone curve. The stiffness will thus vary for different sea states.

5.3 Response analysis

The nonlinear dynamic response analysis and the estimation of long term extreme response are performed as described for the jack up in Section 3; see also Karunakaran (1991). The waves are simulated for each sea state using a JONSWAP model spectrum and assuming long crested sea.

For comparison, a linear analysis of the platform has also been performed, using the same method as described in Section 3.3. In this case the soil is modelled by the standard deviation linear springs described above together with concentrated viscous dampers.

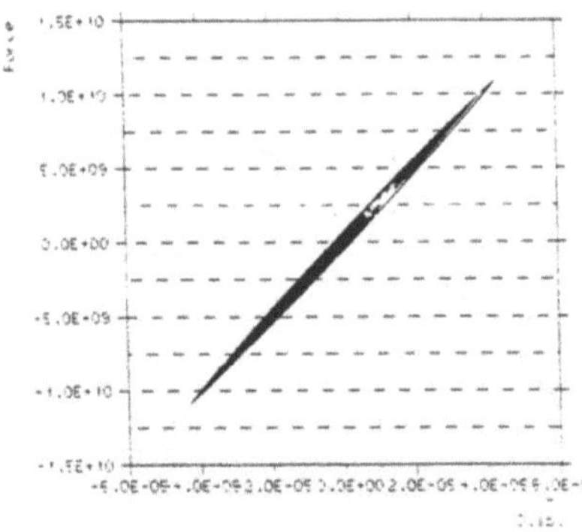

H_S = 2.0 m & T_P = 4.5 sec

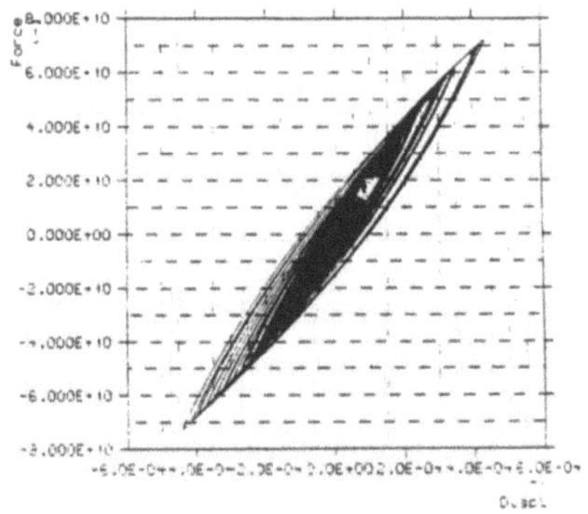

H_S = 15.8 m & T_P = 17.3 sec

Figure 11: Hysteresis curves for overturning moment vs. mudline rotation.

5.4 Results

Dynamic response calculations have been performed for a number of sea states ranging from the low sea state H_S=2 m, T_P=4.5 s to the extreme sea state H_S=15.8 m, T_P= 17.3 s. The first natural period of the platform using standard deviation linear soil stiffness is 4.65 s referred to the lowest sea state and 4.83 s referred to the extreme sea state.

The obtained force–displacement hysteresis curves for two different sea states are shown in Figure 11. It is clearly seen that the hysteresis and thereby the damping increases with increasing sea state.

The nonlinear dynamic response is dominated by resonance for the lowest sea states while for the highest the response is mainly quasi-static. The results are compared with those obtained by linear analysis without additional soil dampers. The comparison shows that the nonlinear analysis systematically reduces the standard deviation of the response. The reductions are most pronounced for sea states with T_P close to resonance (12% for T_P= 4.5 s) while they are almost insignificant for the highest sea states. The standard deviation response in the two cases can however, be matched very well by introducing equivalent linear viscous soil dampers in the linear model.

For the maxima the situation is different. From the sample distribution of the simulated nonlinear response it is seen that the response is close to Gaussian in the lower sea states while in the higher sea states the response becomes non Gaussian resulting in higher maxima than for the linear case. A two parameter Weibull model fits the response very well. Using this model the extreme response during a 6-hour extreme storm with H_S=15.8 m, T_P=17.3 s is 20% higher than in the linear case.

A comparison of long term response corresponding to a 100 year return period shows that the nonlinear analysis increases the maximum response with up to 20%.

6. CONCLUDING REMARKS

From the case studies presented here, the following conclusions can be drawn:

- To optimize inspection planning for jacket structures, nonlinear stochastic collapse analysis can be combined with probabilistic fatigue/fracture analysis. These methods have been used to reduce inspection costs for many North Sea platforms.
- For flexible jack-up structures, nonlinearities arise both in wave kinematics and drag force mechanisms. For these structures, conventional linear extreme and fatigue analysis is often found to be non-conservative.
- Nonlinear wave diffraction forces should be considered for both slow-drift and springing motions of the TLP. Resulting force spectra show large oscillations near typical structural frequencies. Uncertainty in these frequencies and associated dampings may thus have significant impact.
- Proper modelling of the nonlinear soil behavior may in certain cases be important in evaluating the extreme response of large concrete gravity platforms.

These case studies show that nonlinear stochastic methods can play an important role in the design and maintenance of offshore structures. These methods will be increasingly useful as we move into deeper water and search for more economic structures. Continued development of more simple and robust methods, both analytical and simulation-based, will promote their increasing use.

REFERENCES

Almar-Næss, A., Editor, (1985): "Fatigue Handbook for Offshore Steel Structures", *Tapir Publishers*, Trondheim, Norway, 520 pages.

Bjerager, P., A. Karamchandani and C.A. Cornell (1987): "Failure Tree Analysis in Structural System Reliability", *Proc. of 5th International Conference on Applications of Statistics and Probability in Soil and Structural Engineering, ICASP5*, May 25-29, 1987, Vancouver, Canada.

Bjerager, P., Loseth, R., Winterstein, S., and Cornell, C.A. (1988): "Reliability Method for Marine Structures under Multiple Environmental Load Processes", *Proceedings of 5th International Conference on the Behaviour of Offshore Structures*, Vol. 3, Trondheim, Norway, pp. 1239–1253.

Bysveen, S., A.G. Kjelås, J. Lereim and T. Marthinsen (1990): "Experience from Application of Probabilistic Methods in Offshore Field Activities", *Proc. of OMAE'90*, Houston, Texas, pp. 143–149.

Carr, P., Clayton, M., Busby, P.L., and Dobson, J. (1986): "A Probabilistic Strategy for Subsea Inspection of Steel Structures", *Society of Petroleum Engineers, European Petroleum Conference*, Paper No. SPE 15868, London, October, 1986.

Curtis, L.B. (1984): "How Conoco Developed the Tension Leg Platform", *Ocean Industry*, Vol. 19, No. 9, pp. 35–46.

De, R.S. and C.A. Cornell (1990): "Offshore Structural System Reliability: Wave-load Modeling, System Behaviour and Analysis", PhD-Thesis, Report No. RMS-6, Department of Civil Engineering, Stanford University, California.

Guenard, Y.F. (1984): "Application of System Reliability Analysis to Offshore Structures", Report No. 71, John A. Blume Earthquake Engineering Center, Department of Civil Engineering, Stanford University, California.

Holm, C.A., P. Bjerager and H.O. Madsen (1989): "Long Term System Reliability of Offshore Jacket Structures", *Proceedings of 2nd IFIP Working Conference on Reliability and Optimization of Structural Systems*, London, September 26-28, 1989, ed. by P. Thoft-Christensen, Lecture Notes in Civil Engineering, Springer Verlag.

Farnes, K.-A. (1990): "Long Term Statistics of Response in Nonlinear Marine Structures", Dr. ing. thesis, Norwegian Institute of Technology, No. 1990:60, Trondheim, Norway.

Haver, S. and K.A. Nyhus (1986): "A Wave Climate Description for Long Term Response Calculations", *Proc., 5th Intl. Offshore Mech. Arctic Eng. Symp.*, ASME, Vol. IV, pp. 27–34.

Hellan, Ø. (1990): "USFOS: A Computer Program for Progressive Collapse Analysis of Steel Offshore Structures", SINTEF Report No. STF71 A90001, Trondheim, Norway.

Kac, M. and A.J.F. Siegert (1947): "On the Theory of Noise in Radio Receivers with Square Law Detectors" *Journal Appl. Phys.*, Vol. 18, pp. 383–400.

Karamchandani, A., Dalane, J.I., and Bjerager, P. (1991a): "A Systems Approach to Fatigue of Structures". To appear in *Journal of Structural Engineering*, ASCE.

Karamchandani, A.K., Dalane, J.I., and Bjerager, P. (1991b): "Systems Reliability of Offshore Structures including Fatigue and Extreme Wave Loading". To appear in *Journal of Marine Structures.*

Karsan, D.I., and Kumar, A. (1990): "Fatigue Failure Paths for Offshore Platform Inspection". *Journal of Structural Engineering*, ASCE, Vol. 116, No. 6, pp. 1679–1695.

Karunakaran, D. (1990): "Long Term Response Analysis of Dynamically Sensitive Jack-up Platforms", SINTEF Report STF71 A90028, Trondheim, Norway.

Karunakaran, D. (1991): "Procedure for Nonlinear Dynamic Response Analysis of Offshore Structures–Both for Extreme and Fatigue Response", SINTEF Report STF71 A91016, Trondheim, Norway.

Karunakaran, D., N. Spidsøe and O.T. Gudmestad (1991a): "Selection of Design Sea States for Jack-up Platforms with Dynamic Behaviour", *Proc. of ISOPE*, Edinburgh, Scotland.

Karunakaran, D., B.J. Leira, N. Spidsøe, and T. Moan (1991b): "Fatigue Life Estimation for Dynamically Sensitive Slender Offshore Structures Exposed to Non-Gaussian Wave Loading", To appear in *Int. Conf. on Fatigue and Fracture in Steel and Concrete Structures*, Madras, India.

Kirkemo, F. (1988): "Applications of Probabilistic Fracture Mechanics to Offshore Structures", *Proceedings of the Seventh Offshore Mechanics and Arctic Engineering Conference*, Houston, Texas, and also in *Applied Mechanics Review*, Vol. 41, No. 2, February 1988.

Leira, B.J. (1987): "Multi-dimensional Stochastic Linearization of Drag Forces", *Applied Ocean Research*, July 1987.

Leira, B.J., D. Karunakaran, G. Svanø and E. Skomedal (1992): "Linear and Nonlinear Dynamis of the Troll Gravity Platform", submitted to the ISOPE Conference, San Francisco, USA, June 1992.

Lloyd, J.R. and W.C. Clawson (1983): "Reserve and Residual Strength of Pile Founded Offshore Platforms", *Proc. of Symp. on the Role of Design, Inspection and Redundancy in Marine Structural Reliability*, November 1983, National Research Council, pp. 157-196.

Lotsberg, I. and F. Kirkemo (1989): "A Systematic Method for Planning In-Service Inspection of Steel Offshore Structures", *Proc. of OMAE'89*, March 18-23, 1989, The Hague, The Netherlands.

Madsen, H.O. (1987): "Model Updating in Reliability Theory". *Reliability and Risk Analysis in Civil Engineering 1. ICASP5*, ed. by N.C. Lind, University of Waterloo, Ontario, Canada, 564-577.

Madsen, H.O., and J.D. Sørensen (1990): "Probability-Based Optimization of Fatigue Design, Inspection and Maintenance", *Proc. of Conference on Integrity of Offshore Structures*, Glasgow, June, 1990.

Marthinsen, T. (1989): "Hydrodynamics in TLP Design", *Proc., 8th Intl. Offshore Mech. Arctic Eng. Symp.*, ASME, Vol. I, pp. 127–133.

Marthinsen, T. (1991): "Second-Order Hydrodynamic Load and Response Statistics", *Proc., 1st Intl. Workshop on Very Large Floating Structures,* Honolulu, pp. 209–220.

Moan, T., J. Amdahl, A.G. Engseth, and T. Granli (1985): "Collapse Behaviour of Trusswork Steel Platforms", *Proceedings of BOSS'85*, Elsevier Publishers, Amsterdam, The Netherlands.

Molin, B. and X.B. Chen (1990): *Calculation of Second-Order Sum-Frequency Loads on TLP Hulls.* Institute Français du Petrole Report, Paris, France.

Naess, A. (1985): "Statistical Analysis of Second-Order Response of Marine Structures", *Journal of Ship Research*, Vol. 29, pp. 270–284.

Naess, A. (1986): "The Statistical Distribution of Second-Order Slowly-Varying Forces and Motions", *Applied Ocean Research*, Vol. 8, No. 2, pp. 110–118.

Neal, E. (1972): "Second-Order Hydrodynamic Forces Due to Stochastic Excitation", *Proc., 10th ONR Symp.*, Cambridge, MA, pp. 517–539.

Prevost, J-H. (1978): "Anisotropic Undrained Stress-Strain Behaviour of Clays", *Journal of Geotechnical Engineering*, *ASCE*, Vol 104, No 8, pp. 1075-1090.

PROBAN (1989): The PROBabilistic ANalysis Program, Version 2. Theory Manual: VR Report 89-2022, Users Manual: VR Report 89-2024, Example Manual: VR Report 89-2025, Distribution Manual: VR Report 89-2026, and Command Reference: VR Report 89-2027, A.S Veritas Research, Norway.

Skjong, R., and R. Torhaug (1991): "Rational Methods for Fatigue Design and Inspection Planning of Offshore Structures", To appear in *Journal of Marine Structures.*

Stewart, G., M. Efthymiou and J.H. Vugts (1988): "Ultimate Strength and Integrity Assessment of Fixed Offshore Platforms", *Proceedings of the 5th International Conference on the Behaviour of Offshore Structures*, Vol. 3, Trondheim, Norway, pp. 1205–1221.

Svanø, G., D. Karunakaran, B.J. Leira and O. Skjåstad (1991): "TROLL PHASE I. Design Basis for Soil Dynamics—Nonlinear Dynamic Analysis", SINTEF Report STF69 F91005, Trondheim, Norway.

Torsethaugen, K. (1987): "Range of Validity of the JONSWAP-Spectrum" (in Norwegian), Report No. 85.115, Oceanographics Centre, Trondheim, Norway.

Ueda,,, Y. and S.M.H. Rashed (1974): "An Ultimate Transverse Strength Analysis of Ship Structures", *Journal of the Society of Naval Arch. of Japan*, Vol. 139.

Vinje, T. (1984): "On the Statistical Distribution of Second-Order Forces and Motions", *Int. Shipbuilding Prog.*, Vol. 30, pp. 58–68.

Wheeler, J.D. (1970): "Methods for Calculating Forces Produced on Piles in Irregular Waves", *Journal of Petroleum Technology*, March 1970.

Winterstein, S.R. (1988): "Nonlinear Vibration Models for Extremes and Fatigue", *Journal of Engineering Mechanics*, ASCE, 114(10), 1772-1790.

Winterstein, S.R. and T. Marthinsen (1991a): "Nonlinear Effects on TLP Springing Response and Reliability", *Computational Stochastic Mechanics* (Proc., 1st Int. Conf., Corfu), Elsevier, pp. 765–776.

Winterstein, S.R. and T. Marthinsen (1991b): *Springing Response of a Tension Leg Platform: Reliability against Extremes and Fatigue*, Report No. RMS–9, Reliability of Marine Struc. Prog., Civ. Eng. Dept., Stanford Univ.

Wirsching, P. (1984): "Fatigue Reliability of Offshore Structures", *Journal of Structural Engineering*, ASCE, Vol. 110, pp. 2340–2356.

Estimation of Structural Reliability of Nonlinear Systems under Seismic Action

V.V.Bolotin
USSR Academy of Sciences
Institute of Mechanical Engineering
101830 Moscow, USSR

Summary

Response of structures with strong nonlinearities under seismic actions treated as nonstationary random processes is analysed using the numerical simulation. Various nonlinearities such as elasto-plastic behaviour, sliding bearings, and dry friction are included into consideration. The central point of the study is the prediction of structural reliability measures under the seismic action. The problem of the extrapolation of results of the numerical simulation into the area of small probabilities is considered. The properties of "tails" of distributions for essentially nonlinear systems are demonstrated with numerical examples, and approaches to facilitate the engineering solution of the "tails" problem are discussed.

Introduction

Safety and reliability problems for structures and buildings under seismic action attract now the special attention both of engineers and applied mathematicians. There are two main direction of analysis [1]: (1) introduction of more adequate models for structures and actions; (2) development of more fundamental approaches to the assessment of structural seismic risk. Difficulties in the first direction can be easily overcomed due to the progress of computer technology and numerical methods. Difficulties in the second direction are of principal nature. Among them there are the lack of statistical data on structures, soils, and especially ground motion during the strong earthquakes. The essential feature of the seismic safety assessment problems for structures of high responsibility is a number of uncertainties that are always connected with the extrapolation into the area

N. Bellomo · F. Casciati (Eds.)
Nonlinear Stochastic Mechanics
IUTAM Symposium Turin

of rare events. The Monte Carlo simulation with application to large systems produce rather limited samples of structural responses. On the other hand, the real behaviour of "tails" of empirical distributions in fact remains always unknown. This behaviour may be predicted only in the simplest cases when the asymptotic distributions of extrema may be used [2]. The situation is more complicated for essentially nonlinear systems. Firstly, their analysis requires much more computer time, and the problem of the sample size becomes acute. Secondly, the behaviour of nonlinear systems subjected to strong ground motions, as a rule, can vary even qualitatively both in time and from one sample to another. For example, new structural components enter in turn into the plastic stage. That complicates additionally the problem of extrapolation into the area of rare events.

The aim of this paper is to clarify, using the numerical experimention mainly, the substance of difficulties arisen in the assessment of structural seismic risk for nonlinear systems. Another point is to demonstrate the use of apriori disributions for those parameters which are given with uncertainty, and to study how that influences on the empirical distributions of parameters responsible for the final failure. We try to show that the randomization of parameters of seismic actions and structural properties facilitates the solutions of the "tail" problem. At last, we demonstrate the application of the concept of the integral seismic risk for optimization purposes.

Several types of structures are considered. The first one is a classical multi storey building with a metal or reinforced concrete framework (Fig.1,a). Significant plastic deformations are permissible in such structures after strong earthquakes if the damage does not produce danger for people [3]. The second type of structures discussed in this paper is a building on the "floating" foundation. An example is presented in Fig. 1,b where the building of a nuclear power plant is depicted schematically [4]. The nonlinearity of the system is born, in particular, by the possibility of relative sliding of the parts of the foundation and by the involvment of dry friction forces on the sliding surfaces. The third type of nonlinear structural system is depicted in Fig. 1,c. It is a building containing small gaps. During a strong earthquake the response of the structure becomes impact-vibrational and, in fact, strongly nonlinear.

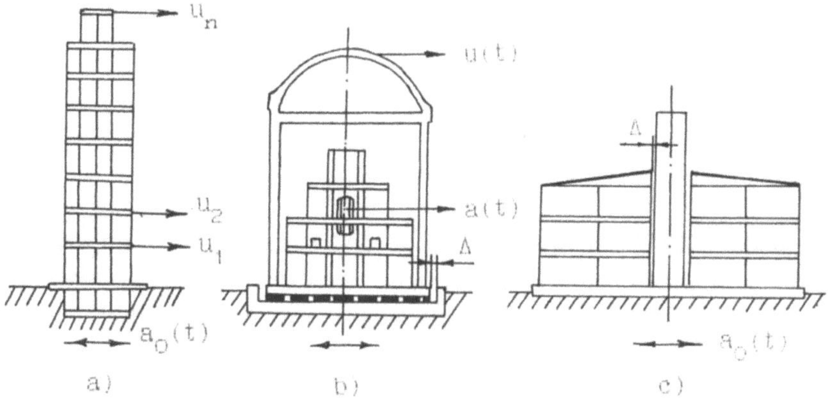

Fig. 1. Typical nonlinear systems: (a) multi-storey building with elasto-plastic deformations; (b) reactor building on a set of metal-elastomer bearings including dry friction; (c) elongated building with gaps.

Seismic Input Schematization with Account of Uncertainties

In risk assessment procedures based on the Monte Carlo simulation, artificial ground motion processes are used which are supposed similar to those expected during strong earthquakes on the site. A number of kinds of artificial accelerogramms were suggested during the last three decades beginning from our paper [5] (the Russian version of this paper appeared a year earlier). Practically all the proposed schematizations present the ground acceleration as a non-stationary random process, and most frequently - as a stationary random process modulated with a slowly varying time function which, generally, may be a random time function, too.

Following paper [5] assume the ground acceleration in the form

$$a_0(\tau) = A_0 c\tau \exp(1-c\tau)\varphi(\tau), \qquad \tau > 0. \tag{1}$$

Here τ is the "fast" time, A_0 is the maximum of the pseudo-envelope $A(\tau) = A_0 c\tau \exp(1-c\tau)$, $\tau_c=1/c$ is the time from the beginning of the ground motion at $\tau = 0$ up to the attainment of this maximum, and $\varphi(\tau)$ is a realization of a stationary random process with zero mean and unity mean square. The term "pseudo--envelope" is used since the real peaks of samples of the process given with Eq. (1) can be higher than A_0.

We introduce the incompleteness of information and the uncertainties by means of apriori distributions of relevant parameters, such as peak accelerations, dominant frequencies, duration of the strong stage, etc. The simplest apriori distributions are uniform and log-uniform distributions within the expected intervals of variation of parameters. In particular, we assume that the values of A_0 are log-uniformly distributed on the segments corresponding to intensities I = 7, 8 and 9 in MSK scale. The average accelerations are taken 1.0, 2.0 and 4.0 m s^{-2}, respectively.

The simplest presentation for $\varphi(t)$ is a normal process given with a two-parameter spectral density depending on the characteristic frequency θ and the bandwidth parameter α. A typical, and not the only possible, equation is [6]

$$S_\varphi(\omega)= \frac{2\alpha\theta^2}{\pi}\;\frac{1}{(\omega^2-\theta^2)^2+4\alpha^2\omega^2}\,. \tag{2}$$

Parameters θ and α depend on the geological conditions in the region and the properties of grounds on the site. We assume them as random variables. Because of the lack of information, apriori distributions are to be used, say, uniform distributions on the segments $[\theta_1, \theta_2]$ and $[\alpha_1, \alpha_2]$. Within a group of ground motions with a given intensity I all parameters are independent, but they became mutually dependent in nonconditional distributions, i.e. for a randomly chosen strong earthquake.

Conditional and Integral Components of the Structural Seismic Risk

To facilitate assessment of the structural risk, a hybrid, analitical-numerical technique is useful. This technique is based on the "splitting" of the risk into parts. As a result, the Monte Carlo simulation is needed only for assessment of conditional probabilities which are not very close to unity (or to zero). Denote $v(\tau|I)$ the response of a structure to a ground motion with intensity I. Let a region Ω is given of admissible values of components of the process $v(\tau|I)$. Introduce the conditional structural risk H(I) equal to the probability that after an earthquake of intensity I the condition $v(\tau|I) \in \Omega$ is violated at least once. Then

$$H(I) = 1 - P\{v(\tau|I) \in \Omega(\tau) \; ; \; \tau \in T_E\} \qquad (3)$$

where the "fast" time τ is used, and T_E is duration of the intensive stage of the ground motion.

Then an assumption is to be introduced on the distribution of ground motion in the "slow" time t. Assume ground motions on the site form a set of independent Poisson point processes in the "slow" time t with parameters $I_1, \ldots, I_m$ and return periods $1/\lambda_1, \ldots, 1/\lambda_m$. Then the integral (unconditional) structural seismic risk is [1]

$$H(t) = 1 - \exp\left[-\sum_{j=1}^{m} \lambda_j t H(I_j)\right]. \qquad (4)$$

Here t is the "slow" time, and all events assumed to be independent and non-hereditary (e.g., the effect of partial damage from previous earthquakes is not taken into account). This simplified approach can be improved in several ways, in particular, taking into account the scatter of properties of materials, structural components, joints, etc. [1].

Elasto-Plastic Response of Buildings to Strong Ground Motions

Consider a multi-storey frame building subjected to a horizontal ground motion (Fig. 1,a). For simplification, assume the storey girders are rigid and carrying all the mass of the adjoining storeys. The columns are clamped in the girders (in the foundation also) and are deformed elasto-plastically. These assumptions correspond to the common "shear" model of tall buildings [3]. For n-storey building vibrating in a plane-mode we come to a system with n degrees of freedom and generalized coordinates equal to horizontal displacements $u_1(t),\ldots,u_n(t)$ of the storeys. The horizontal displacement of the foundation is $u_0(t)$. The elasto-plastic behaviour of columns is described in terms of relationships between the moments M_k in the clamped sections of the columns, and the averaged angles of inter-storey "shear", i.e. $\varphi_k = (u_{k+1}-u_k)/h_k$. Here h_k is the length of the columns. In Fig. 2 the case of translatory linear hardening is presented. Compared with the conventional elasto-plastic analysis, the ultimate ductility angle φ_* is introduced in Fig.2. If inequality $|\varphi| < \varphi_*$ is violated, the corresponding columns fail, and it means the complete collapse of the storey, and, usually, the final failure of the building.

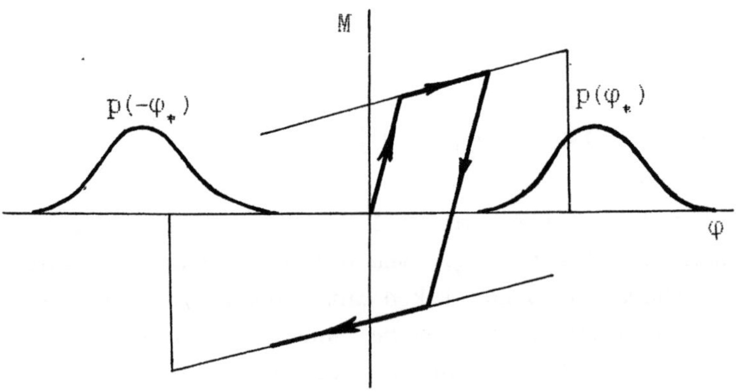

Fig. 2. Elasto-plastic behaviour with linear translatory hardening and randomly distributed ultimate deformation φ_*.

Later on some numerical results for a 16-storey building are presented obtained by Monte Carlo simulation. The seismic input was divided into three groups according to the intensities equal to 7, 8 and 9. The characteristic acceleration A_0 was considered as a log-unformely distributed in corresponding segments. Safety restrictions are put on the inter-storey deflections, absolute accelerations and displacements of each storey. To diminish the dimension of the space of parameters, the region Ω from Eq. (3) was taken as

$$\Omega = \{|\Delta u_1| < \Delta u_*,\ |a_{16}| < a_*,\ |u_{16}| < u_*\} \qquad (5)$$

where $\Delta u_1 = u_1 - u_0 = \varphi_1 h$, $a_{16} = a_0 + \ddot{u}_{16}$; Δu_*, a_*, and u_* are corresponding values limited from the reliability considerations.

Some numerical data, including 20 samples of maximum response for each intensity interval, are presented in Fig. 3 where the functional scale corresponding to the Gumbel distribution is used. The general tendency is evident. When ground motions are comparatively weak, empirical distributions follow the asymptotic distribution of maxima quite satisfactory. Extrapolation into the area of rare events seems here quite sensible. But if a ground motion is strong, there is no universal distribution of maximum response at all. The explanation is very simple: during strong earthquakes some groups of structural components begin to deform elasto-plastically, and that changes the general picture fundamentally. Moreover, each change in samples behaviour can be explained with inelastic behaviour of one or several rows of columns, including the general collapse of the all inter-storey columns. Of course, there is always a possibility to apply the importance sampling, stressing the extrapolation on the right-hand side sample points, as shown in Fig. 3 with the dashed lines. But one cannot be sure that adding new realizations, new tendencies of distributions would not appear.

The input into the integral seismic risk of strong but seldom, and weaker but more frequent earthquakes becomes more or less comparable. As an illustrations the partial seismic risks are

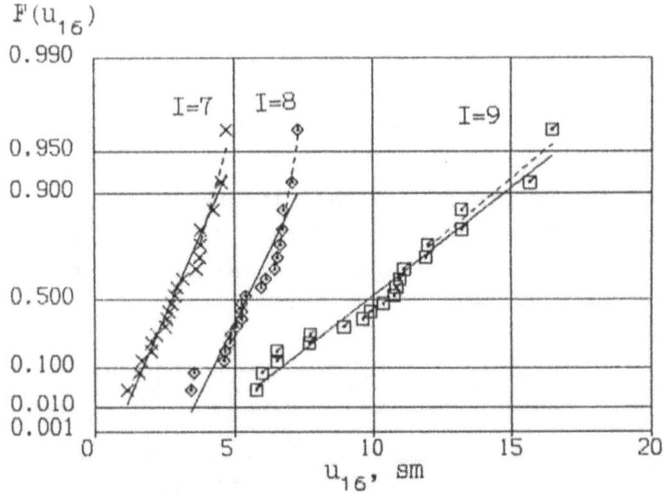

Fig. 3. Sample distributions of the displacement $\max|u_{16}(\tau)|$ on the top of the building.

presented in Table 1 corresponding to violation of one of the restrictions given with Eq. (5). Table 1 contents also the estimates of the seismic risk originated from each group of ground motion. The inequality

$$\max_{\alpha} H_{\alpha}(I) \leqslant H(I) \leqslant \min \{1, \sum_{\alpha} H_{\alpha}(I)\} \tag{6}$$

provides for seldom events a rather narrow-band approximation. The final values of the integral seismic risk are computed using Eq. (4). With respect of the restrictions given with Eq. (6), they are situated in the interval $0.028 \leqslant H(T) \leqslant 0.050$ for the life-time of the building $T = 50$ years.

Randomization of Resistance Parameters

The randomization of structural properties, in addition to ac-

Table 1

I	λ_j	$H_\alpha(I_j)$			$H(I_j)$	
		$\lvert\varphi_1\rvert<\varphi_*$	$\lvert a_{16}\rvert < a_*$	$\lvert u_{16}\rvert < u_*$	min	max
7	10^{-1}	$3.3\cdot10^{-6}$	$6.3\cdot10^{-8}$	$5.8\cdot10^{-15}$	$3.3\cdot10^{-6}$	$3.3\cdot10^{-6}$
8	10^{-2}	$7\cdot10^{-4}$	$8.2\cdot10^{-4}$	$1.6\cdot10^{-10}$	$8.2\cdot10^{-4}$	$1.5\cdot10^{-3}$
9	10^{-3}	$5.7\cdot10^{-1}$	$3.6\cdot10^{-1}$	$9.1\cdot10^{-2}$	$5.7\cdot10^{-1}$	1
				$H(t)$	0.028	0.050

count of the their natural randomness, facilitates the solution of the "tails" problem since it produces more representative and well-mixed samples of responses. Discuss briefly the simulation of the building's responses in the assumption that the limit angles φ_* characterizing ductility of the columns are random values. Suppose that φ_* takes the same magnitude for all the columns of k-th storey independently distributed among the storeys. Safety of the building will be violated if even for one of the storeys the equality $|\varphi_k(\tau)| = \varphi_k^*$ is attained during an earthquake. Then the admissible region takes the form

$$\Omega = \{ \ |\varphi_k(\tau)| < \varphi_k^*; \ \alpha = 1,\dots,n, \ \tau \in T_E \ \}. \tag{7}$$

For sampling it seems more convenient to deal with the risk factors for each storey $z_k = \max|\varphi_k(\tau)|/\varphi_k^*$, $k = 1,\dots, n$. The risk factor for the building as a whole is $z = \max_k z_k$. All these factors of course, are random variables. Denote their conditional distribution functions $F_k(z_k|I)$ and $F(z|I)$, respectively. With account of Eqs. (5) and (7) the conditional risk is $H(I) = 1 - F(1|I)$.

Computations were realized for a building regularly designed against ground motions with I = 8. It was assumed that the limit angles φ_* follow the Weibull distribution for minima, and that ground motions correspond to I = 9. The empirical distributions $F_k(z_k)$ for k = 1, 8, 16 and F(z) for all k = 1, ..., 16 are plotted in Fig.4 in the Gumbel functional scale. The conditional seismic risks $H(z_k|9)$ may be estimated by extrapolation up to the limit value $z_k = 1$, although no extrapolation is needed for z_1 as well as for $z = \max_k z_k$. The apparent irregularities of empirical distributions close to the safety margin are born by elasto-plastic behaviour and collapse of lower storeys' columns.

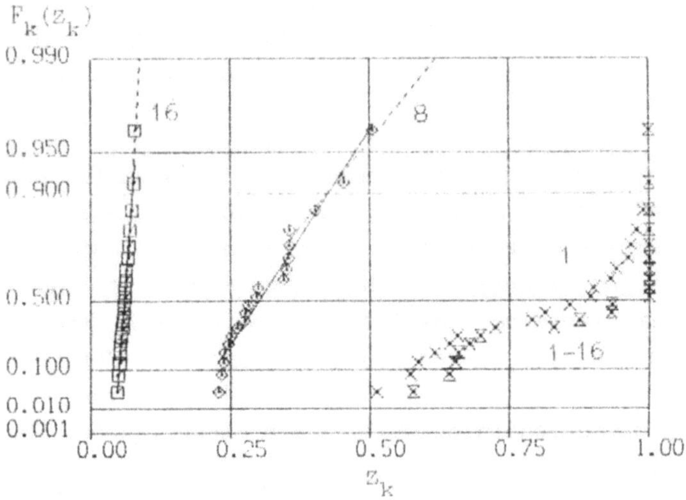

Fig. 4. Sample distribution of the risks factors for a building with random ultimate inter-storey shear.

Buildings on Floating Foundations.

Characteristic frequencies of strong earthquakes are situated in a comparatively narrow band, say 1 - 30 Hz, and only for very weak ground conditions this band is shifted to the lower frequencies. If a building is sufficiently rigid, the installation of floating foundation using, for example, metal-elastomer bearings apears to be effective [4].

Usually, the simple approach is applied to find the appropriate parameters of the foundation: to chose such the horisontal compliance that the eigenfrequency of the building treated as a rigid body were of order of 1 Hz. But on the whole the situation is more complicated. Some parameters (such as displacements) increase with decreasing of the fundamental eigenfrequency, others (such as accelerations) decrease. The problem is to minimize the integral seismic risk with respect to a large amount of controversal requirements.

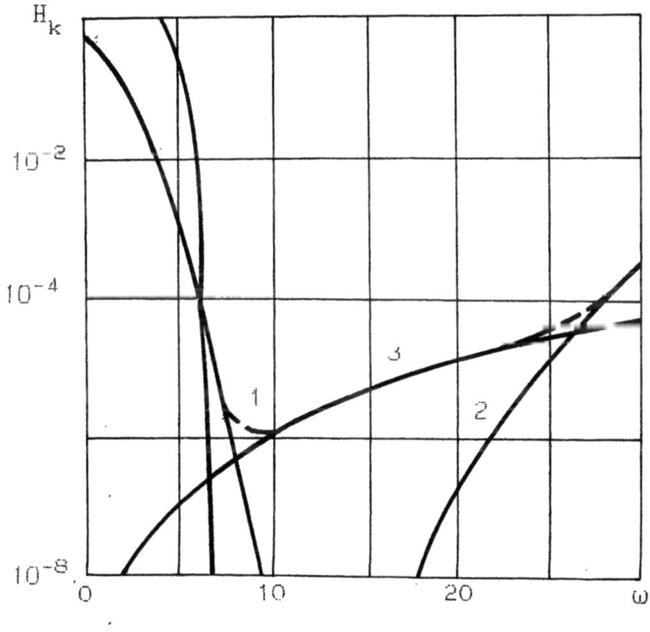

Fig. 5. Integral and partial seismic risk measures versus the fundamental natural frequency ω.

Some numerical results for a reactor building are presented in Fig. 5. There magnitudes of the partial seismic risks (continuous lines) and those of the integral seismic risk (dashed line) are plotted against the characteristic frequency ω equal to the fundamental natural frequency of the rigid building in the horizontal mode. Figures at the curves denote the partial limitations put on: (1) displacement of the upper foundation with respect to the lower one; (2) horizontal acceleration on the top level of the equipment; (3) maximum tensile stress in the concrete containment shell; (4) relative displacement of the sliding parts of the bearings. The optimal solution which is a compromise between the controversal requirements is approximately $\omega = 8\ s^{-1}$, i.e. rather close to 1 Hz. Two-side estimates (6), as it is seen from Fig. 4, present a quite satisfactory approximation for the integral seismic risk.

References

1. Bolotin, V.V.: Prediction of service life for mashines and structures. New York: ASME Press 1989.

2. Bolotin, V.V.: Structural integrity under stochastic loading in the area of small probabilities. In: Nonlinear Stochastic Dynamic Engineering Systems. IUTAM Symposium Innsbruck. Berlin: Springer - Verlag 1988 269-284.

3. Augusti, G.; Baratta, A.; Casciati, F.: Probabilistic Methods in Structural Engineering. London, New York: Chapman and Hall 1984.

4. Ushida, K. et al.: Seismic response analysis of nuclear plant founded on sliding elastomer bearing pads. In: Trans. of SMIRT-9. Rotterdam: Balkema R2 (1987) 667-672.

5. Bolotin, V.V.: Statistical theory of the aseismic design of structures. In: Proc. of 2nd World Conference on Earthquake Engineering. Tokyo: WCEE 1960 1365-1374.

6. Bolotin, V.V.: Random vibrations of elastic systems. The Hague: Martinus Nijhoff 1984.

Engine Probabilistic Structural Analysis Methods Reliability/Certification

C. C. CHAMIS

NASA Lewis Research Center

Cleveland, Ohio, U.S.A.

ABSTRACT

Development of reliability and risk methods for structural components and structures is a major activity at Lewis Research Center. It consists of six program elements: (1) probabilistic loads, (2) probabilistic finite element analysis, (3) probabilistic material behavior, (4) assessment of reliability and risk, (5) probabilistic structural performance evaluation, and (6) extension to other applications. Recent progress includes: (1) the evaluation of the various uncertainties in terms of cumulative distribution functions for various structural response variables based on known or assumed uncertainties in primitive structural variables, (2) evaluation of the failure probability, (3) reliability and risk-cost assessment, and (4) an outline of an emerging approach for eventual certification of man-rated structures by computational methods. Collectively, the results demonstrate that the structural durability/reliability of man-rated structural components and structures can be effectively evaluated by using formal probabilistic methods.

INTRODUCTION

It is becoming increasingly evident that deterministic structural analysis methods will not be sufficient to properly design critical components in future structures in general and aerospace structures in particular. These structural components are subjected to a variety of complex, and severe cyclic loading conditions, including high temperatures and high temperature gradients. Most of these are quantifiable only as best engineering estimates. These complex loading conditions subject the material to complex and coupled nonlinear behavior which depends on stress, temperature, and time. Complex and coupled nonlinear material behavior is nonuniform, is very difficult to determine experimentally, and perhaps impossible to describe deterministically. In addition, hot rotating structural components for aerospace propulsion (engines) are

N. Bellomo · F. Casciati (Eds.)
Nonlinear Stochastic Mechanics
IUTAM Symposium Turin

relatively small. Fabrication tolerances on these components, which in essence are small thickness variations, can have significant effects on the component structural response. Fabrication tolerances by their very nature are statistical. Furthermore, the attachment of components in the vehicles integrated structural system generally differs by some unknown amount from that assumed for designing the component. In summary, the fundamental aspects -- (1) loading conditions, (2) materials behavior, (3) geometric configuration, and (4) supports (attachments) to integrated structures inherently include a variety of uncertainties of unknown magnitude.

There are generally two approaches to handle this wide variety of uncertainties: (1) current practice, and (2) probabilistic evaluation. Current practice is adequate where the new engine structure is *not* very different from an existing one. However, this approach is costly and requires long time-schedules for future structures which will be entirely different from any existing ones. The second approach is to develop probabilistic structural analysis methods where the uncertainties in all the parameters of the four fundamental aspects are described by appropriate probability functions.

Development of the Probabilistic Structural Analysis Methodology (PSAM) is an on-going activity at NASA Lewis and is a joint program of in-house and sponsored research (ref. 1). Theoretical considerations, computer codes, and other relevant applications are described in papers presented in conferences (refs. 2 to 6). Activities and progress up to June 1989 are summarized in reference 7. The objectives of this brief paper are (1) to summarize the fundamental aspects of PSAM and (2) to demonstrate the application of this methodology to a specific example (the reliability/risk of turbine blade components of rocket propulsion systems). The specific example includes the four fundamental aspects (key elements) required in probabilistic structural analysis of future structures, namely: (1) probabilistic loads, (2) probabilistic finite element analysis, (3) probabilistic description of complex coupled nonlinear material behavior, and

(4) evaluation of reliability and risk. Throughout the specific example discussion, appropriate comments are included to illustrate the generality of the method and its application to aerospace and other structures in general. Extensions to other applications are cited to illustrate the versatility of the method.

FUNDAMENTAL CONSIDERATION

Central to the probabilistic structural analysis is the fundamental consideration that: Uncertainties observed in the structural performance (displacements, frequencies, buckling, global fracture toughness, stresses/strains) of structures can be quantified in terms of corresponding uncertainties in basic parameters (primitive variables). The primitive variables are those which are used to describe the structure and its respective environment. For example: (1) structural configuration,
(2) boundary conditions (attachments), (3) loading conditions, and
(4) material thermomechanical nonlinear behavior.

The uncertainties in these primitive variables are then integrated through structural mechanics to quantify the uncertainties in the global structural responses (displacements) and are decomposed to quantify the uncertainties in local responses (stresses/strains). The concept is schematically illustrated in figure 1. The structural component is the blade which is modeled for finite element analysis. The input uncertainties are the blade loads (centrifugal, pressure, and temperature), geometry and material variables. The output is quantification of corresponding uncertainties in structural responses or in local stresses for probable fracture initiation. Brief descriptions are given in subsequent sections on each of these as it is applied to the specific example.

PROBABILISTIC SIMULATION OF LOADS

The fundamental assumption for the probabilistic simulation of loads is that each individual load condition can be probabilistically synthesized from four primitive parts:
(1) steady state, (2) periodic, (3) random, and (4) spike. Each of these parts, except random, is described by a nominal or

deterministic portion and a probabilistic perturbation about this nominal portion. The resulting distribution is similar to the schematic, in figure 1 upper left and as further described in reference 7. One justification for synthesizing each loading condition in terms of primitive parts is that experts, over the years, have developed good judgment of the ranges of perturbations about nominal or deterministic conditions. A computer code CLS (Composite Load Spectra) has been developed to synthesize the four parts of each load condition by using (1) available data from various sources of past experience, (2) probability theory, and (3) a dedicated expert system, which includes the information supplied by the experts. A schematic of the features of the computer code is shown in figure 2.

The results from the application of the Composite Load Spectra computer code to probabilistically simulate loads for two blades are summarized in table 1. The comparisons with the measured data are in very good agreement considering the large number of primitive variables (47) required to synthesize these loads. The conclusion is that methods can be developed and are available to probabilistically synthesize complex load conditions for hot aerospace structures and structures in general.

PROBABILISTIC FINITE ELEMENT STRUCTURAL ANALYSIS

The fundamental assumption for developing Probabilistic Finite Element Methods (PFEM) for structural analysis is that the uncertainties in each primitive structural variable can be represented by an assumed probabilistic distribution. Primitive structural variables are those which are used to describe a structure such as: (1) stiffness, (2) strength, (3) thickness and tolerance, (4) spatial location, (5) attachment, and (6) various nonlinear material dependencies (temperature, stress, time, etc) as is schematically illustrated in figure 1 upper right. See also reference 7. Subsequently, the uncertainties in the load conditions (synthesized by the CLS computer code and the uncertainties in the primitive structural variables are computationally synthesized by performing probabilistic structural analysis to simulate uncertainties in the structural response of a specific structural component or structure. The structural response is generally described in terms of usual quantities such as displacement, frequencies, buckling loads, and structural fracture toughness as was already mentioned. The probabilistic structural analysis is generally performed by using Probabilistic Finite Element Methods (PFEM).

PFEM has been formalized and integrated into a computer code identified as NESSUS (Numerical Evaluation of Stochastic Structures Under Stress). A block diagram of NESSUS is shown in figure 3. NESSUS contains a library of finite elements and is driven by an expert system. It can be used to probabilistically evaluate all types of structures. Representative results obtained using NESSUS on various structures are given in references 1 to 7 and for the specific strength calculations are given in reference 8. The combined effects of the primitive variable uncertainties on structural response are generally shown as probability distributions, figure 1 bottom right (stress from NESSUS). The information generated for these probability distributions can also be used to evaluate the sensitivities which influence these distributions. The results of a sample case are shown in figure 7 where the loads were simulated by CLS using the information in Table 2. The significant point is the PFEM yields a wealth of information which can be used to evaluate: (1) the uncertainties in the structural response, and (2) the sensitivities which can be used to adjust the design for enhanced probability of success. The important conclusion is that probabilistic finite element methods can be developed and are available to quantify uncertainties in the structural performance of a variety of structures. In addition the sensitivities that influence this performance can be evaluated and ranked.

PROBABILISTIC SIMULATION OF MATERIAL BEHAVIOR

The fundamental assumptions to probabilistically simulate complex nonlinear material behavior are: (1) a relationship for material behavior can be developed in terms of primitive variables affecting this behavior and (2) the uncertainties in the primitive variables can be described by assumed distributions (ref. 7). A MultiFactor

Interaction Model (MFIM) for this relationship has been developed and incorporated in the NESSUS computer code. The form of MFIM is shown in Figure 5. It is used to develop resistance probability functions as shown in figure 1, bottom right (strength).

This MFIM is applied to specific structures to probabilistically determine: (1) the resistance curve for damage (crack) initiation and (2) damage propagation and its effects on global structural response. The results for the most probable point for damage initiation, the most probable path, and the degradation in structural integrity can then be determined for specified probabilities. The coupled CLS/NESSUS/MFIM have been used to generate the results shown in Figure 6. The input to MFIM for these results are summarized in Table 3.

The important observation is that the uncertainties in damage initiation, propagation and subsequent effects on structural performance can be probabilistically simulated by the methodology described herein. It is noteworthy that this methodology, in general, is applicable to a variety of structures, and can readily be incorporated to monitor the in-service health of structures in general and aerospace structures in particular.

RISK-COST ASSESSMENT

The methodology described previously has been extended to perform reliability and risk-cost assessments. In order to accomplish this, (1) the cost for component/structure service readiness needs to be quantified and (2) the cost as a consequence of component/structure failure must be established. Both of these have been integrated into the probabilistic structural analysis methodology (ref. 9). The results from the application of this methodology to the specific example blade are summarized in figure 7 in terms of fatigue cycles to failure.

The important observation from the aforementioned discussion is that the reliability and risk-cost of structures in general and man-rated structures in particular can be assessed using probabilistic methods of the type described herein. The implications are far-reaching because these methods are primarily computational and can be applied to existing structures to evaluate their risk for continuing service as well as those on the design board and those still in the conceptual phase.

RELIABILITY/CERTIFICATION -- AN EMERGING APPROACH

The collective observations from the previous discussion led to an emerging approach to computationally simulate: (1) structural reliability, (2) risk, (3) component qualification, and (4) eventually vehicle structure certification. The general steps for this emerging approach are outlined as follows:

1. Develop a coarse structure or structural component/vehicle (global) analysis model.
2. Conduct Probabilistic Structural Analysis (PSA) of the types described herein.
3. Identify the critical component/structure areas from the results of PSA.
4. Perform global/local PSA's to evaluate nonlinear effects and to locate probable sites of damage initiation.
5. Determine the most probable damage propagation path.
6. Evaluate probable structural degradation along this path.
7. Establish probable path extent for violation of specified structural performance criteria (for example, 10-percent increase in displacement or 5 percent reduction in the frequency of the first vibration mode).
8. Assess the corresponding reliability and risk and decide on their acceptability.
9. Schedule inspection intervals and retirement for cause criteria based on the results of items 5, 6, and 7.
10. Verify with probabilistically selected (using respective sensitivities) critical structural components and prototype structure tests.
11. Design a suitable in-service health monitoring system using the results from items 8 and 9 above in order to ascertain that the component/structure will meet the acceptable reliability and risk requirements.

EXTENSION TO OTHER APPLICATIONS

Recently PSAM has been extended to other applications to illustrate the versatility of the method.

1. Application of the method to aerodynamic uncertainties in propulsion systems is described in reference 10. The results from this application show that PSAM can be used to describe the uncertainties in the performance of the rotor as represented by the flow coefficient and by the flow presurvival parameters.
2. Application of the method to space trusses is described in reference 11. The results from these studies show that the uncertainties in truss-member force and truss vibration frequencies can be determined in terms of uncertainties in spatial geometry and loads. The method is currently extended to probabilistically assess the progressive buckling of the truss.
3. Application of the method to probabilistically simulate wind loads and building response is described in reference 12. The results show that the uncertainties of several primitive variables (building height, wind load--pressure side and leeward side) contribute to maximum building's displacement and overturning moment at low probabilities, while only the wind load on the pressure side is the most significant primitive variable at high probabilities.

Although these are limited applications, they do illustrate that the PSAM methods described are not limited to aerospace engine components.

CONCLUSIONS

A methodology has been developed for the formal Probabilistic quantification of uncertainties in the structural performance and subsequent reliability and risk of man-rated structures. The key elements in this methodology are: (1) probabilistic load simulation, (2) probabilistic finite element analysis, (3) probabilistic simulation of complex nonlinear material behavior, and (4) risk-cost assessment. This methodology is described in terms of fundamental aspects and application to a specific structural component which is a turbopump blade of the Space Shuttle Main Engine (SSME) and which was selected for its complexity in order to demonstrate what can be done by using this methodology. The specific example illustrates how the uncertainties in all the basic parameters (primitive variables) for loads, structure and material behavior are incorporated in order to probabilistically simulate the uncertainties in the structural response (global and local). Also, the example illustrates how the reliability and risk-cost can be assessed. Collectively, the summary of the fundamental considerations and the results from the specific example demonstrate that a formal methodology is available to evaluate the reliability and risk-cost of man-rated structures in aerospace environments as well as structures in general. An emerging approach is outlined which can be used to computationally qualify and eventually certify future structures. In addition, extension to other applications is cited to illustrate the versatility of the method.

REFERENCES

1. C. C. Chamis: Probabilistic Structural Analysis Methods for Space Propulsion System Components. NASA TM 88861, 1986.
2. Anon.: AIAA/ASME/ASCE/AHS 28TH Structures, Structural Dynamics and Materials Conference, April 6-8, 1987 (papers 87-0764, -0765, -0766), pp. 252-274.
3. Anon.: Structural Integrity and Durability of Reusable Space Propulsion Systems. NASA CP 2471, 1987, pp. 117-200.
4. Anon.: AIAA/ASME/ASCE/AHS 29th Structures, Structural Dynamics and Materials Conference, April 18-20, 1988 (papers 88-2371 to 88-2376), pp. 1262-1288.
5. AIAA/ASME/ASCE/AHS/ASC 30th Structures, Structural Dynamics, and Materials Conference, April 3-5, 1989 (papers 89-1368 to 89-1372), pp. 1832-1910.
6. Anon.: AIAA/ASME/ASCE/AHS/ASC 31st Structures, Structural Dynamics and Materials Conference, April 2-4, 1990 (papers 90-1098 to 90-1102), pp. 1039-1080.
7. C. C. Chamis and D. A. Hopkins: Probabilistic Structural Analysis Methods of Hot Engine Structures. NASA TM 102091, 1989.

8. C. C. Chamis and T. A. Cruse: Probability Approach for Strength Calculations. AGARD-R-772, 1990.

9. M. C. Shiao and C. C. Chamis: Probability of Failure and Risk Assessment of Propulsion Structural Components. NASA TM 102323, 1990.

10. A. Hamed: Probabilistic Modeling for Simulation of Aerodynamic Uncertainties in Propulsion Systems. NASA TM 102472, 1989.

11. S. S. Pai: Probabilistic Structural Analysis of a Truss Typical for Space Station. NASA TM 103277, 1990.

12. A. R. Shah and C. C. Chamis: Simulation of Probabilistic Wind Loads and Building Analysis. NASA TM 103744, 1991.

TABLE 1

HPOTP & HPFTP Parameters
Phase II Engine Calculated vs Measured

	HPOTP				HPFTP			
	Speed		Turbine Discharge Temp		Speed		Turbine Discharge Temp	
Condition	Calc (rpm)	Measured (rpm)	Calc °R	Measured °R	Calc (rpm)	Measured (rpm)	Calc °R	Measured °R
Hardware — 2σ random	294	—	53	—	288	—	65	—
Test — 2σ random	210	—	157	—	398	—	20	—
Total random	360	—	165	—	554	—	70	—
Low NPSP — det	620	—	225	—	56	—	52	—
High NPSP — det	−317	—	−219	—	−94	—	−62	—
Range = random + det	1660	1500	475	400	1260	1000	114	150
Max	29090	29100	1630	1650	35742	35750	1740	1760
Nom	28100	28200	1374	1380	35130	35300	1688	1690
Min	27430	27500	1155	1250	34482	34750	1625	1610

Measured — Measured variation for phase II test set
Hardware — Variations in engine hardware
Test — Initial test conditions — Inlet temperatures & mixture ratio
Det — Duty cycle effects of inlet pressures plus correlated Ro variations of cavitation

TABLE 2

Random Variables Considered and Their Statistics

No.	Random Variable	Type	Affected FEM Quantities	Mean	Standard Deviation
1	Material axis Z	Material	Anisotropic	−0.087266 radian	0.067544
2	Material axis Y	orientation	material	−0.034907	0.067544
3	Material axis X	Effects	Orientation angles	−0.052360	0.067544
4	Elastic modulus	Material	Elastic constants	18.38E6 psi	0.4695E6
5	Poisson's ratio	properties		0.386	0.00965
6	Shear modulus			18.63E6 psi	0.046575E6
7	Geometric lean	Geometrical	Node coordinates	0 deg	0.14 deg
8	Geometric tilt	variations		0 deg	0.14 deg
9	Geometric twist			0 deg	0.30 deg
10	Mixture ratio	System	Pressure, temperature,	6.0	0.02
11	Fuel Inlet pressure	Independent	centrifugal force	30.0 psi	5.00
12	Oxidizer Inlet pressure	loads		100.00 psi	26.00
13	Fuel Inlet temperature			37 °R	0.50
14	Oxidizer Inlet temperature			164 °R	1.33
15	Pump efficiency	Component	Pressure, temperature,	1.00	0.008
16	Head coefficient	Independent loads	centrifugal force	1.024	0.008
17	Coolant seal leakage	Local effects	Temperature	1.0	0.10
18	Hot gas seal leakage			1.0	0.05

TABLE 3

PRIMITIVE VARIABLE PROBABILITY DISTRIBUTIONS FOR PROBABILISTIC MATERIAL PROPERTY MODEL

VARIABLE	DISTRIBUTION TYPE	MEAN	STANDARD DEVIATION (VALUE)	STANDARD DEVIATION (% OF MEAN)
T_F	NORMAL	2750 °F	51.4 °F	2.0
T_0	NORMAL	68 °F	2.04 °F	3.0
S_F	NORMAL	212.0 ksi	10.6 ksi	5.0
σ_0	CONSTANT	0	0	0
N_{MF}	LOGNORMAL	10^8	5×10^6	5.0
N_{MO}	LOGNORMAL	10^3	50	5.0
n	NORMAL	0.25		3.0
p	NORMAL	0.25		3.0
q	NORMAL	0.25		3.0

FIGURE 1

Component Response Analysis Using CLS Coupled With PSAM

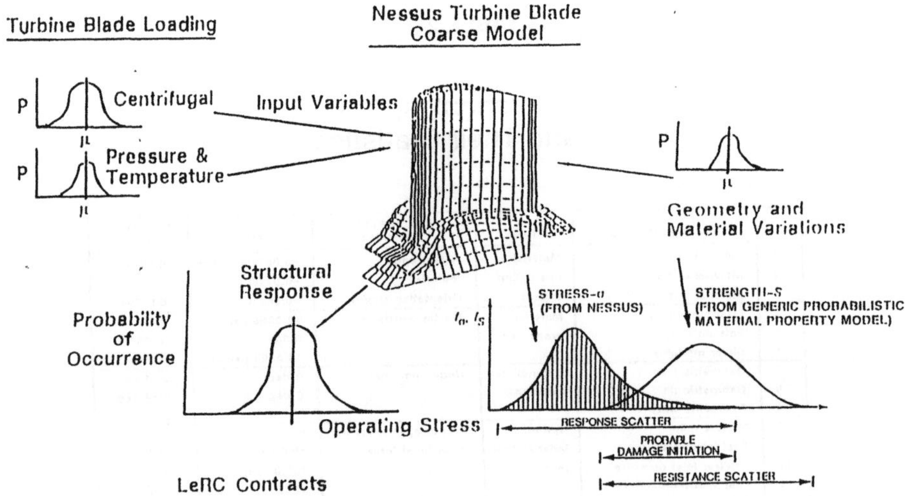

LeRC Contracts

CLS - Composite Loads Spectra
PSAM - Probabilistic Structural Analysis Methods - SWRI

FIGURE 2

COMPOSITE LOAD SPECTRA SIMULATION USING EXPERT SYSTEMS

COMPOSITE LOAD SPECTRA EXPERT SYSTEM

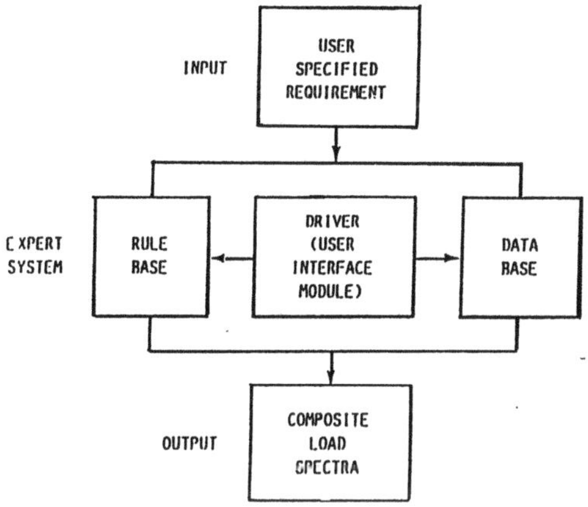

LDEXPT DESIGN

LOAD EXPERT SYSTEM DESIGN PHILOSOPHY

- RULE-BASE PRODUCTION SYSTEM
  - IF THEN RULES
- SIMPLE INFERENCE SCHEME
  - INFERENCE NET (DECISION TREE)
- SOPHISTICATED PROBABILISTIC METHODS
  - DISCRETE PROBABILITY DISTRIBUTION
  - MONTE CARLO
  - BARRIER CROSSING
- POWERFUL KNOWLEDGE BASE
  - INFLUENCE COEFFICIENT
  - SCALING COEFFICIENTS
  - DUTY CYCLE LOAD PROFILES
  - ENGINE CONFIGURATION AND GEOMETRY DATA
  - RAW ENGINE FLIGHT AND TEST DATA

FIGURE 3

COMPUTATIONAL SIMULATION OF PROBABILISTIC STRUCTURAL RESPONSE USING NESSUS

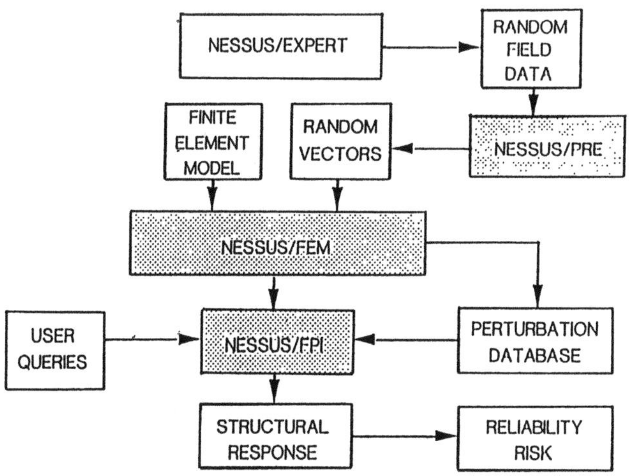

FIGURE 4

PROBABILISTIC SIMULATION OF UNCERTAINTIES IN BLADE STRESSES (SSME HPFTP)

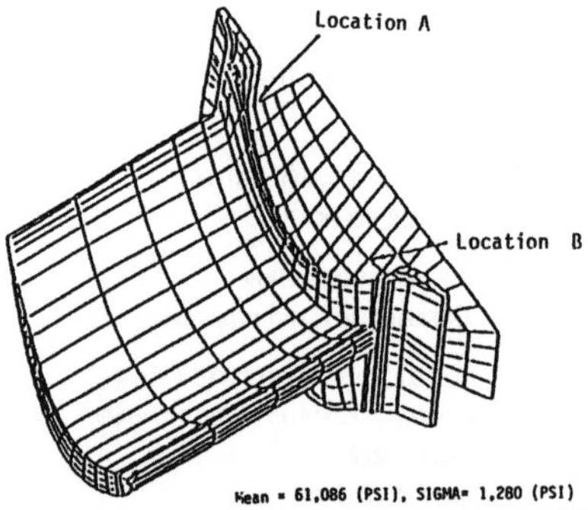

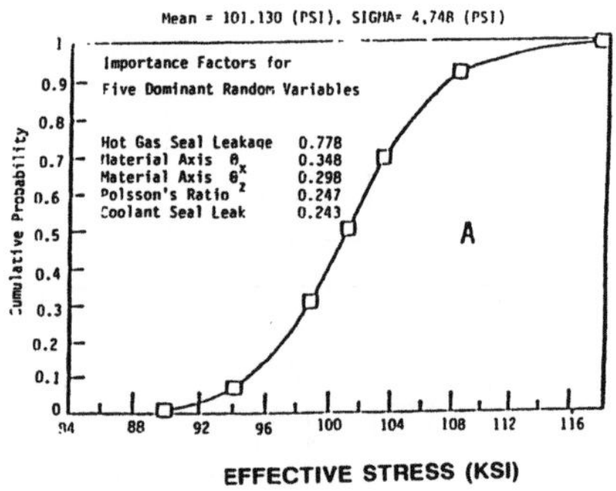

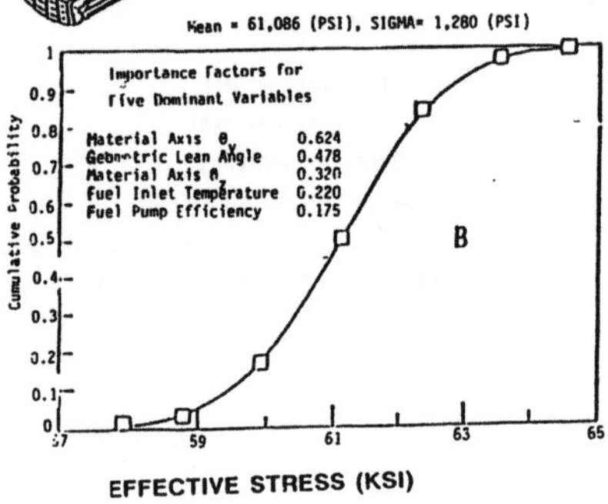

THE METHODOLOGY IS AVAILABLE TO FORMALLY QUANITIFY UNCERTAINTIES AND SENSITIVITIES IN HOT ENGINE STRUCTURES

FIGURE 5

GENERIC PROBABILISTIC MATERIAL PROPERTY DEGRADATION MODEL IN TERMS OF PRIMITIVE VARIABLES

$$M_P = M_{PO}\left[\frac{T_F - T}{T_F - T_O}\right]^n \left[\frac{S_F - \sigma}{S_F - \sigma_O}\right]^p \left[\frac{\log N_{MF} - \log N_M}{\log N_{MF} - \log N_{MO}}\right]^q$$

PRIMITIVE VARIABLES

M_P = DEGRADED MATERIAL PROPERTY
T = TEMPERATURE
S = STRENGTH
σ = STRESS
N_M = MECHANICAL CYCLES

SUBSCRIPTS

F = FINAL CHARACTERISTIC VALUE
O = REFERENCE PROPERTY

FIGURE 6

NATURAL FREQUENCIES DECREASES AS FRACTURE PROGRESSES

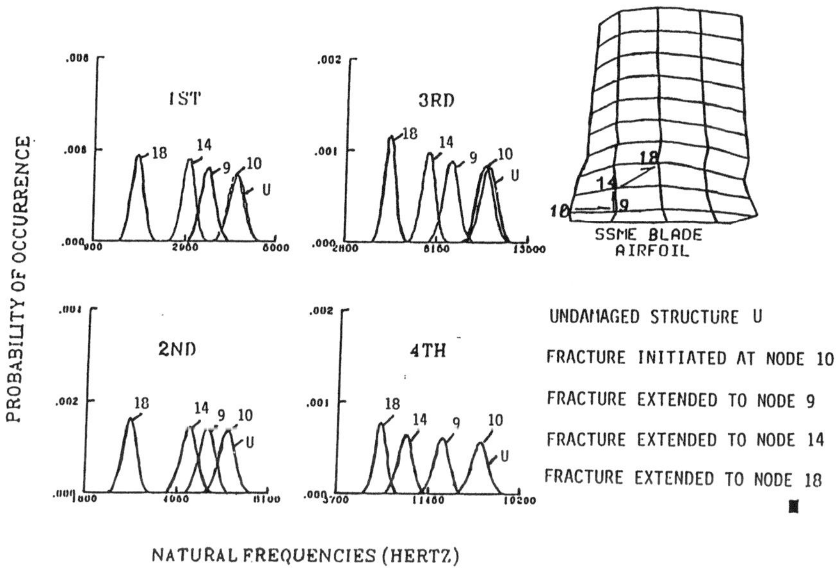

FIGURE 7

PROBABILISTIC RISK-COST ASSESSMENT

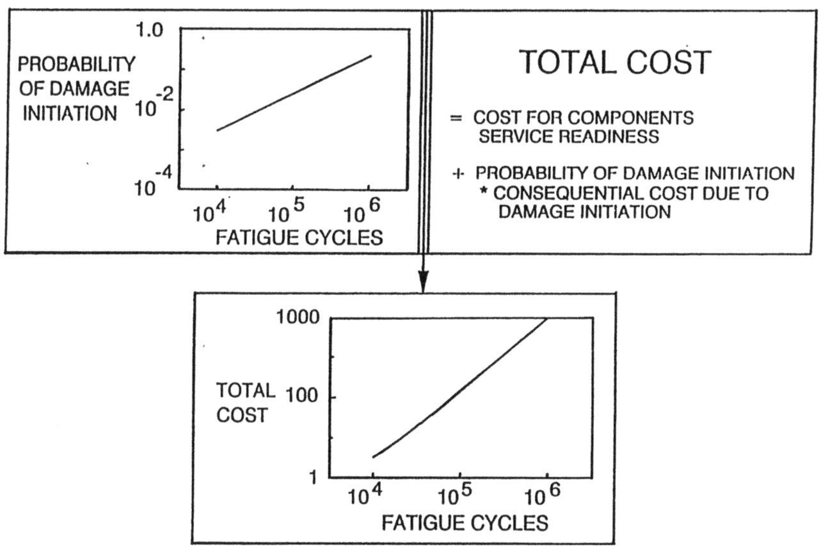

Seismic Reliability of Simple R/C Frames Designed According with the 1976 and 1987 Mexico City Codes

M. CHAVEZ, D. GONZALEZ
INSTITUTO DE INGENIERIA, UNAM, MEXICO
APDO 70-569, 04510 MEXICO DF

ABSTRACT

A methodology is applied to estimate the probability of failure, **Pf**, of simple R/C frames designed by following Mexico's City Code in its 1976 and 1987 versions. The methodology uses Monte Carlo simulation techniques, step by step dynamic analysis, and reliability concepts. Additionally a bayessian hazard model has been applied to generate detailed ground motions expected in Mexico City soft soil. The behaviour of the frames under seismic loading was represented by the elastoplastic hysteretic type (as it is implied in the mentioned Codes) and defined by their initial lateral stiffness, K, and yielding force, Vy. As K and Vy depend on several material and geometrical random variables (i.e. concrete elastic modulus, yielding stress of the steel, etc.) they were considered to be lognormally distributed and statistically independent. Taking this into consideration, by Monte Carlo simulation, families of structural systems with initial natural periods of 0.5, 1.5, and 2.5s were generated.
The ground motions expected in Mexico's City soft soils were also randomly generated from the scaling of 50 accelerograms, which were recorded in Mexico City lake zone during the occurrence of 15 earthquakes with magnitudes larger than 5.5 in the last 30 years. Most of the accelerograms were recorded after 1980. The scaling factors were obtained from the probability distribution of the maximum ground intensity in Mexico's City soft soils, estimated with the seismic hazard model, and the data recorded in the mentioned lapse time.
The ductility demands (ratio of the maximum horizontal deformation by the yielding deformation) expected from the considered systems, under the seismic loads associated to the ground montions mentioned above, were obtained by step by step dynamic analysis by using the DRAIN code. The results of those analyses provided the data from which the complementary accumulated probability distribution of the ductility demands were estimated. The resistance of the systems (their nominal ductilities) was assumed to possess a lognormal distribution. Finally the **Pf** of the structural systems of interest were computed from the two mentioned distributions.
The main results of the study are the following: 1) the probability distributions of the ductility demands for the considered structural systems are of the extreme type, 2) the **Pf** of the frames designed with the 1976 Mexico City Code are three

N. Bellomo · F. Casciati (Eds.)
Nonlinear Stochastic Mechanics
IUTAM Symposium Turin

times larger, than the ones designed with the 1987 version of the Code, 3) the **Pf** of the systems with a 2.5s of initial natural period are up to three orders of magnitude larger than the ones of 0.5s, 4) the **Pf** implicit in both codes for the systems considered is of the order of 10^{-3}.

INTRODUCTION

The estimation of the reliability of structural systems subjected to seismic loads needs to take into account the nonlinear behaviour of the systems under dynamic loads. Also the characteristics of the future ground motions which will be acting on the systems, and the uncertainties on the stuctural parameters should be included. In recent studies [1,2] this problem was studied and a methodology was proposed and applied to estimate the seismic safety of simple R/C frames designed accordingly with the 1976 version of the Mexico City Code. Since then a large number of accelerograms have been recorded in the about 100 instruments installed in Mexico's City upgraded accelerograph network, also new statistical data has been published in relation with the mechanical and geometrical parameters of R/C elements designed with the 1976 and 1987 versions of the mentioned Code. The present paper investigates the impact of this new information on the estimation of the probability of failure, **Pf**, of simple R/C systems designed accordingly with those Codes. The methodology we applied is basically the same as the one proposed in [1], however, in the present research the DRAIN code was utilized to compute the seismic response of the structural systems.

STRUCTURAL MODEL

The structural systems analyzed in this work are ductile R/C frames as the one shown in Fig 1. The behaviour of the columns under flexural cyclic loading is assumed to be of the elastoplastic hysteretic type, Fig 2, and the beam element is considered to be infinitely rigid. The failure mode of the frame due to the horizontal ground acceleration, u_g , is assumed to be unique and due to the simultaneous formation of plastic hinges at the ends of the columns, Fig 1.

The resistance, R, of the frame is associated to the variable δ, where $\delta = \mu-1$, and μ is a random variable representing the ductility available in the frame. Here μ is defined as the ratio of the maximum deformation to the yield deformation, The variable δ is assumed to possess a lognormal distribution. In Fig 3 we present the probability density functions of δ (that in what follows it will be called f_R) for several values of $V\mu$, and a characteristic value of μ of 4. The latter value is suggested in [3,4] for the structural systems under study. As it is shown later, the calculation of **Pf** requires the computation of the probability distributions of the ductility demands on the systems of interest, F_S, when subjected to the ground motions, u_g, expected in the lake zone of Mexico City. Therefore, it is necessary to obtain samples of the expected seismic responses of the mentioned systems. First we will present how the structural systems of interest can be simulated, and later on we will discuss about the simulation of the u_g. The parameters that characterize the behaviour of the frames are the initial lateral stiffness, K, and the yield force Vy, Fig 2. K is a function of the elasticity modulus of the concrete, Ec, the inertia moment of the columns cross section, I, and the length of the columns, L. Vy is a function of the concrete strenght, fc, the yielding stress of the reinforcing steel, fy, the cross section of the reinforcing steel, As, the effective depth of the column cross section, d, the width of the column cross section, b, and a formulae error parameter ϕ. Accordingly with [5], Ec, fc, fy, I, d, and ϕ can be considered as random variables, and As, L, and b as deterministic ones. From this it follows that K and Vy are random variables whose first two statistical moments can be expressed as a function of the first two moments of their associated random variables. It can be shown [6] that, for two columns, rigidly clamped,

$$\overline{K} = 24\ \overline{E}\ \overline{I}/\ L^3 \tag{1}$$

$$\sigma_K^2 = (24\ \overline{I}/\ L^3)^2\ \sigma_E^{\ 2} + (24\ \overline{E}/L^3)^2 \sigma_I^2 \tag{2}$$

$$\overline{V}_Y = 4\ \overline{M}_Y/\ L \tag{3}$$

$$\sigma^2_{Vy} = 16\ \sigma^2_{My} /\ L^2 \tag{4}$$

where the bar over a variable and σ^2 represent its mean and variance respectively, and M_y is the yielding moment of the columns, which can be computed with the following expression [3,4]:

$$My = \phi\ As\ fy\ d\ (\ 1 - As\ fy/\ 2\ b\ d\ fc\ ) \tag{5}$$

If we asume that K and Vy are statistically independent and lognormally distributed, we can randomly simulate pairs of Ki and Vyi values with i = 1,m, where m represents the size of the sample.

SEISMIC RELIABILITY OF SIMPLE R/C FRAMES

If R and S are associated to the random resistance and the random action on the frame under study, respectively, and if they are considered to be statistically independent, the probability of failure of the structural system can be calculated as follows:

$$\mathbf{Pf} = P(S>R) = \int_{\alpha} \bar{F}_S(\alpha)\ f_R(\alpha)\ d\alpha \tag{6}$$

in eq. (6) P(S>R) means the probability that S will be larger than R at a certain time during the lifetime of the frame. $\bar{F}_S$ is the complementary accumulated probability distribution of S, and f_R is the probability density function of R which was described above. In order to calculate $\bar{F}_S$, it is required to have samples of the seismic responses of the frames under the expected ground motions at the site of interest, in our case the so called Lake zone of Mexico City. To obtain those samples, Monte Carlo simulation techniques can be utilized to generate samples of structural systems, by following what was suggested in the previous chapter. The simulation of the ground motions expected in the mentioned zone can be performed as it is described in the next chapter. When the samples of structural systems and of expected accelerograms have been simulated, they are randomly paired and by step by step dynamic analysis, the maximum responses of the structural systems can be computed, therefore Fs can be calculated, as it is shown in the numerical application.

SEISMIC INPUT

The procedure to generate the expected ground motions at the Lake zone of Mexico City can be synthesized as follows : 1) compute the probability distribution of the maximum ground intensities at the site for the time period of interest; 2) select a set of recorded accelerograms which includes the main ground motion characteristics at the site of interest, such as amplitudes, frequency content, and duration; 3) generate the sample of the expected accelerograms by scaling the ordinates of the accelerograms selected in the step 2. The resulting ground motions take the seismic hazard at the site into consideration through the probability distribution mentioned in step 1. Details of the methodology can be found in [1,2], therefore in what follows we will discusse the enlarged data set which was used in this work. In this work we utilized data about the maximum ordinate of the velocity response spectra for a 10% of critical damping, Sv, associated to events with surface wave magnitudes larger than 5.5, which have been recorded in Mexico's City Lake zone in the last 30 years. In Fig. 4 we show the annual rate of exceedance of the Sv data, with squares the data up to 1980 (used in [1] and [2]) and with crosses the complete set which covers the period from 1961 to 1990.
If we assume that the occurrence of earthquakes follows a Poisson process, such that the intensities and the detailed ground motion associated to any two different events are statistically independent and identically distributed, the probability P, that a particular Sv is exceeded can be expressed as

$$P(Sv) = \exp(-\nu(Sv)To) \tag{7}$$

where To is the lapse of interest and the rate of occurrence $\nu(Sv)$ can be calculated with the expression

$$\nu(Sv) = k(Sv^{-q} - Sv_1^{-q}) \qquad (8)$$

here k and q depend on the seismicity of the region where the site is located, and Sv_1 is the maximum Sv which may occur in the site. The estimation of these parameters for the 30 years data of Fig 4 was performed by applying bayesian statistics, the resulting expected posterior values of k and q were 25 and 2 respectively. The Sv_1 value for this case is 400cm/s, that was obtained based on the statistical analysis of historical data and instrumental recordings in the zone of interest . In Fig 4 we present the $\nu(Sv)$ curve corresponding to the mentioned k, q, and Sv_1 values. This curve is very similar to the one propossed in 1984 in [1], notice that the larger value of Sv_1 of the data set, was recorded during the 1985 Mexico earthquake and was equal to 214cm/s. For the present study the value of To equal to 50 years is of interest, therefore the resulting P(Sv) accordingly with equation 7 is shown in Fig 5.

A sample of 50 accelerograms associated to the earthquakes included in the study were selected as representative of the ground motions at the site, the peak ground accelerations varied from 6.2 to 167cm/seg^2, and their duration varied from 50 to 180s.

The scaling of the accelerograms was performed as follows: sets of Sv values were simulated from its distribution (eq 7), and for each simulated Sv one of the 50 accelerograms was randomly chosen and scaled so as to produce the corresponding value of the mentioned Sv. Tests were carried out in order to compare the response spectra of the simulated accelerograms with the response spectra of the recorded ones, the comparison was satisfactory.

NUMERICAL APPLICATION

A numerical application of the methodology is performed in order to evaluate the **Pf** of simple R/C systems designed according with the 1976 and 1987 Mexico City Codes. The systems studied have an initial natural period of 0.5, 1.5 and 2.5s, and a percentage of critical damping of 0.05 (as recommended in those codes for the

R/C constructions, [3,4]). The values of the parameters utilized to simulate the families of the systems of interest, will be presented elsewhere, the sample sizes used were 40 and 80. In Fig 6 the ductility demands of some of the structural systems studied are plotted versus their probability of exceedance, which as in the previous studies corresponds to an extreme type distribution [1,2]. Notice that for the systems designed according with the 1987 Code and initial natural periods of 0.5 and 1.5s, their ductility demands were less than one (remained in the elastic range) and that the systems with a 2.5s initial natural period the ductility demands were larger than that value. The same type of behaviour was observed for the systems designed according with the 1976 Code. In Fig 7 an example is presented of the extreme type distributions, Fs,fitted to the ductility demands data obtained from the simulations.

Once the Fs distributions are computed for the systems of interest, eq (6) is applied and results of the type shown in Fig 8 are obtained. In this figure the **Pf** of systems with initial natural periods of 2.5s are presented in the vertical axis versus the coefficient of variation of the nominal ductility. The values of the later varied from 0.05 to 2. Notice that the values of the computed **Pf** diminish when the value of the coefficient of variation increases (as this represents an increase in the nominal ductility available in the systems). Remark as well that the values of the **Pf** for the sample size of 40 are very close to the ones computed for a sample size of 80. In the same figure we can see that the **Pf** of the systems designed with the 1976 Code are larger than the ones designed with the 1987 Code, i.e. there is an apparent improvement in the seismic behaviour of the structural systems designed with the later Code. The range of values of the **Pf** varied from $6*10^{-2}$ to $9*10^{-4}$. Incidently, these values are very similar to the ones reported in 1984 in [1].

CONCLUSIONS

The main conclusions of the study are the following:

1) The probability distributions of the ductility demands for the considered structural systems are of the extreme type.

2) The probabilities of failure of the R/C frames designed accordingly with the 1976 Mexico City Code are about three times larger than the **Pf** of the ones designed with the 1987 version of the same code.
3) The **Pf** implicit in both codes is of the order of 10^{-3}.

REFERENCES

1. Chavez, M., de Leon, D., Reliability of nonlinear systems with uncertain parameters and random seismic excitation. VIII WCEE, Vol. IV, San Francisco, (1984) 435-441.
2. Chavez, M., Reliability of nonlinear infilled frame systems with uncertain properties under random seismic loading. In F. Ziegler, G. I. Schueller, (Eds.), Nonlinear Stochastic Dynamic Enginnering Systems, IUTAM Symposium, Innsbruck/Igls, Austria, June 21- 26, 1987, Springer-Verlag Berlin Heidelberg, (1988) 507-515.
3. Reglamento de Construcciones para el Distrito Federal, Depto. del Distrito Federal, Mexico D.F., (1976).
4. Reglamento de Construcciones para el Distrito Federal, Depto del Distrito Federal, Mexico D.F., (1987).
5. Meli R., Bases para los criterios de diseño estructural del proyecto de reglamento de construcciones del D.F., Pub. 375, Instituto de Ingenieria, UNAM, (1976).
6. De Leon, D., Confiabilidad de marcos de concreto con parametros estructurales inciertos ante excitacion sismica, Tesis de Maestria, Fac. de Ing., UNAM, Mexico, 1983.

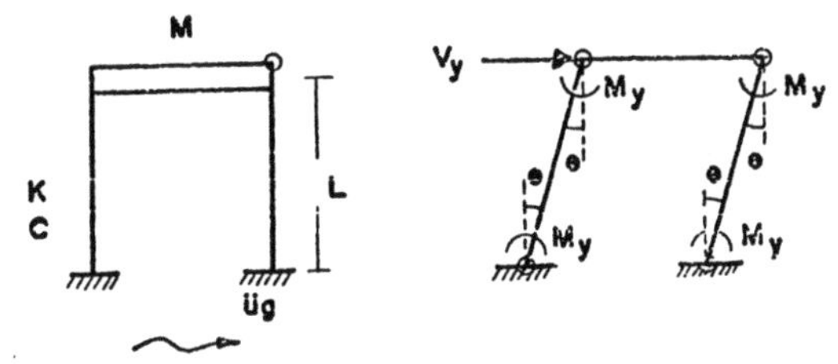

FIG 1. SIMPLE STRUCTURAL SYSTEM AND ITS COLLAPSE MECHANISM

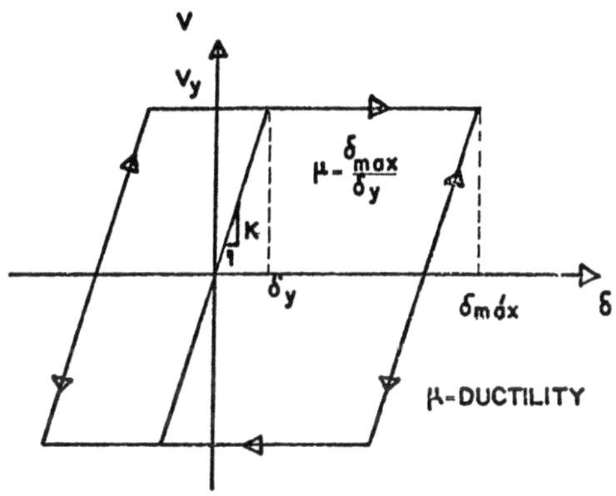

FIG 2. STRUCTURAL SYSTEM ELASTOPLASTIC HYSTERETIC BEHAVIOUR

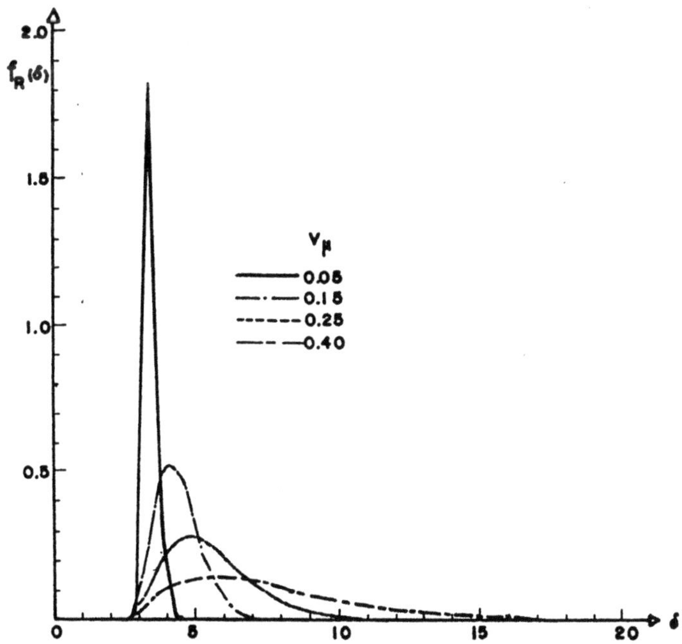

FIG 3. PROBABILITY DENSITY FUNCTION OF NOMINAL DUCTILITY FOR SEVERAL COEFFICIENTS OF VARIATION, $V\mu$, FOR $\mu = 4$ AND $\delta = \mu - 1$

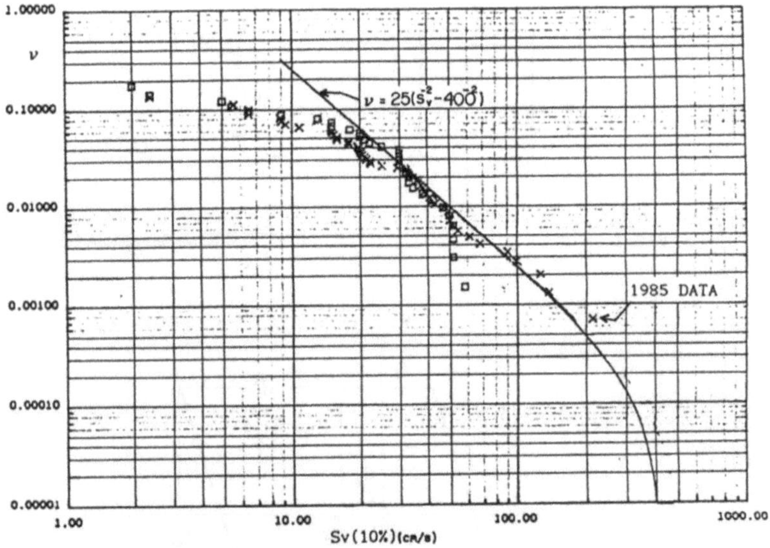

FIG 4. ANNUAL RATE OF EXCEEDANCE, ν, OF THE PEAK VALUES OF THE SEUDOVELOCITY SPECTRA, Sv(10%), OBSERVED IN MEXICO CITY LAKE ZONE FOR A 30 YEARS SAMPLE

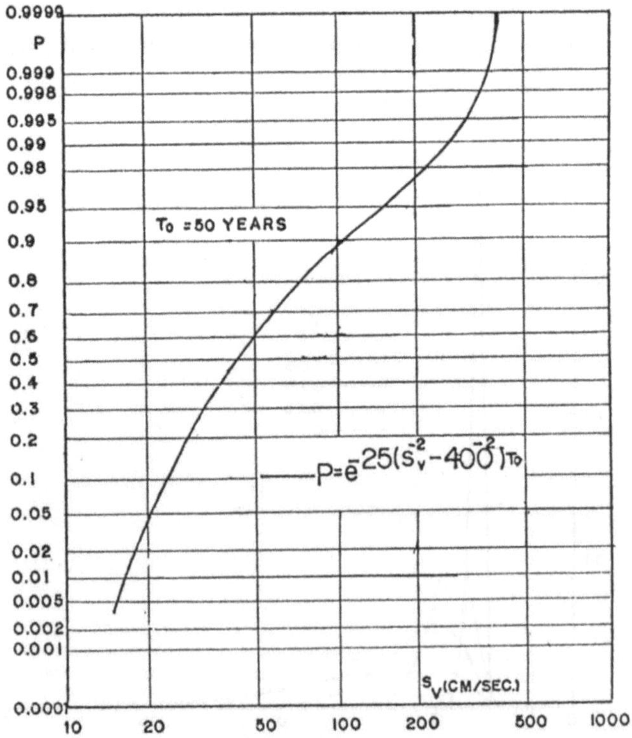

FIG 5. ACCUMULATED PROBABILITY DISTRIBUTION OF THE PEAK VALUES OF THE SEUDOVELOCITY SPECTRA, Sv(10%), OF MEXICO CITY LAKE ZONE

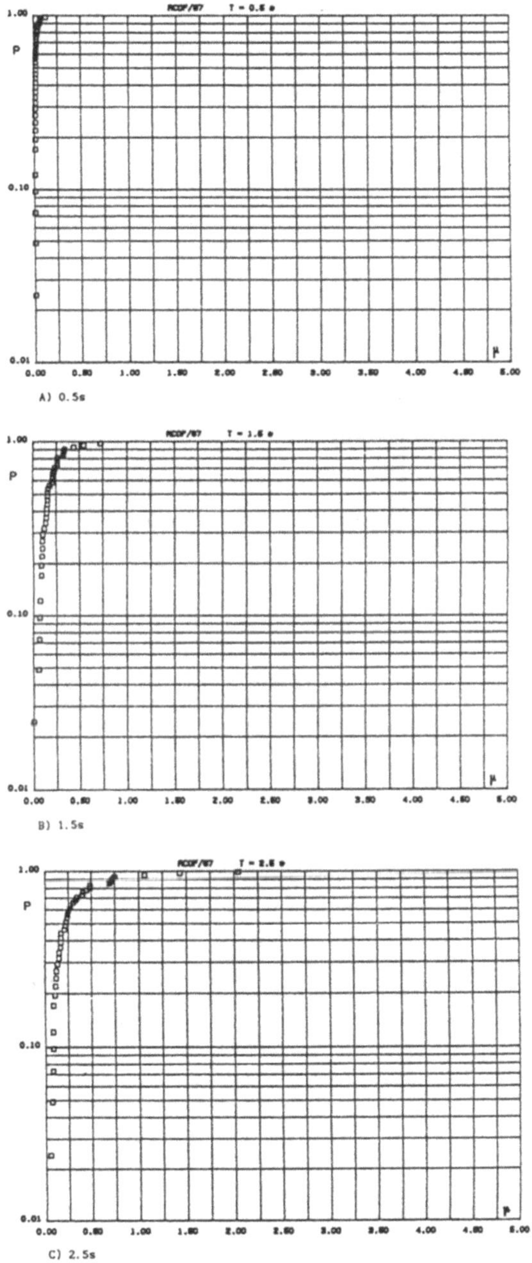

FIG 6. SAMPLES OF DUCTILITY DEMANDS FOR STRUCTURAL SYSTEMS WITH AN INITIAL NATURAL PERIOD OF: A) 0.5s, B) 1.5s, C) 2.5s

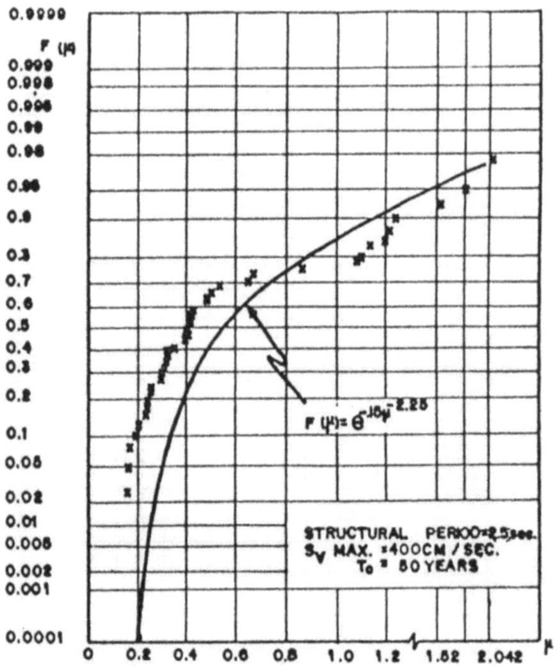

FIG 7. FITTED ACCUMULATED PROBABILITY DISTRIBUTION OF DUCTILITY DEMANDS FOR A STRUCTURAL SYSTEM WITH AN INITIAL NATURAL PERIOD OF 2.5s

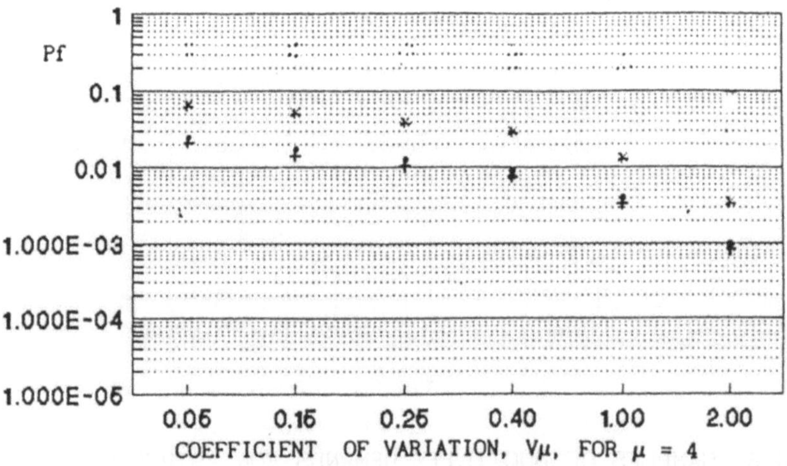

FIG 8. PROBABILITY OF FAILURE, Pf, FOR STRUCTURAL SYSTEMS WITH AN INITIAL NATURAL PERIOD OF 2.5s, DESIGNED ACCORDING WITH THE 1976 AND 1987 MEXICO CITY CODES

Stochastic Finite Element Analysis of Uncertain Nonlinear Trusses under Random Loadings

Rwey-Hua Cherng and Yi-Kwei Wen
Department of Civil Engineering,
University of Illinois at Urbana-Champaign, Urbana, IL 61801, U.S.A.

ABSTRACT

A method of analysis is developed of the responses of nonlinear trusses with uncertain system parameters under random loadings. The structural nonlinear behaviors may arise from large deflection and/or inelastic deformation. Based on a total Lagrangian finite element formulation, a stochastic equivalent linearization method in conjunction with a perturbation method is developed to solve for the response statistics. Responses of a structure under Gaussian filtered white noise ground excitation are obtained and compared with Monte-Carlo simulation results. The stochastic equivalent linearization method was found to yield reasonably accurate results with much less computational time. Depending on the amount of system uncertainties, their contribution to the total response statistics can be quite significant, i.e., more than that due to the randomness of the loading time history. A non-dimensional response sensitivity coefficient is also defined to identify the most dominant system parameters in terms of the contribution to the response variability.

INTRODUCTION

Structural system parameters such as stiffness, damping, and material constants are seldom completely known and loading environments are generally random and dynamic in nature; it is often necessary to perform stochastic response analyses in the assessment of the reliability of structures. Moreover, under severe loadings such as those caused by earthquakes, sea waves, or winds, structures generally exhibit nonlinear behaviors due to inelastic and/or large deformation. For such nonlinear, uncertain structures under random loadings, the response statistics are difficult to obtain except via Monte-Carlo simulations, which can be very costly. Many approximate methods have been proposed. The stochastic equivalent linearization method has been successfully applied to a large class of nonlinear discrete (Wen [9]) and continuous (e.g., Simulescu, Mochio, and Shinozuka [8]) systems under random excitations. For systems with spatial uncertainties, the mean-centered perturbation method (Liu, Besterfield, and Belytschko [7]) has been proven to be one of the most effective methods.

N. Bellomo · F. Casciati (Eds.)
Nonlinear Stochastic Mechanics
IUTAM Symposium Turin

This study considers the simple problem of plane trusses subjected to random excitations; however, nonlinearities due to both large deflection and inelasticity as well as the spatial system parameter uncertainties are included in the formulation. The constitutive law is based on the differential equation model in Wen [9] with the parameters determined in terms of the material constants. The nonlinearities are treated by a stochastic equivalent linearization method, and the system uncertainties are included by a perturbation method.

ELEMENT EQUATIONS OF MOTION

Constitutive Law

Dynamic response of an inelastic system formulated based on classical plasticity constitutive laws are generally difficult to solve. A step-by-step numerical integration method may be used for this purpose; it is, however, usually time consuming, particularly in Monte-Carlo simulations where a large number of time histories are required. The solution procedure can be greatly facilitated if the hysteretic constitutive laws can be described explicitly by a set of differential equations which represent the cyclic stress-strain relationship accurately and are also mathematically tractable.

The constitutive law in this study is governed by the following differential equations:

$$\sigma = \hat{a}E\varepsilon + (1-\hat{a})Ez \tag{1}$$

$$\dot{z} = \kappa\dot{\varepsilon} - \gamma|\dot{\varepsilon}|\,|z|^{n-1}z - \beta\dot{\varepsilon}|z|^{n} \tag{2}$$

in which σ is the second Piola-Kirchhoff axial stress, ε is the Green-Lagrange axial strain, E is the Young's modulus, $\hat{a}$ is the hardening ratio, z is the element auxiliary variable, and n, κ, β, and γ are control parameters of the hysteresis loops. A modification of the one dimensional version of Karray and Bouc's constitutive law of plasticity shows that (Cherng and Wen [2]) $\kappa = 1$ and $\beta = \gamma = E^n/2\sigma_y^n$ where σ_y is the yielding stress of the material. Note that n is the only free parameter in the model; it controls the transition from elastic to plastic state. Also larger values of n give more abrupt transition but satisfy better the Drucker's stability postulate.

Total Lagrangian Formulation

The typical element considered is a prismatic two-node plane truss element; its nodal displacement vector $\underline{\mathbf{u}}$, nodal internal force vector $\underline{\mathbf{f}}$, second Piola-Kirchhoff axial stress σ, and Green-Lagrange axial strain ε in the local coordinate system are defined in Fig. 1. The element may undergo large displacement and large rotation; its strain–displacement relationship is given by (Zienkiewicz and Nayak [10])

$$\varepsilon = \left(\mathbf{B}^{l} + \frac{1}{2}\mathbf{B}^{nl} \right) \underline{\mathbf{u}} \tag{3}$$

in which

$$\mathbf{B}^{l} = \frac{1}{l} \left[-1 \;\; 0 \;\; 1 \;\; 0 \right] \tag{4}$$

$$\mathbf{B}^{nl} = \frac{1}{l^2} \left[-(u_3 - u_1) \;\; -(u_4 - u_2) \;\; (u_3 - u_1) \;\; (u_4 - u_2) \right] \tag{5}$$

Based on Eqs. (1) through (5), the element equations of motion are obtained with the aid of a total Lagrangian formulation (Cherng and Wen [2]).

STOCHASTIC EQUIVALENT LINEARIZATION METHOD

Linearized Global Equations of Motion

Consider the stationary responses of the element under zero-mean stationary Gaussian loadings. Their mean values are generally not zero because of the geometric nonlinearity. It has been shown that (Cherng and Wen [2]), one can derive the linearized equations of motion in global coordinates by applying the stochastic linearization method on the element equations of motion and following a standard assembling procedure used in finite element analyses. The equations are given in matrix forms as follows:

$$\mathbf{M}\ddot{\underline{\mathbf{U}}} + \mathbf{C}\dot{\underline{\mathbf{U}}} + \mathbf{K}\underline{\mathbf{U}} + \mathbf{P}\underline{\mathbf{Z}} + \underline{\mathbf{Q}} = \underline{\mathbf{f}}(\mathbf{t}) \tag{6}$$

$$\dot{\underline{\mathbf{Z}}} = \mathbf{A}\underline{\mathbf{Z}} + \mathbf{B}\dot{\underline{\mathbf{U}}} \tag{7}$$

in which $\underline{\mathbf{U}}$ is the relative displacement vector to the ground, $\underline{\mathbf{Z}}$ is the element auxiliary variable vector, $\mathbf{M}$ is the mass matrix, $\mathbf{C}$ is the viscous damping matrix, $\underline{\mathbf{f}}(\mathbf{t})$ is the random loading vector, and $\mathbf{K}$, $\mathbf{P}$, $\underline{\mathbf{Q}}$, $\mathbf{A}$, and $\mathbf{B}$ can be expressed in terms of explicit functions of system parameters and response statistics.

Governing Equations for Response Statistics

Hereinafter, without loss of generality, it is assumed that the structure is subjected to random ground excitations $\underline{\mathbf{f}}(t)$ given by

$$\underline{\mathbf{f}}(t) = \mathbf{M}\,\underline{\mathbf{r}}a(t) \tag{8}$$

in which $a(t)$ is a ground acceleration random process and $\underline{\mathbf{r}}$ is an on/off vector indicating the influence of $a(t)$ on degree–of–freedoms. In this study $a(t)$ is modeled as Kanai-Tajimi (Kanai [6]) filtered Gaussian white noise, i.e., $2\zeta_g\omega_g\dot{x}_g + \omega_g^2 x_g$, where x_g satisfies the following equation:

$$\ddot{x}_g + 2\zeta_g\omega_g\dot{x}_g + \omega_g^2 x_g = w(t) \tag{9}$$

in which ω_g and ζ_g are the natural frequency and damping ratio of the filter representing the spectral characteristics of the ground excitation and $w(t)$ is a Gaussian white noise with power spectral density S_0.

By defining a response state vector $\underline{\mathbf{Y}} = (x_g\ \dot{x}_g\ \underline{\mathbf{U}}\ \dot{\underline{\mathbf{U}}}\ \underline{\mathbf{Z}})^T$, we can rewrite Eqs. (6) through (9) as:

$$\frac{d\underline{\mathbf{Y}}}{dt} = \mathbf{G}\underline{\mathbf{Y}} + \underline{\mathbf{F}} \tag{10}$$

in which

$$\mathbf{G} = \begin{bmatrix} 0 & 1 & \underline{\mathbf{o}}^T & \underline{\mathbf{o}}^T & \underline{\mathbf{o}}^T \\ -\omega_g^2 & -2\zeta_g\omega_g & \underline{\mathbf{o}}^T & \underline{\mathbf{o}}^T & \underline{\mathbf{o}}^T \\ \underline{\mathbf{o}} & \underline{\mathbf{o}} & \mathbf{O} & \mathbf{I} & \mathbf{O} \\ \underline{\mathbf{r}}\omega_g^2 & \underline{\mathbf{r}}2\zeta_g\omega_g & -\mathbf{M}^{-1}\mathbf{K} & -\mathbf{M}^{-1}\mathbf{C} & -\mathbf{M}^{-1}\mathbf{P} \\ \underline{\mathbf{o}} & \underline{\mathbf{o}} & \mathbf{O} & \mathbf{B} & \mathbf{A} \end{bmatrix} \tag{11}$$

$$\underline{\mathbf{F}} = \begin{bmatrix} 0 & w(t) & \underline{\mathbf{o}} & -\mathbf{M}^{-1}\underline{\mathbf{Q}} & \underline{\mathbf{o}} \end{bmatrix}^T \tag{12}$$

in which $\mathbf{I}$ is a NxN identity matrix, $\mathbf{O}$ is a null matrix, and $\underline{\mathbf{o}}$ is a null vector.

Taking the first and second statistical moments of Eq. (10) respectively and considering stationary responses, we obtain one vector equation for $\overline{\underline{\mathbf{U}}}$, the mean stationary nodal displacement vector, and one matrix equation for $\mathbf{S}$, the one-time covariance matrix of the state vector, as follows:

$$\mathbf{K}\underline{\overline{\mathbf{U}}} + \underline{\mathbf{Q}} = \underline{\mathbf{o}} \tag{13}$$

$$\mathbf{G}\mathbf{S} + \mathbf{S}\mathbf{G}^T + \mathbf{D} = \mathbf{O} \tag{14}$$

in which

$$\mathbf{D} = \begin{bmatrix} 0 & 0 & \underline{\mathbf{o}}^T & \underline{\mathbf{o}}^T & \underline{\mathbf{o}}^T \\ 0 & 2\pi S_0 & \underline{\mathbf{o}}^T & \underline{\mathbf{o}}^T & \underline{\mathbf{o}}^T \\ \underline{\mathbf{o}} & \underline{\mathbf{o}} & \mathbf{O} & \mathbf{O} & \mathbf{O} \\ \underline{\mathbf{o}} & \underline{\mathbf{o}} & \mathbf{O} & \mathbf{O} & \mathbf{O} \\ \underline{\mathbf{o}} & \underline{\mathbf{o}} & \mathbf{O} & \mathbf{O} & \mathbf{O} \end{bmatrix} \tag{15}$$

Solution for Response Statistics

Applying an iterative solution method (Cherng and Wen [2]) to Eqs. (13) and (14) we obtain $\underline{\overline{\mathbf{U}}}$ and $\mathbf{S}$, which contains the first two statistical moments of displacements, velocities, and element auxiliary variables. The statistical moments of other response quantities, such as element stresses, can be derived from $\underline{\overline{\mathbf{U}}}$ and $\mathbf{S}$.

The approximate solutions of the extreme statistics of a response can be also calculated (Davenport [5]) based on the first two moments of the response and its time derivative.

PERTURBATION METHOD

Perturbed Governing Equations for Response Statistics

In Eqs. (13) and (14) the system parameters are assumed to be deterministic. The uncertainties of system parameters, however, may be substantial and contribute significantly to the total response uncertainties and the probability of structural failure. They can be incorporated into the foregoing formulation by applying the mean-centered perturbation method as follows.

The uncertain system parameters are represented by a random vector $\underline{\lambda}$ whose mean vector and covariance matrix are assumed to be either known or derivable by discretizing the prescribed random fields. $\underline{\lambda}$ is then transformed into a non-dimensional zero-mean basic random variable vector $\underline{\epsilon}$ via the following:

$$\epsilon_i = \frac{\lambda_i}{\bar{\lambda}_i} - 1 \; ; \; i = 1, \; NB \tag{16}$$

in which the subscript i denotes the i-th component of a vector; the bar over a variable denotes the mean value; NB is the dimension of $\underline{\lambda}$. Now Eqs. (13) and (14) are rewritten below to indicate that they are conditional on $\underline{\epsilon}$:

$$\mathbf{K}|_{\underline{\epsilon}} \, \underline{\bar{\mathbf{U}}}|_{\underline{\epsilon}} + \underline{\mathbf{Q}}|_{\underline{\epsilon}} = \underline{\mathbf{o}} \tag{17}$$

$$\mathbf{G}|_{\underline{\epsilon}} \, \mathbf{S}|_{\underline{\epsilon}} + \mathbf{S}|_{\underline{\epsilon}} \, \mathbf{G}^T|_{\underline{\epsilon}} + \mathbf{D}|_{\underline{\epsilon}} = \mathbf{O} \tag{18}$$

in which $(.)|_{\underline{\epsilon}}$ designates the dependence of a matrix (or vector) on $\underline{\epsilon}$.

All the matrices in Eqs. (17) and (18) are then expanded into second order Taylor's series of $\underline{\epsilon}$ at $\underline{\epsilon} = 0$. By grouping the terms of the same order, we obtain one set of zero order nonlinear simultaneous equations, NB sets of first order linear simultaneous equations, and $NB(NB + 1)/2$ sets of second order linear simultaneous equations (Cherng and Wen [3]).

The higher order equations may be included in the formulation as well. Nevertheless, if the coefficients of variation of system parameters are not large (say, less than 30%), the computational effort involved which generally increases dramatically in computation of higher order terms may not be justified.

Solution of Conditional Response Statistics

The zero order equations are solved first by the aforementioned iterative method. Based on the zero order solutions, we can proceed to solve for the first order equations. The second order equations are in turn solved by using the zero and first order solutions. Synthesis of these solutions gives us the second order series expansions for $\underline{\bar{\mathbf{U}}}|_{\underline{\epsilon}}$ and $\mathbf{S}|_{\underline{\epsilon}}$, which can be further used to obtain the second order series expansions for other response quantities.

It is important to recognize that the time consuming decompositions of matrices $\mathbf{K}$ and $\mathbf{Q}$ are not necessary in solving the first and second equations since their decomposed forms have already become available when solving the zero order equations. Furthermore, instead of using numerical differentiation schemes, such as in finite difference methods, we has derived explicit expressions for the derivatives of coefficient matrices of Eqs. (17) and (18).

Total Response Statistics

Finally the first two moments of a response R are obtained according to Ang and Tang [1]:

$$\overline{R} = E_{\underline{\epsilon}} \left[\overline{R}|_{\underline{\epsilon}}\right] \tag{19}$$

$$Var[R] = E_{\underline{\epsilon}} \left[Var[R]|_{\underline{\epsilon}}\right] + Var_{\underline{\epsilon}} \left[\overline{R}|_{\underline{\epsilon}}\right] \tag{20}$$

in which $E_{\underline{\epsilon}}$ [.] and $Var_{\underline{\epsilon}}$ [.] denote the mean and variance of the quantity with respect to $\underline{\epsilon}$; $\overline{R}|_{\underline{\epsilon}}$ and $Var[R]|_{\underline{\epsilon}}$ are first two moments of R for given $\underline{\epsilon}$.

Non-dimensional Response Sensitivity Coefficient

To provide a universal measure of the sensitivity of a response moment to a basic random variable, we define a non-dimensional sensitivity coefficient of a response moment μ to a system parameter λ_i as shown below (Chiang [4]):

$$\Omega_i(\mu) \Big|_{\underline{\lambda}=\overline{\underline{\lambda}}} = \frac{\overline{\lambda}_i}{\overline{\mu}} \frac{\partial \mu}{\partial \lambda_i} \Big|_{\underline{\lambda}=\overline{\underline{\lambda}}} \tag{21}$$

$\Omega_i(\mu)|_{\underline{\lambda}=\overline{\underline{\lambda}}}$ can be approximately interpreted as the percentage change in μ due to one percent change in λ_i; it allows us to compare systematically the influence of basic random variables on the response statistics and determine the less significant variables which may be treated as deterministic in subsequent reliability analyses.

NUMERICAL EXAMPLE

The response of a three-tier plane truss structure subjected to a horizontal ground excitation is considered. It is assumed that each plane truss member can be adequately represented by one single element and there is no viscous damping. The basic random variables include material constants of each element (E, σ_y, and $\hat{a}$) and parameters of the ground excitation (S_0, ω_g, and ζ_g of a Kanai-Tajimi filtered Gaussian white noise). They are assumed to be statistically independent of one another; their mean values and coefficients of variation are given in Fig. 2.

The accuracy of the stochastic equivalent linearization method is first examined by analyzing the system with all system parameters set equal to their mean values under the random excitation. Monte-Carlo simulations are carried out; a sample size of 40 is used; the time histories of ensemble r.m.s. displacement at the top and r.m.s. stress of the first story column are shown in Figs. 3 and 4 respectively. Stationary response statistics are obtained by the stochastic linearization method and compared with Monte–Carlo solutions in Table 1. It is seen that the stochastic linearization method gives results which are lower by 4 to 20%. It requires, however, only a small fraction of computational time of Monte-Carlo simulations (approximately 1/30 of one run of 40 sec time history).

When the uncertainties of system parameters are included , the total variance of the maximum displacement increases from 1.31 in^2 (the zero order terms) to 2.94 in^2 (the zero order plus the first order terms); it indicates that the system parameter uncertainties account for 55.4% of the total variance. Meanwhile the non-dimensional sensitivity coefficients of the mean maximum displacement to system parameters are listed in Table 2; it shows that the response is insensitive to the material constant variations in beams and braces but quite sensitive to those in columns and variabilities of ground excitation parameters. In terms of the contribution to the total variance, the ground excitation parameters are most important because of their larger coefficients of variation.

CONCLUSION

A stochastic finite element method is developed for nonlinear trusses. The stochasticities arise from the uncertain system parameters and random loadings; the structural nonlinear behaviors are due to large deflection and inelasticity. A total Lagrangian formulation, with the constitutive law based on a differential equation model, is used to derive the equations of motion. Approximate response statistics are then obtained by a stochastic equivalent linearization method in conjunction with a perturbation method. Numerical examples indicate that the stochastic linearization method yields reasonably accurate results at a small fraction of computational cost of Monte-Carlo simulations, and system uncertainties have a significant contribution to the total response statistics.

The random field modeling of material constants and the related reliability problems will be considered in the future research.

ACKNOWLEDGEMENT

This study is supported by National Science Foundation under Grants NSF ECS 85-11972 and CES 88-22690. The support is gratefully acknowledged. Advises and suggestions from D. A. Pecknold are also greatly appreciated.

	40 Simulations	Linearization
R.M.S. Top Displacement (in)	3.32	2.65
R.M.S. Top Velocity (in/sec)	32.92	31.63
Mean Maximum Top Displacement in 20 sec. (in)	8.96	8.22
Standard Deviation of Maximum Top Displacement in 20 sec. (in)	1.42	1.14
R.M.S. First Story Column Stress (ksi)	22.81	21.82

Table 1 Comparison of Results by Monte-Carlo Simulations and Stochastic Linearization Method

	First Story Column			First Story Brace			First Story Beam			Ground Excitation		
	E	σ_y	$\hat{a}$	E	σ_y	$\hat{a}$	E	σ_y	$\hat{a}$	S_0	ω_g	ζ_g
Sensitivity Coefficients	-.2	.3	-.3 E-1	-.2 E-1	-.1 E-3	.9 E-2	.5 E-3	.2 E-7	.2 E-3	.2	-.3	-.7 E-1

Table 2 Non-dimensional Sensitivity Coefficients of Mean Maximum Top Displacement in 20 sec. to system parameters

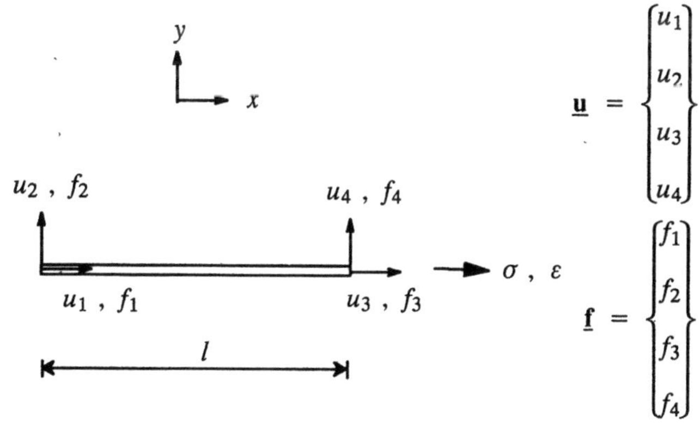

Figure 1 Plane Truss Element in Local Coordinates

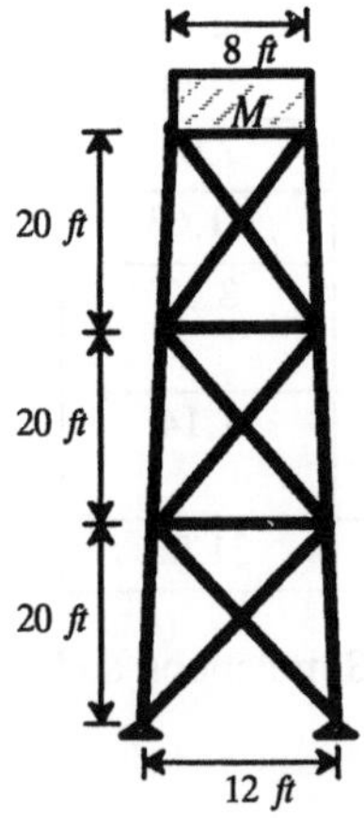

$M = .6\ kp.sec^2/in$

Material Constants (mean values)

$E = 29000\ ksi$, $A = 48\ in^2$

$\sigma_y = 58\ ksi$

$\kappa = 1$, $n = 2$, $\hat{a} = .001$

ϱ (*Density*) $= 7.3\text{x}10^{-7}\ kp.sec^2/in^4$

Horizontal Ground Excitation Parameters

$S_0 = 150\ in^2/\sec^3$

$\omega_g = 19\ rad/sec$, $\zeta_g = .32$

	E	σ_y	$\hat{a}$	S_0	ω_g	ζ_g
C.O.V.	0.10	0.10	0.10	0.30	0.43	0.36

Figure 2 Three-tier Plane Truss Structure

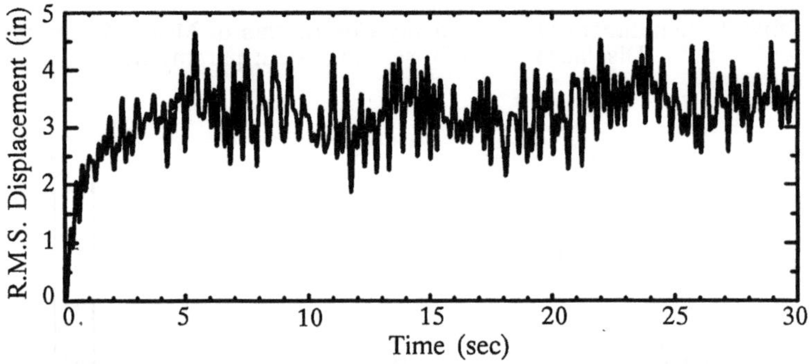

Figure 3 Time History of Ensemble R.M.S. Displacement at the Top of the Structure

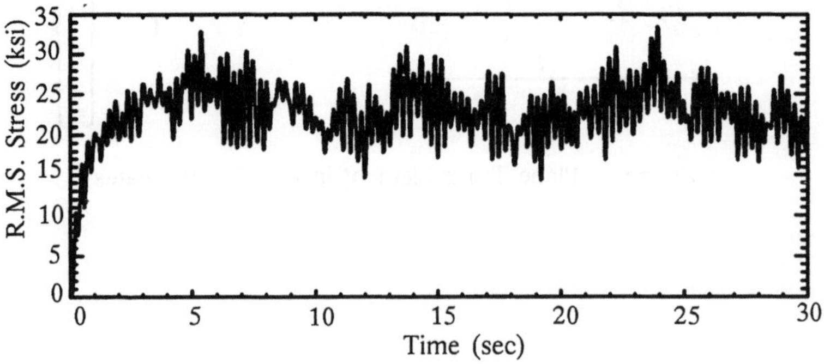

Figure 4 Time History of the Ensemble R.M.S. Stress (the First Story Column)

REFERENCES

1. Ang, A. H-S. and Tang, W. H., Probability Concepts in Engineering Planning and Design, Vol. I, John Wiley & Sons, New York, 1975.
2. Cherng, R.-H. and Wen, Y. K., "Stochastic Finite Element Analysis of Nonlinear Plane Trusses," accepted by *International Journal of Non-Linear Mechanics.*
3. Cherng, R.-H. and Wen, Y. K., "Response of Non-linear Trusses with Spatial Variability under Random Excitations," *Proceedings of the First International Conference on Computational Stochastic Mechanics*, Corfu, Greece, September 1991.
4. Chiang, A. C., Fundamental Methods of Mathematical Economics, McGraw-Hill International Book Company, Tokyo, 1984.
5. Davenport, A. G., "Note on the Distribution of the Largest Value of a Random Function with Application to Gust Loading," Proceedings, Institute of Civil Engineers, London, Vol. 28, 1964.
6. Kanai, K., "Semiempirical Formula for the Seismic Characteristics of the Ground Motion," Bulletin, Earthquake Research Institute, University of Tokyo, Vol. 35, pp. 309–325, 1957.
7. Liu, W. K., Besterfield, G. H., and Belytschko, T., "Variational Approach to Probabilistic Finite Elements" *Journal of Engineering Mechanics Division*, ASCE, Vol. 114, No. 12, pp. 2115–2133, December 1988.
8. Simulescu, I., Mochio, T., and Shinozuka, M., "Equivalent Linearization Method in Nonlinear FEM," *Journal of Engineering Mechanics Division*, ASCE, Vol. 115, No. 3, pp. 475–492, March 1989.
9. Wen, Y. K., "Methods of Random Vibration for Inelastic Structures," *Applied Mechanics Reviews*, Vol. 42, No. 2, pp. 39-52, February 1989.
10. Zienkiewicz, O. C., and Nayak, G. C., "A General Approach to Problems of Large Deformations and Plasticity Using Isoparametric Elements," *Proceedings of Conference on Matrix Methods in Structural Mechanics*, Wright-Patterson AFB, Ohio, 1971.

Stochastic Response on Non-Linear Systems under Parametric Non-Gaussian Agencies

M. Di Paola and G. Falsone
Dipartimento di Ingegneria Strutturale e Geotecnica
Università degli Studi di Palermo,
Viale delle Scienze, I-90120 Palermo, Italy

ABSTRACT

The probabilistic response characterization of non-linear systems subjected to non-normal delta correlated parametric excitation is obtained. In order to do this an extension of both Ito's differential rule and the Fokker-Planck equation is presented, enabling one to account for the effect of the non-normal input. The validity of the approach reported here is confirmed by results obtained by means of a Monte Carlo simulation.

1. INTRODUCTION

For linear or non-linear mechanical systems subjected to normal continuous random excitation the problem of finding the probabilistic characterization of the response process is a well-known subject (see e.g. Stratonovich[1], Lin[2]). For linear systems and normal inputs, the response is also normal and the second moments provide a complete statistical characterization of these processes. Moreover, if the system is driven by a normal white noise process, then the problem may be framed with in the theory of diffusive Markov processes. For such problems the Fokker-Planck-Kolmogorov (FPK) equation, or the Ito's stochastic differential rule[3,4,5], are available for finding useful results.

In some circumstances, however, the excitations are significantly non-normal. Such processes can arise, for example, from the quadratic term commonly used in fluid mechanics, or in cases where the realizations consist of a train of random impulses, Poisson distributed in time. In these circumstances, even for the simplest case of linear systems, the response is non-normal and the probabilistic characterization of the response requires higher order moments. Recently Lutes[6] and Lutes et al.[7] investigated the case of linear systems excited by non-normal delta correlated or filtered processes by means of two methods which permit one to evaluate cumulants up to the fourth order of the response. These methods, however, do not seem easily extendable to non-linear systems. A particular class of linear or non-linear dynamic systems under external non-normal delta correlated random trains of Poisson-distributed impulses has been treated by Roberts[8] who gave a generalization of the Fokker-Planck equation. In this framework Iwankievicz et al.[9] have investigated the dynamic response of a non-linear system driven by a Poisson-distributed train of impulses.

In this paper the most general case of non-linear systems excited by parametric type

N. Bellomo · F. Casciati (Eds.)
Nonlinear Stochastic Mechanics
IUTAM Symposium Turin

non-normal delta correlated processes is treated appropriately by extending the Ito's differential rule and the FPK equation, to account for the non-normal nature of the input. The validity of the approach reported here is confirmed by results obtained by means of a Monte Carlo simulation.

2. PRELIMINARY CONCEPTS

Let the equation of motion be cast in the form

$$dZ = A(Z, t)\,dt + G(Z, t)\,dL \tag{1}$$

where $A(Z, t)$ and $G(Z, t)$ are deterministic functions depending on the process Z and on the time t, while $dL/dt = W(t)$ is a zero-mean stationary non-normal delta correlated process up to the s-th order, (see Stratonovich[1], Grigoriu[10]); that means

$$k_r(W; t_1, t_2, \dots, t_r) = q^{(r)}\,\delta(t_1 - t_2)\,\delta(t_1 - t_3)\dots\delta(t_1 - t_r), \quad r = 2, 3, \dots, s$$

$$k_{s+p}(W; t_1, t_2, \dots, t_r) = 0, \quad p = 1, 2, \dots \tag{2}$$

in which $\delta(\cdot)$ is the Dirac's delta function, $q^{(r)}$ is an intensity coefficient, $k_r(\cdot)$ is the r-th cumulant of W. The latter is related to the log-characteristic function by means of the following relationship:

$$k_r[W; t_1, t_2, \dots, t_r] = k_r[W_1, W_2, \dots, W_r] =$$

$$\left[\frac{1}{(-i)^r}\frac{\partial^r}{\partial\vartheta_1\partial\vartheta_2\dots\partial\vartheta_r}\, ln\, M_{\underset{\sim}{W}}\left(\underset{\sim}{\vartheta}_r; \underset{\sim}{t}_r\right)\right]_{\underset{\sim}{\vartheta}_r = 0} \tag{3}$$

in which $\underset{\sim}{\vartheta}_r$ is a vector of real parameters containing $\vartheta_1, \vartheta_2, \dots, \vartheta_r$, $\underset{\sim}{t}_r$ is the vector containing $t_1, t_2, \dots, t_r$, $W_j = W(t_j)$ and $M_{\underset{\sim}{W}}(\underset{\sim}{\vartheta}_r; \underset{\sim}{t}_r)$ is the characteristic function, related to the probability density function by means of the Fourier transform operator; that is:

$$M_{\underset{\sim}{W}}\left(\underset{\sim}{\vartheta}_r; \underset{\sim}{t}_r\right) = \int_{-\infty}^{\infty}\int_{-\infty}^{\infty} \exp\left(-i\,\underset{\sim}{\vartheta}_r^T\,\underset{\sim}{w}\right) p_{\underset{\sim}{W}}\left(\underset{\sim}{w}; \underset{\sim}{t}_r\right) dw_1\, dw_2 \dots dw_r \tag{4}$$

where the apex T denotes transpose and $p_{\underset{\sim}{W}}(\underset{\sim}{w}; \underset{\sim}{t}_r)$ is the multidimensional probability density function of the random variables $W_1, W_2, \dots, W_r$ all contained in the vector $\underset{\sim}{W}$.

The r-th cumulant of $dL(t)$ is given by the form:

$$k_r[dL_1, dL_2, \ldots, dL_r] = q^{(r)}\delta(t_1 - t_2)\,\delta(t_1 - t_3)\ldots\delta(t_1 - t_r)\,dt_1\,dt_2\ldots dt_r$$

$$k_{s+p}[dL_1, dL_2, \ldots, dL_r] = 0, \quad p = 1, 2, \ldots \tag{5}$$

where $dL_j = dL(t_j)$. Letting $t_1 = t_2 = \ldots = t_r = t$ we obtain

$$k_1[dL(t)] = 0; \; k_r[d(t)] = q^{(r)}dt; \quad r = 2, 3, \ldots, s$$

$$k_{s+p}[dL(t)] = 0, \quad p = 1, 2, \ldots \tag{6}$$

Taking into account the relationships between moments and cumulants, it can be easily seen that, neglecting higher order infinitesimal than dt, the moments of $dL(t)$ coincide with the cumulants; that is:

$$m_r[dL] = E[(dL)^r] = k_r[dL] \tag{7}$$

in which $E[\cdot]$ denotes stochastic average. If $s = 2$ then the process $W(t)$ is a normal white noise; otherwise the process $W(t)$ is a delta correlated one.

3. NON-CLASSICAL STOCHASTIC DIFFERENTIAL RULE

If the probabilistic characterization of the response Z is required, then the moment equation approach can be applied. To do this, it is necessary to write the differential equations governing the moments of the response. In order to obtain these equations it is necessary to extend the Ito's differential rule to take into account the non-normality of the input. For this purpose, we consider an arbitrary scalar real valued function $\phi(Z, t)$, s-times differentiable with respect to Z, and continuously differentiable on t. Then an increment $\Delta\phi$ of ϕ can be written in the following general form:

$$\Delta\phi = d\phi + \frac{1}{2!}d^2\phi + \frac{1}{3!}d^3\phi + \ldots \tag{8}$$

This expression may be found, applied to the case of deterministic function, in a book for students written by Picone and Fichera[11]. Using the chain rule, one can write the first few differentials in the form:

$$d\phi = \frac{\partial \phi}{\partial t} dt + \frac{\partial \phi}{\partial Z} dZ$$

$$d^2 \phi = \frac{\partial^2 \phi}{\partial Z^2} (dZ)^2 + \frac{\partial \phi}{\partial Z} d^2 Z \tag{9}$$

$$d^3 \phi = \frac{\partial^3 \phi}{\partial Z^3} (dZ)^3 + 3 \frac{\partial^2 \phi}{\partial Z^2} dZ\, d^2 Z + \frac{\partial \phi}{\partial Z} d^3 Z$$

where dZ is given by equation (1). It can be easily seen that, on inserting equation (9) in to equation (8), we obtain (neglecting infinitesimals of higher order than dt) the increment of ϕ in the form:

$$\Delta\phi = \frac{\partial \phi}{\partial t} dt + \frac{\partial \phi}{\partial Z} \Delta Z + \frac{1}{2!} \frac{\partial^2 \phi}{\partial Z^2} (\Delta Z)^2 + \ldots + \frac{1}{s!} \frac{\partial^s \phi}{\partial Z^s} (\Delta Z)^s \tag{10}$$

where, according to equation (8), an increment of Z is given as:

$$\Delta Z = dZ + \frac{1}{2!} d^2 Z + \ldots + \frac{1}{s!} d^s Z \tag{11}$$

Equation (10) represents an extension of Ito's rule for the case of non-normal input process, and will be called the non-classical stochastic differential rule (SDR). Taking into account the properties of the process $dL(t)$, it can be easily seen that the following relationships hold:

$$(dZ)^r = G^r(Z, t)\,(dL)^r \tag{12}$$

$$d^r Z = G^{(r)}(Z, t)\,(dL)^r, \quad r = 2, 3, \ldots, s; \quad d^{s+p} Z = 0; \quad p = 1, 2, \ldots \tag{13}$$

where $G^{(r)}(Z, t)$ (with arguments omitted) is given as:

$$G^{(1)} = G; \quad G^{(r)} = \frac{\partial G^{(r-1)}}{\partial Z} G, \quad r = 2, 3, \ldots \tag{14}$$

From the above arguments it follows that an increment of Z can be written as the explicit form:

$$\Delta Z = A(Z, t)\, dt + \sum_{j=1}^{s} \frac{G^{(j)}(Z, t)}{j!} (dL)^j \tag{15}$$

If $s = 2$, then we obtain Ito's stochastic differential equation:

$$\Delta Z = \left(A(Z,t) + \frac{1}{2} \frac{\partial G(Z,t)}{\partial Z} G(Z,t) q^{(2)} \right) dt + G(Z,t)\, dL$$

where the extra term $1/2\ (\partial G(Z,t)/\partial Z)\ G(Z,t)\ q^{(2)}$ is the so called Wong-Zakai or Stratonovich (WZ or S) correction term. Comparing equation (15) with equation (1) the WZ or S correction term is viewed in a new light. It is simply an extra term which compensates for the difference between the derivative of Z ($\Delta Z/dt = \dot{Z}$) and the ratio dZ/dt (only in the case of external excitation does $dZ/dt \equiv \Delta Z/dt$). In the case of non-normal delta correlated processes new coefficients like the WZ or S correction terms appear in the drift coefficient.

4. PROBABILISTIC CHARACTERIZATION OF THE RESPONSE

In order to charecterize the response process Z, the moments of any order of Z have to be evaluated. Setting $\phi(Z,t) = Z$ in equation (10), taking the stochastic average and dividing by dt, we obtain:

$$\dot{m}_r[Z] = r\, E\left[Z^{r-1} \frac{\Delta Z}{dt}\right] + \frac{1}{2} r(r-1)\, E\left[Z^{r-2} \frac{(\Delta Z)^2}{dt}\right] + \ldots + E\left[\frac{(\Delta Z)^r}{dt}\right], \quad r \leq s \tag{16a}$$

$$\dot{m}_r[Z] = r\, E\left[Z^{r-1} \frac{\Delta Z}{dt}\right] + \frac{1}{2} r(r-1)\, E\left[Z^{r-2} \frac{(\Delta Z)^2}{dt}\right] + \ldots.. + \frac{r(r-1)\ldots(r-s+1)}{s!} E\left[Z^{r-s} \frac{(\Delta Z)^s}{dt}\right], \quad r > s \tag{16b}$$

where $m_r[Z] = E[Z^r]$, ΔZ is given in equation (14) and

$$(\Delta Z)^2 = \sum_{j=1}^{s-1} \sum_{k=1}^{s-1} \frac{G^{(j)} G^{(k)}}{j!\, k!} (dL)^{j+k}, \quad j+k = 2, 3, \ldots, s \tag{17a}$$

$$(\Delta Z)^3 = \sum_{j=1}^{s-2} \sum_{k=1}^{s-2} \sum_{l=1}^{s-2} \frac{G^{(j)} G^{(k)} G^{(l)}}{j!\, k!\, l!} (dL)^{j+k+l}, \quad j+k+l = 3, 4, \ldots, s \tag{17b}$$

$$(\Delta Z)^s = G^{(s)} (dL)^s \tag{17c}$$

Hence equation (16) becomes:

$$\dot{m}_r[Z] = r E\left[Z^{r-1} A\right] + r \sum_{j=1}^{s} \frac{1}{j!} E\left[Z^{r-1} G^{(j)}\right] q^{(j)} +$$

$$\frac{1}{2!} r (r-1) \sum_{j=1}^{s-1} \sum_{k=1}^{s-1} \frac{1}{j!\,k!} E\left[Z^{r-2} G^{(j)} G^{(r)}\right] q^{(j+k)} + \ldots + E\left[G^{(r)}\right] q^{(r)}, \quad r \leq s \qquad (18a)$$

$$\dot{m}_r[Z] = r E\left[Z^{r-1} A\right] + r \sum_{j=1}^{s} \frac{1}{j!} E\left[Z^{r-1} G^{(j)}\right] q^{(j)} +$$

$$\frac{1}{2} r (r-1) \sum_{j=1}^{s-1} \sum_{k=1}^{s-1} \frac{1}{j!\,k!} E\left[Z^{r-2} G^{(j)} G^{(k)}\right] q^{(j+k)} + \ldots..$$

$$+ \frac{r (r-1) \ldots (r-s+1)}{s!} E\left[Z^{r+s} G^{(s)}\right] q^{(s)}, \quad r > s \qquad (18b)$$

In the latter equations, according to Ito's stochastic calculus, the following relationships have been used:

$$E\left[g(Z)\,(dL)^r\right] = E[g(Z)] E\left[(dL)^r\right] = E[g(Z)] q^{(r)} dt\,; \quad r = 2, 3, \ldots, s$$

$$E\left[g(Z)\,(dL)^{s+p}\right] = 0, \quad p = 1, 2, \ldots \qquad (19)$$

where $g(Z)$ is any scalar real valued function of Z. In the simple case of a non-linear system driven by an external delta correlated process, that is $G(Z, t) \equiv G(t)$, increments and differentials coincide; then equation (18) simplifies to

$$\dot{m}_r[Z] \doteq r E\left[Z^{r-1} A\right] + \frac{1}{2} r (r-1)\, m_{r-2}[Z] G^2 q^{(2)} + \ldots + G^{(r)} q^{(r)}, \quad r \leq s \qquad (20a)$$

$$\dot{m}_r[Z] = r E\left[Z^{r-1} A\right] + \frac{1}{2} r (r-1)\, m_{r-2}[Z] G^2 q^{(2)} + \ldots.$$

$$+ \frac{1}{s!} r (r-1) \ldots (r-s+1)\, m_{k-s}[Z] G^s q^{(s)}, \quad r > s \qquad (20b)$$

Finally, if the system is linear, then $A(Z, t) \equiv A(t) Z$ and equation (20) simplifies to

$$\dot{m}_r[Z] = r\,A\,m_r[Z] + \frac{1}{2}\,r\,(r-1)\,m_{r-2}[Z]\,G^2 q^{(2)} + \ldots + G^r q^{(r)}, \quad r \leq s \tag{21a}$$

$$\dot{m}_r[Z] = r\,A\,m_r[Z] + \frac{1}{2}\,r\,(r-1)\,m_{r-2}[Z]\,G^2 q^{(2)} + \ldots$$

$$+ \frac{1}{s!}\,r\,(r-1)\ldots(r-s+1)\,m_{r-s}[Z]\,G^s q^{(s)}, \quad r > s \tag{21b}$$

From these equations we note that in the linear case the equation governing the moments of order r of Z involves moments of order r and moments of lesser order than r. It follows that, starting from the knowledge of the first order moment, one can obtain the second order one, and so on. In contrast in the non-linear case the problem of finding the probabilistic characterization of the response becomes more complicated; this happens because the average on the right hand side of equations (18) or (20) must be evaluated by starting from the knowledge of the probability density function of Z, which, in turn, depends on a knowledge of all the moments. Then only approximate solutions can be pursued, by means of truncated closure schemes[14,15].

5. A GENERALIZATION OF THE FPK EQUATION

In this section an extension of the FPK equation to the case of non-normal delta correlated type of excitation is presented. Suppose $\phi(Z)$ has the special form

$$\phi(Z) = \exp(-i\vartheta Z) \tag{22}$$

where ϑ is a real parameter. Inserting this quantity into equation (10), taking the stochastic average and dividing by dt, we obtain the differential equation of the characteristic function $M_Z(\vartheta)$, that is:

$$\frac{\partial M_Z(\vartheta, t)}{\partial t} = -i\vartheta\,E\left[\exp(-i\vartheta Z)\,\Delta Z\right] + \frac{(-i\vartheta)^2}{2}\,E\left[\exp(-i\vartheta Z)\,(\Delta Z)^2\right] + \ldots$$

$$+ \frac{(-i\vartheta)^s}{s!}\,E\left[\exp(-i\vartheta Z)\,(\Delta Z)^s\right] \tag{23}$$

Applying the inverse Fourier transform of this equation, we obtain:

$$\frac{\partial p_Z(z,t)}{\partial t} = -\frac{\partial}{\partial z}\left(p_Z(z,t)\,A(z,t)\right) - \sum_{j=1}^{s}\frac{q^{(j)}}{j!}\frac{\partial}{\partial z}\left(p_Z(z,t)\,G^{(j)}(z,t)\right) +$$
$$\frac{1}{2}\sum_{j=1}^{s-1}\sum_{k=1}^{s-1}\frac{q^{(j+k)}}{j!\,k!}\frac{\partial^2}{\partial z^2}\left(p_Z(z,t)\,G^{(j)}(z,t)\,G^{(k)}(z,t)\right) + \ldots.$$
$$+\frac{(-1)^s}{s!}q^{(s)}\frac{\partial^s}{\partial z^s}\left(p_Z(z,t)\,G^{(s)}(z,t)\right) \tag{24}$$

This equation, which gives the evolution of the probability density function, is an extension of the FPK equation to the case of non-normal delta correlated input. Note that if $q^{(r)} = 0$, $\forall r > 2$; that is for a normal white noise input process, equation (24) exactly coincides with the classical FPK equation.

6. NUMERICAL EXAMPLE

In order to verify the validity of the extension of Ito's rule presented in the previous section, the results obtained by solving the differential equation of moments are compared with some experimental numerical data obtained by means of Monte Carlo simulation. The equation of motion is given in the form:

$$dZ = -A\,Z\,dt + C\,dt + G\,Z\,dL; \quad A > 0 \tag{25}$$

were A, C and G are constants. The system is considered to be at rest with probability one at the time $t = 0$. The choice of this particular system has been made because its moment equations do not constitute an infinite hierarchy; in this way the effect of the new correction term which appears in the equation of moments can be seen clearly, without the masking effects due to closure schemes. The non-normal delta correlated process dL has been chosen to have the following moments:

$$E\left[(dL)^{2r+1}\right] = 0, \quad r = 0,1,2,\ldots$$
$$E\left[(dL)^{2r}\right] = q^{(2r)}dt = \frac{q^{(2)}dt}{2r-1}, \quad r = 1,2,\ldots \tag{26}$$

Moreover, the coefficient G is chosen in such a way that the new correction terms of higher order make significant contributions up to fourth order. Lastly, the coefficient A is chosen in such a way that we do not have stability problems. The equation of moments up to third order of the response are:

$$\dot{m}_1[Z] = \left(-A+\frac{1}{2}G^2q^{(2)}+\frac{1}{24}G^4q^{(4)}\right)m_1[Z]+C \tag{27a}$$

$$\dot{m}_2[Z] = \left(-2A+2G^2q^{(2)}+\frac{2}{3}G^4q^{(4)}\right)m_2[Z]+2C\,m_1[Z] \tag{27b}$$

$$\dot{m}_3[Z] = \left(-3A+\frac{9}{2}G^2q^{(2)}+\frac{27}{8}G^4q^{(4)}\right)m_3[Z]+3C\,m_2[Z] \tag{27c}$$

In order to bring out the influence of the new correction term of higher order we write the same equations by setting $G^{(3)} = G^{(4)} = 0$; that is:

$$\dot{m}_1[Z] = \left(-A+\frac{1}{2}G^2q^{(2)}\right)m_1[Z]+C \tag{28a}$$

$$\dot{m}_2[Z] = \left(-2A+2G^2q^{(2)}+\frac{1}{4}G^4q^{(4)}\right)m_2[Z]+2\,C\,m_1[Z] \tag{28b}$$

$$\dot{m}_3[Z] = \left(-3A+\frac{9}{2}G^2q^{(2)}+\frac{9}{4}G^4q^{(4)}\right)m_3[Z]+3\,C\,m_2[Z] \tag{28c}$$

The solutions of equations (28b) and (28c) are plotted in figs. 2 and 3, respectively, letting $q^{(4)} = 0$ in order to show the non-normality of the input. The solutions of equations (27) and (28) are plotted in figs. 1-3 and are compared with those obtained by means of Monte Carlo simulation (that is directly integrating equation (25)). In order to do this a segment of a stationary process has been generated in the form:

$$W(t) = \frac{dL}{dt} = \sum_{k=1}^{N} R_k\, w_k(t) \tag{29}$$

where R_k are independent random variables characterized by zero mean and an assigned probability distribution $p_R(r)$, and $w_k(t)$ is a window function defined as $w_k(t) = 1\ \forall(t_k - \Delta t/2 \le t < t_k + \Delta t/2)$ and zero elsewere. In this way the duration of the process $W(t)$ is $T = N\,\Delta t$. As $\Delta t \to 0$ the process becomes delta correlated. It will be noted that, starting from a large class of distribution of the random variables R, at the limit the process $W(t)$ becomes a normal white noise. However, a strong distortion of the probability density function of R can produce delta-correlated processes up to higher order. For example, assume a random variable has a probability density function given in the form:

$$p_R(r) = \begin{cases} \dfrac{\alpha}{r^2} & \text{for } -a \le r \le -b \text{ and } b \le r \le 0 \\ 0 & \text{elsewhere} \end{cases} \tag{30}$$

where α is a normalization factor given by $\alpha = ab/2(a-b)$, then setting $a = 1/\Delta t$ and $b = q^{(2)}$ for very small Δt, we have:

$$E\left[R^2\right] = q^{(2)}/\Delta t\,;\;\; E\left[R^4\right] = q^{(2)}/3\,(\Delta t)^3\,;\ldots;\; E\left[R^{2s}\right] = q^{(2)}/(2S-1)\,(\Delta t)^{2s-1} \tag{31}$$

This means that, starting from a distribution given in equation (30), at the limit when $\Delta t \rightarrow 0$, the process $W(t)$ approaches a delta-correlated process up to infinite order. We emphasize that, due to the particular way of constructing a delta input process, when the equation of motion is integrated by means of any numerical method the value of the input at each temporal step must be considered as constant.

The parameters chosen for the analysis are summarized in Table I.

Table I

Fig.	A	G	$q^{(2)}$	Samples
1	7.0	2.5	1	10000
2	5.0	1.5	1	10000
3	3.5	1.0	1	50000

In Figs. 1-3 the solution of equations (27) and (28) are plotted. In order to show the effect of the non-normality the solution of equation (27) is also plotted for $q^{(4)} = 0$. The results are compared with those obtained by means of the Monte Carlo simulation procedure. The time step selected for the analysis is $\Delta t = 0.05$.

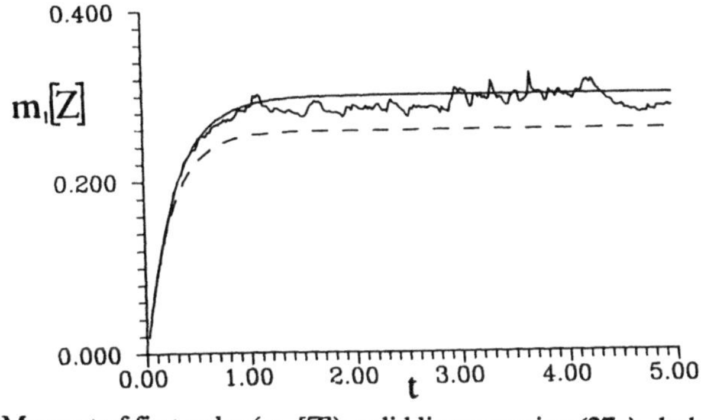

Fig. 1. Moment of first order (m_1 [Z]); solid line: equation (27a), dashed line: equation (28a).

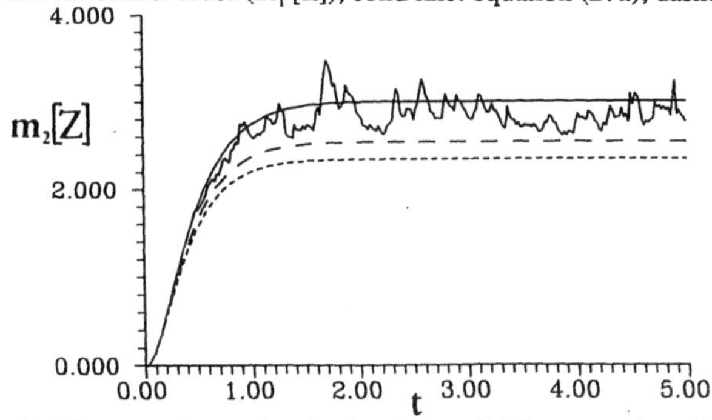

Fig. 2. Moment of second order (m_2 [Z]); solid line: equation (27b), dashed line (– –): equation (28b), dashed line (- -): normal response.

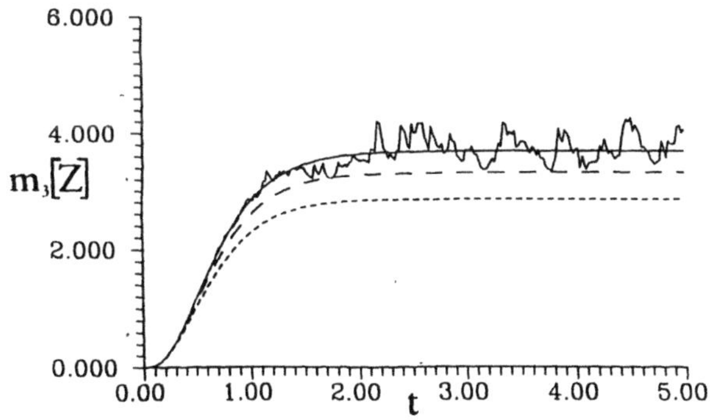

Fig. 3. Moment of third order (m_3 [Z]); solid line: equation (27c), dashed line (– –): equation (28c), dashed line (- -): normal response.

These numerical results show the influence of the new correction term and that the non-normality of the higher order input makes an important contribution. The comparison between the theoretical and the simulation results confirm the validity of the new approach presented here for the extension of stochastic differential calculus to the non-normal input case.

7. REFERENCES

1. Stratonovich, R.L., 1963, "Topics in the theory of random noise", Gordon and Breach, New York, N.Y.

2. Lin, Y.K., 1967, "Probabilistic theory of structural dynamics", McGraw-Hill, New York, N.Y.

3. Ito, K., 1951, "On a formula concerning stochastic differentials, Nagoya Mathematical Journal, Vol. 3, pp. 55-65.

4. Ito, K., 1961, "Lectures on stochastic processes", Tata Institute Fundamental Research, Bombay, India.

5. Jazwinski, A.H., 1973, "Stochastic processes and filtering theory", Academic Press, New York, N.Y.

6. Lutes, L.D., and Hu, S.L.J., 1986, "Non-normal stochastic response of linear systems", Journal of Engineering Mechanics, ASCE, Vol. 112, No. 2, pp. 127-141.

7. Lutes, L.D., 1986, "Cumulants of stochastic response for linear systems", Journal of Engineering Mechanics, ASCE, Vol. 112, no. 10, pp. 1062-1075.

8. Roberts, J.B., 1972, "System response to random impulses", Journal of Sound and Vibration, Vol. 24, pp. 23-34.

9. Iwankiewicz, R., Soren, R.K., and Thoft-Cristensen, P., 1990, "Dynamic response of non-linear systems to Poisson-distributed pulse-tram Markov approach", Structural Safety, Vol. 8, pp. 223-238.

10. Grigoriu, M., 1987, "White noise processes", Journal of Engineering Mechanics Division, ASCE, Vol. 113, No. 5, pp. 757-765.

11. Picone, M., and Fichera, G., 1975, "Corso di analisi matematica", Libreria eredi Virgilio Veschi, Roma (in italian).

12. Wong, E., and Zakaj, M, 1965, "On the relation between ordinary and stochastic differential equations", Int. Journal of Engineering Sciences, Vol. 3, pp. 213-229.

13. Stratonovich, R.L., 1986, "A new form of representing stochastic integrals and equations", Journal SIAM Control, Vol. 4, pp. 362-371.

14. Wu, W.F., and Lin, Y.K., 1986, "Cumulant neglect closure for non-linear oscillators under random parametric and external excitation", Int. Journal of Non-Linear Mechanics, Vol. 19, No. 4, pp. 349-362.

15. Ibrahim, R.A., Somdararajan, A., and Heo, H., 1985, "Stochastic response of non-linear dynamic system based on a non-Gaussian closure", Journal of Applied Mechanics, Vol. 52, No. 12, pp. 965-970.

On the Cross-Correlation Method for Identification of Modal Restoring Force Nonlinearity from Random Vibration Data

M.F.Dimentberg, A.A.Sokolov
Institute for Problems in Mechanics, USSR Academy of Sciences, Moscow, USSR

Abstract

A version of cross-correlation approach is proposed for estimating nonlinearity level in a modal restoring force of a structure from its measured response to a white-noise excitation. This version is based on cross-correlating squared amplitudes of band-pass-filtered response components of main resonance and its integer multiples. The algorithm of signal processing provides effective decomposition of "linear" and "nonlinear" parts of the higher-frequency response component, thus leading to a direct estimate of the desired nonlinearity level in terms of the measured normalized cross correlation factor. The results of numerical simulation are presented.

Identification of nonlinearity in modal restoring force of a structure from its measured response may be important for various engineering applications. The examples include structural components with "flapping" cracks; systems with lumped or distributed parameters with rigid barriers and possible transitions between nonimpact and vibroimpact motions; structures with assembling defects, such as backlashes; etc. This problem may be far from trivial in case of small nonlinearities, since the desired identification usually should be made on the background of other modal responses' contributions into the measured signal and/or measurement noises.

The first identification problem is just to detect the nonlinearity in the modal restoring force $f(X) = \underline{n}^2X + \lambda g(X)$, where X is the modal displacement, that is to check whether $\lambda = 0$ or $\lambda \neq 0$. The desired answer may be sufficient for some applications. However, in certain cases it may be desirable also to estimate the overall level of nonlinearity (parameter λ) with the functional form of the nonlinear component g(X) of f(X) being presumably known. Thus in the example of a flapping crack g(X) is linear and proportional to $\underline{n}^2 - \underline{n}_1^2$ where $\underline{n}_1$ is a reduced slope of the restoring force-displacement characteristic of the cracked element.

These problems are considered here for the case of a sustained motion of the nonlinear element due to a stationary white-noise Gaussian random excitation $\xi(t)$:

$$X + 2aX + f(X) = \xi(t), \quad a \ll \underline{n} \qquad (1)$$

As for the first of the above problems, an efficient method of detecting nonlinearity in the f(X) has been presented in [1], which is based on cross-correlating amplitude of the main harmonic of the steady-state response X(t) with that of the higher response harmonics. The basic idea of the approach is as follows. In lightly damped system (1) each response cycle should

N. Bellomo · F. Casciati (Eds.)
Nonlinear Stochastic Mechanics
IUTAM Symposium Turin

be close to the corresponding damped free vibration cycle (with the same energy). The latter is definitely nonsinusoidal, and therefore should contain sinusoidal components not only with the resonant frequency Ω but with its integer multiples 2Ω, 3Ω, etc. as well. It is the presence of these high-order spectral peaks of X(t) that is used to discriminate between linear and nonlinear systems. Of course, direct detection of these peaks on the background of the nonresonant response components at $\omega \tilde{} 2\Omega$, $\omega \tilde{} 3\Omega$, etc. (as well as possible other modal responses) may be difficult, these peaks may in fact be completely masked. However, the above "extraneous" components of the X(t) may be rejected efficiently by cross-correlating amplitudes of the band-pass-filtered response components with center frequencies Ω and one of integer multiples of Ω. Roughly speaking, in the linear system ($f(X)=\Omega^2 X$ or $\lambda=0$) zero-mean parts of these amplitudes should be uncorrelated; whereas significant cross correlation implies the presence of nonlinearity, which generates higher-order harmonics in the response X(t).

Whilst this effect of cross correlation had been definitely verified by computer simulation [1], its direct use for quantitative estimation of the level of nonlinearity seems not clear because of a "nonadditive" interaction of the "resonant" (correlated) and nonresonant parts of the higher-order response components. Thus, an alternative version of the approach is presented in the following, which is based on cross-correlating zero-means parts of the squared amplitudes.

The undamped periodic free vibration of the system (1) (with $\alpha=0$, $\xi(t)\equiv 0$) may be in case of a small nonlinearity expressed by so-called "improved first approximation" of the Krylov-Bogoliubov method [2] as

$$X(t) = A \cos \psi + \sum_{k=2}^{\infty} A_k \cos k\psi,$$

$$A_k = C_k \psi(A)/[\Omega^2 (k^2-1)] \tag{2}$$

$$C_k(A) = \frac{1}{2\pi A} \int_0^{2\pi} f(A \cos \psi) \cos k\psi \, d\psi$$

The same expression may be used for the full quasiconservative system (1), with A being in this case a slowly varying random process, provided nonresonant response components are added; they are definitely small at frequencies $\omega \tilde{} \Omega$ but in case of small nonlinearity level may be comparable with "resonant components (terms with A_k, $k \geq 2$) at frequencies $\omega \tilde{} 2\Omega$, $\omega \tilde{} 3\Omega$, etc. Thus, the following expression is used for the response X(t), governed by eq.(1), which corresponds to the first basic processing operation (the signal X(t) is passed simultaneously through a pair of narrow band-pass filters, one of which is tuned to the resonant frequency Ω, the other one - to $k\Omega$, where k=2 or 3)

$$X(t) = A \cos \psi + \sum_{k=2}^{\infty} A_k \cos k\psi + X_{nk}(t) \tag{3}$$

$$X_{nk}(t) = X_{nkc}(t) \cos k\psi + X_{nks}(t) \sin k\psi \tag{4}$$

Here $X_{nk}(t)$ is a nonresonant component of the response within a frequency range of the band-pass filter, tuned to $k\Omega$. We assume it to be the same as in the linear system ($\lambda=0$), so that its inphase and quadrature components $X_{nkc}(t)$, $X_{nks}(t)$ are uncorrelated zero-mean Gaussian processes with identical variances

$$\sigma_k^2 \sim \int_{k\Omega-\Delta/2}^{k\Omega+\Delta/2} \frac{D\ \xi d\omega}{(\omega^2-\Omega^2)^2} \simeq \frac{D\xi\Delta}{\Omega^4(k^2-1)^2} \tag{5}$$

where is the filter bandwidth.
It is seen, that squared amplitude of the k-th term of the series (3) - that is, of the output signal of the band-pass filter - is

$$V_k = (A_k + X_{nkc})^2 + X_{nks}^2 \tag{6}$$

where A_k is defined in (2) in terms of $f(X) = \Omega^2 X + \lambda\, g(X)$. Now zero-mean part of $V_k(t)$ is cross-correlated with that of $V(t)=A^2(t)$. The corresponding normalized cross-correlation factor may be calculated by using well-known [3] solution for stationary probability density $p(A)$ of $A(t)$ (or $w(V)$ of $V(t)$; for the case of small nonlinearity λ, though, one may use approximately the well-know Reyleigh (or exponential) distributions

$$\begin{aligned} p(A) &= (A/\sigma^2)\ \exp\ (-A^2/2\sigma^2) \\ w(V) &= (1/2\sigma^2)\ \exp\ (-V/2\sigma^2) \end{aligned} \tag{7}$$

where σ^2 is a total variance of $X(t)$. The second of two distributions (7) has been used actually in the expectation operator (denoted by angular brackets) in the formula for cross-correlation factor r_{1k}

$$r_{1k} = \frac{\langle (V-\langle V\rangle)(V_k-\langle V_k\rangle)\rangle}{[\langle (V-\langle V\rangle)^2\rangle\langle (V_k-\langle V_k\rangle)^2\rangle]^{\frac{1}{2}}} \tag{8}$$

It is seen from (6) and (8), that this version of cross-correlation method does permit indeed to decompose resonant and nonresonant parts in the output of the higher-frequency filter; in fact, the value of r_{1k} provides a direct estimate of the distribution of the resonant part (this estimate, though, is approximate only, since V_k is a complex nonlinear function of A, according to (2), and therefore, in the absence of a nonresonant part r_{1k} is not exactly unity; however, as calculations show, it is sufficiently close to unity for the above estimate to be valid approximately).
The above relations have been used to derive an analytical expression for $r_{1,2}$ (k=2) for the case of a bilinear f(X), i.e.

$$f(X) = \begin{cases} \Omega^2 X & X<X_0 \\ \Omega_1^2 X + (\Omega_1^2-\Omega^2)X_0, & X>X_0 \end{cases} \tag{9}$$

From this lengthly expression the nonlinearity level $\lambda=1-\Omega_1/\Omega$ may be estimated directly in terms of quantities, calculated by processing the measured response X(t), namely, r_{12}, σ^2, $\sigma_k^2 = \sigma_2^2$, etc.

The above procedure has been implemented in computer simulation, with response X(t) being calculated numerically by Runge-Kutta integration of eq.(1) with $\xi(t)$ being supplied by a random number generator. Fig.1 illustrates the relation between preassigned and estimated values of $\lambda=1-\Omega_1/\Omega$, namely, λ_{pr} and λ_{est} respectively. The agreement is seen to be rather poor, but not hopeless. The basic reason for the discrepancies at small λ_{pr} is a very large difference between values of V and $\mathbf{V}_2$ (about four or even five orders of magnitude), which leads to high statistical errors of estimated r_{12} and thus λ_{est}. On the other hand, at high λ_{pr} (about 0.3) the discrepancy is also rather high, probably because the nonlinearity level is not sufficiently small. At moderate levels of nonlinearity, however, within the range $0.12\leq\lambda_{pr}\leq0.26$, the agreement seems reasonable. Thus the level of nonlinearity of the modal restoring force may be estimated by this version of the cross-correlational approach provided it isn't too small and too high. The latter restriction, though, may be relaxed for piecewise-linear f(X) by replacing Krylov-Bogoliubov approximation by the exact generating solution for undamped free vibration; the exact expressions for response probability densities [3] may also be used instead of approximate ones (7) to improve the estimates for the case where the nonlinearity is not small.

References

[1] DIMENTBERG M.F., SOKOLOV A.A. Identification of Restoring Force Nonlinearity from System's Response to a White-Noise Excitation. Intern. Journal of Nonlinear Mechanics (in press)

[2] BOGOLIUBOV N.N., MITROPOLSKY JU.A. Asymptotic Methods in the Theory of Nonlinear Oscillations (in Russian). Moscow, Nauka, 1974

[3] DIMENTBERG M.F. Statistical Dynamics of Nonlinear and Time-Varying Systems. Research Studies Press, Taunton, England, 1988.

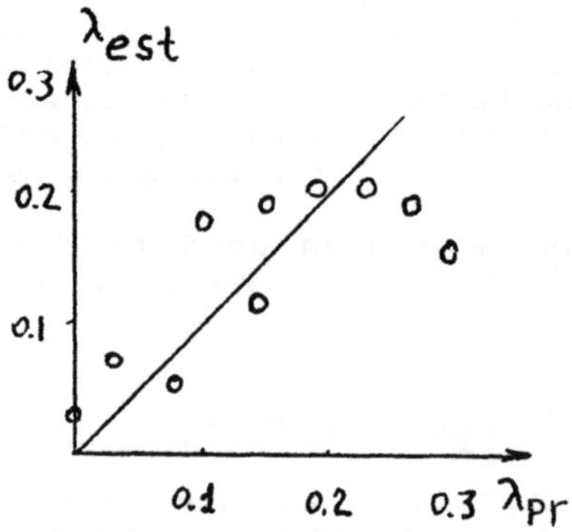

Fig.I

Stochastic Growth of Fatigue Crack under Constant Amplitude Loading

KRZYSZTOF DOLIŃSKI

Institute of Fundamental Technological Research
00-049 Warsaw, ul. Świętokrzyska 21, POLAND

INTRODUCTION

The scatter of experimental fatigue crack paths obtained even in very well-controlled constant amplitude loading tests is very large [Virkler et al. 1979, Ghonem & Dore 1986]. The only origin of this uncertainty is some randomness of the material properties which affect the crack propagation process. The fatigue crack growth phenomenon, however, appears to be physically and mechanically so complicated that pure theoretical considerations alone do not suffice to set up a reliable model which could be used to predict the fatigue crack propagation. Nevertheless, the theoretical analysis of the possible mechanisms which are observed to be present during fatigue crack growth points out some relations between crack growth features, material and load characteristics. In order to verify the theoretical investigations and specify the parameters of the proposed models experimental results have to be used. They, moreover, enable us to identify some random characteristics of the model parameters and help us to explain the random nature of the fatigue crack growth phenomenon.

The main purpose of this contribution is to point out an interdependence and accordance of the deterministic and stochastic modeling of the fatigue crack growth process. It is shown that the stochastic modeling has to originate by the mechanics of the process rather than being proposed *ad hoc* by means of some mathematical assumptions which might assure the model to be very handy but physically misleading.

Firstly, some results of mechanical analysis of the fatigue crack propagation are briefly quoted and a model is proposed in which some random variables account for the scatter of material properties between specimens (elements). Additionally, a random process is admitted to account for the effect of local random nonhomogeneity of material on the fatigue crack growth process. Some illustrative results of statistical analysis based on the experimental data obtained by Virkler et al. and Ghonem & Dore on aluminium alloys are given.

N. Bellomo · F. Casciati (Eds.)
Nonlinear Stochastic Mechanics
IUTAM Symposium Turin

CRACK GROWTH LAW

One of the approaches to derive the governing equation for fatigue crack growth is based on the energy balance. In [Short & Hoeppner 1989] the global/local energy balance is considered for a simplified microstructural material model. It leads to the equation

$$\dot{a} = \frac{W^{(p)} - \Phi}{\Gamma - J} \tag{1}$$

where $\dot{a}$ denotes the fatigue crack propagation (FCP) rate and the other symbols possess the following meaning:

$W^{(p)}$ is the rate of increase of "plastic work" due to a load cycle;

Φ represents the rate of heat generation due to internal friction within crack tip zone of volume $\mathcal{V}$;

Γ may be regarded as the amount of fracture energy which must be added to $\mathcal{V}$ per unit crack advance;

J denotes the path independent contour J-integral introduced into fracture mechanics by Rice, cf. [Rice 1968], and represents the increase in fracture energy within $\mathcal{V}$ per unit crack advance.

In order to specify the quantities in eq.(1) the analysis of a Dugdale-Barenblatt fatigue crack experiencing closure presented previously by Budiansky & Hutchinson [1978] is used by Short & Hoeppner [1989]. It leads to a very nice FCG rate equation which, unfortunately, does not fit the experimental data of either Virkler et al. [1978] or of Ghonem & Dore [1986]. Very restrictive assumptions made in derivation of the energy rate terms, $W^{(p)}$, Φ, Γ and J, in eq.(1) do not meet apparently the real conditions of the process. Especially the derivation of $W^{(p)}$ which strongly depends on the assumed mechanism and stress-strain field configuration in the crack tip zone would require some generalizations to make the crack growth law more flexible being able to fit a wider spectrum of experimental data. Nevertheless, the assumption that the averaged "plastic work" over the plastic zone around the crack tip is solely relevant to the energy balance leads to the proportionality of $W^{(p)}$ to the fourth power of the stress range intensity factor ΔK, i.e.

$$W^{(p)} \sim \Delta K^4 \tag{2}$$

where

$$\Delta K = K_{max} - K_{min} = \Delta S \cdot Y(a) \cdot \sqrt{\pi \cdot a} = (S_{max} - S_{min}) \cdot Y(a) \cdot \sqrt{\pi \cdot a} \tag{3}$$

$\Delta S = S_{max} - S_{min}$ is the nominal stress range and Y(a) denotes the function of the crack length, a, depending on the specimen and crack geometry.

Another assumption, e.g. [Tomkins 1968], that the most crack tip deformation is concentrated along the plastic de-cohesion planes where maximum strain gradient occurs leads to the rate $W^{(p)}$ which is proportional to the second power of the stress range intensity factor, i.e.

$$W^{(p)} \sim \Delta K^2 \tag{4}$$

Comparing these arguments with eq.(1) it is clear that depending on the assumed degree of concentration of the plastic work the parameter $W^{(p)}$ can be considered to be proportional to the fourth or second power of ΔK. However, in the real crack growth process some plastic deformation spreads continuously over a crack tip zone and simultaneously some planes of maximum shear strain gradient develop where plastic de-cohesion effects can be observed, [Tomkins 1968]. Thus, the stress range intensity factor, ΔK, with exponent either 2 or 4 separately cannot be expected to describe sufficiently the crack growth process. This can also be one of the reasons why in the crack growth equations commonly applied to engineering problems an exponent "m" is used as a free parameter to be determined by fitting the experimental data. Such a procedure extends the range of validity of the equation but makes any interpretation of doubtful value.

Keeping in mind two possible modes of concentration of the plastic work the rate of its increase due to a load cycle may be written as a sum

$$W^{(p)} = C_2 \cdot \Delta K^2 + C_4 \cdot \Delta K^4 \tag{5}$$

where C_2 and C_4 weight corresponding to the contribution of the respective modes.

We leave the coefficients C_2 and C_4 undetermined before the statistical analysis of experimental data is carried out. There is, however, one modification which accounts for the crack closure effect and results in reduction of the stress range intensity factor in eq.(6). The crack may remain closed below some values K_{open} and K_{close} when the load increases and decreases, respectively, during a cycle. The mechanism of the crack closure was very carefully investigated by Budiansky & Hutchinson [1978] for a Dugdale-Barenblatt crack model. The results coincide well with the approximation based on experimental observations given by Elber [1971]. These results and many other experimental data show that the values K_{open} and K_{close} are very close to each other and the ratio $q = K_{cont}/K_{max}$, where K_{cont}

denotes the stress intensity factor at crack opening or closing, can be related with the stress asymmetry ratio, $R = K_{min}/K_{max}$. Since there is no universal form of this relationship one has to admit an approximation. Veers [1987] proposed the bi-linear approximation

$$q = \max\left\{q_0 \cdot \left(1 + \frac{R}{|R_0|}\right),\ R\right\} \tag{6}$$

where $q_0 \in [0.2, 0.5]$ and $R_0 \in [-5, -2]$ are the material dependent parameters to be determined from experimental data. This approximation will be adopt in this paper also.

The effective fracture energy added per unit time can be defined as follows

$$W_{eff}^{(p)} - \Phi = C_2 \cdot \Delta K_{eff}^2 + C_4 \cdot \Delta K_{eff}^4 - \Phi \tag{7}$$

where the effective stress range intensity factor, ΔK_{eff}, is given by

$$\Delta K_{eff} = \frac{1 - q}{1 - R} \cdot \Delta K = U(R) \cdot \Delta K \tag{8}$$

The quantity Φ in eq.(1) considered as the heat generation rate by Short & Hoeppner [1989] corresponds formally to the threshold value, ΔK_{th}, which appears often in crack growth rate equations. It is difficult to find in the literature any convincing theoretical arguments which would allow one to define an explicit relation between ΔK_{th} and the material properties and load conditions. It is observed that the threshold strongly depends on alloy strength, grain size and on the load cycle parameters, S_{max} and S_{min}, [Ritchie 1977]. Similarly the quantity Γ in eq.(1) which corresponds to the fracture toughness, K_C, is also dependent on the material parameters and load conditions. The both parameters are also left undetermined and their dependence on the load cycle parameters has to be infered from a statistical analysis of appropriate experimental data.

The FCG rate equation (1) can also be rewritten in the following form

$$\frac{da}{dn} = \frac{C_2 \cdot \Delta K_{eff}^2 + C_4 \cdot \Delta K_{eff}^4 - \Phi(\Delta K_{th})}{K_C^2 - K_{max}^2} \cdot x(a) = g(a; \underset{\sim}{\Theta}) \cdot x(a) \tag{9}$$

where the approximate relation

$$J = \frac{K_{max}^2}{E} \tag{10}$$

is used. The function $\Phi(\Delta K_{th})$ indicates a possible dependence of the heat generation rate on the stress intensity range threshold. There are two coefficients, C_2 and C_4, and two load cycle dependent parameters, Φ and K_C, that should be determined for a given material from statistical analysis of experimental data.

The material dependent parameter vector, $\underset{\sim}{\Theta}$, specified in eq.(9) by the components $[C_2,C_4,\Phi,K_C]$ is taken to be a random vector. The effect of stochastic variability of the material properties within a specimen is modeled by a positive random function of the crack tip position, x(a).

In contrast to eq.(9) it is quite usual to consider the number of cycles, n, or, equivalently, the time, t, as the argument of the stochastic factor x(t). The models with time dependent stochastic process, x(t), in place of x(a) in eq.(9) were investigated in many papers, [Lin & Yang 1983, Sobczyk 1986, Spencer & Tang 1988], where some solutions based on the theory of Markov processes were proposed. The assumption that the function describing the effect of random material nonhomogeneity depends directly on the crack tip position and not on the time has much clearer physical motivation. Such an approach has been considered by Ditlevsen [1986] and Ortiz & Kiremidjan [1986, 1988].

Kozin & Bogdanoff [1981] pointed out that the lifetime, N(a) or T(a), should be considered as the dependent variable. Since the crack growth process is monotonically increasing, $g(a;\underset{\sim}{\Theta}) > 0$ and $x(a) > 0$ almost surely, there is no objection to investigate the reciprocal

$$\frac{dN}{da} = G(a;\underset{\sim}{\Theta}) \cdot X(a) \tag{11}$$

with $G(a;\underset{\sim}{\Theta})$ and X(a) as the reciprocals of $g(a;\underset{\sim}{\Theta})$ and x(a), respectively. Eq.(11) has the dimension units [cycles/mm], say, and may be considered as the "fatigue resistance" of the element with a crack of the length "a".

MODELING OF NONHOMOGENEITY EFFECTS

The most common approach to model the stochastic function, X(a), in fatigue crack growth equations like eq.(11) is based on the assumption that X(a) is a stationary stochastic process, [Ortiz & Kiremidjan 1986, 1988, Ditlevsen 1986]. The additional assumption that X(a) is a stationary random process, $X(a) = \zeta(a)$, with independent values, i.e.

$$\mathbb{E}[\zeta(a)] = \bar{\zeta} \quad \text{and} \quad \mathbb{C}[\zeta(a_1),\zeta(a_2)] = \sigma_\zeta^2 \cdot \delta(a_2 - a_1) \tag{12}$$

where $\mathbb{C}[\cdot,\cdot]$ denotes the covariance function and $\delta(\cdot)$ is the Dirac function enables one to develop very nice mathematical analysis and leads directly to the approximate inverse Gaussian probability distribution of the crack growth increments [Ditlevsen 1986].

The assumption of the stationarity of the function X(a) means that the "fatigue crack resistance", dN/da, is disturbed during the crack growth process by some stationary, random material properties at the crack tip only or, at most, by these properties in the crack tip neighbourhood of a constant size. If we recall, however, the previous discussion about the derivation of crack growth laws it is clearly seen that the size of the crack tip affected zone which interacts during the crack growth process increases fairly proportionally to the length and/or square length of the crack. Therefore, the random nonhomogeneities of material within these extending zones have to affect the "fatigue crack resistance" as well. The following proposal is an attempt to account for this nonstationary random interdependence in the stochastic model of crack growth process.

Let us consider a crack to propagate in a metal sheet of constant thickness. In every point, (x,y), of the sheet surface a random variable, $\zeta(x,y)$, is prescribed. It defines a random "fatigue crack resistance" factor averaged over the sheet thickness. The random variables $\zeta(x,y)$ are assumed to be independent of each other and their collection over the sheet surface constitutes a random field with the constant mean, $\mathbb{E}[\zeta(x,y)] = \bar{\zeta}$ and the covariance function as follows

$$\mathbb{C}[\zeta(x',y'),\zeta(x'',y'')] = \sigma_\zeta^2 \cdot \delta(x'' - x') \cdot \delta(y'' - y') \tag{13}$$

As indicated above, the "fatigue crack resistance", dN/da, is affected by the material properties of the crack tip neighbourhood. In section II two mechanisms were considered to dominate the fatigue crack growth. The first of these is connected with the plasticity of the crack tip zone. Thus, the properties of material points which are placed within this zone will affect the crack propagation process. In the second mechanism the plastic slip lines are active and the randomness of properties of material points along the lines will disturb the crack growth process. These two features of the material nonhomogeneity effect are assumed to additively constitute the factor X(a), i.e.

$$X(a) = \alpha \cdot X_4(a) + (1 - \alpha) \cdot X_2(a) \tag{14}$$

α is the weight factor which defines to what extent the random nonhomogeneity connected with the respective mechanism contributes in stochastic variability of the fatigue crack growth process. Some general forms of the random functions, $X_4(a)$ and $X_2(a)$, are given by Doliński [1991]. Herein the constant shape functions accounting for the spatial effect of the nonhomogeneity are only considered leading to the following definitions

$$X_4(a) = \frac{1}{W_4(a;,k_{x4},k_{y4})} \cdot \int_{-k_{y4}\cdot a}^{k_{y4}\cdot a} \int_{a}^{k_{x4}\cdot a} \zeta(x,y)\, dx\, dy$$

$$X_2(a) = \frac{1}{W_2(a;,\kappa,k_{y2})} \cdot \int_{0}^{k_{y2}\cdot a} \int_{a+y\cdot ctg\vartheta}^{a+\kappa+y\cdot ctg\vartheta} \zeta(x,y)\, dx\, dy \tag{15}$$

where ϑ is the angle between the crack plane and the slip line and $W_4(a;k_{x4},k_{y4})$, $W_2(a;\kappa,k_{y2})$ denote simply, in the present case, the areas of the respective domains whose properties affect the fatigue crack growth, i.e.

$$W_4(a;k_{x4},k_{y4}) = 2\cdot k_{y4}\cdot(k_{x4}- 1)\cdot a^2$$

$$W_2(a;\kappa,k_{y2}) = 2\cdot\kappa\cdot k_{y2}\cdot a \tag{16}$$

It is assumed that the dimensions of the integration domains are proportional to the crack length. It results directly from the theoretical investigations of the fatigue crack growth process where the range of the plastic zone (e.g. for the Dugdale-Barenblatt model)

$$r_p = \frac{\pi}{8}\cdot\frac{K_{max}^2}{\sigma_y^2} \approx 0.4\cdot\frac{K_{max}^2}{\sigma_y^2} \tag{17}$$

and the length of the plastic slip lines (e.g. given by Sahasakmontri and Horii [1991])

$$l_p = \lambda\cdot\frac{K_{max}^2}{\sigma_y^2} \tag{18}$$

appear to be proportional to the square of the maximum stress intensity factor,

K_{max}. σ_y denotes the yield stress in tension. The proportionality coefficient, λ, in (18) depends on the material and enviromental conditions but its value does not differ very much from $\lambda \approx 0.1$ for the aluminium alloys tested in the experiments considered also in the present paper [Sahasakmontri & Horii 1991]. The width of integration interval in the y-direction in the definition of $X_4(a)$ can be assumed to be equal to the range of plastic zone, i.e. $2 \cdot k_{y4} \cdot a = r_p$. It would define a square area in front of the crack tip as the region whose nonhomogeneity affect the crack growth. Another proposal may result from the investigations of the "fracture process zone" [Short & Hoeppner 1989]. Such a zone can be defined approximately as a rectangular of the length, δ_f, to be equal to the range of the fully reversed plastic zone and of the width, W_f, to be equal to the width of the residual stretch zone in the wake if the growing crack. Hence, the integral limits in the definition of $Z_1(a)$ would be as follows

$$\begin{aligned} (k_{x4} - 1) \cdot a &= \delta_f = \frac{1}{4} \cdot (1 - q) \cdot r_p \\ 2 \cdot k_{y4} \cdot a &= W_f = \frac{J}{\sigma_y} \cdot \sqrt{1 - (1 - q)^2} \end{aligned} \tag{19}$$

where q is defined in eq.(6) and J denotes the Rice integral approximated by (10).

The values of the coefficient k_{y2} in the definition of $X_2(a)$, eq.(15), might be expected to approach the result of the following equation

$$k_{y2} \approx \frac{l_p \cdot \sin \vartheta}{a} \approx 0.1 \cdot \frac{S_{max}^2 \cdot \pi}{\sigma_y^2} \cdot \sin \vartheta \tag{20}$$

OUTLINE OF STATISTICAL VERIFICATION

A comprehensive presentation of statistical verification of the proposed model based on the experimental data provided by Virkler et al. [1979] and Ghonem & Dore [1986] will be given elsewhere [Doliński 1991]. This contribution incorporates only some limited results obtained by extensive statistical analysis.

The analysis of all data sets shows that the term $C_4 \cdot \Delta K_{eff}^4 \ll C_2 \cdot \Delta K_{eff}^2$. Moreover, the correlation between the parameters C_2 and Φ in eq.(9) is very heigh (above 0.8) for all data sets considered. Following this result the relationship

$$\Phi = C_2 \cdot \Delta K_{th}^2 = C_2 \cdot \Delta S^2 \cdot Y^2(A_{th}) \cdot \pi \cdot A_{th} \approx C_2 \cdot \Delta S_{eff}^2 \cdot \pi \cdot A_{th} \tag{21}$$

is admitted with A_{th} as a random variable. Thus, the "fatigue crack resistance" equation (11) takes the following form

$$\frac{dN}{da} = \frac{K_C^2 - K_{max}^2}{C_2 \cdot U^2 \cdot \Delta S^2 \cdot \pi \cdot [Y^2(a) \cdot a - A_{th}]} \cdot X(a) \tag{22}$$

Very strong correlation is also observed between K_C^2 and C_2 (above 0.96) for the Virkler data but the limited range of data does not allow one to estimate this parameter for the Ghonem & Dore results and no general conclusion would be justified.

The effect of the loading condition was investigated for three sets of the Ghonem & Dore data obtained for three different load amplitude parameters. Taking q_0= 0.45 and R_0= -5 in eq.(6) the common C_{2G} parameter may be accepted for all Ghonem & Dore tests.

The probability distributions of the parameters were established from goodness-of-fit tests and for the following ones thc highest significance levels were achived:

- the lognormal probability distribution for C_2;
- the Frêchet probability distribution for K_C^2;
- the Weibull probability distribution for A_{th}.

Tho parameters of the random function $\zeta(x,y)$ are determined from the statistical analysis of the residuals

$$R_{\Delta a}(a) = \frac{\Delta N(a, a+\Delta a)}{\Delta a \cdot G(\bar{a}; \underset{\sim}{\vartheta}_j)} = \frac{1}{\Delta a} \cdot \int_a^{a+\Delta a} X(a)\, da \qquad \text{with } \bar{a} = a + \frac{1}{2} \cdot \Delta a \tag{23}$$

averaged over the ensamble of crack length paths for some given crack length intervals Δa. The analysis of variance of the residuals with the function X(a) as defined by eqs.(14) and (15) shows that the weight coefficient $\alpha = 0$. It accords with the result $C_4 \ll C_2$ obtained before while fitting the probabilistic term, $G(a;\underset{\sim}{\Theta})$, of the model. Thus, the covariance of X(a) can be written as

$$c_X(a',a'') = \begin{cases} \sigma_\zeta^2 \cdot \dfrac{\min(a',a'') + \kappa - \max(a',a'')}{\kappa^2 \cdot k_{y2} \cdot \max(a',a'')} & \text{for } \left\{ \kappa > \max(a',a'') - \min(a',a'') \right. \\ 0 & \text{otherwise} \end{cases} \tag{24}$$

and the variance of residuals is determined as follows

$$\mathrm{Var}[R_{\Delta a}(a)] = \frac{1}{\Delta a^2}\cdot\int\int_{a}^{a+\Delta a} c_X(a',a'')\, da'\, da'' =$$

$$= \sigma_\zeta^2\cdot\frac{1}{4\cdot\Delta a^2\cdot\kappa^2\cdot k}\cdot \tag{25}$$

$$\cdot\begin{cases}\Delta a\cdot\left(a + 2\cdot\kappa - \frac{\Delta a}{2}\right) - a\cdot\left(a + 2\cdot\kappa\right)\cdot\ln\left(\frac{a + \Delta a}{a}\right) & \text{for } \Delta a \le \kappa \\ \kappa\cdot\left(a + \frac{3}{2}\cdot\kappa\right) + \kappa^2\cdot\ln\left(\frac{a + \Delta a}{a}\right) - \left(a + \kappa\right)^2\cdot\ln\left(\frac{a + \kappa}{a}\right) & \text{for } \Delta a > \kappa\end{cases}$$

where $k = k_{y2}$. The averaged residuals from the Virkler and Ghonem & Dore data, respectively, for $\Delta a = 0.2$, 0.4 and $\Delta a = 0.1$, 0.2 are used to determined the parameters σ_ζ and κ in eq.(25). In the figures the best fitted curves are shown. The standard deviation, σ_ζ, given $\lambda = 0.1$ in eq.(18) and the active slip line band width, κ, are obtained to be equal, respectively, to 0.04 and 0.25 mm for the Virkler data and for the Ghonem & Dore data slightly differ for various load amplitude ratios: $\sigma_\zeta = 0.023$ and $\kappa = 0.022$ mm for $R = 0.6$, $\sigma_\zeta = 0.018$ and $\kappa = 0.094$ mm for $R = 0.5$, $\sigma_\zeta = 0.013$ and $\kappa = 0.084$ mm for $R = 0.4$.

CONCLUSIONS

The proposed model originates by the mechanics of the fatigue crack growth process. Statistical and stochastic nonhomogeneity of material properties are considered. The former is accounted for by a random vector of material parameters. The latter is modeled by a stationary random field. The effect of the stochastic nonhomogeneity has an integral form by means of averaging the local fluctuations of the material properties over some regions in the crack tip neighbourhood. These are the same regions which are considered in derivation of the mechanical model to affect the fatigue crack growth rate, namely the plastic zone and fatigue process zone. Comparison with limited experimental data and the results of statistical analysis are encouraging to further research to more precisely specify the parameters, to verify the approach on a broader data basis and, eventually, to propose some directions for application.

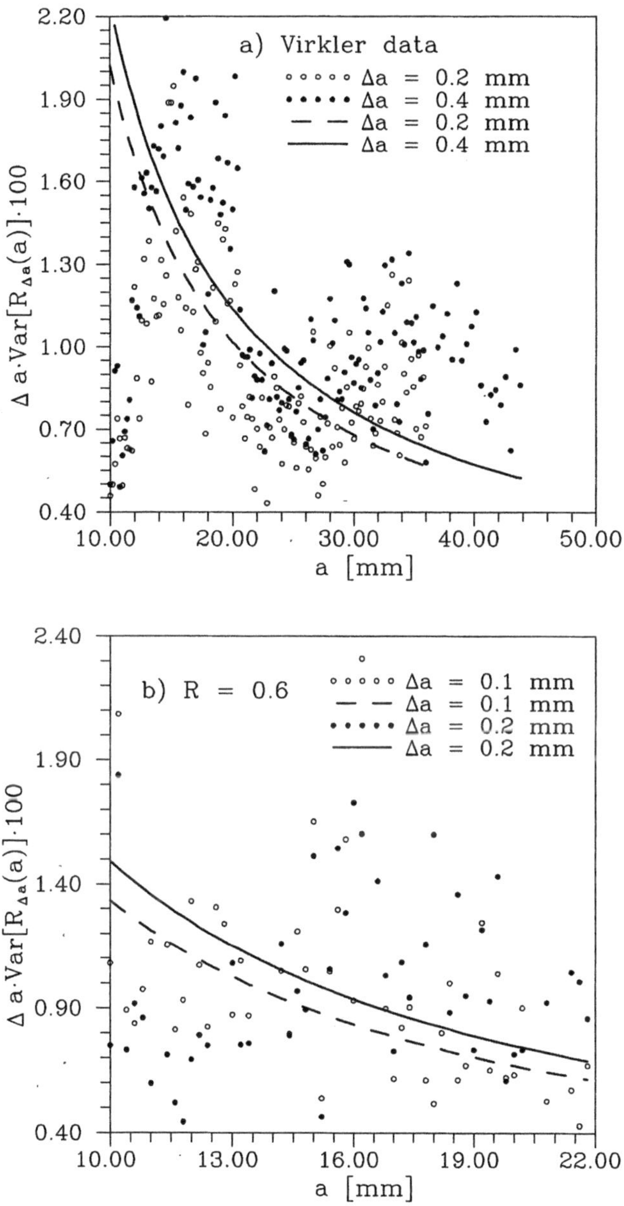
2.20
1.90
1.60
1.30
1.00
0.70
0.40
a) Virkler data
Δa = 0.2 mm
Δa = 0.4 mm
Δa = 0.2 mm
Δa = 0.4 mm
Δ a·Var[R_Δa(a)]·100
10.00
20.00
30.00
40.00
50.00
a [mm]
2.40
1.90
1.40
0.90
0.40
b) R = 0.6
Δa = 0.1 mm
Δa = 0.1 mm
Δa = 0.2 mm
Δa = 0.2 mm
Δ a·Var[R_Δa(a)]·100
10.00
13.00
16.00
19.00
22.00
a [mm]

a)

b)

Fig. 1.

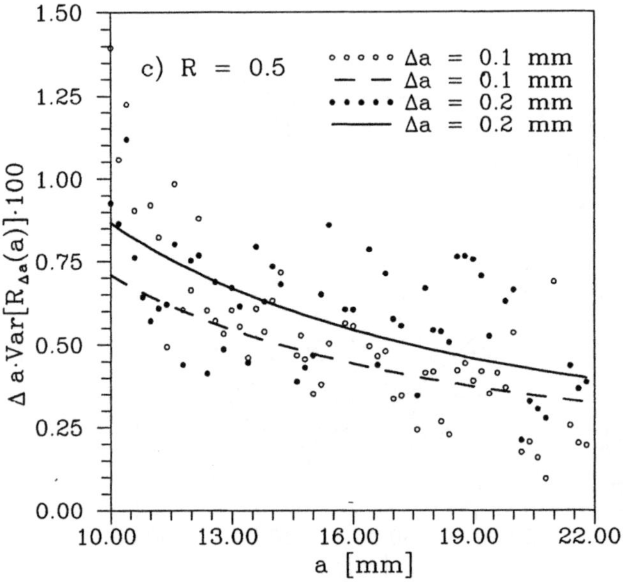

c)

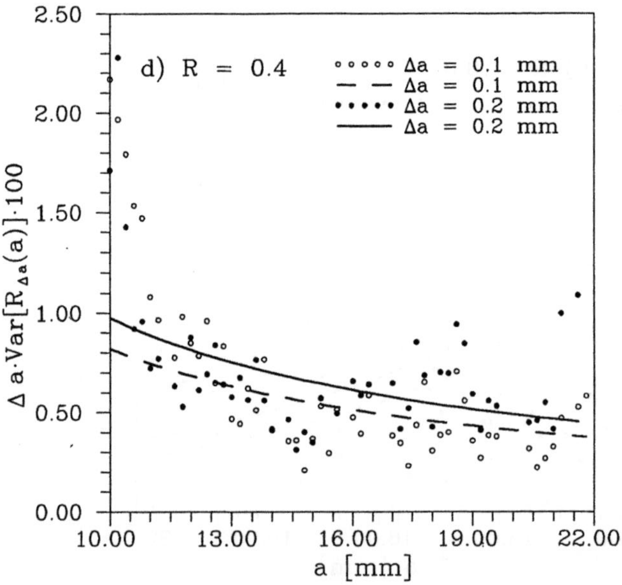

d)

Fig.1.Variance of residuals averaged over the set of test specimens - a) the Virkler data; b) ÷ d) the Ghonem & Dore data.

REFERENCES:

Budiansky B., Huchinson J.W. (1978),Analysis of closure in fatigue crack growth, Trans. ASME, Journal of Applied Mechanics, vol. 45, pp. 267-276;

Ditlevsen O. (1986), Random fatigue crack growth - a first passage problem, Eng. Fract. Mechanics, vol. 23, No. 2, pp. 467-477;

Doliński K. (1991), Stochastic modeling and statistical verification of the crack growth under constant amplitude loading, (in preparation);

Elber W. (1971), The significance of the crack closure in: Damage Tolerance in Aircraft Structures, ASTM STP 486, 230;

Ghonem H., Dore S. (1986) Probabilistic description of fatigue crack growth in aluminum alloys under constant amplitude loading, Report AFOSR-83-0322, Univ. of Rhode Island;

Kozin F., Bogdanoff J.L. (1981), A critical analysis of some probabilistic models of fatigue crack growth, Eng. Fract. Mech., 14, pp. 58-89;

Lin Y.K., Yang J.N. (1983), On statistical moments of fatigue crack propagation, Eng.Fract. Mechanics, vol.18, No. (2), pp. 243-256;

Ortiz K., Kiremidjian A. (1986), Time series analysis of fatigue crack growth data, Eng.Fract. Mechanics, vol. 24, No. 5, pp. 657-675;

Ortiz K., Kiremidjian A. (1988), Stochastic modeling of crack growth, Eng.Fract. Mechanics, vol. 29, No. 3, pp. 317-334;

Rice J.R. (1968), A path independent integral and the approximate analysis of strain concentration by notches and cracks, J.Appl. Mechanics, vol. 35, pp. 379-386;

Ritchie R.O. (1977), Near-threshold fatigue crack propagation in ultra-high strength steel: influence of load ratio and cyclic strength, Trans. ASME, Journal of Engineering Materials and Technology, vol. 99, No. 3, pp. 195-204;

Sahasakmontri K., Horii H. (1991), An analytical model of fatigue crack growth based on the crack-tip plasticity, Eng.Fract. Mechanics, vol. 38, No. 6, pp. 413-437, 1991;

Short J.S. Hoeppner D.W. (1989), A global/local theory of fatigue crack propagation, Eng.Fract. Mechanics, vol. 33, No.2, pp. 175-184;

Sobczyk K. (1986), Modeling of random fatigue crack growth, Eng. Fract. Mech., vol.24, No.4, pp. 609-623;

Spencer Jr. B.F., Tang J. (1988), Markov process model for fatigue crack growth, J.Eng.Mechs, ASCE, vol. 114, No. 12, pp. 2134-2157;

Tomkins B. (1968), Fatigue crack propagation - an analysis, The Philosophical Magazine, vol. 18, No. 155, pp. 1041-1066;

Veers P.S. (1987), Fatigue crack growth due to random loading, SAND87-2039, Sandia National Laboratories;

Virkler D.A., Hillberry B.M., Goel P.K. (1978), The statistical nature of fatigue crack propagation, Technical Report AFFDL-TR-78-43, Wright-Patterson Air Force Base, Ohio;

Equivalent Statistical Quadratization for Multi-Degree-of Freedom Nonlinear Systems

M.G. Donley
Sr. Project Engineer
Structural Dynamics Research Corp.
2000 Eastman Drive
Milford, Ohio 45150

P.D. Spanos
L.B. Ryon Chair in Engineering
Rice University
P.O. Box 1892
Houston, Texas 77251

Summary

The statistical linearization method is often inadequate for estimating spectral properties of random responses of nonlinear systems. This is sometimes due to the fact that the power spectra of responses of linear systems span only the frequency range of the excitation spectrum, whereas significant responses outside this range are possible for nonlinear systems. Recently, the concept of a statistical "quadratization" method was introduced to address this shortcoming of the linearization methods. The effectiveness of statistical quadratization was demonstrated on several single-degree-of-freedom systems. In this paper the method is generalized to multi-degree-of-freedom systems. The statistical quadratization solution procedure involves replacing the nonlinear system with an "equivalent" system with polynomial nonlinearities up to quadratic order. The nonlinear equivalent system has a form whose solutions can be approximated by using the Volterra series method. The non-Gaussian joint response probability distribution is approximated by a third order Gram-Charlier expansion. The method is formulated for systems with general nonlinearities and with nonlinearities of a special form. To demonstrate the method, solutions are obtained for a specific system. The corresponding results compare well with Monte-Carlo simulation data. Further, it is shown that the quadratization method is notably superior to the linearization method for the considered system.

Introduction

In a recent investigation by Donley and Spanos[1], the concept of a statistical "quadratization" method was introduced for determining the response statistics of randomly excited single-degree-of-freedom systems. The effectiveness of the method was demonstrated on several examples and was shown to give more accurate results than the statistical linearization methods. In this paper, the proposed method is extended to multi-degree-of-freedom(mdof) systems.

Loosely speaking, statistical "quadratization" can be viewed as the next level of sophistication beyond statistical linearization. Linearization methods, although they are perhaps the most versatile and popular methods for treating nonlinear systems, see references[2-8], do exhibit shortcomings for certain classes of problems. One is reflected by the fact that the response spectrum obtained by linearization spans only the frequency range of the excitation spectrum. It is well known, however, that for nonlinear systems the response spectra can have significant values outside the frequency range of the excitation. In a recent paper by Donley and Spanos[1], it was demonstrated that in some cases this leads to significant errors in the computed linearized response statistics. The Volterra series method, see Schetzen[9], is an approximate analytical method for treating nonlinear systems and does not suffer from this drawback. This method is best suited for systems with polynomial nonlinearities. However, in many engineering

N. Bellomo · F. Casciati (Eds.)
Nonlinear Stochastic Mechanics
IUTAM Symposium Turin

applications, the nonlinearities can not be described by polynomials and may not even be analytic. In this context, an equivalent statistical "quadratization" method was proposed. The arbitrary system nonlinearity is replaced by "equivalent" polynomials up to quadratic order. The resulting nonlinear "equivalent" system is in a form which is treated by the Volterra series method. The non-Gaussian response probability distribution is approximated by a Gram-Charlier expansion.

In this paper the method is formulated for a general nonlinear multi-degree-of-freedom system subject to a stationary, Gaussian excitation. The computational effort involved in analyzing a system with completely general nonlinearities, however, can be quite expensive. Therefore, a simplified version of the method is also developed for systems which have nonlinearities of a special form. The simplified method is exemplified by considering a system with two degrees of freedom.

Equivalent Quadratic System

Consider the following nonlinear equation of motion of a multi-degree-of-freedom system subject to a stationary, Gaussian random force vector, $\underline{f}(t)$,

$$\mathbf{m}\,\ddot{\underline{x}} + \mathbf{c}\,\dot{\underline{x}} + \mathbf{k}\,\underline{x} + \underline{g}(\underline{x},\dot{\underline{x}}) = \underline{f}(t) \tag{1}$$

The system has N degrees of freedom, so that the vector $\underline{x}$ is an N×1 displacement response vector, and **m**, **c**, and **k** are N×N mass, damping, and stiffness matrices, respectively. The symbol $(\dot{\ })$ denotes differentiation with respect to time, t. The vector $\underline{g}(\underline{x},\dot{\underline{x}})$ is the nonlinear force vector which may contain velocity and displacement terms. The ith term of this vector is assumed to be of the following general form

$$g_i(\underline{x},\dot{\underline{x}}) = \sum_{j=1}^{N} \sum_{k=1}^{N} g_{jk}(x_j, x_k, \dot{x}_j, \dot{x}_k) \tag{2}$$

The force excitation is described mathematically by the sum of a deterministic mean vector, $\underline{\mu}_f$, and a vector of random filtered white-noise processes. The latter part is expressed mathematically by a time-invariant linear transform involving the convolution of white noise with the filter impulse response function $\underline{h}_f(\tau)$. Specifically, $\underline{f}(t)$ can be written as follows

$$\underline{f}(t) = \underline{\mu}_f + \int_{-\infty}^{\infty} \underline{h}_f(\tau)\, w(t-\tau)\, d\tau \tag{3}$$

In this equation, w(t) is a zero-mean, Gaussian, white-noise process. That is, $\langle w(t)w(t+\tau)\rangle = 2\pi\delta(\tau)$, where the symbols $\langle\cdot\rangle$ and δ stand for the operator of mathematical expectation and the Dirac delta function, respectively.

The spectral density matrix of the excitation is given by the equation

$$S_{ff}(\omega) = \frac{1}{2\pi} \int_{-\infty}^{\infty} \langle \underline{f}(t)\, \underline{f}^T(t+\tau)\rangle \exp(-i\omega\tau)\, d\tau \tag{4}$$

where $i = \sqrt{-1}$. Substituting equation (3) into equation (4) yields

$$S_{ff}(\omega) = \mu_f \mu_f^T \delta(\omega) + \underline{H}_f^*(\omega)\, \underline{H}_f^T(\omega)\, S_{ww}(\omega) \tag{5}$$

In this equation, $\underline{H}_f(\omega)$ is the filter transfer function vector and is related to filter impulse response function vector, $\underline{h}_f(\tau)$, by the following Fourier transform

$$\underline{H}_f(\omega) = \int_{-\infty}^{\infty} \underline{h}_f(\tau) \exp(-i\omega\tau)\, d\tau \tag{6}$$

The symbol $S_{ww}(\omega)$ denotes the spectral density of w(t) and is identically equal to one because w(t) is delta-correlated.

In general, the nonlinear force vector, $\underline{g}(\underline{x},\dot{\underline{x}})$, is non-symmetric. Hence, the solution of equation (1) may not have a zero mean. Therefore, the stationary response is written in the following form

$$\underline{x}(t) = \mu_x + \hat{\underline{x}}(t) \tag{7}$$

where μ_x is the mean displacement vector, and $\hat{\underline{x}}(t)$ is a vector of stationary, zero-mean processes, which in general are not Gaussian. Substituting equation (7) into equation (1) gives

$$\mathbf{m}\,\ddot{\hat{\underline{x}}} + \mathbf{c}\,\dot{\hat{\underline{x}}} + \mathbf{k}\,(\mu_x + \hat{\underline{x}}) + \underline{g}(\mu_x + \hat{\underline{x}}, \dot{\hat{\underline{x}}}) = \underline{f}(t) \tag{8}$$

Ensemble averaging equation (8) leads to the following equation for the mean response

$$\mu_x = \mathbf{k}^{-1}[\mu_f - \langle \underline{g}(\mu_x + \hat{\underline{x}}, \dot{\hat{\underline{x}}})\rangle] \tag{9}$$

This equation represents an exact relation, which can be evaluated if the joint probability distributions are known. However, the exact response probability distribution are generally not known and have to be estimated. Statistical linearization offers one such method for approximating the response probability distributions, however, this method suffers from the previously stated drawbacks. Thus, the idea is to replace the original nonlinear mdof system by an "equivalent quadratic" system. Although this system cannot be treated exactly as an "equivalent linear" system could, it can be investigated approximately by the Volterra series method. The equivalent quadratic system is constructed as follows

$$\begin{aligned}\mathbf{m}\,\ddot{\hat{\underline{x}}} &+ \mathbf{c}\,\dot{\hat{\underline{x}}} + \mathbf{k}\,\hat{\underline{x}} + \boldsymbol{\alpha}_1\hat{\underline{x}} + \boldsymbol{\alpha}_2\dot{\hat{\underline{x}}} + \boldsymbol{\alpha}_3\{[\hat{\underline{x}},\hat{\underline{x}}] - \langle[\hat{\underline{x}},\hat{\underline{x}}]\rangle\} \\ &+ \boldsymbol{\alpha}_4\{[\dot{\hat{\underline{x}}},\dot{\hat{\underline{x}}}] - \langle[\dot{\hat{\underline{x}}},\dot{\hat{\underline{x}}}]\rangle\} + \frac{1}{2}\boldsymbol{\alpha}_5\{[\hat{\underline{x}},\dot{\hat{\underline{x}}}] + [\dot{\hat{\underline{x}}},\hat{\underline{x}}] - \langle[\hat{\underline{x}},\dot{\hat{\underline{x}}}]\rangle - \langle[\dot{\hat{\underline{x}}},\hat{\underline{x}}]\rangle\} = \underline{f} - \mu_f\end{aligned} \tag{10}$$

where the $\boldsymbol{\alpha}_i$'s are the quadratization coefficient matrices. The special notation $[\cdot,\cdot]$ denotes a vector obtained by the following operation

$$[\hat{\underline{x}}_1, \hat{\underline{x}}_2] = \sum_{i=1}^{N} \hat{\underline{x}}_1 \hat{\underline{x}}_2^T \underline{e}_i \tag{11}$$

where the vector $\underline{e}_i$ is the ith column of the N×N identity matrix. It is noted that the equivalent stochastic quadratization method reduces to the equivalent stochastic linearization method for the special case when $\boldsymbol{\alpha}_3 = \boldsymbol{\alpha}_4 = \boldsymbol{\alpha}_5 \equiv \mathbf{0}$.

The quadratization coefficient matrices are computed such that

$$\langle \underline{\varepsilon}^T \underline{\varepsilon} \rangle = \text{minimum} \tag{12}$$

where the error vector, $\underline{\varepsilon}$, is defined by the equation

$$\begin{aligned} \underline{\varepsilon} \equiv \mathbf{k}\underline{\mu}_x + \underline{g}(\underline{\mu}_x+\underline{\hat{x}},\underline{\dot{\hat{x}}}) - \underline{\mu}_f - \boldsymbol{\alpha}_1\underline{\hat{x}} - \boldsymbol{\alpha}_2\underline{\dot{\hat{x}}} - \boldsymbol{\alpha}_3\{[\underline{\hat{x}},\underline{\hat{x}}]-\langle[\underline{\hat{x}},\underline{\hat{x}}]\rangle\} \\ - \boldsymbol{\alpha}_4\{[\underline{\dot{\hat{x}}},\underline{\dot{\hat{x}}}]-\langle[\underline{\dot{\hat{x}}},\underline{\dot{\hat{x}}}]\rangle\} - \frac{1}{2}\boldsymbol{\alpha}_5\{[\underline{\hat{x}},\underline{\dot{\hat{x}}}] + [\underline{\dot{\hat{x}}},\underline{\hat{x}}] - \langle[\underline{\hat{x}},\underline{\dot{\hat{x}}}]\rangle - \langle[\underline{\dot{\hat{x}}},\underline{\hat{x}}]\rangle\} \end{aligned} \tag{13}$$

Performing a mean square minimization on the error, and accounting for the stationarity of the responses it can be shown that

$$\langle \underline{\tilde{x}}\underline{\tilde{x}}^T \rangle \underline{\alpha}_i = \langle g_i \underline{\tilde{x}} \rangle \qquad i = 1 \text{ to } N \tag{14}$$

where

$$\begin{aligned} \underline{\tilde{x}}^T = \Big(\underline{\hat{x}}^T \mid \underline{\dot{\hat{x}}}^T \mid \{[\underline{\hat{x}},\underline{\hat{x}}]-\langle[\underline{\hat{x}},\underline{\hat{x}}]\rangle\}^T \mid \{[\underline{\dot{\hat{x}}},\underline{\dot{\hat{x}}}]-\langle[\underline{\dot{\hat{x}}},\underline{\dot{\hat{x}}}]\rangle\}^T \mid \\ \frac{1}{2}\{[\underline{\hat{x}},\underline{\dot{\hat{x}}}]+[\underline{\dot{\hat{x}}},\underline{\hat{x}}]-\langle[\underline{\hat{x}},\underline{\dot{\hat{x}}}]\rangle-\langle[\underline{\dot{\hat{x}}},\underline{\hat{x}}]\rangle\}^T\Big) \end{aligned} \tag{15}$$

and

$$\underline{\alpha}_i^T = \underline{e}_i^T (\alpha_1 \mid \alpha_2 \mid \alpha_3 \mid \alpha_4 \mid \alpha_5) \tag{16}$$

The notation, $\mid$, denotes matrix partition. The N linear systems of 5N equations given by equation (14) must be solved to obtain the quadratization coefficients. The left hand side matrix, whose elements consist of the joint central moments of the response, is the same for each of the N system of equations, but the right hand side vector is different.

The expectation on the left hand side of equation (14) involves joint central moments of $\underline{\hat{x}}$ and $\underline{\dot{\hat{x}}}$ up to fourth order. Towards this end, the representation of the response of the nonlinear equivalent system in the time and frequency domain is pursued in the next section.

Representation of the Equivalent Quadratic System

Time Domain Formulation

The joint central moments are evaluated from the response of the nonlinear equivalent system. Following the standard procedure for using the Volterra series, see Schetzen(1980), the scalar coefficient λ is introduced for bookkeeping purposes; for numerical computation its value is set equal to one. In this context, the zero-mean force $\underline{f}(t) - \underline{\mu}_f$ is replaced by $\lambda(\underline{f}(t) - \underline{\mu}_f)$ without altering the equivalent quadratic system. The nonlinear equivalent system in equation (10) is thus rewritten as

$$\begin{aligned} \mathbf{m}\,\underline{\ddot{\hat{x}}} + \mathbf{c}_{eq}\underline{\dot{\hat{x}}} + \mathbf{k}_{eq}\underline{\hat{x}} = \lambda(\underline{f}(t) - \underline{\mu}_f) - \boldsymbol{\alpha}_3\{[\underline{\hat{x}},\underline{\hat{x}}]-\langle[\underline{\hat{x}},\underline{\hat{x}}]\rangle\} \\ - \boldsymbol{\alpha}_4\{[\underline{\dot{\hat{x}}},\underline{\dot{\hat{x}}}]-\langle[\underline{\dot{\hat{x}}},\underline{\dot{\hat{x}}}]\rangle\} - \frac{1}{2}\boldsymbol{\alpha}_5\{[\underline{\hat{x}},\underline{\dot{\hat{x}}}] + [\underline{\dot{\hat{x}}},\underline{\hat{x}}] - \langle[\underline{\hat{x}},\underline{\dot{\hat{x}}}]\rangle - \langle[\underline{\dot{\hat{x}}},\underline{\hat{x}}]\rangle\} \end{aligned} \tag{17}$$

where

$$\mathbf{k}_{eq} = \mathbf{k} + \boldsymbol{\alpha}_1 \tag{18}$$

$$\mathbf{c}_{eq} = \mathbf{c} + \boldsymbol{\alpha}_2 \tag{19}$$

Following the Volterra series method, it is assumed that the solution to equation (17) can be written as an infinite series in the following form involving powers of the parameter, λ, and unknown functions of time

$$\hat{\underline{x}}(t) = \sum_{j=1}^{\infty} \lambda^j \hat{\underline{x}}^{(j)}(t) \tag{20}$$

For numerical computation, this series must be truncated. Clearly, terms up to at least j=2 must be included to account for the effect of the quadratic terms. Therefore, terms after j=2 are neglected. Substituting the truncated form of equation (20) into equation (17) and equating like powers of λ gives two linear differential equations: one operating on $\hat{\underline{x}}^{(1)}$ and the other one operating on $\hat{\underline{x}}^{(2)}$. The differential operator in both equations is the same; hence, due to linearity the two equations can be combined into one equation as

$$\mathbf{m}\,\ddot{\hat{\underline{x}}} + \mathbf{c}_{eq}\dot{\hat{\underline{x}}} + \mathbf{k}_{eq}\hat{\underline{x}} = \hat{\underline{f}}(t) = \hat{\underline{f}}^{(0)} + \hat{\underline{f}}^{(1)} + \hat{\underline{f}}^{(2)} \tag{21}$$

where

$$\hat{\underline{f}}^{(0)} = \alpha_3<[\hat{\underline{x}}^{(1)},\hat{\underline{x}}^{(1)}]> + \alpha_4<[\dot{\hat{\underline{x}}}^{(1)},\dot{\hat{\underline{x}}}^{(1)}]> + \frac{1}{2}\alpha_5\{<[\hat{\underline{x}}^{(1)},\dot{\hat{\underline{x}}}^{(1)}]> + <[\dot{\hat{\underline{x}}}^{(1)},\hat{\underline{x}}^{(1)}]>\} \tag{22}$$

$$\hat{\underline{f}}^{(1)} = \underline{f}(t) - \mu_f \tag{23}$$

$$\hat{\underline{f}}^{(2)} = -\alpha_3[\hat{\underline{x}}^{(1)},\hat{\underline{x}}^{(1)}] - \alpha_4[\dot{\hat{\underline{x}}}^{(1)},\dot{\hat{\underline{x}}}^{(1)}] - \frac{1}{2}\alpha_5\{[\hat{\underline{x}}^{(1)},\dot{\hat{\underline{x}}}^{(1)}] + [\dot{\hat{\underline{x}}}^{(1)},\hat{\underline{x}}^{(1)}]\} \tag{24}$$

Comparing equations (3) and (23), it is apparent that the force $\hat{\underline{f}}^{(1)}(t)$ is the zero-mean part of the linear filtered white-noise process. It follows from equation (24) that the force $\hat{\underline{f}}^{(2)}(t)$ is a quadratic transform of the white-noise process. Thus, the force $\hat{\underline{f}}(t)$ can be viewed as a second order Volterra series in which white noise is the input function. This is expressed mathematically by the equation

$$\hat{\underline{f}}(t) = \hat{\underline{f}}^{(0)} + \int_{-\infty}^{\infty} \underline{h}_f^{(1)}(\tau_1)\, w(t-\tau_1)\, d\tau_1 + \iint_{-\infty}^{\infty} \underline{h}_f^{(2)}(\tau_1,\tau_2)\, w(t-\tau_1)w(t-\tau_2)\, d\tau_1 d\tau_2 \tag{25}$$

where the functions $\underline{h}_f^{(1)}(\tau)$ and $\underline{h}_f^{(2)}(\tau_1,\tau_2)$ are respectively linear and quadratic Volterra kernel vectors for force. The forces $\hat{\underline{f}}^{(1)}$ and $\hat{\underline{f}}^{(2)}$ are referred to as the linear and quadratic Volterra forces respectively.

It is interesting to note that the equivalent nonlinear quadratic system defined in equation (10) has now been replaced by a linear system given by equation (21). This has been accomplished by approximating the nonlinear terms by a higher order excitation force. This transformation is important because solutions of linear systems can now be used. Consequently, the steady-state response of equation (21) can be written as

$$\hat{\underline{x}}(t) = \int_{-\infty}^{\infty} \mathbf{h}(\tau)\, \hat{\underline{f}}(t-\tau)\, d\tau \tag{26}$$

where the function $\mathbf{h}(\tau)$ is the impulse response function matrix for the linear differential operator. Substituting equation (25) into equation (26) leads to the following Volterra series representation for the system response

$$\hat{\underline{x}}(t) = \hat{\underline{x}}^{(0)} + \int_{-\infty}^{\infty} \underline{h}_{\hat{x}}^{(1)}(\tau_1)\, w(t-\tau_1)\, d\tau_1 + \iint_{-\infty}^{\infty} \underline{h}_{\hat{x}}^{(2)}(\tau_1,\tau_2)\, w(t-\tau_1)w(t-\tau_2)\, d\tau_1 d\tau_2 \tag{27}$$

where

$$\hat{\underline{x}}^{(0)} = \hat{\underline{f}}^{(0)} \int_{-\infty}^{\infty} \mathbf{h}(\tau)\, d\tau \tag{28}$$

$$\underline{h}_{\hat{x}}^{(1)}(\tau_1) = \int_{-\infty}^{\infty} \mathbf{h}(\tau)\, \underline{h}_{\hat{f}}^{(1)}(\tau_1-\tau)\, d\tau \tag{29}$$

$$\underline{h}_{\hat{x}}^{(2)}(\tau_1,\tau_2) = \int_{-\infty}^{\infty} \mathbf{h}(\tau)\, \underline{h}_{\hat{f}}^{(2)}(\tau_1-\tau,\tau_2-\tau)\, d\tau \tag{30}$$

It follows from equations (22) and (24) that the zero order force, $\hat{\underline{f}}^{(0)}$, is the negative of the expected value of the second order force, $\hat{\underline{f}}^{(2)}$. Therefore, $<\hat{\underline{f}}> = \underline{0}$,. Further, due to linearity of the differential operator, $< \hat{\underline{x}} > = \underline{0}$ as formulated.

From equations (2) and (23) it follows that the linear kernel is the same as the linear impulse response function of the white-noise filter. The quadratic kernel of the Volterra force is found by substituting the linear component of equation (27) into equation (24) and taking into consideration equation (25). Specifically, it is found that

$$\underline{h}_{\hat{f}}^{(2)}(\tau_1,\tau_2) = -\alpha_3 [\underline{h}_{\hat{x}}^{(1)}(\tau_1), \underline{h}_{\hat{x}}^{(1)}(\tau_2)] - \alpha_4 [\underline{h}_{\dot{\hat{x}}}^{(1)}(\tau_1), \underline{h}_{\dot{\hat{x}}}^{(1)}(\tau_2)]$$

$$- \frac{1}{2}\alpha_5 \{ [\underline{h}_{\hat{x}}^{(1)}(\tau_1), \underline{h}_{\dot{\hat{x}}}^{(1)}(\tau_2)] + [\underline{h}_{\dot{\hat{x}}}^{(1)}(\tau_1), \underline{h}_{\hat{x}}^{(1)}(\tau_2)] \} \tag{31}$$

The kernel $\underline{h}_{\dot{\hat{x}}}^{(1)}(\tau)$ is the linear Volterra kernel for $\dot{\hat{\underline{x}}}(t)$ and is the time derivative of $\underline{h}_{\hat{x}}^{(1)}(\tau)$.

Frequency Domain Formulation

In the ensuing analysis, all numerical computations are performed in the frequency domain. Thus, for the Volterra series in equations (25) and (27), only the transfer functions need to be computed.

As it has already been stated, the force transfer function vector $\underline{H}_{\hat{f}}^{(1)}(\omega)$ is the same as the transfer function of the white-noise filter transfer function. Thus, it is related to the linear kernel by the one dimensional Fourier transform relation defined in equation (6). Using the linear transform and equation (29) the transfer function for the linear displacement is derived as

$$\underline{H}_{\hat{x}}^{(1)}(\omega) = \mathbf{H}(\omega)\underline{H}_{\hat{f}}^{(1)}(\omega) \tag{32}$$

In equation (32), the transfer function matrix $\mathbf{H}(\omega)$ is the frequency domain counterpart of $\mathbf{h}(\tau)$ and is computed by the equation

$$H(\omega) = [-\omega^2 m + i\omega c_{eq} + k_{eq}]^{-1} \tag{33}$$

Linear systems theory also yields the following equation for the velocity response transfer function vector

$$\underline{H}_{\dot{x}}^{(1)}(\omega) = i\omega\, \underline{H}_{x}^{(1)}(\omega) \tag{34}$$

The quadratic transfer functions and quadratic kernels are related by a two dimensional form of the Fourier transform. Specifically, the quadratic Volterra force transfer function and kernel are related by the equation

$$\underline{H}_{f}^{(2)}(\omega_1,\omega_2) = \iint_{-\infty}^{\infty} \underline{h}_{f}^{(2)}(\tau_1,\tau_2) \exp(-i\omega_1\tau_1) \exp(-i\omega_2\tau_2)\, d\tau_1 d\tau_2 \tag{35}$$

The quadratic transfer functions for displacement, $\underline{H}_{x}^{(2)}(\omega_1,\omega_2)$, and velocity, $\underline{H}_{\dot{x}}^{(2)}(\omega_1,\omega_2)$, are related to their corresponding kernels by the same Fourier relation.

The quadratic transfer function for the Volterra force is found by substituting equation (31) into equation (35) and taking into consideration equation (34). This procedure leads to the following expression

$$\underline{H}_{f}^{(2)}(\omega_1,\omega_2) = [\,-\alpha_3 + \alpha_4\,\omega_1\omega_2 - \tfrac{i}{2}\alpha_5(\omega_1+\omega_2)][\underline{H}_{x}^{(1)}(\omega_1).\underline{H}_{x}^{(1)}(\omega_2)] \tag{36}$$

The quadratic transfer function for the displacement is found by transforming equation (30) into the frequency domain using the two dimensional Fourier transform. After some mathematical manipulation this procedure yields

$$\underline{H}_{x}^{(2)}(\omega_1,\omega_2) = H(\omega_1+\omega_2)\underline{H}_{f}^{(2)}(\omega_1,\omega_2) \tag{37}$$

It can be shown that the quadratic transfer functions for displacement and velocity are related by the following equation

$$\underline{H}_{\dot{x}}^{(2)}(\omega_1,\omega_2) = i(\omega_1+\omega_2)\, \underline{H}_{x}^{(2)}(\omega_1,\omega_2) \tag{38}$$

Inspecting equation (36), it is seen that the force quadratic transfer function possesses the following symmetries

$$\underline{H}_{f}^{(2)*}(\omega_1,\omega_2) = \underline{H}_{f}^{(2)}(-\omega_1,-\omega_2) \tag{39}$$

$$\underline{H}_{f}^{(2)}(\omega_1,\omega_2) = \underline{H}_{f}^{(2)}(\omega_2,\omega_1) \tag{40}$$

where (*) denotes complex conjugation. It follows from equations (33), (37) and (38) that the response quadratic transfer functions have the same properties. This fact greatly reduces the requisite computational effort and memory capacity for applying the proposed method to a specific problem.

Response Statistics Computation

In general, the joint probability distribution of $\hat{\underline{x}}(t)$ is not Gaussian. Approximate joint distributions can be obtained from a Gram-Charlier expansion which utilizes the partial derivatives of a joint Gaussian distribution, as described in Johnson and Kotz[10]. Specifically, the joint probability distribution, $p(\underline{\xi})$, between the N displacement components and their corresponding velocities of the equivalent system is approximated by the following relation

$$p(\underline{\xi}) = \sum_{j_1=0}^{\infty} \sum_{j_2=0}^{\infty} \cdots \sum_{j_{2N}=0}^{\infty} r_{j_1 j_2 \cdots j_{2N}} \frac{\partial^{j_T} \Phi(\underline{\xi})}{\partial \xi_1^{j_1} \partial \xi_2^{j_2} \cdots \partial \xi_{2N}^{j_{2N}}} \tag{41}$$

where $j_T = j_1 + j_2 + \cdots + j_{2N}$. The symbol $\underline{\xi}$ denotes the state space vector

$$\underline{\xi}^T = (\hat{x}_1 \mid \dot{\hat{x}}_1 \mid \hat{x}_2 \mid \dot{\hat{x}}_2 \mid \cdot \mid \cdot \mid \hat{x}_N \mid \dot{\hat{x}}_N) \tag{42}$$

The function $\Phi(\underline{\xi})$ is a joint Gaussian distribution given by the equation

$$\Phi(\underline{\xi}) = (2\pi)^{-N} |S|^{-1/2} \exp[-\tfrac{1}{2}\underline{\xi}^T S^{-1} \underline{\xi}] \tag{43}$$

where the matrix S is the covariance matrix for the state space vector.

For a Gram-Charlier expansion, the coefficients $r_{j_1 j_2 \cdots j_{2N}}$ are related to the cumulants of the distribution by the following equation taken from Beaman and Hedrick[7]

$$r_{j_1 j_2 \cdots j_{2N}} = \begin{cases} 1 & j_T = 0 \\ 0 & 0 < j_T < 3 \\ (-1)^{j_T} \dfrac{\kappa_{j_1 j_2 \cdots j_{2N}}}{j_1!\, j_2! \ldots j_{2N}!} & j_T \geq 3 \end{cases} \tag{44}$$

The cumulants are related to the joint central moments through the joint characteristic function, see Nigam[11]. The central moments of $\underline{\xi}$ are defined by the equation

$$\mu_{j_1 j_2 \cdots j_{2N}} = \langle \prod_{i=1}^{2N} \xi_i^{j_i} \rangle \tag{45}$$

They are computed by substituting the Volterra series given by equation (27) into equation (45) and transforming into the frequency domain. For example, the relation for an arbitrary second moment is obtained as

$$\langle \xi_i \xi_j \rangle = \int_{-\infty}^{\infty} H_{\xi_i}^{(1)*}(\omega) H_{\xi_j}^{(1)}(\omega)\, d\omega + 2 \iint_{-\infty}^{\infty} H_{\xi_i}^{(2)*}(\omega_1,\omega_2) H_{\xi_j}^{(2)}(\omega_1,\omega_2)\, d\omega_1 d\omega_2 \tag{46}$$

where $H_{\xi_i}^{(1)}$ and $H_{\xi_i}^{(2)}(\omega_1,\omega_2)$ are linear and quadratic transfer functions of the state space variable ξ_i. In a similar manner, an arbitrary third order central moment is obtained by the equation

$$\begin{aligned}<\xi_i\xi_j\xi_k> = {} & 2\iint_{-\infty}^{\infty} H_{\xi_i}^{(1)*}(\omega_1)\, H_{\xi_j}^{(1)*}(\omega_2)\, H_{\xi_k}^{(2)}(\omega_1,\omega_2)\, d\omega_1 d\omega_2 \\ & + 2\iint_{-\infty}^{\infty} H_{\xi_i}^{(1)*}(\omega_1)\, H_{\xi_k}^{(1)*}(\omega_2)\, H_{\xi_j}^{(2)}(\omega_1,\omega_2)\, d\omega_1 d\omega_2 \\ & + 2\iint_{-\infty}^{\infty} H_{\xi_j}^{(1)*}(\omega_1)\, H_{\xi_k}^{(1)*}(\omega_2)\, H_{\xi_i}^{(2)}(\omega_1,\omega_2)\, d\omega_1 d\omega_2 \\ & + 8\iiint_{-\infty}^{\infty} H_{\xi_i}^{(2)*}(\omega_1,\omega_2)\, H_{\xi_j}^{(2)}(\omega_1,-\omega_3)\, H_{\xi_k}^{(2)}(\omega_2,\omega_3)\, d\omega_1 d\omega_2 d\omega_3 \end{aligned} \tag{47}$$

Since the computation of moments higher than third order becomes quite costly, the distribution in equation (41) is truncated beyond terms of order three. Although response moments higher than third order are not needed for the truncated joint probability distribution, they are needed to solve for the quadratization coefficient matrices using equation (14). These higher order moments are computed by using the truncated joint distribution. From the characteristic function of this joint distribution it can be shown that a general fourth order central moment is given by the equation

$$<\xi_i\xi_j\xi_k\xi_l> = <\xi_i\xi_j><\xi_k\xi_l> + <\xi_i\xi_k><\xi_j\xi_l> + <\xi_i\xi_l><\xi_j\xi_k> \tag{48}$$

In addition to the response statistics, the response spectral density matrix, $S_{\hat{\underline{x}}\hat{\underline{x}}}(\omega)$, is often of interest. Using the expression for $\hat{\underline{x}}(t)$ given by equation (27) and the definition of spectral density given by equation (4) results in the following well known equation for linear systems

$$\mathbf{S}_{\hat{\underline{x}}\hat{\underline{x}}}(\omega) = \mathbf{H}^*(\omega)\, \mathbf{S}_{\hat{\underline{f}}\hat{\underline{f}}}(\omega)\, \mathbf{H}^T(\omega) \tag{49}$$

where $S_{\hat{\underline{f}}\hat{\underline{f}}}(\omega)$ is the spectral density matrix of the force $\hat{\underline{f}}(t)$. The force spectral density matrix is computed similarly by using equation (25) for $\hat{\underline{f}}(t)$. This leads to

$$S_{\hat{\underline{f}}\hat{\underline{f}}}(\omega) = \underline{H}_{\hat{f}}^{(1)*}(\omega)\underline{H}_{\hat{f}}^{(1)T}(\omega) + 2\int_{-\infty}^{\infty} \underline{H}_{\hat{f}}^{(2)*}(\omega_1,\omega-\omega_1)\underline{H}_{\hat{f}}^{(2)T}(\omega_1,\omega-\omega_1)\, d\omega_1 \tag{50}$$

Thus, the force spectral density matrix consists of two parts. The first part is the spectral density of the linear filtered white-noise force, $\hat{\underline{f}}^{(1)}(t)$. The second part is the spectral density of the quadratic force, $\hat{\underline{f}}^{(2)}(t)$, and involves a frequency domain convolution. The convolution introduces spectral contributions in the force $\hat{\underline{f}}(t)$ which are at frequencies that are the difference and the sum of the frequencies of the filtered white noise.

<u>Simplified Solution for Systems of a Special Form</u>

The most computationally intensive part of the described method is in the computation of the joint third order moments required for equation (14). However, for systems with nonlinearities of a special form the requisite number of joint third order moments to be computed can be greatly reduced. Specifically, the case in which the terms of the nonlinear force vector, $\underline{g}(\underline{x},\dot{\underline{x}})$, have the following simplified form is considered

$$g_i(\underline{x},\dot{\underline{x}}) = g_i(x_i,\dot{x}_i) \tag{51}$$

For the simplified form of the nonlinearity given by equation (51), the equivalent quadratic system is again given by equation (10), except that now the notation [·,·] denotes a vector obtained by the following operation

$$[\hat{\underline{x}}_1,\hat{\underline{x}}_2] = \sum_{i=1}^{N} \mathrm{Diag}(\hat{\underline{x}}_1\hat{\underline{x}}_2^T)\,\underline{e}_i \tag{52}$$

where the notation Diag( ) represents an matrix operation in which all off-diagonals terms are made equal to zero. The unknown coefficient matrices, $\boldsymbol{\alpha}$'s, are now diagonal matrices. In this case, the linear system of equations represented by equation (14) reduces from a 5Nx5N system to a 5x5 system where the vector $\underline{\tilde{x}}$ is now given by the equation

$$\underline{\tilde{x}}^T = \left(\hat{x}_i \mid \dot{\hat{x}}_i \mid (\hat{x}_i^2-<\hat{x}_i^2>) \mid (\dot{\hat{x}}_i^2-<\dot{\hat{x}}_i^2>) \mid \hat{x}_i\dot{\hat{x}}_i\right) \tag{53}$$

and the vector $\underline{\alpha}$ is

$$\underline{\alpha}^T = (\alpha_{1i} \mid \alpha_{2i} \mid \alpha_{3i} \mid \alpha_{4i} \mid \alpha_{5i}) \tag{54}$$

where α_{ji} refers to the ith diagonal term of the $\boldsymbol{\alpha}_j$ matrix. Thus, only moments of the ith degree of freedom are needed to determine the quadratization coefficients associated with that degree of freedom. Joint moments of different degrees of freedom may still be of interest, but would not need to be computed until after the iteration process has converged.

The following Gram-Charlier expansion for the joint probability distribution of the displacement and velocity of the ith degree of freedom is used to solve for the expectations on the right hand side of equation (14) and the mean displacement given by equation (9)

$$p(\hat{x}_i,\dot{\hat{x}}_i) = \left\{ 1 - \frac{1}{6}\mu_{\hat{x}_i^3}\frac{\partial^3}{\partial\hat{x}_i^3} - \frac{1}{6}\mu_{\dot{\hat{x}}_i^3}\frac{\partial^3}{\partial\dot{\hat{x}}_i^3} - \frac{1}{2}\mu_{\hat{x}_i\dot{\hat{x}}_i^2}\frac{\partial^3}{\partial\hat{x}_i\partial\dot{\hat{x}}_i^2}\right\}\phi(\hat{x}_i)\,\phi(\dot{\hat{x}}_i) \tag{55}$$

Here the joint central moments, $\mu_{\hat{x}_i^m\dot{\hat{x}}_i^n}$, are defined as

$$\mu_{\hat{x}_i^m\dot{\hat{x}}_i^n} = <\hat{x}_i^m\,\dot{\hat{x}}_i^n> \tag{56}$$

and $\phi(\hat{x}_i)$ and $\phi(\dot{\hat{x}}_i)$ are the following one dimensional Gaussian distributions

$$\phi(\hat{x}_i) = \frac{1}{\sqrt{2\pi\mu_{\hat{x}_i^2}}}\exp\left(-\frac{\hat{x}_i^2}{2\mu_{\hat{x}_i^2}}\right) \tag{57}$$

$$\phi(\dot{\hat{x}}_i) = \frac{1}{\sqrt{2\pi\mu_{\dot{\hat{x}}_i^2}}}\exp\left(-\frac{\dot{\hat{x}}_i^2}{2\mu_{\dot{\hat{x}}_i^2}}\right) \tag{58}$$

Note that the distribution in equation (55) has been simplified to reflect the fact that for stationary responses $\mu_{\hat{x}_i^2\dot{\hat{x}}_i}$ and $\mu_{\hat{x}_i\dot{\hat{x}}_i} = 0$.

Solution Procedure

Clearly, the implementation of the preceding solution scheme for either the general or special case requires an iterative procedure since the quadratization coefficient matrices are initially unknown. The procedure starts by assuming initial values for the quadratization coefficient matrices. Next the central moments required for the solution of equation (14) are computed using equations (46) and (47). Solving equation (14) yields new quadratization matrices. Using the updated the quadratization coefficient matrices, new values for the moments of the system response are computed. This procedure is repeated until a preset solution convergence criterion is satisfied.

Example of Application

The discussed quadratization procedure has been used for the two-degree-of-freedom system with energy dissipating nonlinearities shown in Figure 1. The nonlinear damping is given by the following equation

$$g_i(\dot{x}_i) = \gamma_i |a_i + \dot{x}_i|(a_i + \dot{x}_i) \qquad i = 1, 2 \tag{59}$$

For this kind of nonlinearity the simplified solution procedure can be used. The following numerical values are used: $k_1 = 1$, $k_2 = 1$, $m_1 = 1$, $m_2 = 1$, $c_1 = .1$, $c_2 = .1$, $\gamma_1 = 1$, $\gamma_2 = 1$, and $a_1 = a_2 = a$ where a is a parameter to be varied. In this example, the white-noise filter of the excitation force vector, $\underline{f}(t)$, is given by the equation

$$H_{f_i}(\omega) = \begin{cases} \dfrac{1}{\sqrt{2(\omega_b - \omega_a)}} & \omega_a \le |\omega| \le \omega_b \\ 0 & \text{everywhere else} \end{cases} \tag{60}$$

for i = 1, 2. The elements of the covariance matrix for the excitation are all equal to one. The mean force vector, μ_f, is assumed to be equal to zero.

According to the described quadratization procedure, the mean response is computed by the equation

$$\mu_x = -\begin{pmatrix} 2 & -1 \\ -1 & 1 \end{pmatrix}^{-1} \begin{pmatrix} \langle |a + \dot{x}_1|(a + \dot{x}_1)\rangle \\ \langle |a + \dot{x}_2|(a + \dot{x}_2)\rangle \end{pmatrix} \tag{61}$$

and the equivalent quadratic system is given by the equation

$$\mathbf{m}\,\ddot{\hat{\underline{x}}} + \mathbf{c}\,\dot{\hat{\underline{x}}} + \mathbf{k}\,\hat{\underline{x}} + \alpha_2 \dot{\hat{\underline{x}}} + \alpha_4\{[\dot{\hat{\underline{x}}},\dot{\hat{\underline{x}}}] - \langle[\dot{\hat{\underline{x}}},\dot{\hat{\underline{x}}}]\rangle\} = \underline{f}(t) \tag{62}$$

Analytical expressions for the expectations in equation (61) and the right hand side vector in equation (14) can be derived. For further details the reader is referred to Donley and Spanos[1].

Numerical results have been obtained for the two-degree-of-freedom system using both equivalent statistical linearization and quadratization. Further, Monte Carlo simulation has been performed to assess the accuracy of these results. The simulation is performed by using the sum of harmonics method, see Borgman[12], to produce 300 realizations. The linear filter has the frequency parameters: $\omega_a = 4$ and $\omega_b = 8$. Note that if there were no nonlinear forces in this system (i.e. $\gamma_1 = \gamma_2 = 0$), its natural frequencies would be $\omega_1 = .618$ rad/sec and $\omega_2 = 1.618$ rad/sec. Both of these frequencies are well away from the chosen excitation frequency range.

For the velocity offset, a, equal to .3, the displacement auto-spectral densities of mass 1 and mass 2 in one-sided form are shown in Figures 2 and 3, respectively. Both of these response spectral densities exhibit two peaks. This result is best explained in terms of the spectral density of the Volterra series force given by equation (50). The linear part of the spectral density is due entirely to the input excitation and thus spans the frequency range of 4 to 8 rad/sec. The spectral density of the quadratic force is obtained by convolutions of the filtered white-noise spectrum. This produces low frequency contributions from 0 to 4 rad/sec and high frequency contributions from 8 to 16 rad/sec. In general, the magnitude of the quadratic force is several orders less than the linear force. However, in this case the low frequency response is significant relative to the linear response because of the first resonance frequency. A resonance peak is not seen at the second natural frequency because the magnitude of the quadratic force has tapered off considerably. It can be seen that the spectra obtained by simulation and quadratization are in quite good agreement over the entire frequency range. Linearization on the other hand yields good results for the linear response but can not predict the quadratic order response.

The probability distributions for both masses are shown in Figures 4 and 5. The vertical lines mark the mean displacements. The distributions derived by simulation and quadratization are in reasonably good agreement and are clearly non-Gaussian as evidenced by their skewness. The distribution obtained by the linearization solution is Gaussian and differs significantly from the simulation distribution. It is noted that the correlation between the distributions obtained by simulation and quadratization is better near the mean than in the tails of the distribution. In fact, the tail of the distributions obtained by quadratization becomes negative over a small range. This of course is physically impossible, but can be produced mathematically from the Gram-Charlier expansion. This result is attributable to an insufficient number of moments used to accurately describe the tail of the distributions rather than inaccuracies in the computed moments. From a design viewpoint, this means that the analytical distribution would not be appropriate for reliability analyses, but is suitable for fatigue analyses. In any case, the non-Gaussian distributions derived by statistical quadratization are a substantial improvement over the Gaussian distributions obtained by linearization.

In Figures 6 to 8, the mean, standard deviation, and skewness of the displacements of mass 1 and 2 are plotted for a range of the velocity offset parameter. The mean responses obtained by all methods are in good agreement. On the other hand, linearization underpredicts the standard deviation by as much as 100% in some cases because it does not account for quadratic effects. It is noted that the linearization and quadratization statistics coincide for the value of a = 0. At this value the nonlinearity is perfectly symmetric. Consequently, the equivalent quadratic system has only linear terms. It is noted that the analytical standard deviations at a = 0 disagree slightly with the simulation standard deviations. This is perhaps due to higher order responses which are not considered.

The quadratization method yields estimates for the skewness which have the correct trend but are somewhat high. Linearization fails to exhibit any skewness since it induces a Gaussian probability distribution.

Summary and Conclusions

A stochastic quadratization method has been developed for stationary response analysis of nonlinear multi-degree-of-freedom systems subject to filtered Gaussian white-noise excitation. The stationary response is written as the sum of a mean time-independent response and a stationary zero-mean response which is assumed to be non-Gaussian. The mean response is determined from expectations involving the original nonlinear equation. The zero-mean response is obtained as the solution to an "equivalent" quadratic system which is constructed by replacing the nonlinearity with polynomials up to quadratic order. The equivalent system has a form whose solution can be approximated by using the Volterra series method. The coefficients of the polynomials are determined by invoking a mean square minimization criterion. A third order Gram-Charlier expansion is used to describe the joint non-Gaussian response probability distribution, and the system response statistics are determined in an iterative manner.

The proposed method, developed for systems with general nonlinearities, can be cumbersome since many higher order moments need to be computed. Therefore, a version which requires less computational effort is developed for nonlinearities of a special form. This version is applied on a system with two degrees of freedom. It is shown that significant resonance responses which can not be accounted for by linearization can be accurately predicted by quadratization. The quadratization results are obtained quite efficiently and compare well with time domain Monte Carlo simulation data.

References

1. M.G. Donley and P.D. Spanos, *Dynamic Analysis of Non-Linear Structures by the Method of Statistical Quadratization*, Lecture Notes in Engineering **57**, Springer-Verlag, New York, (1990).

2. T.K. Caughey, Equivalent linearization techniques, *J. Acoust. Soc. of Am.* **35**, 1706-1711 (1963).

3. W.D. Iwan and I.M. Yang, Application of statistical linearization techniques to nonlinear multidegree-of-freedom systems, *J. Appl. Mech.* **39**, 545-550 (1972).

4. T.S. Atalik and T. Utku, Stochastic linearization of multi-degree-of-freedom non-linear systems, *Earthquake Engng. Struct. Dyns.* **4** , 411-420 (1976).

5. P-T.D. Spanos, Formulation of stochastic linearization for symmetric or asymmetric m.d.o.f. nonlinear systems, *J. Appl. Mech.* **47**, 209-211 (1980).

6. P-T.D. Spanos, Monte Carlo simulations of responses of a non-symmetric dynamic system to random excitations, *Computers and Structs.* **13,** 371-376 (1981).

7. J.J. Beaman and J.K. Hedrick, Improved statistical linearization for analysis and control of nonlinear stochastic systems: Part I: An extended statistical linearization technique, *J. Dyn. Sys., Meas. and Con.* **102**, 14-21 (March 1981).

8. J.B. Roberts, and P.D. Spanos, *Random Vibrations and Statistical Linearization*, John Wiley, New York (1990).

9. M. Schetzen, *The Volterra and Wiener Theories of Nonlinear Systems,* John Wiley, New York, (1980).

10. N.L. Johnson and S. Kotz, *Distributions in Statistics: Continuous Multivariate Distributions*, 1972, John Wiley, New York (1972).

11. N.C. Nigam, *Introduction to Random Vibrations,* MIT Press, Cambridge, Massachusetts, (1983).

12. L.E. Borgman, Ocean wave simulation for engineering design, *ASCE J. Water. & Harb. Div.*, 557-583 (1969).

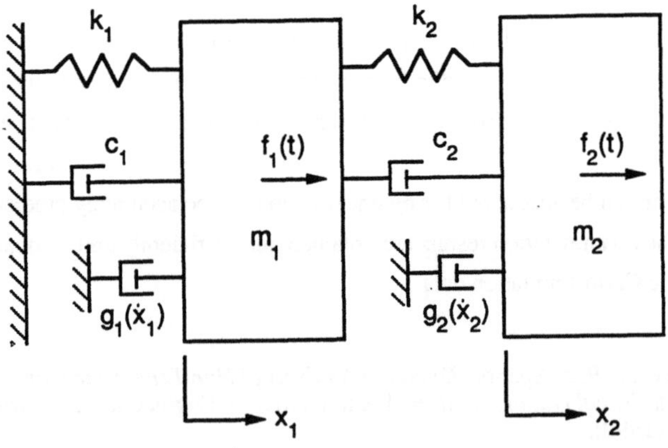

Figure 1. Nonlinear two-degree-of-freedom system.

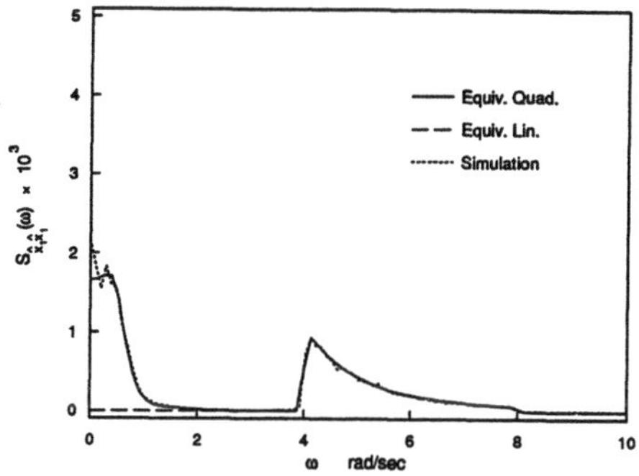

Figure 2. Displacement response spectral density for mass 1, a =.3.

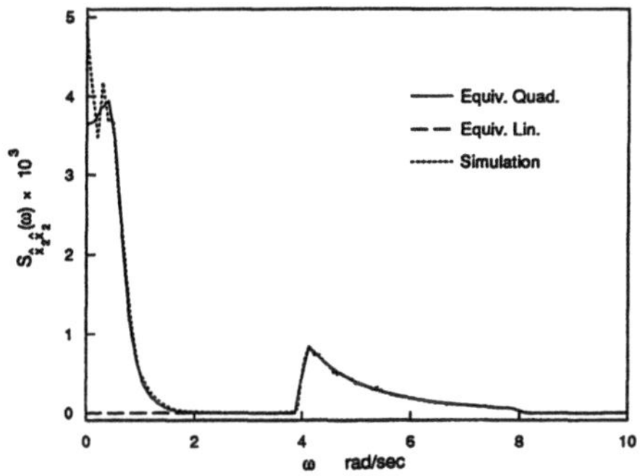

Figure 3. Displacement response spectral density for mass 2, a =.3.

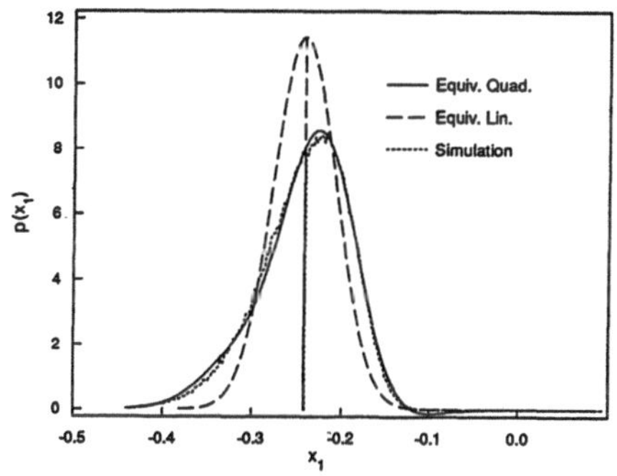

Figure 4. Displacement response probability distribution for mass 1 a = .3.

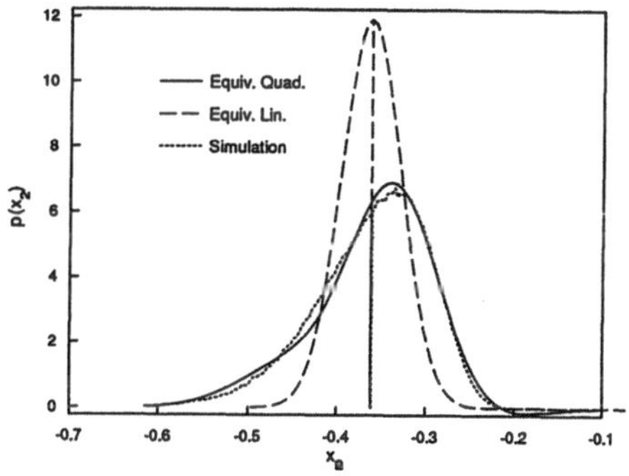

Figure 5. Displacement response probability distribution for mass 2 a = .3.

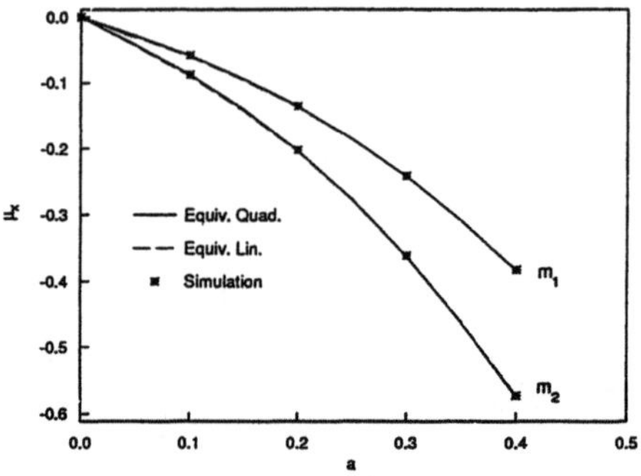

Figure 6. Mean displacement response versus velocity offset, a.

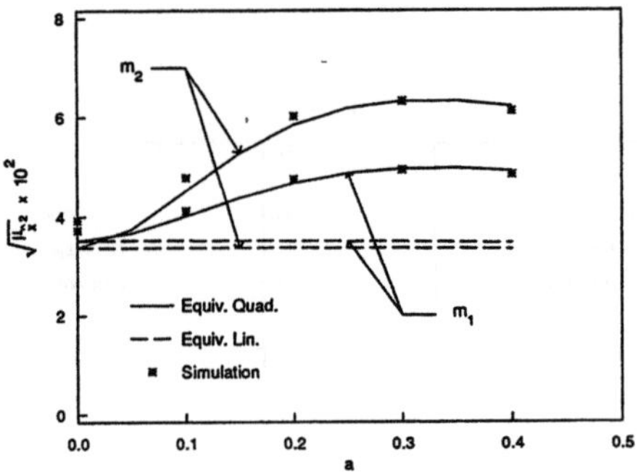

Figure 7. Standard deviation of displacement response versus velocity offset, a.

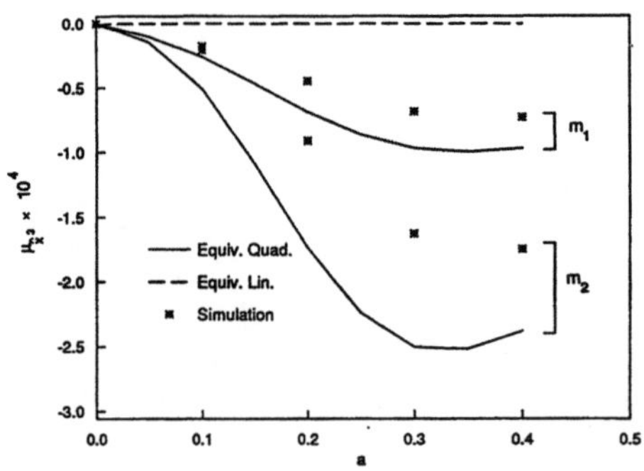

Figure 8. Skewness of displacement response versus velocity offset, a.

Comparison of the New Energy-Based Versions of the Stochastic Linearization Technique

Isaac Elishakoff and Ruichong Zhang

Center for Applied Stochastics Research
and Department of Mechanical Engineering
Florida Atlantic University
Boca Raton, FL 33431-0991, U.S.A

Summary

In this study a generalization of the stochastic linearization method is proposed; namely the nonlinear system is suggested to be replaced by a linear system equivalent to the original one in the following sense: The two systems should share common mean-square values of potential energies, as well as have coincident mean square values of energy dissipation function. An example of a system with nonlinear damping and nonlinear stiffness is numerically evaluated, to elucidate the proposed method.

Introduction

In order to elucidate new versions of the stochastic linearization method, it is instructive first to give a brief overview of its conventional version [1,2]. To do this, consider a single-degree-of-freedom system characterized by the following differential equation

$$\ddot{X} + g(X,\dot{X}) = F(t) \tag{1}$$

where $g(X,\dot{X})$ is a nonlinear deterministic function of the displacement X and velocity $\dot{X}$, dot denotes differentiation with respect to time; F(t) is a random excitation resulting in stochastic response X(t). A problem consists in determining the probabilistic characteristics of X(t) once those of F(t) are given.

Within the conventional stochastic linearization scheme, this differential equation is replaced by the equivalent linear equation

$$\ddot{X} + \beta_{eq}\dot{X} + k_{eq}X = F(t) \tag{2}$$

where the coefficients β_{eq} and k_{eq} are determined through some suitable equivalence criteria. Almost universally, the equivalence criterion consists of the requirement of minimum of the mean-square value of the error defined as

N. Bellomo · F. Casciati (Eds.)
Nonlinear Stochastic Mechanics
IUTAM Symposium Turin

$$\varepsilon = g(X,\dot{X}) - \beta_{eq}^{(1)}\dot{X} - k_{eq}^{(1)}X \tag{3}$$

resulting in two conditions that

$$\frac{\partial}{\partial\beta_{eq}^{(1)}}E(\varepsilon^2) = 0, \qquad \frac{\partial}{\partial k_{eq}^{(1)}}E(\varepsilon^2) = 0 \tag{4}$$

For F(t) we assume that it is a stationary, Gaussian random process with zero mean. Further analysis is based on the assumption that the displacement X and velocity $\dot{X}$ are Gaussian too. In addition, we assume that X and $\dot{X}$ are jointly stationary. These lead to the following expressions for β_e and k_{eq}, respectively:

$$\beta_{eq}^{(1)} = \frac{E[\dot{X}g(X,\dot{X})]}{E(\dot{X}^2)}, \qquad k_{eq}^{(1)} = \frac{E[Xg(X,\dot{X})]}{E(X^2)} \tag{5}$$

For the case of separable stiffness and damping in Eq. (1), namely, for

$$g(X,\dot{X}) = f(X) + h(\dot{X}) \tag{6}$$

we arrive at

$$\beta_{eq}^{(1)} = \frac{E[\dot{X}h(\dot{X})]}{E(\dot{X}^2)}, \qquad k_{eq}^{(1)} = \frac{E[Xf(X)]}{E(X^2)} \tag{7}$$

There exists a large amount of studies which implement this scheme of stochastic linearization.. For the detailed account one may consult with the articles by Spanos [3] and Crandall [4] and the recent monograph by Roberts and Spanos [5], which also contain pertinent bibliography.

Recently, for the case of nonlinear stiffness, i.e. when function h(X) is linear, but the function f(X) is nonlinear, two new criteria have been proposed: Zhang, Elishakoff and Zhang [6] suggested that the coefficient k_{eq} should be found from the requirement that the mean-square difference of potential energies of the original system, governed by Eq. (1), and of the imitating linear system, characterized by Eq. (2), be minimal, i.e. that

$$\frac{\partial}{\partial k_e^{(2)}}E[U(X) - \frac{1}{2}k_{eq}^{(2)}X^2]^2 = 0 \tag{8}$$

where

$$U(X) = \int_0^X f(u)du \tag{9}$$

In addition, Elishakoff and Zhang [7] suggested another new stochastic linearization scheme based on the requirement that the mean-square values of the potential energy in the original system and of its linear counterpart be equal:

$$E\{[U(X)]^2\} = E\left\{[\frac{1}{2}k_{eq}^{(3)}X^2]^2\right\} \tag{10}$$

In this study these two new criteria are generalized to include a system with nonlinear damping. These generalizations are elucidated on an example studied earlier in the literature via the perturbation method by Khabbaz [8]. Namely, the system governed by the following differential equation

$$m\ddot{X} + c_0\dot{X}(1+b_2\dot{X}^2 + b_4\dot{X}^4)+k_0X(1+a_2X^2 + a_4X^4) = F(t) \tag{11}$$

is investigated. In Eq. (11) m is the mass, c_0 and k_0 are coefficients modelling the damping and spring stiffness of the linear system which is obtained by a formal substitution: $a_2 = b_2 = a_4 = b_4 = 0$. Here coefficients a_2, b_2, a_4 and b_4 are assumed to be positive, and specified, F(t) is a stationary Gaussian white noise. The exact solution of Eq. (11) is unavailable. We derive approximate solutions through three stochastic linearization criteria. The results are compared both with those yielded by the perturbation method and by the Monte Carlo simulation.

Conventional Linearization

Condition (4) yields respectively,

$$k_{eq}^{(1)} = \frac{k_0}{E[X^2]}\left[E(X^2) + a_2E(X^4) + a_4\, E(X^6)\right] \tag{12}$$

$$c_{eq}^{(1)} = \frac{c_0}{E(\dot{X}^2)}\left[E(\dot{X}^2) + b_2E(\dot{X}^4) + b_4E(\dot{X}^6)\right] \tag{13}$$

The mean square displacement of the linear system (with $a_2 = b_2 = a_4 = b_4 = 0$) equals

$$\sigma_{x_0^2} = \frac{\pi S}{c_0 k_0} \tag{14}$$

where S is the spectral density of F(t). For the equivalent linear system the mean-square displacement is

$$\sigma_x^2 = \frac{\pi S}{c_{eq} k_{eq}} \tag{15}$$

In addition the mean-square velocity equals

$$\sigma_{\dot{x}}^2 = \frac{k_{eq}}{m}\sigma_x^2 = \frac{\pi S}{c_{eq} m} \tag{16}$$

Utilizing the postulated jointly normal probability density of the displacement and velocity for evaluation of Eqs. (12) and (13) results in

$$k_{eq}^{(1)} = k_0(1+3a_2\sigma_x^2 + 15a_4\sigma_x^4) \tag{17}$$

$$\begin{aligned} c_{eq}^{(1)} = c_0\big[1 &+ 3b_2(k_o/m)(1 + 3a_2\sigma_x^2 + 15a_4\sigma_x^4)\sigma_x^2 \\ &+15b_4(k_0^2/m^2)(1 + 3a_2\sigma_x^2 + 15a_4\sigma_x^4)^2\sigma_x^4\big] \end{aligned} \tag{18}$$

Substitution of Eqs. (17) and (18) into Eq. (15) yields a polynomial equation

$$\sum_{i=0}^{9} A_i(\sigma_x^2)^i = 0 \tag{19}$$

where coefficients A_i read

$$A_0 = -\sigma_{x_0}^2, \qquad A_1 = 1, \qquad A_2 = \frac{3}{m}(a_2 m + k_0 b_2)$$

$$A_3 = \frac{3}{m^2}(5a_4 m^2 + 6mk_0 a_2 b_2 + 5k_0^2 b_4)$$

$$A_4 = \frac{9k_0}{m^2}(3ma_2^2 b_2 + 10ma_4 b_2 + 15k_0 a_2 b_4)$$

$$A_5 = \frac{135k_0}{m^2}(2ma_2a_4b_2+3k_0a_2^2b_4+5k_0a_4b_4)$$

$$A_6 = \frac{135k_0}{m^2}(5ma_4^2b_2+3k_0a_2^3b_4+30k_0a_2a_4b_4)$$

$$A_7 = 2025k_0^2a_4b_4(3a_2^2+5a_4)/m^2 \tag{20}$$

$$A_8 = 30375\frac{k_0^2a_2a_4^2b_4}{m^2}, \qquad A_9 = 50625\frac{k_0^2a_4^3b_4}{m^2}$$

Here $\sigma_{x_0}^2$ is the mean-square response of the corresponding

linear system. According to the Descartes' rule of signs, the polynomial equation (19) has a single positive root for σ_x^2, since the sequence of coefficients in Eq. 19 has only one change in sign. The evaluation of the mean-square displacement is performed numerically. The results are depicted in Fig. 1, as curve 1.

Generalization of Zhang-Elishakoff-Zhang Criterion [6]

Eq. (9) becomes, for the potential energy of the system

$$U(X) = k_0\left\{\frac{1}{2}X^2 + \frac{1}{4}a_2X^4 + \frac{1}{6}a_4X^6\right\} \tag{21}$$

For the energy dissipation function we obtain

$$\begin{aligned}\Phi(\dot{X}) &= \int_0^{\dot{X}} c_0\dot{u}(1 + b_2\dot{u}^2 + b_4\dot{u}^4)d\dot{u} \\ &= c_0\left\{\frac{1}{2}\dot{X}^2 + \frac{1}{4}b_2\dot{X}^4 + b_4\frac{1}{6}\dot{X}^6\right\}\end{aligned} \tag{22}$$

The expression for $k_{eq}^{(2)}$ reads

$$k_{eq}^{(2)} = 2k_0\, K_{eq}^{(2)} \tag{23}$$

where

$$K_{eq}^{(2)} = \frac{1}{2} + \frac{5}{4}a_2\sigma_x^2 + \frac{35}{6}a_4\sigma_x^4 \tag{24}$$

To obtain an expression for the equivalent linear damping we require that the mean square deviation between the energy dissipation functions in the original system and in its linear analog be minimal, i.e. that

$$E\left[\Phi(\dot{X}) - \frac{1}{2}c_{eq}^{(2)}\dot{X}^2\right]^2 \rightarrow \min \tag{25}$$

This is achieved by requiring that

$$\frac{\partial}{\partial c_{eq}^{(2)}} E\left[\Phi(\dot{X}) - \frac{1}{2}c_{eq}^{(2)}\dot{X}^2\right]^2 = 0 \tag{26}$$

yielding

$$c_{eq}^{(2)} = \frac{2E[\dot{X}^2\Phi(\dot{X})]}{E(\dot{X}^4)} \tag{27}$$

or

$$c_{eq}^{(2)} = 2c_0\left[\frac{1}{2} + \frac{5}{4}b_2\sigma_{\dot{X}}^2 + \frac{35}{6}b_4\sigma_{\dot{X}}^4\right] = 2c_0C_{eq}^{(2)} \tag{28}$$

where

$$\begin{aligned} C_{eq}^{(2)} = \frac{1}{2} + \frac{5}{4}b_2\frac{2k_0}{m}\left(\frac{1}{2} + \frac{5}{4}a_2\sigma_x^2 + \frac{35}{6}a_4\sigma_x^4\right)\sigma_x^2 \\ + \frac{35}{6}b_4\frac{4k_0^2}{m^2}\left(\frac{1}{2} + \frac{5}{4}a_2\sigma_x^2 + \frac{35}{6}a_4\sigma_x^4\right)^2\sigma_x^4 \end{aligned} \tag{29}$$

The mean-square displacement equals

$$\sigma_x^2 = \frac{\pi S}{c_{eq}^{(2)}k_{eq}^{(2)}} = \frac{\sigma_{X_0}{}^2}{4C_{eq}^{(2)}K_{eq}^{(2)}} \tag{30}$$

Substitution of Eqs. 24 and 29 into Eq. 30 leaves us with a polynomial equation

$$\sum_{i=0}^{9} B_i\,(\sigma_x^2)^i = 0 \tag{31}$$

where the coefficients B_i are defined as

$$B_0 = -\sigma_{X_0}^{\ 2}, \qquad B_1 = 1, \qquad B_2 = \frac{5}{2m}(a_2 m + k_0 b_2)$$

$$B_3 = \frac{5}{6m^2}(14a_4 m^2 + 15mk_0 a_2 b_2 + 14k_0^2 b_4)$$

$$B_4 = \frac{25}{24}\frac{k_0}{m^2}(15a_2^2 b_2 m + 56a_4 b_2 m + 84k_0 a_2 b_4)$$

$$B_5 = \frac{175}{12}\frac{k_0}{m^2}(10a_2 a_4 b_2 m + 15k_0 a_2^2 b_4 + 28k_0 a_4 b_4) \tag{32}$$

$$B_6 = \frac{875}{72}\frac{k_0}{m^2}(28a_4^2 b_2 m + 15k_0 a_2^3 b_4 + 168k_0 a_2 a_4 b_4)$$

$$B_7 = \frac{6125}{36}\frac{k_0^2}{m^2}a_4 b_4(15a_2^2 + 28a_4)$$

$$B_8 = \frac{214375}{18}\frac{k_0^2 a_2 a_4^2 b_4}{m^2}, \qquad B_9 = \frac{1500625}{81}\frac{k_0^2 a_4^3 b_4}{m^2}$$

Again, Eq. 31, like Eq. 19 has a single positive root. The results of this version of stochastic linearization are depicted in Fig. 1 as curve 2.

Generalization of Elishakoff-Zhang Criterion [7]

Within this new version of the stochastic linearization, we demand that the mean-square values of the potential energies of the original non-linear system and its linear imitation be equal (see Eq. 10).

In addition for the system with nonlinear damping we require that the mean-square values of the energy dissipation functions of both systems be equal, i.e.

$$E\left\{[\Phi(\dot{X})]^2\right\} = E\{[\frac{1}{2}c_{eq}^{(3)}\dot{X}^2]^2\} \tag{33}$$

These requirements yield equivalent stiffness $k_{eq}^{(3)}$ and equivalent damping coefficient $c_{eq}^{(3)}$, respectively

$$k_{eq}^{(3)} = 2\sqrt{\frac{E[U^2(X)]}{E(X^4)}}, \qquad c_{eq}^{(3)} = 2\sqrt{\frac{E[\Phi^2(\dot{X})]}{E(\dot{X}^4)}} \tag{34}$$

Substituting equation (34) into Equations (15) and (16) we arrive at

$$k_{eq}^{(3)} = 2k_0 K_{eq}^{(3)} \tag{35}$$

where

$$K_{eq}^{(3)} = \left[\frac{1}{4} + \frac{5}{4}a_2\sigma_x^2 + 35\left(\frac{1}{16}a_2 + \frac{1}{6}a^4\right)\sigma_x^4 + \frac{105}{4}a_2a_4\sigma_x^6 + 105a_4\sigma_x^8\right]^{1/2} \tag{36}$$

Also

$$C_{eq}^{(3)} = 2c_0 C_{eq}^{(3)} \tag{37}$$

where

$$C_{eq}^{(3)} = \left\{\frac{1}{4} + \frac{5}{4}\frac{b_2k_{eq}^{(3)}}{m}\sigma_x^2 + 35\left(\frac{1}{16}b_2 + \frac{1}{6}b_4\right)\left[\frac{k_{eq}^{(3)}}{m}\right]^2\sigma_x^4 + \frac{105}{4}b_2b_4\left[\frac{k_{eq}^{(3)}}{m}\right]^3\sigma_x^6 + 105b_4\left[\frac{k_{eq}^{(3)}}{m}\right]^4\sigma_x^8\right\}^{1/2} \tag{38}$$

This yields an equation for σ_x^2:

$$\sigma_X^2 - \frac{\pi S}{c_{eq}^{(3)}k_{eq}^{(3)}} \equiv \sigma_X^2 - \frac{\sigma_{X_0}^2}{4K_{eq}^{(3)}C_{eq}^{(3)}} = 0 \tag{39}$$

The explicit form of the resulting polynomial equation in terms of σ_x^2 is cumbersome and is not reproduced here. One can show, however, that the resulting equation also possesses a single root, as in previous cases. Curve 3 in Fig. 1 portrays the results of this stochastic technique.

Numerical Simulation

Since the theoretical solutions are not available for the nonlinear system under consideration, Monte-Carlo simulations have been performed to estimate the standard deviation of the displacement and compare the results with those obtained by the proposed stochastic linearization criteria. The displacement of the system was determined numerically by using a fourth order Runge-Kutta algorithm. The zero-mean Gaussian white-noise excitation F(t) was digitally simulated as a sequence of pulses with random heights with identical durations

Δt [9,10] as follows

$$F(t) = \sum_{N}^{k=1} Y_k \{U[t-(k-1)\Delta t] - U[t-k\Delta t]\} \tag{40}$$

where Y_k are independent normally distributed random variables with zero mean and a standard deviation equal to $2\pi S/\Delta t$, and U[t] denotes a unit step function.

The spectral density of such a simulated random process tends to be a constant value, S, over the entire frequency range as Δt approaches zero. To provide sufficient accuracy, Δt is chosen to be small. Here Δt was fixed at 0.01 sec. Sample of $N = 10^5$ numerical realizations of the loading function was created by choosing various realizations of Y_k. For each realization $f_j(t)$ of the loading function, the displacement $x_j(t)$ was numerically determined, ($j = 1,2...,N$). Subsequently, a statistical analysis was performed to calculate the mean-square displacement.

The results of numerical simulation of four cases are shown in figures by triangles. For hardening spring vibration system, the coefficients were fixed at $a_2 = b_2 = a_4 = b_4 = 1.0$, $k_0/m = 1.0$ rad^2/sec^2 and $c_0/m = 0.4$ rad/sec. As is seen, the results of simulation (denoted by triangles) depict large differences with the conventional stochastic linearization. Performance of generalized Zhang-Elishakoff-Zhang criterion (given by Equations 8 and 26) and generalized Elishakoff-Zhang criterion (given by Equations 10 and 33) is much better, especially for smaller values of $\sigma_{x_0}^2$.

Vibration System with Nonlinearity of Softening Type

Consider now a softening vibration system. Namely, the functions f and h in Eq. 8 are expressed as

$$f(X) = k_0 \frac{sgn(X)(1-e^{-\varepsilon|x|})}{\varepsilon}, \qquad h(\dot{X}) = c_0\dot{X} \tag{41}$$

It can be shown that the right hand side of Eq. 1 will approach k_0 when ε approaches zero (Fig.2). Numerical computations were performed and compared with the results of the Monte Carlo method. Fig. 3 portrays the dependence of the mean-square displacement versus coefficient ε. Fig. 3 reveals that for very small values of ε, all three criteria yield results converging to the numerical simulation. For values of ε up to unity the generalized Elishakoff-Zhang criterion yields results in best agreement with the simulation. In the range $1 \le \varepsilon \le 2$, the generalized Zhang-Elishakoff-Zhang criterion is superior to two other criteria.

Note, that the results of simulation cross the two curves based on considerations of energy and the dissipation function. Fig. 4 depicts the dependence of the ratio $\sigma_x^2/\sigma_{x_0}^2$ versus $\sigma_{x_0}^2$. Again, two proposed generalized criteria of stochastic linearization turn out to be advantageous over the conventional version of stochastic linearization in this problem.

Acknowledgment

This paper is partially supported by a grant, AFOSR-91-0073 from the Air Force Office of Scientific Research, Air Force Systems Command, USAF, monitored by Dr. Spencer Wu. The U.S. Government is authorized to reproduce and distribute reprints for governmental purposes, notwithstanding any copywright notation thereon.

References

1. Y. K. Lin, Probabilistic Theory of Structural Dynamics, McGraw Hill, New York, 1967 (Second Edition: Robert Krieger Company, Malabar, FL, 1976)

2. V. V. Bolotin, Random Vibration of Elastic Systems, Martinus Nijhoff Publishers, The Hague, 1984.

3. P. D. Spanos, Stochastic Linearization in Structural Dynamics, Applied Mechanics Reviews, 1981, Vol. 34, No.1, pp 1-8.

4. S. H. Crandall, Heuristic and Equivalent Linearization Techniques for Random Vibration of Nonlinear Oscillators, Proceedings of the 8th International Conference on Nonlinear Oscillation, Academia, Prague, 1979, Vol. 1, pp. 211-226.

5. J. B. Roberts and P. D. Spanos, Random Vibration and Statistical Linearization, John Wiley and Sons, Chichester, 1990.

6. X. Zhang, I. Elishakoff and R. Zhang, A Stochastic Linearization Technique Based on Minimum Mean Square Deviation of Potential Energies, in "Stochastic Structural Dynamics-New Theoretical Developments " (Y.K. Lin and I. Elishakoff, eds.), Springer Verlag, Berlin, 1991, pp.

7. I. Elishakoff and X. Zhang, An Appraisal of Stochastic Linearization Techniques, Journal of Sound and Vibration, 1991, to appear.

8. G.R. Khabbaz, Power Spectral Density of the Response of a Nonlinear System to Random Excitation, Journal of the Acoustical Society of America, Vol 38, 1964, pp. 847-850.

9. R.W. Glough and J. Penzien, Dynamics of Structures, McGraw Hill, Auckland, 1975.

10. G.Q. Cai and Y.K. Lin, Hysteretic Columns under Random Excitations, Technical Report NCEER-89-0003, Center for Applied Stochastics Research, Florida Atlantic University, Jan. 1989.

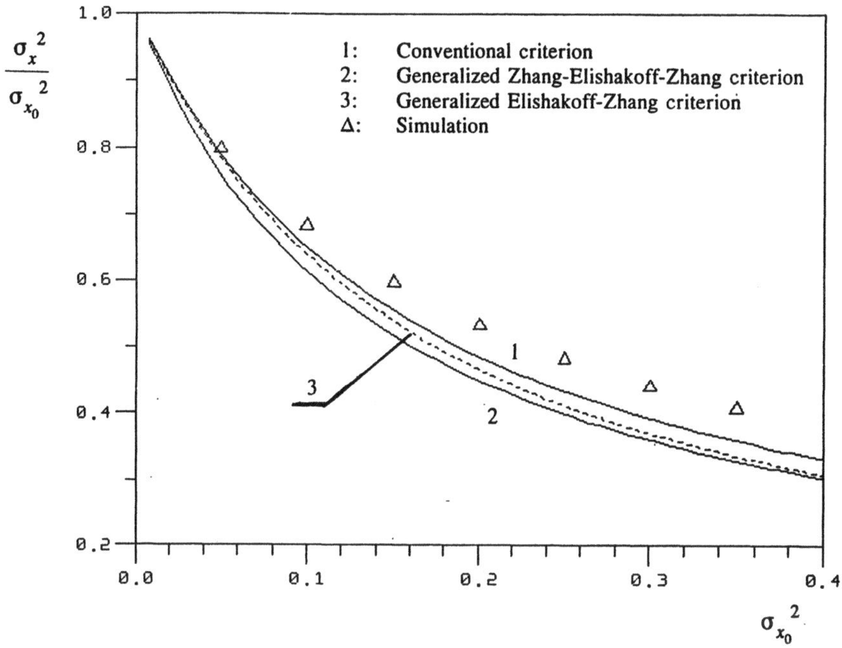

Fig. 1 Comparison of three stochastic linearization criteria with the Monte Carlo solution: hardening nonlinearity

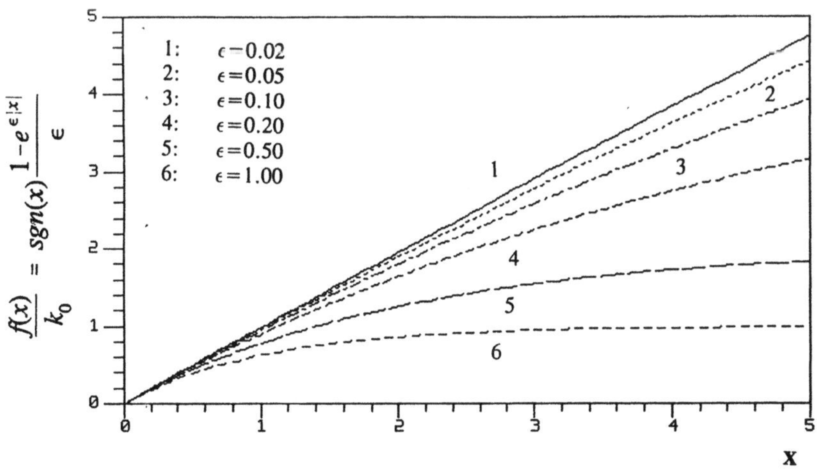

Fig. 2 Restoring force versus displacement for softening nonlinearity

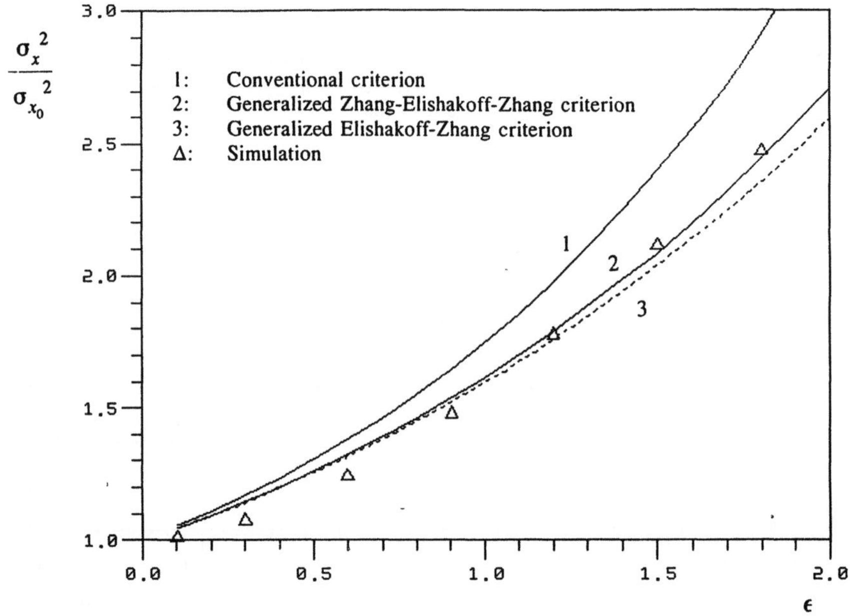

Fig. 3 Dependence of the nondimensional mean-square displacement versus ϵ for softening nonlinearity ($\sigma_{x_0}^2 = 0.4$ m^2)

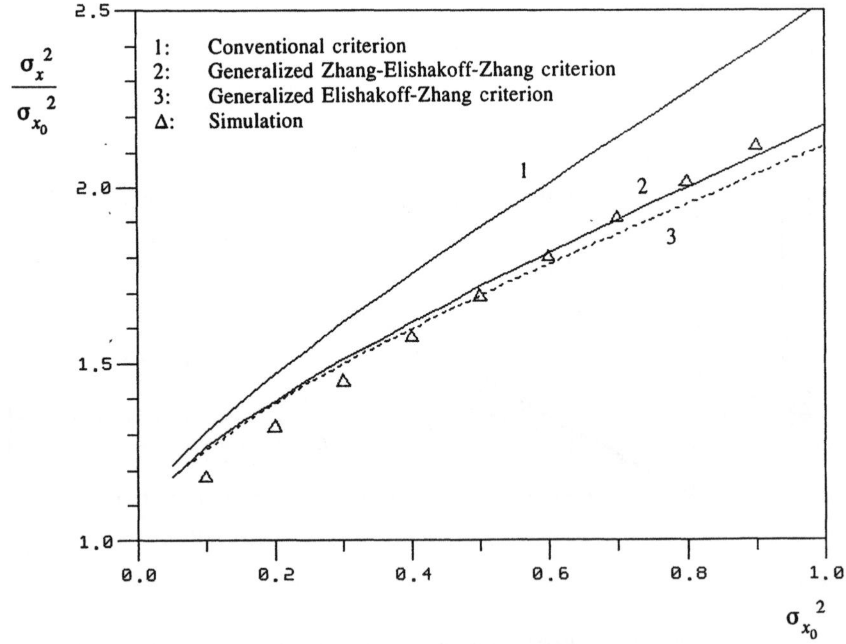

Fig. 4 Dependence of the nondimensional mean-square displacement versus $\sigma_{x_0}^2$ for softening nonlinearity ($\epsilon = 0.4$)

Structural Reliability via Response Surface

L. Faravelli

Institute of Energetics, University of Perugia, I06100 Perugia, Italy

SUMMARY

Problems of nonlinear stochastic mechanics (both static and dynamic) are governed by equations whose solution must generally be pursued in a numerical way. Stochasticity may lie in the nature of the excitation and/or in the nature of distributed system properties. The absence of analytical expressions and the presence of stochastic processes and random fields prevent one from using classical reliability methods. A general - purpose approach for assessing the reliability of a nonlinear mechanical system under stochastic conditions is presented in this paper. It combines the skeleton of classical reliability methods with the adoption of a response surface scheme.

INTRODUCTION

The study of any nonlinear mechanical problem is generally characterized by the impossibility of obtaining analytical solutions. Results are generally achieved by an intensive use of numerical algorithms and computer time.

Nonlinear problems become stochastic when one or more random components are met in formulating the governing equations [1][2]:

- stochastic processes could be the best idealization for the external excitations to be considered;

- random fields or random vectors could be appropriately introduced to model continuous or distributed properties of the system;

- random variables or stochastic processes could express initial and boundary conditions, respectively, for a set of differential equations.

Given a nonlinear stochastic system and one or more performance criteria, the reliability of the system is the probability that it does not leave the intersection of the performance (safe) domains. For its assessment, classical methods of structural reliability [3] [4] [2] require an analytical form

$$g(\underline{\mathbf{X}}) > 0 \tag{1}$$

for the performance domain intersection, where $\underline{\mathbf{X}}$ is a vector of random quantities and $g(\;)$ is the so called performance function. Stochastic processes and random fields are excluded.

For the formulation of a general approach, therefore, the problem is twofold; one must be able:

i) to study systems whose behaviour can only be analysed in a numerical way. For this purpose a finite element code is coupled, in Ref.[5], with an iterative calculus of the safety

N. Bellomo · F. Casciati (Eds.)
Nonlinear Stochastic Mechanics
IUTAM Symposium Turin

index. The code is activated whenever the value of $g(\ )$ and/or the value of its gradient in the space of the variables $\mathbf{X}$ is required by the reliability assessment process.

ii) to incorporate the spatial and temporal dependency of the randomness. Temporal variability has been considered in Ref. [6] by adding a further auxiliary random variable to the vector $\underline{\mathbf{X}}$. More generally, one can make use of mean-centered perturbation techniques [7] and/or expansions [8], introduced inside the numerical solving algorithm. It is assumed here that the numerical operator which governs the system is too complex to be decomposed and must be regarded as a black-box. Repeated experiments have, therefore, to be carried out. As an alternative to Monte Carlo techniques [9], [10], response surface methodology (see Ref.[2] for a comprehensive list of references) has been investigated [11]-[15]. Applications and extensions [16]-[21] cover the study of nuclear reactor vessels, the assessment of seismic local amplification and crashworthiness analyses.

This extensive research activity has led to the general-purpose approach to the problem of assessing the reliability of stochastic nonlinear systems which is developed in this paper. Unlike previous different attempts by other authors [22], the resulting procedure presented here:

1) makes use of second order polynomial forms for modelling the response surface;

2) builds the experiments plans according to experiment design theory [23][2];

3) iterates the response surface, which holds over the corresponding experiment region, in a space of reduced dimension, compared to the one of the space of the design variables $\mathbf{X}$;

4) establishes objective conditions of convergence;

5) incorporates stochastic processes and random fields.

GOVERNING RELATIONS

Let $\underline{\mathbf{X}}$ denote the random quantities in the mechanical system under investigation. It can be partitioned into:

$$\underline{\mathbf{X}} = \left\{ \begin{array}{c} \underline{\mathbf{X}}_v \\ \underline{\mathbf{X}}_s \end{array} \right\} \tag{2}$$

where $\underline{\mathbf{X}}_v$ is a vector of random variables and $\underline{\mathbf{X}}_s$ groups those random quantities whose spatial or temporal distribution cannot be neglected. Any component $\underline{X}_{si}$ of $\underline{\mathbf{X}}_s$ can be written in one of the two forms:

$$\begin{aligned} \underline{X}_{si}(t,x,y,z) &= \underline{X}_{svi} + \underline{X}_{ssi}(t,x,y,z) \\ \underline{X}_{si}(t,x,y,z) &= \underline{X}_{svi} \cdot \underline{X}_{ssi}(t,x,y,z) \end{aligned} \tag{3}$$

where $\underline{X}_{svi}$ is any random central value of $\underline{X}_{si}$ and $\underline{X}_{ssi}$ denotes the deviations of $\underline{X}_{si}$ from the central value $\underline{X}_{svi}$. Discretization could lead one to regard $\underline{X}_{ssi}$ as a random vector rather than a random field.

The extended response surface model formulated by the author [14][15][2] is based on the relationship

$$g(\mathbf{X}) = f(\mathbf{X}_v, \mathbf{X}_{sv}, \boldsymbol{\theta}) + \varepsilon(\mathbf{X}_{ss}) \tag{4}$$

where, as usual, $f(\ )$ is a polynomial form of the coefficients $\boldsymbol{\theta}$ and ε is a random term which incorporates both the lack of fit and the randomness of the terms which do not appear explicitly in Eq.(4). The reader is referred to the literature [2] for the more important aspects met in modelling Eq.(4). They include:

- the way of designing the numerical experiments which will form the basis for estimating $\boldsymbol{\theta}$ and ε;

- the procedure for computing the parameters;
- the possibility of working on transformations of g and $\mathbf{X}_v, \mathbf{X}_{vs}$ rather than on the original variables.

In the following the variables $\mathbf{X}_v$ and $\mathbf{X}_{sv}$ are mapped in the standardized space $\mathbf{Z}$ where all the variables are uncorrelated, and have zero mean and unit variance. This can be done by any transformation of classical reliability theory. It follows that Eq.(4) is mapped in the space $\mathbf{Z}$ by the parametrized form

$$g_z(\mathbf{Z}|\mathbf{X}_{ss}) = f(\mathbf{Z}, \boldsymbol{\theta}_z) + \varepsilon_z(\mathbf{X}_{ss}) \tag{5}$$

Note that a single function $f(\ )$ cannot cover with accuracy all the space of the variables $\mathbf{Z}$. Moreover, one must work in the central region of the space of all the variables, until the actual location of $g_z(\ )$, in the $\mathbf{Z}$ space, has been identified.

A GENERAL-PURPOSE RELIABILITY ASSESSMENT APPROACH

Consider now a reliability problem governed by the positiveness of the performance function $g(\mathbf{X})$. Let $g_z(\mathbf{Z}|\mathbf{X}_{ss})$ be the performance function in the space of the standardized variables $\mathbf{Z}$. For the purpose of acquiring information with minimal computational effort, one selects:
- a linear polynomial form for the response surface;
- an appropriate fractional replicate of a factorial design centered at the origin of the space $\mathbf{Z}$ as an experiment plan.

Table I provides the number of the experiments N required for different dimensions $n(2 < n < 99)$ of the space $\mathbf{Z}$. The number N is suitably increased with respect to the mere n+1 one in order to also obtain an estimate of the model error. Note that all the selected designs of experiments are fractional factorials [2]. They are a class of orthogonal designs: an orthogonal design has the important property that it is a minimum variance design for estimating the coefficients of a first order equation [24]. The identification problem is summarized in the relation:

$$\bar{\mathbf{g}}_z = \mathbf{A}\boldsymbol{\theta}_z + \boldsymbol{\varepsilon}_z \tag{6}$$

in which the Nx$(n+1)$ matrix $\mathbf{A}$ takes the form:

$$\mathbf{A} = \begin{bmatrix} 1 & Z_{11} & \dots & Z_{n1} \\ 1 & Z_{12} & \dots & Z_{n2} \\ & & \dots & \\ 1 & Z_{1N} & \dots & Z_{nN} \end{bmatrix} \tag{7}$$

Z_{ij} being the level of input of the i-th variable at the j-th experimental point. Orthogonality means that

$$\mathbf{A}^T\mathbf{A} = n\mathbf{I} \tag{8}$$

where $\mathbf{I}$ is the identity matrix.

To the author's knowledge, many computer codes are available for building fractional factorial designs, but no code gives a design with $(\frac{1}{2})^8$ and less replicate. This led the author to write specific software incorporating the requirements of Table I. It is presently operative within a main program which, on the basis of information on the variables (number, type

Table 1: Experiments required for fitting a linear response surface in the space of all the variables **Z** (in the last column plus 1 denote an additional central point experiment)

number of variables	coefficients to be determined	minimum required experiments	type of $(\frac{1}{2})^P$ replicate	number of experiments $N = 2^{n-P} + 1$
$n < 25$	$n+1$	$2n+1$		
2	3	5	full factorial	$2^2+1=5$
3	4	7	"	$2^3+1=9$
4	5	9	"	$2^4+1=17$
5	6	11	$(\frac{1}{2})$ - replicate	$2^4+1=17$
6	7	13	$(\frac{1}{2})^2$ - replicate	$2^4+1=17$
....	....	....	....	....
9	10	19	$(\frac{1}{2})^4$ - replicate	$2^5+1=33$
....	....	....	....	....
17	18	35	$(\frac{1}{2})^{11}$ - replicate	$2^6+1=65$
....	....	....	....	....
$n \geq 25$	$n+1$	$26+n$		
25	26	51	$(\frac{1}{2})^{19}$ - replicate	$2^6+1=65$
....	....	....	....	....
40	41	66	$(\frac{1}{2})^{33}$ - replicate	$2^7+1=129$
....	....	....	....	....
99	100	125	$(\frac{1}{2})^{92}$ - replicate	$2^7+1=129$

of the distributions, their means and standard deviations), builds the appropriate design in the space of the transformed variables **Z**. Moreover the code transforms the design point coordinates back into the original coordinates $\mathbf{X}_v, \mathbf{X}_{sv}$. This is required by the computer code which develops the mechanical analysis of the system under investigation. The components of the l.h.s. vector of Eq.(6) are obtained one for each of the required numerical experiments. A regression scheme then provides

$$\boldsymbol{\theta}_z = (\mathbf{A}^T\mathbf{A})^{-1}\mathbf{A}^T\bar{\mathbf{g}}_z = \frac{1}{N}\mathbf{A}^T\bar{\mathbf{g}}_z \tag{9}$$

due to the selection of an orthogonal design (Eq.(8)). The analysis of variance (ANOVA) provides an estimation of the variance of $\underline{\varepsilon}_z$, whose mean is assumed to be zero in the model. This variance should be very low to certify a correct modelling. If so, its contribution can be neglected and Eq.(5) is now an analytical form $g_z^{(1)}(\mathbf{Z})$ for the performance function $g_z(\mathbf{Z})$. A classical level II reliability analysis scheme can be used for solving the reliability problem.

This first response surface application (where the contribution of the quantities $\mathbf{X}_{ss}$ was neglected) leads therefore the following intermediate results:

- estimation of the design point $\mathbf{Z}^*_{(1)}$ for the linear model. It is the point of $g_z^{(1)}(\mathbf{Z})$ closest to the origin and is the most likely failure point [2]. Its distance from the origin, $\beta_{(1)}$ is the reliability index and represents a measure of the reliability of the system.

- the coefficients of the linear approximation $g_z^{(1)}(\mathbf{Z})$, which, after appropriate normalization, provide the direction cosine $\boldsymbol{\alpha}_{(1)}$ of the straight line between $\mathbf{Z}^*_{(1)}$ and the origin. The following relation holds:

$$\beta_{(1)} = \mathbf{Z}^{*T}_{(1)}\boldsymbol{\alpha}_{(1)} \tag{10}$$

The three values ($\beta_{(1)}, \mathbf{Z}^*_{(1)}$ and $\boldsymbol{\alpha}_{(1)}$) differ from the exact ones ($\beta, \mathbf{Z}^*$ $\boldsymbol{\alpha}$) since:

1) a mere linearization of the actual performance function $g_z(\mathbf{Z})$ has been introduced;

2) the performance function $g_z(\mathbf{Z})$ has been fitted by experiments conducted in the central region ($\mathbf{Z} = 0$) rather than in the neighbourhood of $\mathbf{Z}^*$.

The true values of $\beta, \mathbf{Z}^*$ and $\boldsymbol{\alpha}$ are therefore pursued by a sequence of planned experiments, conceived for both fitting second order polynomial response surfaces and incorporating the terms $\mathbf{X}_{ss}$ according to Eq.(6):

$$g_v^{(i)}(\mathbf{V}_{(i)}) = f(\mathbf{V}_{(i)}, \boldsymbol{\theta}_{(i)}) + \varepsilon_{(i)}(\mathbf{X}_{ss}) \tag{11}$$

where

$$\mathbf{V}_{(i)} = \mathbf{Z} - \mathbf{Z}^0_{(i)} \tag{12}$$

In Eq.(11) $f(\ )$ is a second order polynomial of parameters $\boldsymbol{\theta}$ and $\varepsilon(\mathbf{X}_{ss})$ is a zero mean random term which accounts for both the lack - of -fit and the stochastic terms $\underline{\mathbf{X}}_{ss}$. In Eq.(12), for $i = 2$, $\mathbf{Z}^0_{(i)} = \gamma \mathbf{Z}^*_{(1)}$ where γ is a relaxation coefficient. Previous considerations of the inaccuracy of $\beta_{(1)}$, $\mathbf{Z}^*_{(i)}$ and $\boldsymbol{\alpha}_{(1)}$ suggest a value of γ not greater than 0.5 to ensure convergence to the actual values.

Since the computational effort may turn out to be excessive, one can decide to assume γ close to 1 and accept a lower accuracy in the final solution.

A further result of the first iteration can fruitfully be exploited in order to reduce the dimension of the space where the response surface is written and reduce to a reasonable level the number of experiments to be carried out. For this purpose note that the values of $\boldsymbol{\alpha}_{(1)}$ are often referred to as the "sensitivity factors": it can be shown that $\beta_{(i)}$ is increased by a factor approximatively $(1/(1 - \alpha^2_{(i)j}))$ if the variance of the corresponding variable is set equal to zero [3]. Therefore, for basic variables Z_j with small numerical values $\alpha_{(i)j}$, no great error is introduced by considering these variables as fixed and equal to their mean values (equal to zero in the case of standardized variables). Let n_t be the number of these variables: from now on, all analysis will be developed in the reduced space of the

$$n_m = n - n_t \tag{13}$$

variables $\mathbf{V}_m$. Nevertheless, the effect of neglecting the variables $\mathbf{V}_t$ may become more significant as the iteration proceeds. As a consequence their effect will be considered in the error term

$$\underline{\varepsilon}_m = \varepsilon(\underline{\mathbf{X}}_{ss}, \underline{\mathbf{V}}_t) \tag{14}$$

together with the one of the terms $\underline{\mathbf{X}}_{ss}$. The response surface is pursued in the reduced space of $\mathbf{V}_m$. There, Eq. (12) relates the variables $\mathbf{V}_m$ to the ones of the reduced space $\mathbf{Z}_m$ by the shift $\mathbf{Z}^0_m$: note that both $\mathbf{V}_m$ and $\mathbf{Z}^0_m$ depend on the iteration, as well as $\underline{\varepsilon}_m$ does. If the standard deviation of $\underline{\varepsilon}_m$ becomes too big, as the iteration proceeds, the dimension of the working space must be re-arranged.

Operative reasons suggest that one should work, rather than in the space $\mathbf{V}_m$, in a particular space $\mathbf{U}$, obtained from it by rotation. The new first axis U_1 coincides with the direction, from the origin of $\mathbf{Z}$, to $\mathbf{Z}^*_{m(1)}$; i.e. the direction of direction cosines $\boldsymbol{\alpha}_{(1)}$ $/(\sum_{j=1}^{n_m}(\alpha_{(1)j})^2)^{1/2}$

The motivations can be summarized as follows:

1) The experiment along axis U_1 in the direction of the failure set gives a better estimation of $g(\ )$ when this function approaches zero or becomes negative;

2) Probabilistic computations in the space of variables $\mathbf{U}$ are very easy, since all the variables remain with zero mean and unit variance, except the first, U_1, whose mean is $-\gamma\beta_{(1)}$ for $i = 2$ and so on;

3) Convergence is easily detected from the fact that $\alpha_{(i)1}$ must be as close as possible to unity and the other direction cosines are close to zero.

The variables $\mathbf{U}$ and $\mathbf{V}_m$ are related by the orthogonal transformation

$$\mathbf{U} = \mathbf{Q}\mathbf{V}_m \tag{15}$$

The first column of $\mathbf{Q}^T$ is formed with $\boldsymbol{\alpha}$. The remaining columns of $\mathbf{Q}^T$ are obtained by the Gram-Schmidt algorithm [24]. Let $\mathbf{Q}_0$ be the identity matrix which transforms $\mathbf{V}_m$ in itself, $\mathbf{q}$ the matrix of column vectors already transformed, $\mathbf{q}_i$ the next column of $\mathbf{Q}_0^T$ to be transformed and $\mathbf{q}_{iT}$ the transformed vector orthogonal to those vectors already in $\mathbf{q}$:

$$\mathbf{q}_{iT} = \mathbf{q}_i - \mathbf{q}(\mathbf{q}^T\mathbf{q})^{-1}\mathbf{q}^T\mathbf{q}_i \tag{16}$$

After Eq.(16) is used the sum of the squares of the components of each row of $\mathbf{Q}$ must be forced to be zero. A similar transformation was also suggested in [25].

Once Eq.(11), in the form

$$g_u^{(i)}(\mathbf{U}_{(i)}) = g_u(\mathbf{U}_{(i)}, \boldsymbol{\theta}_{(i)}) + \varepsilon_{(i)}(\mathbf{X}_{ss}, \mathbf{V}_{t(i)}) \tag{17}$$

is fitted, the reliability analysis is performed in the space of the $n_m + 1$ variables $\mathbf{U}_{(i)}$ and $\varepsilon_{(i)}$ and leads to an assessment of the quantities $\beta_{(i)}$, $\mathbf{U}^*_{(i)}$ and $\boldsymbol{\alpha}_{(i)}$.

As a consequence one can compute

$$\mathbf{V}^*_{m(i)} = \mathbf{Q}^T_{(i)}\mathbf{U}^*_{(i)} \tag{18}$$

and

$$\mathbf{Z}^*_{m(i)} = \mathbf{Z}^0_{m(i)} + \mathbf{V}^*_{m(i)} \tag{19}$$

Moreover, the rotation of the space to make the first axis coincident with the new minimal distance, leads one to compute by the procedure of Eq.(16) the transformation matrix $\boldsymbol{\Delta}\mathbf{Q}_{(i)}$.

For the next step one uses:

$$\mathbf{U}_{(i+1)} = \boldsymbol{\Delta}\mathbf{Q}_{(i)}\mathbf{Q}_{(i)}\mathbf{V}_{m(i+1)} = \mathbf{Q}_{(i+1)}\mathbf{V}_{m(i+1)} \tag{20}$$

$$U^0_{(i+1)1} = (\gamma(\mathbf{Z}^*_{(i)} - \mathbf{Z}^0_{(i)}) + \mathbf{Z}^0_{(i)})^T\boldsymbol{\alpha}_{(i)} \tag{21}$$

$$U^0_{(i+1)k} = 0 \qquad fork > 1 \tag{22}$$

$$\mathbf{Z}^0_{m(i+1)} = \mathbf{Q}^T_{(i+1)}\mathbf{U}^0_{(i+1)} \tag{23}$$

The iteration procedure stops when convergence between $\beta_{(i)}$ and $\beta_{(i-1)}$ and $\boldsymbol{\alpha}_{(i)}$ and $\boldsymbol{\alpha}_{(i-1)}$ is reached.

A final refinement (step 2) is eventually conceived as follows:

$$\mathbf{U}_{(l)} = \mathbf{Q}_{(l)}\mathbf{V}_{(l)} \tag{24}$$

$$U^0_{(l)1} = \beta_{(l-1)} \tag{25}$$

This leads to a final response surface with its center at the design point. Since the axis U_1 is normal to the line $g(\circ) = 0$, one is interested in a better approximation of the response

Table 2: Iteration process for the r.c. cross section problem for both the classical reliability assessment approach (C) and response surface scheme (RS) proposed in this paper. The last row gives the result of a fast RS (FRS) approach, centering the second experiment plan at the design point estimated at the end of the first iteration

Iteration	method	β	X_1	X_2	X_3	X_4	X_5	X_6	X_7
1	C	3.31	2.401	-0.977	-1.843	-0.921	0.055	-0.028	-0.083
1	RS	3.35	2.431	-0.940	-1.861	-0.935	0.062	-0.031	-0.090
2	RS	3.39	2.605	-0.835	-1.857	-0.746	-	-	-
3	RS	3.40	2.617	-0.830	-1.868	-0.786	-	-	-
4	RS	3.41	2.611	-0.828	-1.868	-0.789	-	-	-
5	C	3.41	2.615	-0.831	-1.866	-0.790	0.037	-0.019	-0.056
2	FRS	3.40	2.321	-0.840	-1.840	-0.775	-	-	-

surface over the plane normal to U_1. This suggest a distortion of the experiment plan along the first axis, so that all the experiments contribute to the definition of $g(\circ) = 0$. This is simply obtained by a reduction of scale (say 1/10) along axis U_1. The resulting coefficients in $g(\;)$ will then be appropriately scaled (multiplied by 10 or 10^2) in the reliability calculation. In this way the maximum accuracy in modelling $g(\circ) = 0$ is reached, i.e. the maximum accuracy in the reliability assessment.

A summary of the whole procedure is given in Figures 1 and 2.

NUMERICAL EXAMPLES

Two different examples illustrate the main features of the reliability assessment procedure proposed in this paper:

1) a reinforced-concrete cross section analysis in seven random variables, already solved by a classical approach in Ref.[3];

2) a plane-stress finite-element problem, for which the propagation of uncertainty has already been investigated in Ref.[7].

R.C. Cross Section

In Ref.[3] the following vector of Gaussian mutually uncorrelated random variables is introduced when dealing with a reinforced concrete cross section subjected to pure bending:

$$\underline{\mathbf{X}}^T = \{\underline{M}_B, \underline{D}, \underline{\sigma}_{YS}, \underline{A}_S, \underline{K}, \underline{B}, \underline{\sigma}_{YC}\}$$

where M_B = sectional bending moment; D = distance of the reinforcement from the compressed end; σ_{YS} = steel yielding stress; A_S = reinforcement area; K = factor related to the stress-strain relation of concrete(0.5 for an ideal plastic stress-strain curve); B = cross section width and σ_{YC} = maximum compressive strength of the concrete. The safety domain can then be written as:

$$g(\underline{\mathbf{X}}) = \underline{X}_2\underline{X}_3\underline{X}_4 - \frac{\underline{X}_5\underline{X}_3{}^2\underline{X}_4{}^2}{\underline{X}_6\underline{X}_7} - \underline{X}_1 > 0$$

For given mean values and standard deviations (for which the reader is referred to reference [3]), the classical reliability approach starts at the origin of the standardized space $\mathbf{Z}$ and ends its first iteration with the values of $\beta_{(1)}$ and $\mathbf{X}^*_{(1)}$ (obtained from $\mathbf{Z}^*_{(1)}$) given in the first row of Table 2.

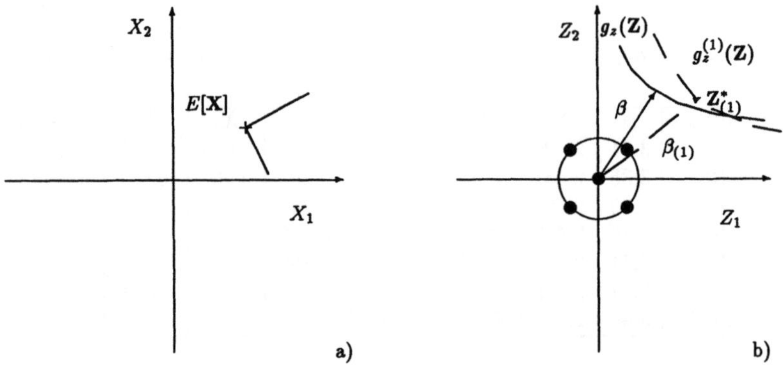

Figure 1 - a) Space of the original correlated variables **X**; b) Standardized variables **Z**; first iteration.

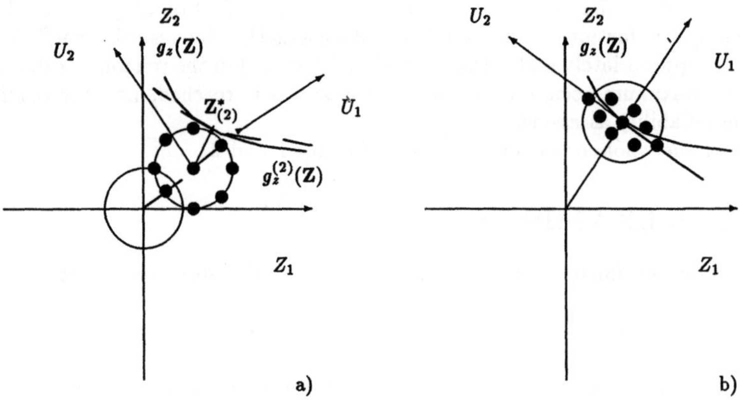

Figure 2 - a) Second iteration; introduction of the rotated **U** space; b) Last iteration.

Table 3: Probabilistic model for the plane-stress FEM problem. The autocorrelation function holds for different vertical layers. Each variable is strictly correlated within the same layer. All the variables are Gaussian

Quantity	Variable	Mean Value	Stand. Dev.	Autocorrelation function
$\sigma_y(x)$[p/sqi] =	$\sigma_{y1} \equiv X_1$	25 10^3	1.77 10^3	-
$\sigma_{y1} + \sigma_{y2}(x)$	σ_{y2}	0	1.77 10^3	$\exp(- \mid x_i - x_j \mid /.5L)$
$E_T(x)$[p/sqi] =	$E_{T1} \equiv X_2$	30 10^4	2.12 10^4	-
$E_{T1} + E_{T2}(x)$	E_{T2}	0	2.12 10^4	$\exp(- \mid x_i - x_j \mid /.5L)$
w[p/inch]	$w \equiv X_3$	10.24 10^2	2.12 10^2	-
Δ_u[inch]	$\Delta_u \equiv X_4$	1.	.1	-

The response surface scheme proposed in this paper starts by building a response plan of 17 experiments (according to Table 1, for $n = 7$) centered at the origin of the standardized space $\mathbf{Z}$. The results of Table 2 shows an estimation of β (3.352) close to its actual final value: this is the contribution for the additional nine experiments that the fractional factorial design requires; redundancy means a better approximation and, hence, a better estimation of β. The first iteration provides the following vector $\boldsymbol{\alpha}^T_{(1)}$: {-0.7249, 0.2964, 0.5548, 0.2788, -0.0184, 0.0092, 0.0269 }. The last three elements are lower, in modulus, than 0.05 by the modulus of the maximum component (.7249). Considering 0.05 times the maximum as a significance gate, the next steps will consider only $n_m = 4$ variables, the last three being accounted for directly in the error term.

A second order polynomial form is now adopted; and the experiment design is made with 31 points [2]. The reference system is rotated in such a way that the first axis coincides with the straigth line joining the origin of $\mathbf{Z}$ with $\mathbf{Z}_{(1)}{}^*$ in the reduced space of four variables. The origin of the new reference system is in $\beta_{(1)}/2$, i.e $U^0_{(2)1} = 1.676$. At the end of the third iteration ($U^0_{(3)1} = 2.5319$) $\alpha_{(3)1}$ is found to be .9994 and $\beta_{(4)} = 3.40$: since the errors are .06% and .029%, respectively, both of them are lower than 1% and the process is considered to convergence. The 4-th iteration is the refinement obtained with the experiment plan distorted along axis U_1, while the origin has $U^0_{(4)1} = 3.40$. The coincidence of the results with the ones obtained after 5 steps by the classical method of Ref.[3] is complete.

The last row of Table 2 shows the results achievable at the second iteration by centering the second experiment plan in the design point estimated at the end of the first iteration. The accuracy for engineering purposes is quite satisfactory in this case.

Plane-stress FEM Problem

The elastic-plastic cantilever-beam subject to the distributed load w, which was studied in Refs.[26] and [7] is considered. The cantilever length is $L = 40$"(101.6 cm) and its height is $H = 10$"(25,4 cm). A plane stress analysis must be conducted since the length-height ratio is not large enough for the use of beam theory to be reasonable. Therefore the system is discretized into 16 vertical layers, each made with 4 plane-stress finite elements. The main parameters of the constitutive law are Young's modulus E, the post yielding modulus E_T and the yielding stress σ_y. The reader is referred to Ref. [17] for a complete description of the problem. The probabilistic modelling adopted is summarized in Table 3.

The analysis is static, but nothing prevents one from conducting a dynamical study. The considered performance function is

$$g(\mathbf{X}) = \Delta_U - \Delta(\sigma_y, E_T, w)$$

where Δ is a function whose analytical form is unknown. It can only be computed, for given

Table 4: Iteration sequence for the plane-stress finite element problem

Iteration	β	X_1^*	X_2^*	X_3^*	X_4^*	ε^*
1	3.094	-0.8758	-0.0349	-2.6168	-1.3957	-
2	1.941	-0.5813	-	-1.7872	-0.4183	-0.2431
3	1.889	-0.6685	-	-1.6878	-0.3621	-0.3771
4	1.896	-0.6779	-	-1.7115	-0.3672	0.2655

realizations of the fields σ_y and E_T and of the variable w, by running a nonlinear analysis finite element code.

The results of the reliability assessment following the response surface procedure proposed in this paper are given in Table 4

CONCLUSIONS

This paper illustrates and discusses a procedure for the reliability assessment of nonlinear stochastic systems. The procedure makes use of a response surface scheme and is particularly suitable when the analysis of the system can only be conducted in a numerical way. The main features of the proposed approach are that:

1) any numerical algorithm, finite element code or boundary element program can be coupled with the response surface scheme;

2) no modification must be introduced in the structural analysis algorithm; only a repeated use of this algorithm is required.

ACKNOWLEDGEMENT

This research was supported by a grant of the Italian National Research Council (CNR)

References

[1] Lin Y.K., Probabilistic Theory of Structural Dynamics, Krieger, Huntington, N.Y., 1976

[2] Casciati F. and Faravelli L., Fragility Analysis of Complex Structural Systems, Research Studies Press, Taunton, U.K., 1991

[3] Madsen H.O., Krenk S., Lind N.C., Methods of Structural Safety, Prentice Hall Ellingwod Cliffs, NS, 1986

[4] Shinozuka M., Basic Analysis of Structural Safety, Journal of Structural Division, ASCE, 109(3), 1983, 721-740

[5] Liu P-L., Lin H-Z. and Der Kiureghian A., CALREL - Program for Structural Reliability Analysis, University of California, Berkeley, 1989

[6] Wen Y.K. and Chen H.C., Reliability of Structural Systems under Time Varying Loads, Proc. ICASP 5, Vancouver, 1987, vol. I, 366-373

[7] Liu W.K., Besterfield G.H. and Belytschko T., Variational Approach to Probabilistic Finite Element, Journal of Eng.Mechanics, ASCE, 114(12), 1988, 2115-2133

[8] Deodatis G. and Shinozuka M., Response Variability and Reliability of Stochastic Systems Using the Weighted Integral Method, Proc. of 1991 ASCE Eng. Mech. Specialty Conf., Columbus, Ohio, 1991, vol. 1, 263-267

[9] Rubinstein R.Y., Simulation and Monte Carlo Method, J. Wiley & Sons, New York, 1981

[10] Bjerager P., Probability Integration by Directorial Simulation, Journal of Eng. Mech., ASCE, 114(8), 1988, 1285-1301

[11] Veneziano D., Casciati F. and Faravelli L., Method of Seismic Fragility for Complicated Systems, Proc, 2nd Specialist Meeting on Probabilistic Methods in Seismic Risk Assessment for Nuclear Power Plants, Livermore, 1983, 67-88

[12] Rackwitz R., Reliability Analysis of Structural Components and Systems, in Thoft-Christensen P.(ed), Reliability Theory and Its Applications in Structural and Soil Mechanics, M. Nijhoff Publ., 1983, 171-214

[13] Schueller G.I., Basic Principles of Probability, Stochastic Processes and Reliability Methods, in Schueller G.I. and Shinozuka M.(eds.), Stochastic Methods in Structural Dynamics, M.Nijhoff Publ, 1987

[14] Faravelli L., A Response Surface Approach for Reliability Analysis, Jour. of Eng. Mech., ASCE, 115(12), 1989, 2763-2781

[15] Faravelli L., Finite-Element Analysis of Stochastic Nonlinear Continua, in Liu W.K. and Belitschko T. (eds.), Computational Mechanics of Probabilistic and Reliability Analysis, Elmepress Int., Lausanne, 1989, 263-280

[16] Faravelli L. and Lucia A.C., Stochastic Finite Element Analysis of Nozzle-Corner Response, Proc. 9th SMiRT, Lausanne, 1987, vol. M, 327-332

[17] Faravelli L., Stochastic Finite Elements by Response Surface Technique, in Computational Probabilistic Methods, ASME-AMD, 1988, 197-203

[18] Faravelli L., Response Variable Correlation in Stochastic Finite Element Analysis, Meccanica, 22(2), 1988, 102-106

[19] Faravelli L. and Bigi D., Stochastic Finite Elements for Crash Problems, Structural Safety, 7 (3-4), 1990, 113-130

[20] Faravelli L., Marcellini A. and Franceschina L., Local Amplification in Ground Acceleration Synthesis, Proc. of ICASP6 (Int. Conf. on Appl. of Stat. and Prob. in Civ. Eng.), Mexico, 1991, vol. 1, 565-571

[21] Bohm F. and Bruckher-Foint A., On a Criterion for Accepting a Response Surface Model, 1991, submitted for pubblication

[22] Bucher C.G. and Borgund U., A Fast and Efficient Response Surface Approach for Structural Reliability Problems, Structural Safety, 7, 1990, 57-66.

[23] Petersen R.G., Design and Analysis of Experiments, Marcel Dekker, NY, 1985

[24] Draper N. and Smith H., Applied Regression Analysis, 2nd Edition John Wiley & Sons, 1991, p. 277

[25] Der Kiureghian A., Lin H.Z. and H.Wang S.J., Second Order Reliability Approximations, Journal of Eng. Mechanics, ASCE, 113(8), 1987, 1208-1225

[26] Liu W.K. et al., Application of Probability Finite Element Method in Elastic-Plastic Dynamics, J. of Eng. for Industry (ASME), 109, 1987, 1-8

Stochastic Analysis of a Beam on Random Foundation with Uncertain Damping Subjected to a Moving Load

L. FRYBA

Institute of Theoretical and Applied Mechanics of the Czechoslovak Academy of Sciences, Prague, Czechoslovakia

S. NAKAGIRI and N. YOSHIKAWA

Institute of Industrial Science, University of Tokyo, Tokyo, Japan

Summary

The steady state vibration of the rail on the foundation with uncertain stiffness and damping is investigated. The stochastic finite element method is employed to the analysis. Basic formulation is reviewed first. Relevant matrices for the analysis are newly formulated. The variances of deflection and bending moment of the rail caused by the uncertainties are evaluated by the first-order perturbation technique and first-order second-moment method. The validity of the method is demonstrated by numerical examples.

1. Introduction

The theoretical model of the rail as an infinitely long beam resting on elastic foundation and subjected to a moving load was introduced by S. P. Timoshenko [1] and refined by J. T. Kenney [2] and L. Fryba [3] for all cases of the speed of the moving load and viscous damping. This deterministic model has been widely accepted for the calculation of static and dynamic stress in rail.

The foundation is in fact a variable medium with more or less random properties, however. L. Fryba therefore attempted to solve this problem in stochastic conditions using approximate analytical methods, i.e. the perturbation method and successive approximations of the integro-differential equation [4].

N. Bellomo · F. Casciati (Eds.)
Nonlinear Stochastic Mechanics
IUTAM Symposium Turin

In the same time, S. Nakagiri and T.Hisada developed the stochastic finite element method capable to solve the problem of random foundation [5]. The uncertainty in damping is also possible to be taken into account [6]. The damping of both foundation and rail is to be dealt with as spatial stochastic process. This paper demonstrates applicability of the stochastic finite element method to the problem of the said rail. The basic formulation of the steady state vibration of the beam on elastic foundation and of the stochastic finite element method is reviewed first. Some numerical analyses verify the validity of the proposed method.

2. Formulation of the Problem

Suppose that an infinitely long beam rests on elastic foundation of Winkler type, and that the foundation stiffness $k(x)$ is a random, stationary and ergodic function of the length coordinate x, see Fig. 1, as well as the coefficient of viscous damping $\omega_b(x)$.

The beam is subjected to a single constant load F which moves at constant speed c from the left to the right in Fig.1. Under the assumption that the load moves from minus infinity to plus infinity, the boundary conditions are prescribed in the following manner : the deflection, slope and bending moment in infinitely long distance from the load in both directions hold zero in every moment.

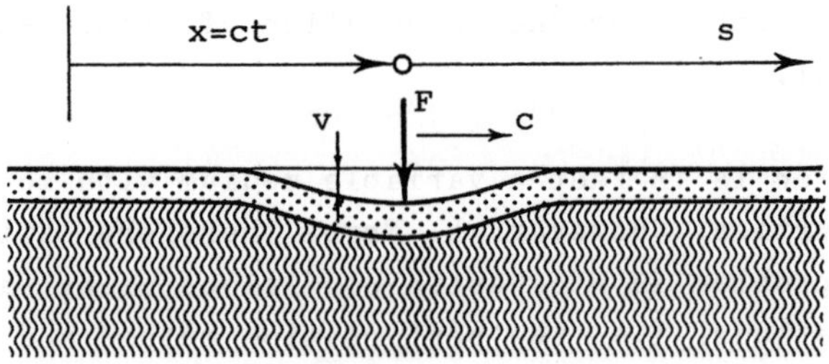

Fig.1 Movement of a load F along a beam on elastic foundation

The deflection of the beam is described by the following Bernoulli-Euler differential equation,

$$EI\frac{\partial^4 v(x,t)}{\partial x^4}+\mu\frac{\partial^2 v(x,t)}{\partial t^2}+2\mu\omega_b(x)\frac{\partial v(x,t)}{\partial t}+k(x)v(x,t)$$

$$=\delta(x-ct)F \qquad (1)$$

where it is denoted :

$v(x,t)$ - vertical deflection of the beam at point x and time t,

EI - constant flexural rigidity of the beam,

$\mu = \rho A$ - constant mass per unit length of the beam with the cross-sectional area A and mass density ρ and

$\delta(x)$ - Dirac's delta function.

The foundation stiffness (force per unit length that causes a unit deflection) can be decomposed into its constant mean value $\bar{k}$ and stochastic value $\varepsilon(x)$ as given below.

$$k(x)=\bar{k}[1+\varepsilon(x)] \qquad (2)$$

The coefficient of viscous damping can be given similarly as

$$\omega_b(x)=\bar{\omega}_b[1+\gamma(x)] \qquad (3)$$

with constant mean value $\bar{\omega}_b$. The stochastic variables are assumed small as follows.

$$\varepsilon(x)\ll 1 \quad , \quad \gamma(x)\ll 1 \qquad (4)$$

The following formulation is developed to investigate only the steady state vibration of the beam irrespective of any initial conditions. We introduce a new independent variable,

$$s=x-ct \qquad (5)$$

whose origin is moving together with the load F, see Fig.1. An ordinary differential equation is derived from substitution of Eq.(5) into the basic partial differential equation (1).

$$EI\frac{\partial^4 v(s)}{\partial s^4}+\mu c^2\frac{\partial^2 v(s)}{\partial s^2}-2\mu c\omega_b(s)\frac{\partial v(s)}{\partial s}+k(s)v(s)=\delta(s)F \qquad (6)$$

Equation (6) can be simplified by introducing such parameters as λ in Eq.(7) for dimensionless coordinate, v_0 in Eq.(8) and M_0 in Eq.(9) for the deflection and bending moment of the beam under the load F at x = s = 0 used in the static deterministic condition [3], respectively.

$$\lambda=\left(\frac{\bar{k}}{4EI}\right)^{\frac{1}{4}} \tag{7}$$

$$v_0=\frac{F}{8\lambda^3 EI}=\frac{F\lambda}{2\bar{k}} \tag{8}$$

$$M_0=\frac{F}{4\lambda} \tag{9}$$

The following dimensionless parameters are further introduced,

speed parameter

$$\alpha=\frac{c}{c_{cr}}=\frac{c}{2\lambda}\left(\frac{\mu}{EI}\right)^{\frac{1}{2}} \tag{10}$$

damping parameter

$$\beta=\bar{\omega}_b\left(\frac{\mu}{\bar{k}}\right)^{\frac{1}{2}} \tag{11}$$

and critical speed

$$c_{cr}=2\lambda\left(\frac{EI}{\mu}\right)^{\frac{1}{2}} \tag{12}$$

which are used as deterministic characteristics governed by the mean value of $\bar{k}$ and $\bar{\omega}_b$.

The assumption of stationarity and ergodicity of k(x) and $\omega_b(x)$ enables us to replace the variable x with the new variable s in the expressions (2) and (3) and to apply k(s) and $\omega_b(s)$ in what follows. The solution v(s) of Eq.(6) is the particular integral of Eq.(1) and expresses the steady state vibration which seems to be stationary random process to the observer moving together with the moving load. The solution does not satisfy any initial conditions, but it takes place shortly after the moving load starts to act. The deterministic solution of Eq.(6), i.e. for $\varepsilon = \gamma = 0$, may be found in [3], Eqs.

(13.46) and (13.47). The results of the present numerical solution in deterministic cases are in good agreement with this analytical solution.

In the undamped case of $\gamma = 0$, Eq.(6) expresses the behavior of a beam on elastic foundation subjected to a load F and an axially induced compressive force $S = -\mu c^2$. For such a beam, the critical force is $S_{cr} = -4EI\lambda^2$, and the expressions (10) and (12) imply that S_{cr} is attained by $\alpha = 1$. Thus, the critical speed of the moving load is related to elastic stability of a compressed bar.

3. Stochastic Finite Element Method

The beam is divided into N beam elements of equal length ℓ. The deflection along the element is written in the matrix form of Eq.(13),

$$v(s) = \lfloor N(s) \rfloor \{u\} \tag{13}$$

where the row matrix $\lfloor \; \rfloor$ denotes the displacement (interpolation) function and the column matrix {u} expresses the nodal displacements of an element(see Fig.2).

$$\lfloor N(s) \rfloor = \lfloor 1-3\left(\frac{s}{l}\right)^2+2\left(\frac{s}{l}\right)^3, s\left(1-\frac{s}{l}\right)^2, 3\left(\frac{s}{l}\right)^2-2\left(\frac{s}{l}\right)^3, \frac{s^2}{l}\left(\frac{s}{l}-1\right) \rfloor \tag{14}$$

$$\{u\} = \begin{Bmatrix} v_1 \\ \theta_1 \\ v_2 \\ \theta_2 \end{Bmatrix} \tag{15}$$

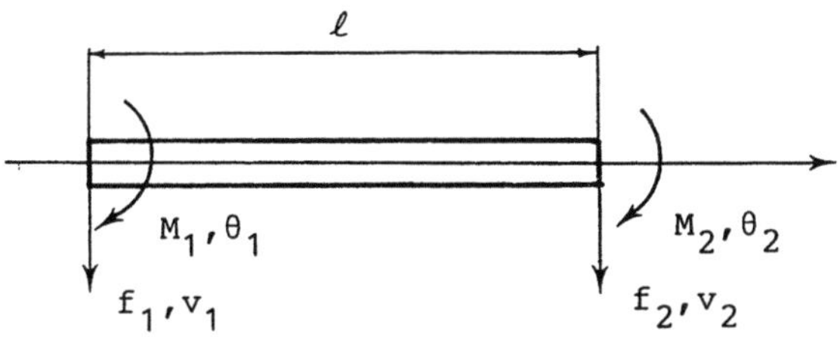

Fig.2 Finite element of the beam

Applying the principle of virtual work after substituting Eq.(13) into Eq.(6), we obtain the stiffness equation of an element in the form of Eq.(16), where {f} is external force vector.

$$\Big([K]-[D]+[V]+[R]\Big)\{u\}=\{f\} \tag{16}$$

In the above, the matrices for an element are defined as follows.

stiffness matrix

$$[K]=\frac{2EI}{l^3}\begin{bmatrix} 6 & 3l & -6 & 3l \\ 3l & 2l^2 & -3l & l^2 \\ -6 & -3l & 6 & -3l \\ 3l & l^2 & -3l & 2l^2 \end{bmatrix} \tag{17}$$

dynamic matrix

$$[D]=\frac{c^2\rho A}{30l}\begin{bmatrix} 36 & 3l & -36 & 3l \\ 3l & 4l^2 & -3l & -l^2 \\ -36 & -3l & 36 & -3l \\ 3l & -l^2 & -3l & 4l^2 \end{bmatrix} \tag{18}$$

viscous matrix

$$[V]=\frac{2cb\omega_b}{60}\begin{bmatrix} -30 & -6l & -30 & 6l \\ 6l & 0 & -6l & l^2 \\ 30 & 6l & 30 & -6l \\ -6l & -l^2 & 6l & 0 \end{bmatrix} \tag{19}$$

reaction matrix

$$[R]=\frac{blk}{420}\begin{bmatrix} 156 & 22l & 54 & -13l \\ 22l & 4l^2 & 13l & -3l^2 \\ 54 & 13l & 156 & -22l \\ -13l & -3l^2 & -22l & 4l^2 \end{bmatrix} \tag{20}$$

The matrices [K] and [D] remain deterministic, while the matrices [V] and [R] are not deterministic in terms of the stochastic variables k and ω_b.

The stochastic distribution of the foundation stiffness and damping is discretized together with the finite element division. The stochastic variables ε and γ are assigned to each element in the following form,

$$
\begin{aligned}
k &= \bar{k}(1+\varepsilon_n) \\
\omega_b &= \bar{\omega}_b(1+\gamma_n)
\end{aligned}
\tag{21}
$$

where n denotes the element number. The unknown nodal displacement vector {U} is calculated by use of the first-order perturbation technique [7] in the following form,

$$
\{U\} = \{\bar{U}\} + \sum_{n=1}^{N}\{U^I_{\varepsilon_n}\}\varepsilon_n + \sum_{n=1}^{N}\{U^I_{\gamma_n}\}\gamma_n \tag{22}
$$

where the notations used are

- $\{\bar{U}\}$ - mean value (deterministic) of displacements,
- $\{U^I_{\varepsilon_n}\}$ - first-order rate of change for ε ,
- $\{U^I_{\gamma_n}\}$ - first-order rate of change for γ .

Applying the first-order second-moment method [8], we obtain an expression for the variance Var[U] of a nodal displacement U as given below,

$$
\begin{aligned}
\mathrm{Var}[U] &= \sum_{i=1}^{N}\sum_{j=1}^{N} U^I_{\varepsilon_i} U^I_{\varepsilon_j} \mathrm{Cov}[\varepsilon_i \varepsilon_j] + \sum_{i=1}^{N}\sum_{j=1}^{N} U^I_{\gamma_i} U^I_{\gamma_j} \mathrm{Cov}[\gamma_i \gamma_j] \\
&\quad + 2\sum_{i=1}^{N}\sum_{j=1}^{N} U^I_{\varepsilon_i} U^I_{\gamma_j} \mathrm{Cov}[\varepsilon_i \gamma_j]
\end{aligned}
\tag{23}
$$

where Cov[ ] denotes the input covariance function of the pertinent variables.

4. Numerical Analysis

The constants in Eq.(6) are set equal to EI = 558 KN$\cdot$m^2, ρA = 59.7 Kg/m, F = 294 KN. The mean values of stochastic variables are $\bar{k}$ = 14.2 MPa, $\bar{\omega}_b$ = 101 /sec or 505 /sec. The total length of the space investigated is set equal to 20 m, and is divided into 200 elements. The load F is applied to the 101th node at the center of the space.

Figure 3 shows the validity of the perturbation technique for a point lying near the point where the load is applied. The calculated results of v and M are shown in Fig.3 for various ε in the 105th element while $\gamma_i = 0$. The coefficient of damping is fixed so that damping parameter is equal to 0.5 in this case. These verify that the first-order perturbation result agrees with the deterministic solution in the range of $-0.5 < \varepsilon_i < +0.5$.

Figures 4 and 5 show the distribution of the standard deviation of the deflection of the beam. The solid lines show the mean values of deflection for $k = \bar{k}$ in Fig.4 and for $\omega_b = \bar{\omega}_b$ in Fig.5 in the case of $\alpha = 0.5$ and $\beta = 0.5$. The broken lines show the values of $v \pm \sigma_v$(σ_v - standard deviation of the deflection) and the dotted line gives the deflection for $k = 1.2\bar{k}$ in Fig.4 and for $\omega_b = 1.2\bar{\omega}_b$ in Fig.5. These figures indicate that the deflection is affected by the randomness of the foundation stiffness more than that of the damping.

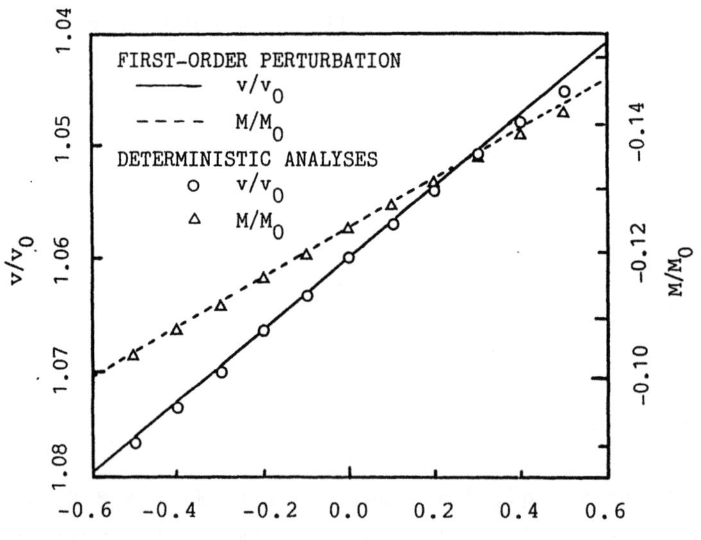

Fig.3 Dependence of the deflection $v(-\ell)$ and bending moment $M(4.5\ell)$ under the moving load on the stochastic increment ε of the coefficient of foundation, $\gamma_i = 0$, $\alpha = 0.5$, $\beta = 0.5$

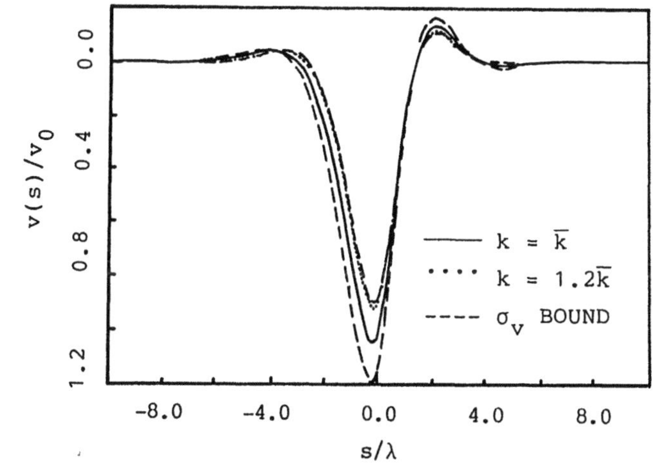

Fig.4 Distribution of the deflection along the length of the beam, effect of foundation stiffness, $\alpha = 0.5$, $\beta = 0.5$, $\gamma = 0$, $Cov[\ \varepsilon_i \varepsilon_j\ ] = 0.2^2$ (full correlation), solid line $k = \bar{k}$, broken lines $v \pm \sigma_v$, dotted line $k = 1.2\bar{k}$

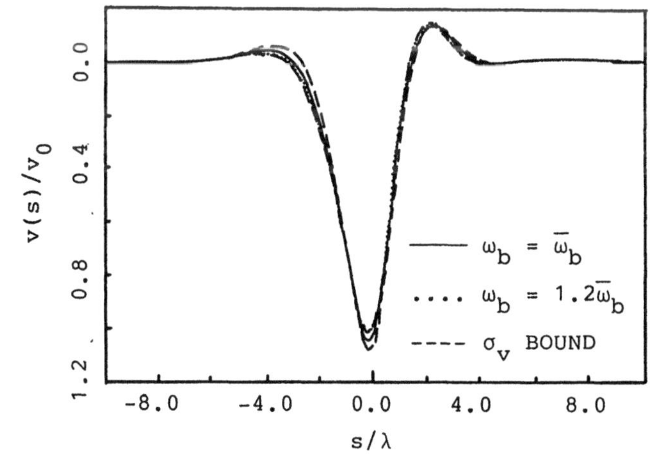

Fig.5 Distribution of the deflection along the length of the beam, effect of damping, $\alpha = 0.5$, $\beta = 0.5$, $\varepsilon = 0$, $Cov[\gamma_i\ \gamma_j] = 0.2^2$(full correlation), solid line $\omega_b = \overline{\omega_b}$, broken lines $v \pm \sigma_v$, dotted line $\omega_b = 1.2\bar{\omega}_b$

The effect of speed of the moving load is worthy to be elucidated from practical point of view. Figure 6 illustrates the effect of speed on the variance of the deflection along the beam. In this figure related to the case of $\gamma = 0$ and $\beta = 0.2$, the randomness of the foundation stiffness is dealt with in the form of cosine type correlation in Eq.(24),

$$\mathrm{Cov}[\varepsilon_i \varepsilon_j] = \sigma^2 \cos 2\pi \frac{s_i - s_j}{D} \tag{24}$$

where σ is the standard deviation of random foundation stiffness taken equal to 0.2, D is the scale of decay taken equal to 10λ and s_i means the centroid coordinate of the i-th element. In addition to the variability of deflection, the variability of bending moment is also dependent on the speed of the moving load and randomness. Figure 7 summarizes the specific coefficient of variation of the deflection v and bending moment M defined as follows,

$$V_v = \frac{\max(\sigma_v/\sigma)}{v_0} \tag{25}$$

$$V_M = \frac{\max(\sigma_M/\sigma)}{M_0} \tag{26}$$

as function of the dimensionless speed parameter α. In the above σ_M stands for the standard deviation of bending moment. It is seen that the variability takes a single peak at the critical speed, and that the peak of the deflection is higher than that of the bending moment.

5. Conclusions

The effect of random foundation on the deflection and bending moment of rail is investigated in problem of random vibration of an infinitely long beam on elastic foundation and subjected to a moving load. The foundation is of Winkler type whose stiffness and damping are spatial stochastic process. The stochastic process is supposed to be ergodic, and the speed of the moving load is taken constant so that steady state vibration is investigated on the basis of the equation of motion of the beam under the Bernoulli-Euler hypothesis.

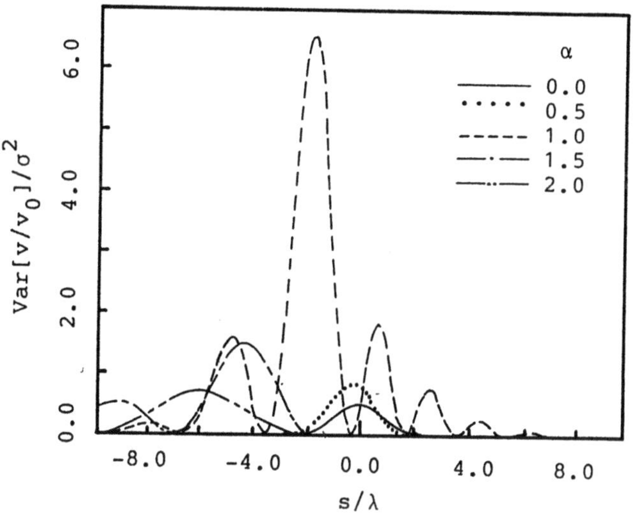

Fig.6 Distribution of variance of the deflection ($Var[v/v_0]/\sigma^2$) along the beam for various dimensionless speed parameter α, $\beta = 0.2$, $\gamma = 0$, $D = 10\lambda$

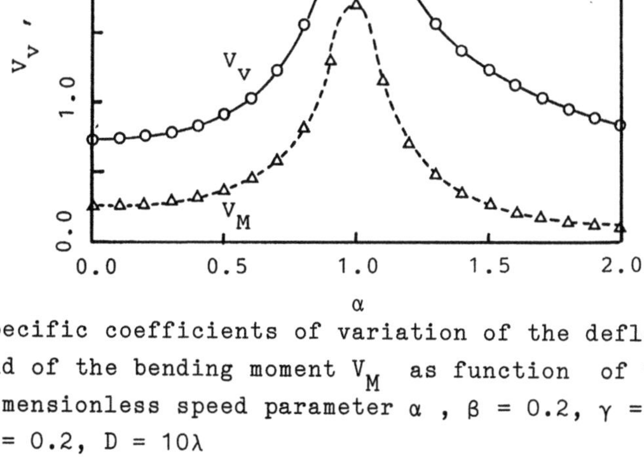

Fig.7 Specific coefficients of variation of the deflection V_v and of the bending moment V_M as function of the dimensionless speed parameter α, $\beta = 0.2$, $\gamma = 0$, $\sigma = 0.2$, $D = 10\lambda$

The perturbation-based stochastic finite element method is employed to solve the problem. The relevant matrices are newly derived. The variances of the deflection and bending moment of the beam are evaluated by means of the first-order perturbation and the first-order second-moment method.

It has appeared that the random foundation stiffness influences the deflection more than the uncertain damping does. The variability of the deflection and bending moment is found to take a single peak at the critical speed of the moving load at which the beam loses its axial stability in undamped case. The variability of the deflection is greater than that of the bending moment. This indicates that the movement of vehicles is more sensitive to random foundation than the stress in rail is.

References

[1] Timoshenko,S.P.: Statical and Dynamical Stresses in Rails. Proc. Intern. Congr. Appl. Mech., Zurich, 1926, pp.407-418.

[2] Kenney,J.T.: Steady-State Vibrations of Beams on Elastic Foundation for Moving Load. J.Appl.Mech., 21 (1954), No.4, pp.359-364, discussion 22 (1955), No.3, p.436.

[3] Fryba,L.: Vibration of Solids and Structures Under Moving Loads. Prague, Academia, Groningen, Noordhoff International Publishing, 1972, Chapters 13 and 23.2.

[4] Fryba,L.: Random Vibration of a Beam on Elastic Foundation Under Moving Force., Elishakoff,I. and Lyon,R.H. (eds.) : Random Vibration - Status and Recent Developments. Amsterdam, Elsevier, 1986, pp.127-147.

[5] Nakagiri,S. and Hisada,T.: A Note on Stochastic Finite Element Method (Part 6) - An Application in Problems of Uncertain Elastic Foundation. Seisan-Kenkyu (Monthly Journal of Institute of Industrial Science, University of Tokyo), 35 (1983), No.1, pp.20-23.

[6] Nakagiri,S., Hisada,T. and Toshimitsu,K.: Stochastic Time-History Analysis of Structural Vibration with Uncertain Damping. Probabilistic Structural Analysis. PVP - Vol.93, pp.109-120. Published by ASME, New York, 1984.

[7] Nakagiri,S. and Hisada,T.: Stochastic Finite Element Method (in Japanese). Tokyo, Baifukan, 1985.

[8] Cornell,C.A.: A Probability-Based Structural Code. ACI Journal, 66 (1969), No.12, pp.974-985.

Stochastic Aspects in Nonlinear Discrete Kinetic Theory

E. GABETTA – L. PARESCHI
Dipartimento di Matematica
Via Machiavelli 35, 44100 Ferrara (Italy)

Summary
A rarefied gas is often studied by means of the discrete velocity models of the Boltzmann equation. Here we emphasize the analogies between the evolution of the density of a lattice discretization of the Broadwell discrete model, in connection with the large time behaviour, and the evolution of a probability vector towards the equiprobable state.

1. Introduction

Recently we introduced a variety of discrete kinetic models of the Boltzmann equation [1], [2], [3], such that the velocity directions are connected with the symmetries of the cube. These models seem in principle very easy to handle, from a numerical point of view, to provide some answers about their behaviour in a bounded domain.

In addition, the probabilistic aspects of the relationships between the discrete kinetic theory and these lattice models were investigated in [4].

In more details, we discovered in [3] that the fully discretized Broadwell model (but surely this result can be extended to systems with a larger set of possible velocities) present the interesting feature of the evolution of the density towards a steady state.

The aim of the present talk is to reformulate, for the previous model, the evolution equation and the problem of the trend to equilibrium. This new approach to the initial–boundary value problems clarifies the analogies between our original problem and the stochastic evolution of a Markov chain.

To start with, we recall the physical motivation and the general formulation of kinetic models.

Kinetic models are introduced in the statistical description of a large number of particles (electrons, molecules, neutrons,...) which are impossible to follow simultaneously from the usual dynamical point of view, the main idea being to describe the *averaged* behaviour of this collection of particles rather than the real behaviour.

N. Bellomo · F. Casciati (Eds.)
Nonlinear Stochastic Mechanics
IUTAM Symposium Turin

A lot of situations can be treated at this level (fluid mechanics and rarefied gases, plasma physics, nuclear physics, semiconductors, ...) and the list of models is rather impressive, also in many cases the answers derived by their use are quite satisfactory and in good agreement with experimental results.

Nevertheless, a crucial point is not clear at this stage, namely how to justify the derivation of these models from first principles (Newton's law, nonrelativistic quantum mechanics,...). Clearly the latter is by no means a trivial subject, and many fundamental questions remain without an answer.

As we said above, the evolution of our collection of particles is described in terms of a statistical quantity which represents the density of particles $f(\mathbf{x}, \mathbf{v}, t)$, where $f(\mathbf{x}, \mathbf{v}, t)d\mathbf{x}, d\mathbf{v}$ is the number of the particles which are in the phase volume $d\mathbf{x}\, d\mathbf{v}$ around the position $\mathbf{x} \in \mathbf{R}^3$ and the velocity $\mathbf{v} \in \mathbf{R}^3$ at time $t \geq 0$.

Our treatment will be confined to models involving only one type of particles.

More general situations can be studied. In these cases one needs to introduce a density function for each species (mixtures), but except for additional complexity in the notations, the results can be easily adapted to those systems.

Of course, f being a density, satisfies $f(\mathbf{x}, \mathbf{v}, t) \geq 0$.

The first type of models to be quoted corresponds to *conservative* systems, where one has only to describe the fact that the density f is constant along particle paths, which for classical particles are recovered by Newton's law.

In particular, if we only consider free particles which do not interact, and move in absence of any external field, we immediately obtain the following infinitesimal version of the above principle

$$\frac{\partial f}{\partial t} + \mathbf{v} \circ \nabla f = 0 \tag{1}$$

for $\mathbf{x}, \mathbf{v} \in \mathbf{R}^3$,$t > 0$. In (1) ∇ denotes the spatial gradient, i.e. the gradient of f with respect to $\mathbf{x}$, and $\mathbf{a} \circ \mathbf{b}$ denotes the scalar product of $\mathbf{a}, \mathbf{b} \in \mathbf{R}^3$.

The second important group of models concerns physical situations in which the particles can interact. Typically, one speaks of collisions that modify the velocities of the colliding particles, and thus modify f. In this case, (1) is replaced by

$$\frac{\partial f}{\partial t} + \mathbf{v} \circ \nabla f = C$$

where C, the so-called collisional operator, depends on f.

The fundamental example is furnished by the Boltzmann equation

$$\frac{\partial f}{\partial t} + \mathbf{v} \circ \nabla f = Q(\mathbf{f},\mathbf{f}) \tag{2}$$

where

$$Q(\mathbf{f},\mathbf{f}) = \int_{\mathbf{R}^d \times S^{n-1}} d\mathbf{v}_* d\mathbf{n}\, B(|\mathbf{v} - \mathbf{v}^*|, \mathbf{n})(\widehat{f}\widehat{f}_* - f f_*)$$

in $\mathbf{R}^d \times \mathbf{R}^d \times [0,\infty)$.

$(\mathbf{v}, \mathbf{v}_*)$ are the pre–collisional velocities

$(\hat{\mathbf{v}}, \hat{\mathbf{v}}_*)$ are the corresponding post–collisional velocities, and

$f = f(\mathbf{v}),\ f_* = f(\mathbf{v}_*),\ \hat{f} = f(\hat{\mathbf{v}}),\ \hat{f}_* = f(\hat{\mathbf{v}}_*)$;

$B(|\mathbf{v} - \mathbf{v}_*|, \mathbf{n})$ is the function characterizing the scattering of the gas particles due to an interaction potential.

In the discrete kinetic theory, the original integrodifferential Boltzmann equation (2) is replaced by a system of m semilinear partial differential equations of hyperbolic type, which defines the evolution of the molecular densities $\{f_i\}$ joined to the finite set of the m allowed possible velocities. In general, only binary collisions between particles are taken into account, so that the general model reads

$$\frac{\partial f_i}{\partial t} + \mathbf{v}_i \circ \nabla \mathbf{f}_i = Q_i(\mathbf{f},\mathbf{f}) \tag{3}$$

where

$$Q_i(\mathbf{f},\mathbf{f}) = \sum_{j=1}^{m} \sum_{(k,l)} \left(A_{ij}^{kl} f_k f_l - A_{kl}^{ij} f_i f_j \right)$$

$\{\mathbf{v}_i\}$, $i = 1,..,m$ is the set of chosen velocities, and

$$\mathbf{f}(\mathbf{x},t) = \{f_1(\mathbf{x},t),\ldots,f_m(\mathbf{x},t)\}$$

is the m-vector whose i-th component represents the density of the molecules with velocity $\mathbf{v}_i \in \mathbf{R}^3$ in the position $\mathbf{x} \in \mathbf{R}^3$ at time $t \in \mathbf{R}^+$.

A wide variety of these models can be built up by different choices of the set of velocities.

2. The trend towards equilibrium

Once the model equation is written, it has to be supplemented with initial conditions and, when the particles are constrained in a bounded domain, with boundary conditions. The mathematical issues are numerous and contain in particular the analysis of the existence and uniqueness of a solution to the initial-boundary value problem, the study of various properties of the solution, like regularity, large-time behaviour, stability and exact solutions, the approximation of solutions, and so on. The understanding of all these questions (even for any of the models), is far from being complete, and this is particularly the case for the large-time behaviour of the discrete velocity models.

To focus the problem, we restrict our attention to the simpler case of a discrete velocity gas which is assumed spatially homogeneous. In this case, the discrete Boltzmann equation (3) reduces to a system of nonlinear differential equations of the form

$$\begin{aligned} \frac{dp_i}{dt} &= Q_i(\mathbf{p},\mathbf{p}) \\ p_i(t=0) &= p_{i0} > 0 \quad i = 1.2,\ldots,m\,. \end{aligned} \tag{4}$$

Let us assume $\sum_{i=1}^{m} p_{i0} = 1$, in such a way that we transform, from an abstract point of view, our original problem into the corresponding evolution problem of the probability densities driven by system (4). Due to the form of the collisional operator, it is a classical matter to derive the usual conservation laws (the collisional invariants), namely the vectors $\phi_k = (\phi_{1k},\ldots,\phi_{mk})$ such that

$$\sum_{i=1}^{m} \phi_{ik} Q_i(\mathbf{p},\mathbf{p}) = 0\,. \tag{5}$$

At least, (5) is verified if ϕ_1 is the vector with all equal components, $\phi_j = (v_{1,j},\ldots,v_{m,j})$, $j = 1,2,3$ and $\phi_5 = (v_1^2,\ldots,v_m^2)$.

The above conservation laws correspond to the conservation of mass, linear momentum and energy. In addition, assuming that the vector $\mathbf{p}(t)$ is a solution of the Cauchy problem (4), the structure of the collisional operator permits to conclude that, if

$$H(\mathbf{p}(t)) = \sum_{i=1}^{m} p_i(t) \log p_i(t)$$

the classical H-theorem follows, i.e. $H(\mathbf{p}(t))$ decreases with time.

Existence and uniqueness of a global solution to (4) are known [5], even in the case of multiple collisions. By means of the H-theorem, the asymptotic behaviour of the solution is found. $\mathbf{p}(t)$ converges to the unique equilibrium solution $\mathbf{p}_*$ (that satisfies $Q_i(\mathbf{p}_*, \mathbf{p}_*) = 0$) with the same moments of the initial value.

In the nonhomogeneous case, the understanding of this kind of questions is far from being complete. Particular results have been obtained in the case of small initial densities, which correspond to a perturbation of the vacuum [6], and for one dimensional in space discrete models by Beale [7].

Concerning with the problem in a bounded domain, the large-time behaviour of the solution has been recently studied by Slemrod [8], in the case of the reduced Broadwell model in a strip. He studied the initial-boundary value problem with specularly reflecting boundary conditions which in addition guaranty the regularity of the global solution. The goal of his paper was to prove that the densities tend, as time goes to infinity, to lead to running waves without interaction.

One can easily argue that, at this stage, the problem of the long-time behaviour of the solution is not settled.

To obtain informations about this specific problem, we will utilize in this paper a model which has been recently introduced in [3] to approximate the Slemrod solution. The model we propose is based on a suitable discretization both in the time and in the space variables, which has the great advantage to be handled easily from a numerical point of view. The introduction of this lattice discretization allows us here to study the problem as a classical (nonlinear) evolution problem for a probability density vector, in such a way that the analogies with a discrete in time Markov process will be apparent. Due to the form of the lattice collisional operator, we will prove that, independently of the initial density, the sequence of the probability vectors possesses a unique accumulation point, which is nothing but the steady state represented by the vector with all equal components.

3. The Broadwell model and the lattice approximation

The Broadwell model is often used in the study of some specific problems related to the evolution of a rarefied gas. Broadwell himself introduced it for a more simplified approach to the shock wave problem [9].

For flows which depend only on one space variable, the model can be reduced to the simplest form

$$\begin{aligned}\frac{\partial u}{\partial t}+\frac{\partial u}{\partial x}&=w^2-uv\\ \frac{\partial v}{\partial t}-\frac{\partial v}{\partial x}&=w^2-uv\\ \frac{\partial w}{\partial t}&=-\frac{1}{2}(w^2-uv)\end{aligned}\tag{6}$$

The discretization we proposed in [3] consists of the lattice model obtained by considering the gas concentrated at the points

$$x_1=\frac{1}{2(2N)}=\frac{1}{2}\delta\ ;\quad x_i=x_{i-1}+\delta,\quad i=2,3,\ldots,2N.$$

The iterative scheme corresponding to (6) is the following

$$\begin{aligned}u_i^{n+1}&=u_{i-1}^n+\delta\left[(w_i^n)^2-u_{i-1}^n v_{i+1}^n\right]\\ v_i^{n+1}&=v_{i+1}^n+\delta\left[(w_i^n)^2-u_{i-1}^n v_{i+1}^n\right]\\ w_i^{n+1}&=w_i^n-\frac{1}{2}\delta\left[(w_i^n)^2-u_{i-1}^n v_{i+1}^n\right]\end{aligned}\tag{7}$$

with $i=2,3,\ldots,2N-1$, $n\geq 0$.

The boundary conditions are

$$\begin{aligned}u_1^{n+1}&=v_i^n+\delta\left[(w_1^n)^2-v_1^n v_2^n\right]\\ v_1^{n+1}&=v_2^n+\delta\left[(w_1^n)^2-v_1^n v_2^n\right]\\ w_1^{n+1}&=w_1^n-\frac{1}{2}\delta\left[(w_1^n)^2-v_1^n v_2^n\right]\end{aligned}\tag{8}$$

for $i=1$, and

$$\begin{aligned}u_{2N}^{n+1}&=u_{2N-1}^n+\delta\left[(w_{2N}^n)^2-u_{2N-1}^n u_{2N}^n\right]\\ v_{2N}^{n+1}&=u_{2N}^n+\delta\left[(w_{2N}^n)^2-u_{2N-1}^n u_{2N}^n\right]\\ w_{2N}^{n+1}&=w_{2N}^n-\frac{1}{2}\delta\left[(w_{2N}^n)^2-u_{2N-1}^n u_{2N}^n\right]\end{aligned}\tag{9}$$

for $i=2N$.

Let us slightly change the notations. Let us put, for $i=1,2,\ldots,2N$, $n\geq 0$

$$\begin{aligned}p_i^n&=\delta u_i^n\\ p_{2N+i}^n&=\delta v_i^n\\ p_{4N+i}^n&=4\delta w_i^n\end{aligned}$$

and let $\mathbf{p}^n=(p_1^n,p_2^n,\ldots,p_{6N}^n)$.

If we choose nonnegative initial data $\{u_i^0,v_i^0,w_i^0\}$, and total initial mass

$$\delta\sum_{i=1}^{2N}\left(u_i^0+v_i^0+4w_i^0\right)=1$$

system (7), with the boundary conditions (8) and (9), is written in compact form as

$$\mathbf{p}^{n+1} = A(\mathbf{p}^n)\,\mathbf{p}^n; \quad n \geq 0 \tag{10}$$

where$A(\mathbf{p})$ is a square matrix whose elements depend linearly on $\mathbf{p}$.

By the conservation of the mass, since $\sum_{i=1}^{6N} p_i^0 = 1$, for any $n > 0$ we obtain, at any subsequent step $\sum_{i=1}^{6N} p_i^n = 1$

The process defined by (10) possesses the following property.

Theorem. *The sequence $\{\mathbf{p}^n\}$ converges to $\mathbf{p}^\infty$, where $\mathbf{p}^\infty$ is the vector with all equal components.*

PROOF: Let F be the normed linear space of the vectors with $6N$ components, equipped with the euclidean norm. First, we start by proving that, if $\mathbf{p}^n$ converges in F, then the limit is $\mathbf{p}^\infty$. Since $w_i^n \to w_i$, $i = 1, \dots, 2N$, from the third equation in (7) we get

$$\lim_{n\to\infty} \left[(w_1^n)^2 - v_{i-1}^n v_{i+1}^n \right] = 0$$

If we now consider the other equations in (7), we see that

$$\lim_{n\to\infty} \left[u_i^{n+1} - u_{i-1}^n \right] = 0; \quad \lim_{n\to\infty} \left[v_i^{n+1} - v_{i+1}^n \right] = 0$$

Because of the boundary conditions (8), we find that the limit values satisfy $u_1 = v_1$,$v_1 = v_2$, which imply $v_i = v$, for all i.

On the other hand, due to the boundary conditions (9), it follows that $u_{2N} = u_{2N-1}$, so that $u_i = u$ for all i. Finally, since the first equation in (8) implies $u = v$, we obtain $w_i = w = u = v$.

Assume now by contradiction that $\mathbf{p}^n$ does not converge to $\mathbf{p}^\infty$.
Then, there exists a subsequence $\{\mathbf{p}^{n_k}\}$ such that

$$\|\mathbf{p}^{n_k} - \mathbf{p}^\infty\| > \epsilon > 0$$

for some $\epsilon > 0$. Since the conservation of the mass ensures that $\|\mathbf{p}^n\| < const.$, by the Bolzano-Weierstrass theorem there exists a subsequence $\{\mathbf{p}^{n_{kj}}\}$ of $\{\mathbf{p}^{n_k}\}$ which converges. Applying now the first part of the proof we conclude that

$$\|\mathbf{p}^{n_{kj}} - \mathbf{p}^\infty\| \to 0$$

and the contradiction arises.

Remark 1: *The result of the theorem can be generalized to more complex models, which correspond to more complex matrices* $A(\mathbf{p})$ *in* (10). *In the case we dealt with, the matrix* A *possesses in each line only two elements different from zero.*

Looking at the proof, it seems that the result is obtained straightforwardly since in (7) *the third equation connects at each step* w_i^{n+1} *to* w_i^n. *This corresponds to a matrix* A *in which the diagonal has not all zero elements.*

It would be interesting, on the light of this result, to investigate for conditions on A *which are sufficient to identify the limit of* $\mathbf{p}^n$ *as* $\mathbf{p}^\infty$.

Remark 2: *In* [3] *we investigated the plane Broadwell model in a square box, introducing a lattice discretization in the square.*

In the presence of boundary conditions of specular reflection *(different from the ones we used here, in which the collisional part is present), numerical computations showed that the system manifests a trend towards running waves.*

Since different boundary conditions (which of course have to be such that the mass is conserved), generate different matrices, perhaps the choice we made in [3] *was not able to eliminate this type of behaviour.*

Acknowledgment

This work has been partially supported by the National Research Council (Gruppo Nazionale per la Fisica Matematica).

References

[1] E. Gabetta: From stochastic mechanics to the discrete Boltzmann equation: the Broadwell model, *Math. Comput. Modelling*, **15**, 1 (1991) 1-10.

[2] E. Gabetta: Probabilistic aspects of the relationships between the discrete kinetic theory and lattice models, *Ann. Univ. Ferrara*, Sez. 7, Vol. XXXV (1991).

[3] E. Gabetta: On a Broadwell-like lattice in a box, *Proceedings of Discrete Models of Fluid Dynamics*, Coimbra 1990, A. Alves Ed. World Scientific Singapore, 218-229 (1991).

[4] E. Gabetta, L. Pareschi: Approximating the Broadwell model in a strip, M^3AS (in press) (1991).

[5] G. Toscani: On the discrete Boltzmann equation with multiple collisions, *Atti Acc. Pelor. Peric. Messina*, Fasc. Spec. (1991)

[6] G. Toscani: Global existence and asymptotic behaviour for the discrete velocity models of the Boltzmann equation, *J. Math. Phys.*, **11**, 2918-2921 (1985).

[7] J.T. Beale: Large-time behaviour of discrete velocity Boltzmann equations, *Comm. Math. Phys.*, **106**, 659-678 (1986).

[8] M. Slemrod: Large-time behavior of the Broadwell model of a discrete velocity gas with specularly reflective boundary conditions, *Arch. Rational Mech. Anal.*, **111**, 323-342 (1990).

[9] J.E. Broadwell: Shock structure in a simple discrete velocity gas, *Phys. Fluids*, **7**, 1243-1247 (1964).

Random Motion of Particles in a Flow with Application to Groundwater Pollution

JOHAN GRASMAN
Department of Mathematics
Agricultural University
Wageningen, the Netherlands

Abstract

Dispersive groundwater flow is studied from the random motion of a particle in the flow. From the Fokker-Planck equation it can be computed at what rate particles enter some area, e.g. a domain where the groundwater is pumped up.

1. Introduction

In this paper we analyse the flow of groundwater that is confined in a layer, called aquifer. The thickness of this layer is assumed to be small and so it suffices to model this physical process as a 2D-flow problem.

In the past the study of groundwater flow mainly dealt with the change of the level of the groundwater because of discharges and with the natural and invoked flow pattern. Such an analysis, based on the potential equation, does not give information on the displacement of individual water particles and soluted substances as only the average flow field at a macroscopic level is computed. By including dispersion one may account for the effects of the random motion of a particle in the flow. The granular structure of the ground is the main source of dispersion, see fig. 1. This so-called macroscopic dispersion is proportional to the size of the velocity of the flow. The (small) dispersion constant depends on the direction: in the direction perpendicular to the flow it has a smaller value than in the longitudinal direction, see Bear and Verruyt [1]. The effect of molecular diffusion can in most cases be neglected.

Fig. 1 The granular structure of the ground and macroscopic dispersion.

* Financial support for this contribution was provided by the European Community under scientific cooperation contract no. ST2J-0316-C.

N. Bellomo · F. Casciati (Eds.)
Nonlinear Stochastic Mechanics
IUTAM Symposium Turin

This contribution deals with an analytical method for solving the advection-dispersion equation for the flow. The development of numerical methods for this equation tends to be quite laborious. Moreover, one has to be cautious that not a scheme is chosen with so much numerical dispersion that it masks the relatively small hydrological dispersion in the problem. An alternative is the use of Monte Carlo methods simulating the random path of particles, see Uffink [3]. A numerical scheme based on the stochastic differential equation is easily formulated. However, a large number of runs is required and has to be carried out for each possible pollution site. The asymptotic method we propose uses the Fokker-Planck equation that corresponds with the random motion. It gives detailed results on the rapid change of the probability that a particle is pumped up, if it starts somewhere in a boundary layer surrounding the domain of flow towards the well.

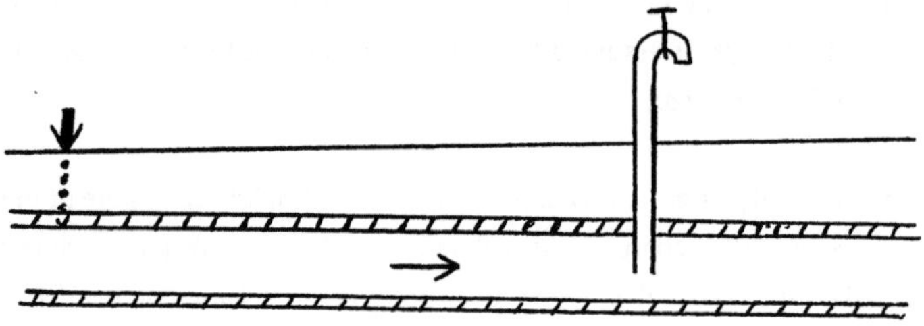

(a) sketch of the situation

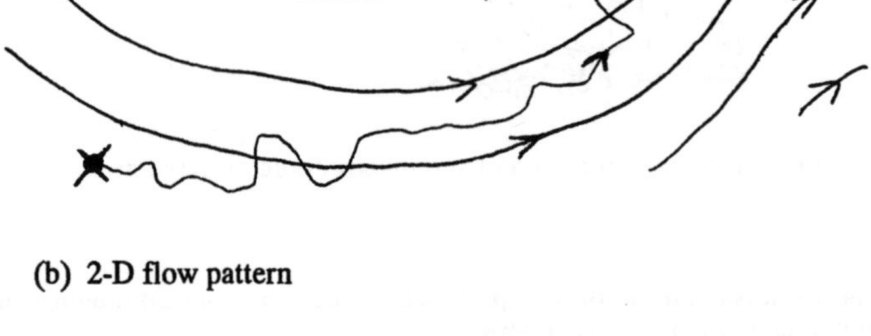

(b) 2-D flow pattern

Fig. 2 Groundwater flow towards a well with an upstream pollution site

In fig. 2 we sketch a typical situation of discharge of (drinking) water from the acquifer. We also consider the possibility that the water is polluted at a site somewhere upstream. When this site is within the domain, from which the water is pumped up, all pollution enters the well. A more refined analysis of the dispersive flow is required when the site is just outside the domain of flow affluent to the well. By dispersion a small, but possibly hazardous, portion of the polluting substance may be pumped up.

2. The Fokker-Planck equation

Let $p(x,y,t)$ be the probability density of finding a particle at a point (x,y) at time t. Then for a velocity field $v = (v_x(x,y), v_y(x,y))$ the function $p(x,y,t)$ satisfies

$$\frac{\partial p}{\partial t} = Mp \tag{1}$$

with

$$M = -\frac{\partial}{\partial x}(v_x\cdot) - \frac{\partial}{\partial y}(v_y\cdot) + \frac{\partial}{\partial x}\left(D_{xx}\frac{\partial\cdot}{\partial x}\right) + \frac{\partial}{\partial x}\left(D_{xy}\frac{\partial\cdot}{\partial y}\right) + \frac{\partial}{\partial y}\left(D_{xy}\frac{\partial}{\partial x}\right) + \frac{\partial}{\partial y}\left(D_{yy}\frac{\partial\cdot}{\partial y}\right),$$

where

$$D = \begin{pmatrix} D_{xx} & D_{xy} \\ D_{xy} & D_{yy} \end{pmatrix}$$

is the dispersion matrix. The entries of this matrix are given by

$$D_{xx} = a_T \mid v \mid + (a_L - a_T) v_x^2 / \mid v \mid,$$
$$D_{xy} = (a_L - a_T) v_x v_y / \mid v \mid,$$
$$D_{yy} = a_T \mid v \mid + (a_L - a_T) v_y^2 / \mid v \mid$$

with a_T and a_L, respectively, the transversal and longitudinal dispersion coefficient.

In the theory of stochastic processes M is called the forward operator, see e.g. Gardiner [2]. The formal adjoint of M is called the backward operator. It is of the form

$$(M^* =)L \equiv v_x\frac{\partial}{\partial x} + v_y\frac{\partial}{\partial y} + \frac{\partial}{\partial x}\left(D_{xx}\frac{\partial\cdot}{\partial x}\right) + \frac{\partial}{\partial x}\left(D_{xy}\frac{\partial\cdot}{\partial y}\right) + \frac{\partial}{\partial y}\left(D_{xy}\frac{\partial\cdot}{\partial x}\right) + \frac{\partial}{\partial y}\left(D_{yy}\frac{\partial\cdot}{\partial y}\right).$$

This backward operator plays an important role in exit problems. Let a particle start at $t = 0$ in a point $(x,y) \in \Omega$ where Ω is a bounded domain and let the boundary $\partial\Omega$ be composed of two parts $\partial\Omega_0$ and $\partial\Omega_1$. Then the probability $q(x,y)$ of leaving Ω at $\partial\Omega_1$ follows from the Dirichlet problem

$$Lq = 0 \text{ in } \Omega, \quad q = 1 \text{ at } \partial\Omega_1, \quad q = 0 \text{ at } \partial\Omega_0. \tag{2}$$

Moreover, the expected arrival time $T(x,y)$ at $\partial\Omega$ satisfies

$$LT = -1 \text{ in } \Omega, \quad T = 0 \text{ at } \partial\Omega. \tag{3}$$

For the expected arrival time $T_1(x,y)$ at $\partial\Omega_1$ we have

$$T_1 = T_q/q \tag{4a}$$

with

$$LT_q = -q \text{ in } \Omega, \quad T_q = 0 \text{ at } \partial\Omega, \tag{4b}$$

see Van Herwaarden and Grasman [5].

3. Exit from a stream domain

In this section we analyse the problem of passage of a separating streamline by a particle. Flow lines with different destinations are separated by a streamline that ends in a stagnation point, see fig. 3. We consider an arbitrary flow near such a streamline given by the velocity vector

$$(v(\rho,n),\ w(\rho,n)),$$

with $\rho > 0$ the coordinate along the streamline and n the one perpendicular to it. The stagnation point is in $(\rho,n) = (0,0)$. First we study the probability $q(\rho,n)$ that a particle starting in $\Omega = \{(\rho,n) \mid \rho, n > 0\}$ will reach the separating streamline $\partial\Omega_1$, see fig. 3. For that purpose we introduce the local coordinate $\eta = n/\sqrt{a_T}$. Switching from the x,y-coordinates to ρ and η we obtain after letting $a_L, a_T \to 0$ the asymptotic approximation

$$v(\rho,0)\frac{\partial q}{\partial \rho} + w_n(\rho,0)\eta\frac{\partial q}{\partial \eta} - v(\rho,0)\frac{\partial^2 q}{\partial \eta^2} = 0, \tag{5a}$$

$$q(\rho,0) = 1. \tag{5b}$$

Its solution reads

$$q(\rho,\eta) = \mathrm{erfc}(\eta s(\rho)) \qquad (6a)$$

with

$$s(\rho) = \left\{ \frac{2}{\nu(\rho,0)^2} \int_0^{\rho} \nu(\bar{\rho},0)^2 d\bar{\rho} \right\}^{-\frac{1}{2}}. \qquad (6b)$$

As a special case that can be handled analytically we take the problem of a well in a uniform background flow, see Van der Hoek [4]. The stream pattern satisfies

$$v_x(x,y) = \frac{x^2 + x + y^2}{(1+x)^2 + y^2}, \quad v_y(x,y) = \frac{-y}{(1+x)^2 + y^2}.$$

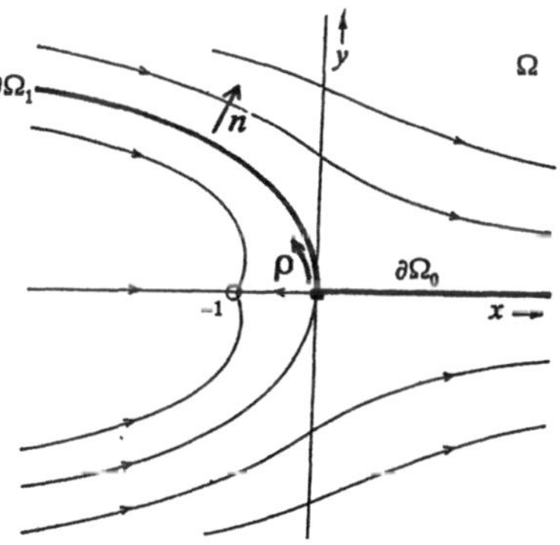

Fig. 3 Probability $q(\rho, n)$ that a particle reaches the separating streamline $\partial\Omega_1$. The probability of being pumped up at the well is approximately $\frac{1}{2}q(\rho, n)$.

The separating streamline $\partial\Omega_1$ (see fig. 3) is given by

$$x = -1 + \frac{y}{\tan y}.$$

A point at $\partial\Omega_1$ is determined by its y-value. For the corresponding ρ-value we get

$$\rho = \int_0^{y} \left\{ 1 + \left(\frac{1}{\tan\bar{y}} - \frac{\bar{y}}{\sin^2\bar{y}} \right)^2 \right\}^{\frac{1}{2}} d\bar{y}.$$

This gives for (6b)

$$s(\rho)=\left\{\frac{2}{Q(y)}\int_0^y Q(\bar{y})R(\bar{y})d\bar{y}\right\}^{-\frac{1}{2}}$$

with

$$Q(y)=\sin^2 y\{1+(\tfrac{1}{y}-\tfrac{1}{\tan y})^2\} \text{ and } R(y)=\left\{1+\left(\frac{1}{\tan y}-\frac{y}{\sin^2 y}\right)^2\right\}^{\frac{1}{2}}.$$

Consequently, we can compute the probability of exit from Ω through $\partial\Omega_1$ from (6).

4. The random walk model

The position $(X(t),Y(t))$ of a particle, for which the probability density function $p(x,y,t)$ of section 2 applies, satisfies the stochastic differential equation

$$dX=\left(v_x+\frac{\partial}{\partial x}D_{xx}+\frac{\partial}{\partial y}D_{xy}\right)dt+\sqrt{\frac{2a_L}{|v|}}v_x dW_L+\sqrt{\frac{2a_T}{|v|}}v_y dW_T,$$

$$dY=\left(v_y+\frac{\partial}{\partial x}D_{xy}+\frac{\partial}{\partial y}D_{yy}\right)dt+\sqrt{\frac{2a_L}{|v|}}v_y dW_L-\sqrt{\frac{2a_T}{|v|}}v_x dW_T,$$

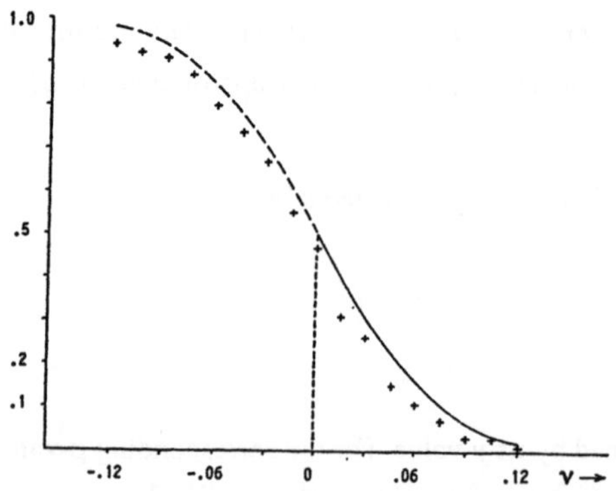

Fig. 4 The probability of reaching the well in a uniform background flow from $N = 100$ simulations at each point (ρ, n) with $\rho = 2{,}958$ for a dispersive flow with $a_T = .001$ and $a_L = .01$. The result is given by (+). The curve represents the asymptotic solution.

where $W_T(t)$ en $W_L(t)$ are independent Wiener processes. From this equation a stochastic difference equation can be derived.

In fig. 4 we give the result for different starting values $(X(0), Y(0))$ at a line normal to the separating streamline for the special problem of section 3. For each starting value a Monte Carlo run ($N = 100$) is made. In this way an approximation is found for the fraction of particles that reaches the well.

5.Some remarks

The method we presented gives an approximation for the function of particles, released in a flow, that leaves or enters some domain. This analytical result (formula (6)) can also be used for a flow that is generated numerically. In that case the integral of (6b) also has to be evaluated numerically from the available data. This computation is less complex than a numerical integration of the complete advection-dispersion equation. The method we presented can also be used to compute the expected arrival time at the well for a particle. For that purpose a nonhomogeneous Dirichlet problem for the backward equation has to be solved, see Van Herwaarden and Grasman [5].

Literature

[1] Bear, J., and A. Verruyt, *Modelling Groundwater Flow and Pollution*, Reidel, Dordrecht, 1987.

[2] Gardiner, C.W., Handbook of Stochastic Methods for Physics, Chemistry and the Natural Sciences, Springer-Verlag, Berlin, 1983.

[3] Uffink, G.J., *Application of Kolmogorov's backward equation in random walk simulations of groundwater contaminant transport*, in *Contaminant Transport in Groundwater*, H.E. Kobus and W. Kinzelbach (eds.), Balkema, 1989, p. 283-289.

[4] Van der Hoek, C.J., *Contamination of a well in a uniform background flow*, Report Dept. of Math., Agricult. Univ. Wageningen Techn. Note 90-03, 1990.

[5] Van Herwaarden, O.A., and J. Grasman, *Dispersive groundwater flow and pollution*, Math. Models and Meth. in Appl. Sci. 1, 1991, p. 61-81.

A Probability Consistent Closure Method

Mircea Grigoriu

School of Civil & Environmental Engineering
Cornell University
Ithaca, New York 14853-3501 U.S.A.

Abstract

A probability-consistent closure method is developed for finding approximate probabilistic characteristics of the response of a nonlinear dynamic system subject to a random input. The method is based on an estimator of the probability density of the system response that consists of a linear combination of specified kernels. The parameters of the linear combination can be determined from the condition that the moment equations of the dynamic system are satisfied in an optimal sense up to a closure level. The optimization involves constraints that assure that the estimator is a probability density fucntion. Several examples are presented to illustrate the proposed closure method.

Introduction

The exact determination of probabilistic characteristics of the state vector of nonlinear systems subject to random dynamic inputs is possible in only a few cases of a limited practical interest (1). Simulation, equivalent linearization, and closure methods are the most common approximate methods for finding probabilistic characteristics of the state vector of a dynamic system (1,6,8). The closure method is based on the hierarchy of moment equations of the state vector and heuristic relationships between moments of order higher and lower than a selected closure level. These relationships can be used to close the hierarchy of moment equations and find moments of the state vector approximately (1).

This paper presents a new closure method that is consistent with the theory of probability. The method is based on an estimator of the probability density function consisting of a superposition of specified kernels weighted by undetermined parameters. The parameters are determined from the condition that the moment equations of the state vector are satisfied in an optimal sense up to the closure level.

The Problem of Moments

Consider a bounded non-decreasing function F such that

$$\int x^n \, dF(x) = \mu_n \quad , \qquad n = 0, 1, 2, \ldots \tag{1}$$

where μ_n are prescribed values. This problem of moments was first formulated by Stieltjes who has also gave necessary and sufficient conditions

N. Bellomo · F. Casciati (Eds.)
Nonlinear Stochastic Mechanics
IUTAM Symposium Turin

for the existence of a solution when F(x) is defined on the positive real line (1). The Stieltjes formulation can be generalized, e.g., by extending the domain of definition of F to the entire real line (1)

The solution of the problem of moments is usually not unique. There generally exist two ore more different distributions satisfying Eq. 1. For example, consider the lognormal probability density

$$f(x) = \frac{1}{\sqrt{2\pi}} x^{-1} e^{-1/2(\log x)^2} \quad , \qquad x > 0 \tag{2}$$

and the density

$$\tilde{f}(x) = f(x)\ [1 + a \sin(2\pi \log x)] \ , \ x > 0, \ -1 \le a \le 1 \tag{3}$$

These densities have the same moments but differ significantly (2).

The Problem of Closure

Consider a random variable $\underline{X}$ with values in R^m, m = 1, 2, ..., and moments $\mu_{s_1 \ldots s_m} = E \prod_{i=1}^{m} X_i^{s_i}$ of order $s = \sum_{i=1}^{m} s_i$, that are assumed to exist up to any desired order. The objective is to estimate the distribution F and the density f of random variable $\underline{X}$ from moments $\mu_{s_1 \ldots s_m}$ of this variable of order s = 0, 1, ..., k, where k is the closure level. The problem does not have a unique solution because it is based on less information than in the problem of moments for $k < \infty$.

Details of the proposed closure method are first presented for random variables. Then, the method is extended to the case of nonlinear dynamic systems subject to random inputs. Examples are used to demonstrate the proposed closure method.

Direct Method. Let m = 1 and assume that the random variable X takes on values in [0, 1] almost surely. Then, moments $\mu_s = EX^s$, s = 0, 1, ..., form a completely monotone sequence because

$$(-1)^r \Delta^r \mu_s \ge 0 \quad , \qquad r = 0, 1, \ldots \tag{4}$$

in which $\Delta\mu_s = \mu_{s+1} - \mu_s$.

Suppose that the first k moments of X are known. Then,

$$p_j^{(k)} = \binom{k}{j} (-1)^{k-j} \Delta^{k-j} \mu_j \quad , \qquad j = 0, 1, \ldots, k \tag{5}$$

are probabilities of a discrete random variable $X^{(k)}$ with values j/k, j = 0, 1, ..., k. Moreover, the cumulative distribution function

$$F^{(k)}(x) = \sum_{j/k \leq x} p_j^{(k)} \tag{6}$$

of $X^{(k)}$ converges to the distribution F(x) of X as $k \to \infty$ at each point of continuity of F (2). These results may suggest that $F^{(k)}$ can be used to approximate F. However, $F^{(k)}$ converges slowly to F and its moments differ from the moments of F.

Alternatively, it is possible to approximate X by another discrete random variable $Y^{(k)}$ taking the values $\{x_j\}$, j = 0, 1, ..., k, with the unknown probabilities $q_j^{(k)}$, j = 0, 1, ..., k. The probabilities $q_j^{(k)}$ can be obtained from the conditions that the first k moments

$$\hat{\mu}_s = \sum_{j=0}^{k} x_j^s \, q_j^{(k)} \tag{7}$$

of $Y^{(k)}$ coincide with the corresponding moments of X. However, the resultant parameters $q_j^{(k)}$ are not probabilities because they can be negative. The approach can be generalized if neither the location nor the probability mass of the atoms of the density of $Y^{(k)}$ are specified. The analysis becomes difficult and results may not be consistent with the theory of probability. A notable exception is the case in which k = 3 and $\mu_1 = EX = 0$ (5).

Optimization Method. Consider an estimator

$$\hat{f}(\underline{x}) = \sum_{q=0}^{n} f_q \, \alpha_q(\underline{x}) \tag{8}$$

of the density $f(\underline{x})$ of $\underline{X}$. The parameters f_q and the specified kernels $\alpha_q(\underline{x})$ of the estimator satisfy the conditions

$$f_q \geq 0, \quad \alpha_q(\underline{x}) \geq 0, \quad \underline{x} \epsilon R^m, \qquad q = 0, 1, \ldots, n$$
$$\int_{R^m} f(\underline{x}) \, d\underline{x} \tag{9}$$

The nodal values f_q of the estimator $\hat{f}$ can be determined from the condition that the differences between the approximate moments

$$\hat{\mu}_{s_1 \ldots s_m} = \sum_{q=0}^{n} f_q \int \underset{R^m}{\cdots} \int x_1^{s_1} \ldots x_m^{s_m} \alpha_q(\underline{x})\, d\underline{x} \tag{10}$$

of the estimator and the exact moments $\mu_{s_1 \ldots s_m} = E \prod_{i=1}^{m} X_i^{s_i}$ be minimized in some sense. These differences can be measured by various objective functions, e.g.,

$$\epsilon_1 = \sum_{s=1}^{k} \left[\hat{\mu}_{s_1 \ldots s_m} - \mu_{s_1 \ldots s_m} \right]^2 \tag{11}$$

and

$$\epsilon_2 = \max_{i \le s \le k} \left| \hat{\mu}_{s_1 \ldots s_m} - \mu_{s_1 \ldots s_m} \right| \tag{12}$$

in which k is the closure level. The optimal nodal values f_q can be determined by minimizing the objective functions ϵ_1 or ϵ_2 with the constraints

$$f_q \ge 0 \quad , \quad q = 0, 1, \ldots, n$$
$$\int \underset{R^m}{\cdots} \int \hat{f}(\underline{x})\, d\underline{x} = 1 \tag{13}$$

The use of ϵ_2 has numerical advantages because linear programming algorithms can be used for solution. Additional constraints can be added to the optimization problem in Eqs. 11-13. This paper considers a Lipschitz-type condition of smoothness for the estimator. It is required that

$$\left| \hat{f}(\underline{x}_1) - \hat{f}(\underline{x}_2) \right| \le \eta \left\| \underline{x}_1 - \underline{x}_2 \right\| \quad , \quad 0 < \eta < 1 \tag{14}$$

The condition limits rapid fluctuations of $\hat{f}(\underline{x})$ that are rarely observed in practice.

The selection of the functional form of the estimator in Eqs. 8 and 9 involves two phases. First, a bounded domain $D \subseteq R^m$ and a partition of it in cells are selected such that most of the probability of $\underline{X}$ is included in D. Second, interpolation functions are selected to represent $\hat{f}(x)$ within each cell of D. For example, consider a partition of D in rectangular cells. Each cell can be mapped into the standard cell $\overset{m}{\underset{i=1}{X}} [-1, 1]$ by translation and scaling. The estimator $\hat{f}(x)$ in the standard cell is

$$\hat{g}(\xi) = \sum_{r=1}^{2^m} f_r \, \psi_r(\xi) \tag{15}$$

in which f_r are nodal values,

$$\psi_r(\xi) = \frac{1}{2^m} \prod_{i=1}^{m} (1 + \zeta_{r,i} \, \xi_i) \tag{16}$$

denote interpolation functions, and $\zeta_{r,i}$ = the coordinates of node r.

Consider the one-dimensional case (m=1) and let $\{x_q\}$, q = 0, 1, ..., n, be nodal points. Examples of shape functions for the estimator in Eqs. 8 and 9 are

$$\alpha_q(x) = \begin{cases} \dfrac{x-x_q-1}{x_q-x_q-1} & , \quad x \in [k_{q-1}, x_q] \\ \dfrac{x_{q+1}-x}{x_{q+1}-x_q} & , \quad x \in [x_q, x_{q+1}] \\ 0 & , \quad \text{otherwise,} \end{cases} \tag{17}$$

$$\alpha_q(x) = \exp[-\zeta_q(x - x_a)^2] \tag{18}$$

or

$$\alpha_q(x) = \delta(x-x_q) \tag{19}$$

in which δ = the delta function.

<u>Extension of Optimization Method to Nonlinear Random Vibration</u>. Let $\underline{X}(t)$ be the stationary response of a nonlinear dynamic system subject to a random excitation. The moments $\mu_{s_1 \ldots s_m} = E\, X_1(t)^{s_1} \ldots X_m(t)^{s_m}$ of $\underline{X}(t)$ satisfy an infinite hierarchy of algebraic equations that can be obtained by using the Itô rule.

Consider as in the previous section an estimator $\hat{f}(\underline{x})$ of the probability density of $\underline{X}(t)$ with moments $\hat{\mu}_{s_1 \ldots s_m}$ in Eq. 10. The optimal values of the parameters $\{f_q\}$, q = 0, ..., n, in Eq. 8 can be obtained from the minimization of the objective functions in Eqs. 11 and 12 in which the differences $\hat{\mu}_{s_1 \ldots s_m} - \mu_{s_1 \ldots s_m}$ are replaced by the moment equations of order s of $\underline{X}(t)$ based on the approximate moments $\hat{\mu}_{s_1 \ldots s_m}$ of the state vector.

The application of this version of the closure method is demonstrated in Example 3 of the next section involving a one-dimensional nonlinear random vibration problem with additive noise.

Application

The optimal estimator in Eqs. 8, 9, and 14 is used to approximate moments and probability density functions of a one-dimensional random variable and diffusion process. The diffusion process represents the response of a one-dimensional nonlinear system subject to white noise. It is assumed that the first k moments or the first k moment equations are available for random variable and process, respectively.

Example 1. Consider a random variable X with the probability density function

$$f(x) = \alpha \frac{1}{\sigma_1} \phi \left[\frac{x-\mu_1}{\sigma_1} \right] + (1 - \alpha) \frac{1}{\sigma_2} \phi \left[\frac{x-\mu_2}{\sigma_2} \right], \quad -\infty < x < \infty \tag{20}$$

in which $\phi(u) = \exp(-0.5\, u^2)/\sqrt{2\pi}$, $\alpha = 0.3$, $\mu_1 = -\mu_2 = -3$, and $\sigma_1 = \sigma_2 = 1.0$. Figure 1 shows the exact density f(x) in Eq. 20 and estimates of it in Eqs. 8 and 9 for D = (-10, 10), n = 40 nodes, closure level k = 8, and the Gaussian kernels in Eq. 18 with $\rho_q = 0.4$. The probability estimates differ significantly depending on the imposed smoothness condition. On the other hand, Table 1 shows that the estimated moments are remarkably stable. Unsatisfactory moments are only obtained when smoothing is excessive.

Example 2. Consider the same random variable X and estimators as in the previous example, the case in which smoothness is adequate. Figure 2 shows the sensitivity of the estimator to the closure level. A very low closure level (k = 1) is insufficient to detect the bimodal feature of the density f(x). However, the bimodal character of the density is identified when the closure level is k = 4. Further improvements can be obtained as the closure level increases. The moments in Table 2 are satisfactory even when their order exceeds the closure level, e.g., moments of order 5, 6, 7, and 8 for closure level k = 4.

Example 3. Consider the scalar diffusion process

$$dX(t) = [a_1\, X(t) + a_2\, X(t)^2 + a_3\, X(t)^3]\, dt + \beta\, dB(t) \tag{21}$$

in which B(t) is the Brownian motion process with Gaussian independent increments dB(t) of mean zero and variance dt. This constitutes a quite general one-dimensional nonlinear random vibration problem with additive

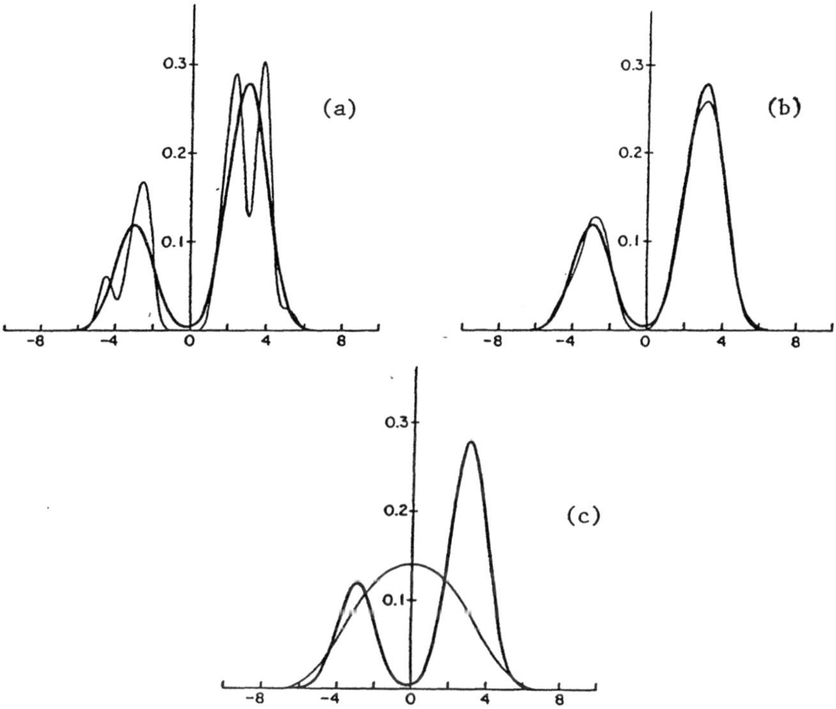

Figure 1. Exact (▬) and estimated (—) densities: (a) no smoothness, (b) adequate smoothness, and (c) excessive smoothness.

Table 1. Exact and Estimated Moments

Order	Moments			
	Exact	Optimal Estimator		
		No Smoothing	Adequate Smoothing	Excessive Smoothing
1	1.20	1.20	1.20	-8.2×10^{-2}
2	1.00×10	1.00×10	1.00×10	6.23
3	1.44×10	1.44×10	1.44×10	-1.45
4	1.38×10^{2}	1.38×10^{2}	1.38×10^{2}	9.26×10
5	2.23×10^{2}	1.23×10^{2}	2.23×10^{2}	-3.28×20
6	2.36×10^{3}	2.36×10^{3}	2.36×10^{2}	1.93×10^{3}
7	4.18×10^{3}	4.18×10^{3}	4.18×10^{3}	$-8.68.10^{2}$
8	4.79×10^{4}	4.79×10^{4}	4.79×10^{4}	4.86×10^{4}

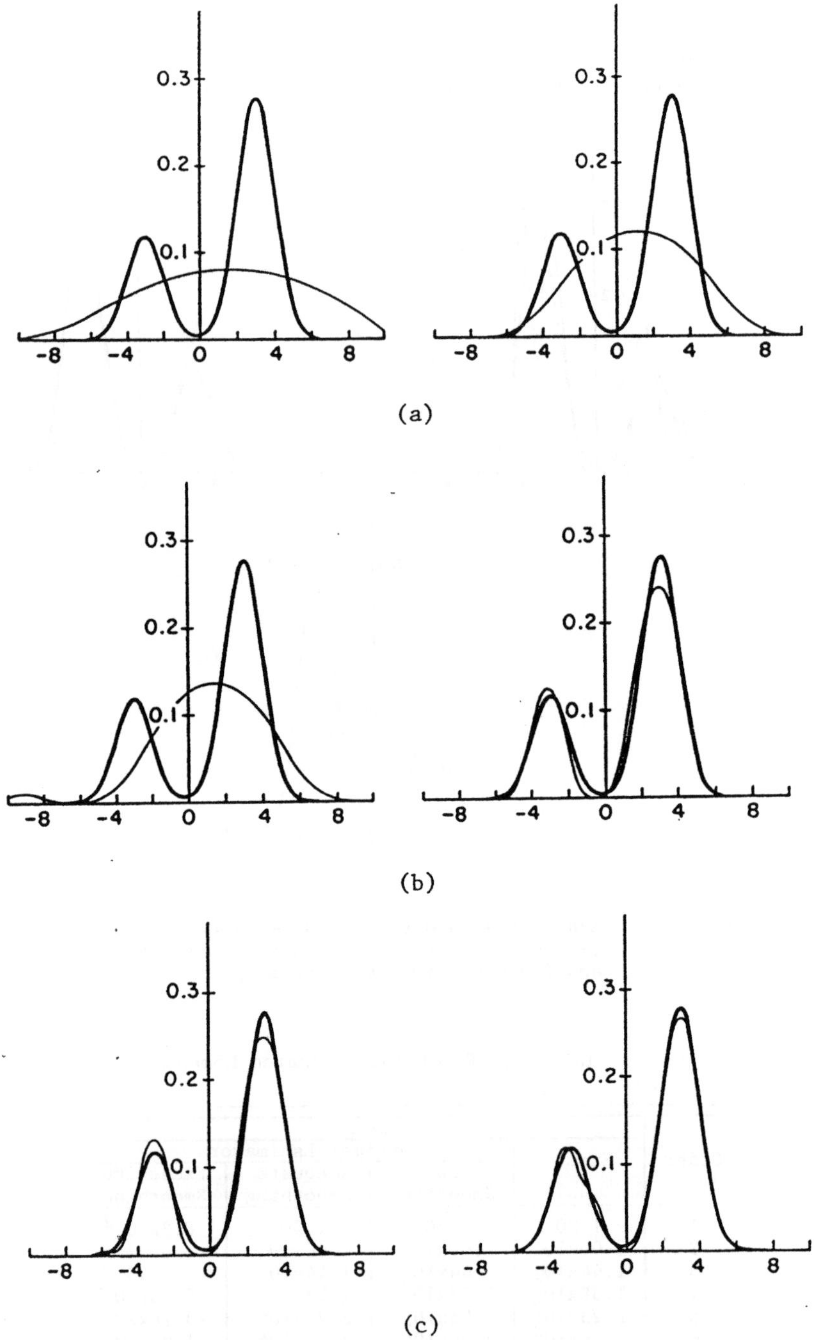

Figure 2. Exact (▬) and estimated (—) densities: (a) k = 1, (b) k = 4, and (c) k = 8.

Table 2. Exact and Estimated Moments

Order	Moments			
	Exact	Optimal Estimator with Adequate Smoothing		
		k=1	k=4	k=10
1	1.20	1.20	1.20	1.20
2	1.00x10	1.98x10	1.00x10	100x10
3	1.44x10	5.94x10	1.44x10	144x10
4	$1.38x10^2$	$8.93x10^2$	$1.38x10^2$	$1.38x10^2$
5	$2.23x10^2$	$3.72x10^3$	$2.40x10^2$	$2.23x10^2$
6	$2.36x10^3$	$5.37x10^4$	$2.32x10^2$	$2.36x10^3$
7	$4.18x10^3$	$2.67x10^5$	$4.82x10^3$	$4.18.10^3$
8	$4.79x10^4$	$3.76x10^6$	$4.48x10^4$	$4.79x10^4$

noise. The stationary density of the process is

$$f(x) = C \exp \left[\frac{1}{\beta^2} \left\{ a_1 x^2 + \frac{2}{3} a_2 x^3 + \frac{1}{2} a_3 x^4 \right\} \right] \qquad (22)$$

where C is a normalization constant. The stationary moments $\mu_s(t) = \mu_s = EX(t)^s$ of the process satisfy the algebraic equations (1,4)

$$\frac{\beta^2}{2} s(s-1) \mu_{s-2} + s[a_1\mu_s + a_2\mu_{s+1} + a_3\mu_{s+2}] \qquad (23)$$

The objective function corresponding to Eq. 11 is

$$\epsilon_1 = \sum_{s=1}^{k} \left[\frac{\beta^2}{2} s(s-1) \hat{\mu}_{s-2} + s(a_1\hat{\mu}_s + a_2\hat{\mu}_{s+1} + a_3\hat{\mu}_{s+2}) \right]^2 \qquad (25)$$

in which $\hat{\mu}_s = \int x^s \hat{f}(x)\, dx$. The procedure of the previous section can be applied to find the optimal estimator $\hat{f}(x)$ and its moments.

Figure 3 shows the exact density f(x) and the estimator $\hat{f}(x)$ with and without smoothing. Results correspond to $a_1 = 4$; $a_2 = 0.5$; $a_3 = -0.5$; $\beta = 4$; a closure level k = 8, and Gaussian kernels. The estimator without smoothing differ significantly from the density of X(t). However, the estimated moments of X(t) are not sensitive to the degree of smoothing used in the analysis (Table 3).

Conclusions

A method was developed for estimating probabilistic characteristics of random variables and processes from their first k moments or moment equations. The method is based on an estimator of the probability density function involving specified kernel functions weighted by undetermined

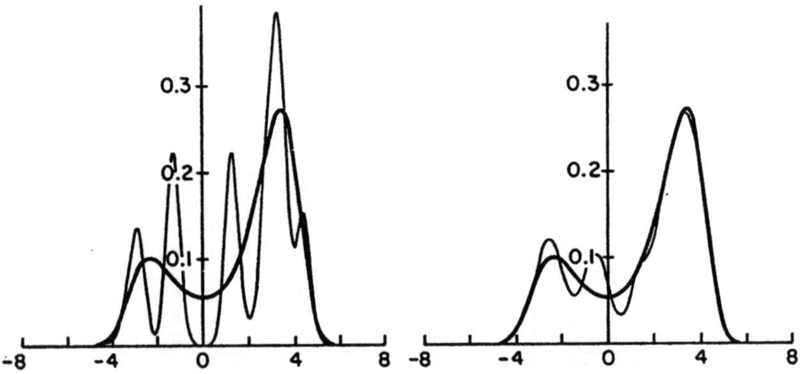

Figure 3. Exact (▬) and estimated (—) densities: (a) no smoothing, and (b) smoothing.

Table 3. Exact and Estimated Moments.

Order	Moments		
	Exact	Optimal Estimator	
		No Smoothing	Smoothing
1	1.46	1.41	1.37
2	1.15	1.00x10	8.03
3	1.99x10	1.94x10	1.90x10
4	$1.01x10^2$	$1.00x10^2$	$1.00x10^2$
5	$3.07x10^2$	$3.01x10^2$	$2.95x10^2$
6	$1.51x10^3$	$1.49x10^3$	$1.47x10^3$
7	$5.23x10^3$	$5.14x10^4$	$5.06x10^3$
8	$2.54x10^4$	$2.51x10^4$	$2.48x10^4$

parameters. The optimal estimator can be obtained from the condition that the estimator matches in an optimal sense a set of moments or moment equations.

The method was first applied to estimate the probability density of a random variable with a bimodal density from its first k moments. Then, estimates were obtained for the moments and the probability density of a scalar diffusion process defined by a stochastic differential equation. Results are satisfactory and show the robustness of the proposed estimator.

Acknowledgements

Discussions with Professor A. M. Hasofer were most helpful in the completion of this paper. Mr. K. Yuan performed the numerical calculations.

References

1. V. V. Bolotin, Random Vibrations of Elastic Systems, Martinus Nijhoff Publishers, 1984, Boston.

2. W. Feller, An Introduction to Probability Theory and Its Applications, Vol. 2, John Wiley, New York, 1971.

3. M. Grigoriu, "A Consistent Closure Method for Non-Linear Random Vibration," International Journal of Non-Linear Mechanics, Vol. 26, 1991.

4. A. H. Jazwinski, Stochastic and Filtering Theory, Academic Press, Inc., New York, 1970.

5. E. Rosenblueth, "Two-point Estimates in Probabilities," Applied Mathematical Modelling, Vol. 5, Oct. 1981, pp. 329-335.

6. M. Shinozuka and G. Deodatis, "Simulation of Stochastic Processes by Spectral Representation," Applied Mechanics Reviews, 1991.

7. J. A. Shohat and J. D. Tamarkin, The Problem of Moments, American Mathematical Society, New York, 1943.

8. P. D. Spanos and W. D. Iwan, "On the Existence and Uniqueness of Solutions Generated by Equivalent Linearization," International Journal of Nonlinear Mechanics, 13, 1978, pp. 71-78.

Probabilistic Methods to Waves in Heterogeneous Media

Henryk Gzyl and Raúl Jiménez

1 Introduction.

This line of work is motivated by our wanting to solve wave equations, to be used for direct modelling in geophysical applications, without having to go thru the wear and tear of numerical grid methods. Here we provide path integral representations for the solution to wave equation (1) which can be the starting point for Montecarlo simulation techniques.

The problem in its simplest versions consists in solving the scalar wave equations

$$u_{tt} = c^2(x)\Delta u + f(t,x) \quad \text{for} \quad t > 0, \quad x \in D \tag{1}$$

where D is a bounded or unbounded domain in R^d, having a smooth boundary ∂D. Equation (1) is suplemented with the initial conditions

$$u(0,x) = \varphi(x), \quad u_t(0,x) = \psi(x) \quad \text{for} \quad x \in D \tag{2}$$

and subject to the boundary conditions of Dirichlet type

$$u(t,x) = g(t,x) \quad \text{for} \quad t > 0, \quad x \in \partial D \tag{3}$$

or to the boundary conditions of Newmann type

$$(\gamma(x), \nabla_x u(t,x)) = -g(t,x) \quad \text{for} \quad t > 0, \quad x \in \partial D \tag{4}$$

where $\gamma(x)$ is an appropriate nontangential vector field on ∂D.

Actually, most of our probabilistic analysis can be carried out for right hand terms with $c^2\Delta$ replaced by an operator $\mathcal{L}$ of the type

$$\mathcal{L} = \sum_{ij} a_{ij}(x)\partial^2_{ij} + \sum_{i=1}^{d} b_i(x)\partial_i \tag{5}$$

where $\partial_i = \partial/\partial x_i$. The coefficients $a_{ij}(x)$, $b_j(x)$ are asssumed to be bounded Hölder continuous, which is not unreasonable from the physical point of view and ensures that the diffusions naturally associated with $\mathcal{L}$ to have the desired properties.

We direct the reader to [4],[5] for a guide to the basic references, and here we only stress that the probability theory we need is well known and, in most cases, has already made it to standard textbooks.

Let $X(t)$ be the diffusion in R^d having the operator $\mathcal{L}$ as generator. We shall assume that with $X(t)$ is associated a transition semigruop P_t defined on the bounded measurable functions f on R^d by means of

$$P_t f(x) = E^x f(X_t)$$

and that a strictly positive density $\rho(x)$ exists such that

$$P_t f(x) = \int p_t(x,y) f(y) \rho(y) dy$$

and furthermore $p_t(x,y)$ is symmetric.

N. Bellomo · F. Casciati (Eds.)
Nonlinear Stochastic Mechanics
IUTAM Symposium Turin

When dealing with the Neumann boundary conditions we shall assume that $\rho(x)$ is defined on $\bar{D}$, $p_t(x,y)$ is defined on $\bar{D}\text{x}\bar{D}$, and satisfies

$$(\gamma,\ \nabla p_t(x,y)) = 0 \text{ for } x\epsilon D.$$

In both cases we assume that $p_t(x,y)$ satisfies

$$\partial_t P_t = \mathcal{L}P_t \quad t>0, \quad x \in D, \quad y \in D$$

Associated to P_t there is a semigroup Q_t defined on the bounded functions on D by

$$Q_t f(x) = E^x[f(X_t;\ t<T] \tag{6}$$

where T is the stopping time $T = inf\{t > 0 : X_t\epsilon\partial D\}$. To this process there is associated a process $\hat{X}_t$ that coincides with X_t up to T and is killed and sent to a cementery point Δ at $t = T$. But this is not our business here.

2 Transmutation approach.

To begin this section, observe that the operator

$$(Tf)(t) = \frac{1}{2\sqrt{\pi t^3}}\int_0^\infty \tau e^{-\tau^2/4t} f(\tau)d\tau. \tag{7}$$

maps a function $f(t)$ having at least two derivatives onto a function $(Tf)(t)$ having at least one derivative such that

$$\frac{\partial}{\partial s}(Tf)(s) = (T(\frac{\partial^2}{\partial t^2}f))(s) \tag{8}$$

As consequence of this we obtain the fact that if $u(t,x)$ satisfies the wave equation with initial condition $u(0,x) = 0$, $u_t(0,x) = \psi(x)$, then $v(s,x) = (Tu(\cdot,x))(s)$ satisfies

$$v_t = \mathcal{L}v + f^* \text{ for, } v(0,x) = \psi(x) \tag{9}$$

where $f^*(s,x) = (Tf(\cdot,x))(s)$. Furthermore, if $u(t,x)$ satisfies the Dirichlet boundary conditions (3) then $v(t,x)$ satisfies the Dirichlet conditions

$$v(t,x) = g^*(t,x) = (Tg(\cdot,x))(t) \tag{10}$$

The solution to (9,10) is given by

$$\begin{aligned} v(t,x) &= Q_t\psi(x) + \int_0^t Q_{t-s}f^*(s,x)ds \\ &+ \int_0^t \int g^*(t-s,y)P^x(X_T\epsilon dy, T\epsilon ds) \end{aligned} \tag{11}$$

The last term is the inverse Laplace transform of

$$E^x[g^*(\alpha, X_T)e^{-\alpha T}] \tag{12}$$

and when the domain D has a smooth boundary and is such that a Green function exists, then

$$P^x(X_T\epsilon dy, T\epsilon ds) = \frac{1}{2}\frac{\partial}{\partial\eta_y}q_s(x,y)\ ds\ \sigma(dy)$$

where $q_s(x,y)$ is essentially the density of the semigroup introduced in (6) and $\sigma(dy)$ is Lebesgue area measure on ∂D. Also $\frac{\partial}{\partial\eta_y}$ denotes the derivative along the inward normal at y (The basic result appears in [7]).

The interest on (11) and (12) is that, according to (6) $Q_s\psi$ can be computed as a path integral

$$Q_s\psi(x) = E^x[\psi(X_s); s < T]$$

wich can be computed by simulating a diffusion X.

The question now is to invert the transmutation operator T introduced in (7). For this, set

$$B(t) = \sum_{n=0}^{\infty} \frac{(-1)^n t^{2n}}{2n!n!} = {}_0\Gamma_2(1, \frac{1}{2}, -\frac{t^2}{4}) \tag{13}$$

and define

$$(Rg)(t) = \int_0^\infty J_0(2\sqrt{\tau t}) \int_0^\infty B(\tau\sqrt{s})F(s)dsd\tau. \tag{14}$$

Using the following identities

$$\int_0^\infty e^{-\alpha s}\frac{te^{-t^2/4s}}{2\sqrt{\pi s^3}}ds = e^{-t\sqrt{\alpha}},$$
$$\int_0^\infty e^{-\alpha\tau}B(\tau\sqrt{s})d\tau = \frac{1}{\alpha}\,e^{-s/\alpha^2},$$
$$\int_0^\infty e^{-\alpha\tau}J_0(2\sqrt{\tau t})dt = \frac{1}{\alpha}\,e^{-\tau/\alpha},$$

it is easy to verify that R defined above and T defined by (7) are inversese to each other. This implies that the solution to (1) is obtained by applying the operator R to (11). See [6] for more.

3 Laplace Inversion Techniques

Standard folklore dictates that initial value problems should be approached by means Laplace transforms. Doing that, recast our original problem as

$$(\mathcal{L} - \alpha^2)\hat{u}(\alpha, x) = -\hat{\phi}(\alpha, x) \tag{15}$$

with $\hat{\phi}(\alpha, x) = \hat{f}(\alpha, x) - \alpha\varphi(X) - \psi(x)$. Here $\hat{F}(\alpha, x)$ is the Laplace Transform of $F(t, x)$. Now the boundary conditions become

$$\hat{u}(\alpha, x) = \hat{g}(\alpha, x) \tag{16}$$

in the Dirichlet case or

$$(\gamma(x), \nabla_x\hat{u}(t, x)) = -\hat{g}(t, x) \tag{17}$$

in the Neumann case.

The easiest case corresponds to the whole space, therefore we begin with (1) plus (2) and we are left with (15) only. In this case we can obtain $\hat{u}(\alpha, x)$ as

$$\hat{u}(\alpha, x) = \int_0^\infty e^{-\alpha^2 t}P_t\hat{\phi}(\alpha, x)dt = E^x\int_0^\infty e^{-\alpha^2 t}\hat{\phi}(\alpha, X_t)dt \tag{18}$$

When we work in a domain D, instead of (18) we obtain for the Dirichlet case

$$\hat{u}(\alpha, x) = E^x\int_0^T e^{-\alpha^2 t}\hat{\phi}(\alpha, X_t)dt + E^x[e^{-\alpha^2 T}\hat{g}(\alpha, X_T), T < \infty] \tag{19}$$

when D is not bounded the solution provided by (19) is not unique unless D^c is recurrent for $X(t)$. See references in [2] and [4].

For for the Neumann case

$$\hat{u}(\alpha, x) = E^x\int_0^\infty e^{-\alpha^2 t}\hat{\phi}(\alpha, X_t)dt + E^x\int_0^\infty e^{-\alpha^2 t}\hat{g}(\alpha, X_t)dL_t \tag{20}$$

where L_t is the Local time at the boundary. See [3] and [5] for details about this.

From representations (18),(19), or (20) we can move on in several directions. One can prove existence results for (1) plus the appropiate boundary conditions as follows: first prove that $\hat{u}(\alpha, x)$ is a Laplace Transform of a function and then that function satisfies the wave equation plus the appropriate boundary conditions.

A different direction consists in obtaining large time behaviour of the solutions to (1) from the behaviour of $\hat{u}(\alpha, x)$ for small α. Preliminary results appears in [4].

Another direction consists in studying **highfrecuency** approximations, that is, to study $\hat{u}(\alpha, x)$ for large values of α. The following results were obtained in [1] and [2].

We shall begin by considering (1) on the whole space, but with $\mathcal{L} = c^2\Delta$, $0 < c_1 < c(x) < c_2 < \infty$. Then we can rewrite (15) as

$$(\frac{1}{2}\Delta - \alpha^2 q(x))\hat{u}(\alpha, x) = -\frac{1}{2}q(x)\hat{\phi}(\alpha, x) \tag{21}$$

with $q(x) = 1/2c^2(x)$. By rescaling coordinates as x/c_0 for some reference speed c_0, putting $\eta^2(x) = q(x)/c_0^2$ we have, instead of (15) the representation

$$\hat{u}(\alpha, x) = E^x \int_0^\infty e^{-\alpha^2 A_t}\eta^2(B_t)\hat{\phi}(\alpha, B_t)dt$$

where B_t is a standard d-dimensional brownian motion and $A_t = \int_0^t \eta^2(B_s)ds$. Assume that there is one smooth solution to

$$(\nabla W)^2(x) = \eta^2(x)$$

and denote by $X_0(t)$ the solution to

$$dX_0(t) = -\nabla W(X_0(t))dt, \quad X_0(0) = x$$

and introduce the time scale

$$\tau(t, x) = \int_0^t \eta^2(X_0(s))ds = W(x) - W(X_0(t))$$

With all this, one can prove that

$$\hat{u}(\alpha, x) = \int_0^\infty e^{-\alpha\tau(t,x)}\eta^2(X_0(t))\hat{\phi}(\alpha, X_0(t))e^{-\frac{1}{2}\int_0^t \Delta W_0(X_0(s))ds}dt + O(1/\alpha) \tag{22}$$

and since η^2 is strictly positive and $d\tau = \eta dt$ we can time change this to obtain

$$\hat{u}(\alpha, x) = \int_0^\infty e^{-\alpha\sigma}\hat{\phi}(\alpha, X_0(t(\sigma)))e^{-\frac{1}{2}\int_0^\sigma \Delta W_0(X_0(t(\sigma')))d\sigma'}d\sigma + O(1/\alpha) \tag{23}$$

When we are within a bounded domain D like the earth, we have a boundary condition that suplements (21), namely $\hat{u}(\alpha, x) = \hat{h}(\alpha, x)$ for $x \epsilon \partial D$. In this case representation (19) reads

$$\hat{u}(\alpha, x) = E^x \int_0^T e^{-\alpha^2 A_t}\hat{\phi}(\alpha, B_t)dt + E^x[e^{-\alpha^2 A_T}\hat{g}(\alpha, B_T)]$$

Procceding as above, we obtain in this case that

$$\begin{aligned}\hat{u}(\alpha, x) \quad &\sim \textstyle\int_0^{T_0(x)} e^{-\alpha\tau(t,x)}e^{-\frac{1}{2}\int_0^t \Delta W(X_0(s))ds}\hat{\phi}(\alpha, X_0(t))dt \\ &+ e^{-\frac{1}{2}\int_0^{T_0(x)} \Delta W(X_0(s))ds}e^{-\alpha\tau(T_0(x),x)}\hat{g}(\alpha, X_0(T_0(x)))\end{aligned}$$

where $T_0(x)$ is the time that it takes to hit ∂D along the trajectory $X_0(t)$.

4 Local times and Waves in Stratified Media

Here we shall consider a one dimensional situation and only assert that the multi-dimensional problem presents no extra difficulty. Also, we shall consider (1) with some preasigned initial conditions, but no boundary conditions. We are again in the one dimensional version of (15), the solution to which is represented by

$$\hat{u}(\alpha,x)=\frac{1}{2}E^x\int_0^\infty e^{-\alpha^2 A_t}\eta^2(B_t)\hat{\phi}(\alpha,B_t)dt \tag{24}$$

Recall that the local time L_t^a at a of the brownian motion is a random increasing functional, continuous in (t,a) such that

$$L_t^a=\int_0^t I_{\{a\}}(B_s)dL_s^a,\text{ and }\int_0^t F(B_s)ds=\int_{-\infty}^\infty F(a)L_t^a da$$

i.e., it is a random occupation density. With this (24) can be rewritten as

$$\hat{u}(\alpha,x)=\frac{1}{2}\int_{-\infty}^\infty\hat{\phi}(\alpha,a)\eta^2(a)(E^x\int_0^\infty e^{-\alpha^2 A(\tau_t^a)}dt)da$$

where τ_t^a is the rigth continuuous inverse of L_t^a. It can be proved that

$$E^x\int_0^\infty e^{-\alpha^2 A(\tau_t^a)}dt=\frac{W_\alpha(x,a)}{J(\alpha,a)}$$

where $W_\alpha(x,a)=W_\alpha^>(x,a)$ or $W_\alpha^<(x,a)$ depending on whether $x>a$ or $x<a$, and in each case satisfies

$$(\frac{d^2}{dx}-\alpha^2\eta^2(x))W_\alpha=0,\ W_\alpha(0,a)=1,\ W\downarrow 0 \text{ as}|x|\to\infty$$

More importantly

$$J(\alpha,x)=\frac{1}{2}\{(\frac{d}{dx}W_\alpha^<)(x,x)-(\frac{d}{dx}W_\alpha^>)(x,x)\}$$

With all this $\hat{u}(\alpha,x)$ becomes

$$\hat{u}(\alpha,x)=\frac{1}{2}\int_{-\infty}^\infty\hat{\phi}(a,\alpha)\eta^2(a)\frac{W_\alpha(x,a)}{J(\alpha,a)}da.$$

Similar formulae can be found for the cases of Dirichlet or Neumann boundaries, and we refer the reader to [3] and [5] for more.

Final Comments: We should mention that probabilistic methods applied to wave equations, especially for Montecarlo simmulation are not new, see reference [8].

References

[1] Gzyl, H. and Jiménez, R.: PROBABILISTIC APPROACH TO WAVE PROPAGATION PROBLEMS, I. Mathl. Comput. Modelling, vol 13, n° 11, pp. 57-65, 1990

[2] Gzyl, H. : PROBABILISTIC APPROACH TO WAVE PROPAGATION PROBLEMS, II: BOUNDARY VALUE PROBLEMS. Mathl. Comput. Modelling, vol 13, n° 11, pp. 67-76, 1990

[3] Gzyl, H. : PROBABILISTIC APPROACH TO WAVE PROPAGATION PROBLEMS, III: BROWNIAN LOCAL TIMES AND WAVES IN LAYERED MEDIA. Mathl. Comput. Modelling, vol 15, n° 7, pp. 56-62, 1991

[4] Gzyl, H. and Jiménez, R.: DIFFUSIONS AND WAVES. To appear in J. Math. Anal Appl.

[5] Betz, C and Gzyl, H. : PROBABILISTIC APPROACH TO WAVE PROPAGATION PROBLEMS, IV: REFLECTING BOUNDARIES. To appear in Mathl. Comput. Modelling.

[6] Gzyl, H. and Jiménez, R.: FROM HEAT EQUATIONS TO WAVE EQUATIONS. MORE ON TRANSMUTATION OPERATORS. To appear in Mathl. Comput. Modelling.

[7] Hsu, P.: BROWNIAN EXIT DISTRIBUTION OF BALL. Seminar Sochastic Processes, Birkhauser, Boston (1985).

[8] Ermakov, S., Nekrutkin, V. and Sipin, A.: RANDOM PROCESSES FOR CLASSICAL EQUATIONS OF MATHEMATICAL PHYSICS. Kluwer Academic Publishers, Dordrecht (1989).

Henryk Gzyl
A.P. 52120
Caracas 1050 – *A*
Venezuela

Raul Jimenez
A.P. 89000
Caracas 1080 – *A*
Venezuela

Solutions of Random Initial Value Problems

D.G. Harlow and T.J. Delph
Lehigh University
Bethlehem, PA 18015, USA

Introduction and Analysis - We consider the random nonlinear initial value problem

$$\frac{d\mathbf{X}}{dt} = \mathbf{g}(t,\mathbf{X},\mathbf{A}) \ ; \ \mathbf{X}(t_0) = \mathbf{X}_0 \tag{1}$$

where $\mathbf{X} = (X_1, X_2, ..., X_n)^T$, $\mathbf{g} = (g_1, g_2, ..., g_n)^T$, and T denotes transpose. The randomness in the problem arises from the vector of initial conditions $\mathbf{X}_0 = (X_{1_0}, X_{2_0}, ..., X_{n_0})^T$, whose components are random variables with joint probability density function (jpdf) $f_{\mathbf{X}_0}(\mathbf{x}_0)$, and the m random parameters $\mathbf{A} = (A_1, A_2, ..., A_m)$, which appear on the right-hand side of eqn. (1) and which have the jpdf $f_{\mathbf{A}}(\mathbf{a})$. In general, $\mathbf{A}$ and $\mathbf{X}_0$ may be jointly distributed; however they are often independent in many applications.

The class of random nonlinear initial value problems to be considered herein consists of those in which the random parameters are time-independent. Specifically, problems involving stochastic processes are excluded from the class. This problem has been the focus of a fairly extensive amount of study, and a number of approximate techniques for its solution have been formulated, e.g., [1-2]. These methods, however, either are not uniformly valid, and/or are not valid for arbitrarily large times. A somewhat more generally valid technique, which utilizes standard numerical methods for solving differential equations, has been proposed by Bellomo and Pistone, [3]. This technique makes use of what is known as Liouville's theorem for dynamic systems, [1], [4]. However this is restricted to the case in which all the initial conditions are random. In the present work, we wish to focus upon the case in which one or more, and perhaps all, of the initial data are degenerate, i.e., deterministic constants. The analysis is presented in more detail in ref. [5], where it is shown that Liouville's theorem may be derived as a special case.

N. Bellomo · F. Casciati (Eds.)
Nonlinear Stochastic Mechanics
IUTAM Symposium Turin

We begin by letting $\mathbf{X}_{00}$ represent the nondegenerate probability subspace contained in $\mathbf{X}_0$, and further by taking the dimension of $\mathbf{X}_{00}$ to be $l < n$, where n is the dimension of $\mathbf{X}_0$. We focus attention upon the direct mapping $\Phi_{t,t_0} : (\mathbf{X}_{00},\mathbf{A}) \rightarrow \mathbf{X} = \mathbf{h}(t,t_0,\mathbf{X}_0,\mathbf{A})$. The calculation of the Jacobian determinant for this mapping will be of great importance in the ensuing development. If $l + m$, the dimension of the product of the probability spaces for $\mathbf{X}_{00}$ and $\mathbf{A}$, is at least n, the elements of this determinant can be calculated very simply by direct differentiation of eqn. (1) with respect to A_i and X_{0_i}, [5], and an interchange of the order of differentiation, assuming that this is valid. Let $\mathbf{X}(t) = \mathbf{h}(t,t_0,\mathbf{X},\mathbf{A})$ be the solution to eqn. (1). Furthermore let $\partial\mathbf{X}/\partial\mathbf{A} = \partial\mathbf{h}/\partial\mathbf{A} = \mathbf{Y}$, $\partial\mathbf{X}/\partial\mathbf{X}_0 = \partial\mathbf{h}/\partial\mathbf{X}_0 = \mathbf{Z}$, and $\mathbf{g}_{\mathbf{X}} = \partial\mathbf{g}/\partial\mathbf{X}$. Then $\mathbf{Y}$ and $\mathbf{Z}$ satisfy, respectively, the following initial value problems:

$$\frac{d\mathbf{Y}}{dt} = \mathbf{g}_{\mathbf{X}}(t,\mathbf{h}(t,t_0,\mathbf{X}_0,\mathbf{A}),\mathbf{A})\mathbf{Y} + \frac{\partial\mathbf{g}}{\partial\mathbf{A}}(t,\mathbf{h}(t,t_0,\mathbf{X}_0,\mathbf{A}),\mathbf{A}); \quad \mathbf{Y}(t_0) = 0. \tag{2}$$

$$\frac{d\mathbf{Z}}{dt} = \mathbf{g}_{\mathbf{X}}(t,\mathbf{h}(t,t_0,\mathbf{X}_0,\mathbf{A}),\mathbf{A})\mathbf{Z}; \quad \mathbf{Z}(t_0) = \mathbf{I}_n \tag{3}$$

Here $\mathbf{I}_n$ is the nxn unit matrix. The solution of eqns. (2-3) yields the elements of K. We consider the following two cases.

Case 1 - We first consider the situation for which $0 \le l \le n$, $1 \le m$, and $l + m \ge n$. Let $q = l + m - n \ge 0$. For the sake of definiteness, we assume that $q > 0$, the case $q = 0$ being a trivial extension. In order to obtain an invertible mapping, we carry out a standard augmentation of the probability space of $\mathbf{X}$. To be specific, let us assume, without loss of generality, that $q < m$, and define

$$B_1 = A_{m-q+1},\ B_2 = A_{m-q+2},\ \dots,\ B_q = A_m \tag{4}$$

We now consider the mapping $\Phi_{t,t_0} : (\mathbf{X}_{00},\mathbf{A}) \rightarrow (\mathbf{X}, \mathbf{B}) = (\mathbf{h}(t,t_0,\mathbf{X}_0,\mathbf{A}), \mathbf{B}(\mathbf{A}))$, which is well-defined. The Jacobian of the mapping is

$$K(t,t_0,\mathbf{X}_{00},\mathbf{A}) = \left| \frac{\partial(\mathbf{h}, \mathbf{B})}{\partial(\mathbf{X}_{00}, \mathbf{A})} \right| \tag{5}$$

The columns of the Jacobian matrix are calculated from eqns. (2-3), and in the present case, eqn. (4) as well. We assume the mapping is invertible in the form $(\mathbf{X}_{00}, \mathbf{A}) = \Phi^{-1}_{t,t_0}(\mathbf{X},\mathbf{B})$. However, it is sufficient to assume that the support of the probability law of $(\mathbf{X}_{00}, \mathbf{A})$ is, up to a negligible set, the union of a finite family of open sets $U_1, \ldots, U_k$, such that, if $V_i = \Phi(U_i)$, the restriction of Φ on each U_i is a diffeomorphism. In this case, the following computation is available on each V_i. In any case, the Jacobian for the inverse mapping satisfies

$$J(t,t_0,\mathbf{X},\mathbf{B}) = \left| \frac{\partial \Phi^{-1}_{t,t_0}}{\partial(\mathbf{X},\mathbf{B})} \right| = \frac{1}{K(t,t_0,(\mathbf{X},\mathbf{B}))} \tag{6}$$

We then may compute the jpdf $f_{\mathbf{X},\mathbf{B}}$ from the relation, [6],

$$f_{\mathbf{X},\mathbf{B}}(t,t_0,\mathbf{x},\mathbf{b}) = f_{\mathbf{A},\mathbf{X}_{00}}(\Phi^{-1}_{t,t_0}(\mathbf{x},\mathbf{b}))\,|J(t,t_0,\mathbf{x},\mathbf{b})|, \tag{7}$$

and the jpdf for $\mathbf{X}$ may then be obtained from eqn. (7) by

$$f_{\mathbf{X}}(t,t_0,\mathbf{x}) = \int \cdots \int f_{\mathbf{X},\mathbf{B}}(t,t_0,\mathbf{x},\mathbf{b})\, db_1 \ldots db_q . \tag{8}$$

We now consider the more complicated case $l + m < n$, i.e., the sum of the dimensions of the nondegenerate initial conditions and the random parameters $\mathbf{A}$ is less than that of $\mathbf{X}$.

Case 2 - Let $0 \le l < n$, $1 \le m < n$, and $l + m < n$. Here the jpdf $f_{\mathbf{X}}(t,t_0,\mathbf{x})$ does not exist. Hence the focus of our attention will be on the marginal pdf for the variable $X_i = h_i(t,t_0,\mathbf{X}_0,\mathbf{A})$, $1 \le i \le n$. We will assume that the solution for X_i is invertible in terms of one of the random parameters $(\mathbf{X}_{00},\mathbf{A})$, which, without loss of generality, we will assume to be A_1. Hence we may write

$$A_1 = g_1(t,t_0,X_i,X_{00_1},\ldots,X_{00_l},A_2,\ldots,A_m). \tag{9}$$

We now perform an augmentation somewhat similar to that outlined previously, and define

$$B_1 = A_2, \ldots, B_{m-1} = A_m, B_m = X_{00_1}, \ldots, B_{m+l-1} = X_{00_l}. \tag{10}$$

By virtue of eqn. (10), it can easily be seen that each row below the first in the Jacobian determinant for the direct mapping $(\mathbf{X}_{00}, \mathbf{A}) \rightarrow (\mathbf{X}_i, \mathbf{B}) = (\mathbf{h}_i(t,t_0,\mathbf{X}_0,\mathbf{A}), \mathbf{B}(\mathbf{X}_{00},\mathbf{A}))$ has a one on the diagonal and zeros elsewhere. Hence, the Jacobian has the simple form

$$K(t,t_0,\mathbf{X}_{00},\mathbf{A}) = \frac{\partial X_i}{\partial A_1}. \tag{11}$$

As before, the Jacobian for the inverse mapping satisfies

$$J(t,t_0,\mathbf{X}_i,\mathbf{B}) = 1/K(t,t_0,h_i(t,t_0,X_i,....),A_2=B_1,........,X_{00_l} = B_{m+l-1}). \tag{12}$$

The jpdf is given by

$$f_{\mathbf{X}_i,\mathbf{B}}(t,t_0,\mathbf{x}_i,\mathbf{b}) = |J|\, f_{\mathbf{X}_{00},\mathbf{A}}(a_1 = g_1, a_2 = b_1,.....,x_{00_l} = b_{m+l-1}) \tag{13}$$

from which the marginal pdf for X_i may be determined by

$$f_{X_i}(t,t_0,x_i) = \int ... \int f_{\mathbf{X}_i,\mathbf{B}}(t,t_0,\mathbf{x}_i,\mathbf{b})\, db_1 \, ...\, db_{m+l-1}. \tag{14}$$

Examples - We now illustrate the method by several examples.

Example 1 - Consider the following two coupled nonlinear differential equations

$$\frac{dY}{dt} = -\kappa\, |Y|\, Y - \Omega \sin(X) \quad ; \quad Y(0) = 0$$

$$\frac{dX}{dt} = Y \quad ; \quad X(0) = 1.3 \tag{15}$$

These equations derive from a second-order equation describing the motion of a pendulum with nonlinear damping. We assume that κ and Ω are independent random parameters, with κ parabolically distributed on [0.075, 0.525] and Ω linearly distributed on [0.10, 0.40]. Thus we have a system with two dependent variables and two random parameters, κ and Ω (Case 1, with $q = 0$). Again, assume that the mapping Φ: $(\kappa, \Omega) \rightarrow (X, Y)$, with Jacobian K, is invertible in the form $(\kappa, \Omega) = \Phi^{-1}(t, X, Y)$, with Jacobian J, so that $J = 1/K$. From eqn. (2),

$$\frac{d}{dt}\begin{Bmatrix}\frac{\partial X}{\partial \Omega}\\ \frac{\partial Y}{\partial \Omega}\end{Bmatrix} = \begin{bmatrix} 0 & 1 \\ -\Omega\cos(X) & -2|Y| \end{bmatrix}\begin{Bmatrix}\frac{\partial X}{\partial \Omega}\\ \frac{\partial Y}{\partial \Omega}\end{Bmatrix} + \begin{Bmatrix} 0 \\ -\sin(X)\end{Bmatrix} \tag{16}$$

with $\partial X/\partial\Omega = \partial Y/\partial\Omega = 0$ at $t = 0$. A similar equation holds for $\{\partial X/\partial\kappa, \partial Y/\partial\kappa\}$. The solution of eqns. (15-16), and the equation for $\{\partial X/\partial\kappa, \partial Y/\partial\kappa\}$ yields the Jacobian K of the direct mapping $(\kappa, \Omega) \rightarrow (X, Y)$, and hence the desired Jacobian J. Then from eqn. (7)

$$f_{X,Y}(x,y,t) = |J|\, f_{\kappa\Omega}(\Phi^{-1}(x,y,t)) \tag{17}$$

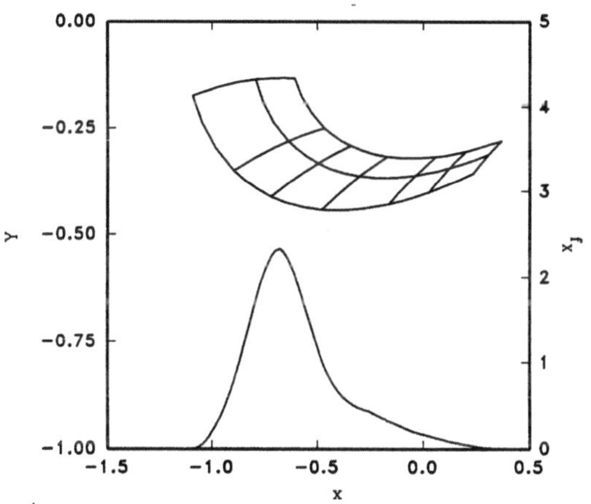

Fig. 1 - Grid in X-Y plane and resulting pdf $f_X(x, 5)$

The marginal pdf $f_X(x,t)$ may be determined by integrating the jpdf $f_{X,Y}(x,y,t)$ with respect to y. This was accomplished numerically by subdividing the rectangular domain of the parameters (κ, Ω) into a grid. In the (X, Y) plane, for a value of $t = 5$, this rectangular grid maps into that shown in the upper portion of Fig. 1. For fixed values of κ and Ω lying at the grid points, the corresponding values of $X(5,\kappa,\Omega)$ and $Y(5,\kappa,\Omega)$ may be calculated, yielding pointwise values of the inverse mapping $(X(5),Y(5)) \rightarrow (\kappa,\Omega)$. Eqns. (15-16) were then solved simultaneously to yield the Jacobian K, and hence $J = 1/K$. This procedure gave the values of the jpdf $f_{X,Y}(x,y,t)$ at the grid points in the (X,Y) plane. To carry out the integration, cubic finite element isoparametric interpolating functions, which are defined by the calculated values of the jpdf at the grid points, were used to interpolate to values of $f_{X,Y}(x,y,t)$ on the interior of the domain. The finite element approximation to $f_{X,Y}(x,y,t)$ was then numerically integrated with respect to y to determine the marginal pdf $f_X(x,t)$. The lower portion of Fig. (1) shows the results.

Example 2 - The Paris-Erdogan equation, which has been used extensively in the analysis of fatigue crack growth, may be written as

$$\frac{dN}{da} = e^{-Y}(\Delta K(a))^{-M} \tag{18}$$

where N is the number of cycles, a is the crack length, and $\Delta K(a)$ is the crack tip stress intensity factor. In the probabilistic formulation, Y and M are taken to be random constants. See ref. [7] for more details and background. To eqn. (18), we append the initial condition $N(a_0) = 0$, where a_0 is the deterministic initial crack length. Based upon experimental data, [8], it has been concluded that Y and M may be considered to be jointly normally distributed, [9]. It is convenient to transform to uncoupled variables X_1 and X_2

$$m = \sqrt{2}\sigma_m x_1 + \eta_m \quad ; \quad y = \sigma_y(\sqrt{2(1-r^2)}\, x_2 + \sqrt{2}r\, x_1) + \eta_y \tag{19}$$

which yields

$$f_{X_1X_2}(x_1, x_2) = \frac{1}{\pi} e^{-x_1^2 - x_2^2} \tag{20}$$

Thus we have a single dependent variable N which depends upon the two random parameters X_1 and X_2 (Case 1). Given the previous assumptions, we suppose that we may write $X_2 = g(N, X_1, a)$. We now consider the mapping $(X_1, X_2) \rightarrow (B = X_1, N)$ with Jacobian determinant K and the inverse mapping $(B = X_1, N) \rightarrow (X_1, X_2)$ with Jacobian determinant J. Given that $\partial B/\partial X_1 = 1$ and $\partial B/\partial X_2 = 0$, we have $K = 1/J = \partial N/\partial X_2$. Hence, from eqn. (7),

$$f_{N,B}(n, b, a) = \left| \left(\frac{\partial n}{\partial x_2}\right)^{-1} \right| \frac{1}{\pi} e^{-b^2 - x_2^2(n,b,a)} \tag{21}$$

and

$$f_N(n, a) = \frac{1}{\pi} \int_{-\infty}^{\infty} |J| \; e^{-b^2 - x_2^2(n,b,a)} \, db \qquad (22)$$

with $\partial n/\partial x_2 = 0$ at $a = 0$. A numerical solution of eqns. (18) and (23), with the given value of X_1 and the value of X_2 determined by the shooting procedure, yields the desired quantity.

In order to compute $f_{N,B}(n, b, a)$, we require both $\partial n/\partial x_2$ and the mapping $x_2 = g(n, b, a)$. The latter may be computed by a simple shooting scheme. For given values of the arguments N and $X_1 = B$, we pick a value of X_2 (typically a quadrature point for the integral in eqn. (22)) and integrate eqn. (18) forward from the given initial condition to calculate a value of N at the crack length a of interest. The value of X_2 is then varied until integration of eqn. (18) yields the given value of N to within desired accuracy. In order to obtain $\partial n/\partial x_2$, we differentiate eqn. (18) with respect to x_2, and, assuming a sufficient degree of continuity, interchange the order of differentiation. Then we have

$$\frac{d}{da}\left(\frac{\partial n}{\partial x_2}\right) = \sigma_y \sqrt{2(1-r^2)} \, \sigma^{-Y} (\Delta K(a))^{-M} \qquad (23)$$

with $\partial n/\partial x_2 = 0$ at $a = 0$. A numerical solution of eqns. (18) and (23), with the given value of X_1 and the value of X_2 determined by the shooting procedure, yields the desired quantity.

Fig. 2 shows the calculated cumulative distribution function (cdf), calculated by numerically integrating the marginal pdf, plotted on Weibull probability paper for various crack lengths. It can be seen that the agreement with the experimental results of Wirkler *et al.*, [8], for crack lengths greater than 20 mm, is excellent. For crack lengths less than this value, the agreement becomes increasingly poorer, however, probably due to the increasing importance of stochastic effects for shorter crack lengths.

Example 3 - We consider the creep deformation of a cantilever beam whose uniaxial creep behavior is given by the power-law creep equation $d\varepsilon^c/dt = C\,\sigma^N$. Delph and Yukich, [10], have considered the more general multiaxial case. Here the material parameters C and N are taken to be dependent random variables, jointly log-normal/normal, with the parameters for the jpdf derived from experimental data, [11]. We let $Y = ln(C)$ and make use of the transformation given by eqn. (19), which yields the jpdf given by eqn. (20).

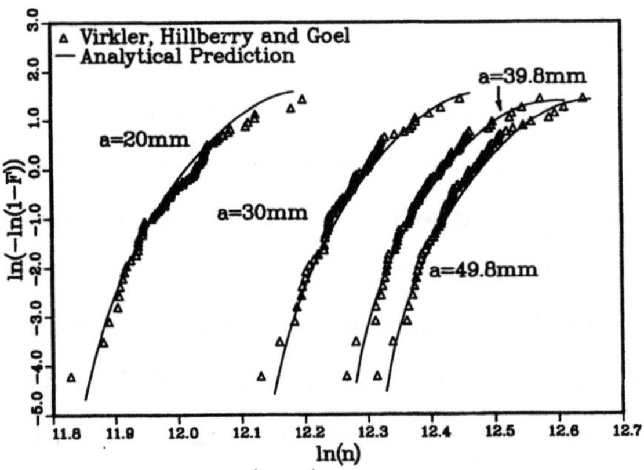

Fig. 2 - Calculated cdf's $F_N(n, a)$ and experimental data.

This example illustrates *Case 2*, in which the number of variables in the problem exceed the number of random parameters. According, we assume that the solution for the quantity of interest A may be expressed in the form $A = h(t,X_1,X_2)$. We further assume that A and X_1 are one-to-one, so that we may make the inversion $X_1 = g(t,A,X_2)$. We now set $X_2 = B$, and consider the mapping $(X_1,X_2) \rightarrow (A, B)$ with Jacobian determinant K and the inverse mapping $(A, B) \rightarrow (X_1,X_2)$ with Jacobian determinant J. Then, as in the previous example,

$$f_{A,B}(t,a,b) = |J| \frac{1}{\pi} e^{-x_1^2(b,a,t)-b^2} \tag{24}$$

and hence

$$f_A(t,a) = \frac{1}{\pi}\int_{-\infty}^{\infty} |J| \; e^{-x_1^2(b,a,t)-b^2} \, db \tag{25}$$

Now

$$K = \left| \frac{\partial(A, B)}{\partial(X_1, X_2)} \right| = \left(\frac{\partial A}{\partial X} \right) = \frac{1}{J} \tag{26}$$

and $\partial A/\partial X = (\partial A/\partial N)(\partial N/\partial X) + (\partial A/\partial Y)(\partial Y/\partial X) = \eta_N \sqrt{2(1-r^2)}\ (\partial A/\partial N)$. To calculate $\partial A/\partial N$ and to carry out the numerical inversion, we make use of finite element techniques.

For a uniform distributed load q, the finite element equations are

$$\mathbf{K}\{w\} = -\frac{q}{EI}\int_0^L \{H\}\,dx - \frac{2b}{I}\int_0^L\int_0^h y\,\varepsilon^c \left\{ \frac{d^2 H}{dx^2} \right\} dy\,dx \tag{27}$$

where $\mathbf{K}$ is the elastic stiffness matrix, $\{H\}$ is the vector of finite element shape functions, $\{w\}$ is the vector of nodal point displacements and slopes, and I is the area moment of inertia of the rectangular cross-section having width b and height $2h$. We have in addition the strain-displacement relation

$$\varepsilon = -y \left\{ \frac{d^2 H}{dx^2} \right\}^T \{w\} \tag{28}$$

and Hooke's law $\sigma = E(\varepsilon - \varepsilon^c)$. The beam was modelled using 10 C^1 cubic elements. The result is a fairly large set of coupled nonlinear equations containing two random parameters. As noted, we are interested in computing the pdf for the variable A, where A could represent a nodal point displacement, the stress at a given location, or any other quantity of interest. The quantity $\partial A/\partial N$, which we need in order to calculate the Jacobian, is obtained as before by direct differentiation of the governing equations, and an interchange of the order of differentiation where appropriate. This leads to

$$\frac{\partial}{\partial t}\left(\frac{\partial \varepsilon^c}{\partial N} \right) = C\,\sigma^N \ln\sigma + N\,C\,\sigma^{N-1}\frac{\partial \sigma}{\partial N} \tag{29}$$

$$\mathbf{K}\left\{ \frac{\partial w}{\partial N} \right\} = -\frac{2b}{I}\int_0^L\int_0^h y \left(\frac{\partial \varepsilon^c}{\partial N} \right) \left\{ \frac{d^2 H}{dx^2} \right\} dy\,dx \tag{30}$$

$$\frac{\partial \varepsilon}{\partial N} = -y \left\{ \frac{d^2 H}{dx^2} \right\}^T \left\{ \frac{\partial w}{\partial N} \right\} \tag{31}$$

$$\frac{\partial \sigma}{\partial N} = E \left[\frac{\partial \varepsilon}{\partial N} - \frac{\partial \varepsilon^c}{\partial N} \right] \tag{32}$$

The solution of eqns. (29-32) allow the determination of the quantity $\partial A/\partial N$, and hence K. An alternate method for calculating $\partial A/\partial N$, which is slightly less accurate, but computationally more efficient, is to estimate $\partial A/\partial N$ during the final stages of the numerical shooting procedure described earlier, which is used to compute the numerical inversion.

More complete details of the solution are given in ref. [10]. We present some of the results here. Figs. 3-4 show, respectively, the calculated pdf and cdf for the displacement of the beam tip after 10,000 hours of creep. Also shown on Fig. 4 are the results of 500 Monte Carlo trials. It can be seen that the agreement between computation and Monte Carlo simulation is excellent.

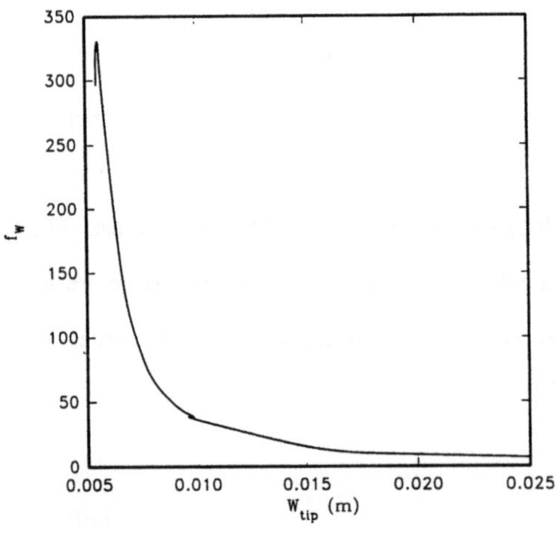

Fig. 3 - Calculated beam tip pdf (t = 10000 hrs.).

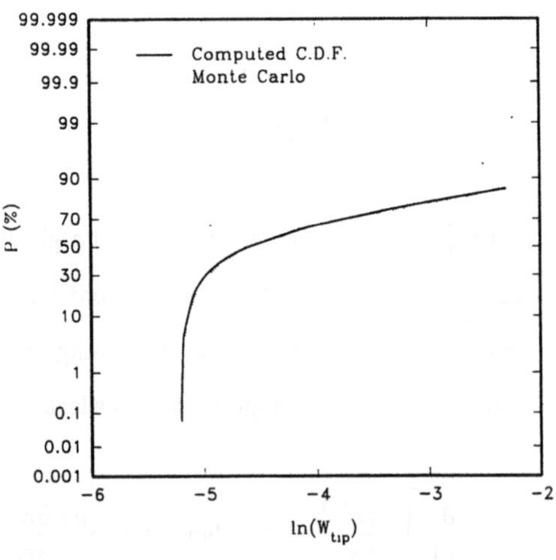

Fig. 4 - Calculated beam tip cdf and results of Monte Carlo simulations (t = 10000 hrs.).

Discussion - We have outlined here a numerical technique for the solution of random initial value problems containing an arbitrary mixture of deterministic initial conditions, random initial conditions, and random parameters. This work is felt to be a significant extension of previous work in the area, e.g. [3-4], which utilized Liouville's theorem to calculate directly the Jacobian determinant J of the inverse mapping, but which was restricted to the case in which all the initial conditions were random. When one or more initial conditions are deterministic, however, Liouville's theorem is no longer applicable, and one must instead calculate the Jacobian determinant K of the direct mapping. The quantity of interest, J, is then just the reciprocal of K.

The technique that we have described here is simple to apply, at least in principle. It is, moreover, based upon exact relations from probability theory, and can be implemented using standard numerical techniques. The two main disadvantages of the technique are that it is invalid for problems involving stochastic processes, and that it may become somewhat computationally intensive for large systems of equations. However it is for such systems, resulting, for example, from finite difference or finite element discretizations, e.g., [10], that the technique presented here is felt to have its most important applications. We close by noting that the mapping theorem upon which the present technique is based is equally applicable to systems of time-independent equations. An example of such an application, in the context of finite element equations, has recently been given by Delph, [12].

References

1. T.T. Soong, *Random Differential Equations in Science and Engineering*, Academic Press, New York (1973).
2. G. Adomian, *Stochastic Systems*, Academic Press, New York (1983).
3. N. Bellomo and G. Pistone, *J. Math. Anal. and Appl.* **77**, 215 (1980).
4. N. Bellomo and G. Pistone, *Mech. Res. Comm.* **6**, 75 (1979).
5. D.G. Harlow and T.J. Delph, *Math. & Computers in Simulation*, to appear.
6. P. Billingsley, *Probability and Measure*, John Wiley & Sons, New York (1986).
7. D. Xiao, D.G. Harlow, and T.J. Delph, *Eng. Fract. Mech.*, to appear.
8. D.A. Virkler, B.M. Hillberry, and P.K. Goel, *J. Eng. Mat. Tech.* **101**, 148 (1979).
9. O. Ditlevsen and R. Olesen, *Eng. Frac. Mech.* **25**, 177 (1986).
10. J.E. Yukich and T.J. Delph, submitted for publication (1991).
11. I. Sprung and V.A. Zilberstein, in *Understanding Variability in Creep and Rupture Behavior*, ed. M Prager and J.D. Parker, Amer. Soc. Mech. Eng., New York (1988).
12. T.J. Delph, submitted for publication (1991).

Large Amplitude Random Vibration of Polygonal Plates

R. HEUER *, H. IRSCHIK ** and F. ZIEGLER *

* Civil Engineering Department, Technical University of Vienna, Wiedner Haupstraße 8-10/E201, Austria, A-1040

** Institute of Technical Mechanics (331), Johannes-Kepler-University Linz, Linz-Auhof, Austria, A-4040

Summary

Geometrical nonlinearity has a major influence on the flexural response of beams and plates in free and forced vibrations. Recently it was found for free vibrations of simply supported polygonal plates that a modal decomposition (a Ritz-Galerkin procedure using the linear eigenfunctions) gives a unifying result for the nonlinear eigenfrequencies that holds independently of the special planform and for all modes when time is properly non-dimensionalized. The present paper gives a proper extension to forced vibrations that are described in a multimode approximation by a set of coupled inhomogeneous Duffing-type equations. The restoring forces can be derived from a potential function, and for random vibrations the FPK-equation can be set up and solved in closed form for the stationary transition probability density with the further assumption of linear modal damping. Numerical studies illustrate the beauty of the analytical solution.

Introduction

Considerable effort has been spent on the development of techniques for the prediction of the response of nonlinear dynamic systems to random excitation. When the excitation is a white noise process, or a field of independent white noises, the response is a Markov process, see Caughey /1/. The transition probability density function that completely describes the response together with a given initial distribution (or distributions), satisfies the Fokker-Planck-Kolmogorov (FPK) equation. The coefficients of the FPK-equation have to be derived from the nonlinear equations of motion of the dynamic system. Exact stationary solutions can be constructed for a class of problems in which the nonlinear restoring forces have a potential function. The theory of the Markov process and its associated FPK-equation has been applied to nonlinear random vibrations of strings, beams, and rectangular plates in the papers of Ariaratnam /2/ and Herbert /3/, /4/ among others. A survey of exact and approximate methods for nonlinear systems subjected to random vibrations is given by Crandall and Zhu /5/.

The purpose of this paper is to investigate the large multimode response of polygonal plates due to white-noise excitation by means of the FPK-equation. The geometrically nonlinear plate problem is treated by employing of Berger's approximation, see /6/, that is suitable for plates with

N. Bellomo · F. Casciati (Eds.)
Nonlinear Stochastic Mechanics
IUTAM Symposium Turin

immovable edges. Plates which are rigid in shear with arbitrarily shaped, polygonal planform are considered under simple support conditions. Furthermore, thermal prestress due to a prescribed spatial distribution of the cross-sectional mean temperature is taken into account, but only vibrations that are symmetric to the horizontal plane are considered.

A multimodal expansion and the Galerkin procedure are applied to the corresponding differential equation of the forced plate vibrations, thereby extending the contribution /7/ where nonlinear natural oscillations have been treated. The forced flexural vibrations are described by a coupled set of inhomogeneous Duffing-type equations. By means of a proper time scaling a nondimensional formulation is obtained that is independent of the special polygonal planform of the plate, and it is the same for all successive orders of the modes. The analysis in the frequency domain of deterministic nonlinear flexural vibrations of homogeneous plates on a two-parameter foundation /8/ and of layered plates /9/ has been generalized to include shear and rotatory inertia.

The FPK-equation of the associated nonlinear MDOF-system has a stationary closed form solution of the N-dimensional transition probability-density function of the first N modal amplitudes by properly selecting the intensities of the forcing white noise processes /2/. In the postbuckling range of thermal prestress that solution excludes unsymmetric vibrations and any chaotic transitions. Following the lines of the approximation of Herbert /3/, an analytical integration of the nonlinear modal variances is performed. Numerical integration of the defining integrals of the nonlinear modal variances, taking into account an increasing number of modes and their second-order moments, shows qualitatively convergence, and the results are compared to solutions derived within a linear analysis. A wide range of the forcing parameter is considered.

While the method of FPK-equation for nonlinear problems is well discussed from the point of probability theory /10/, its application to complex structures seems to be a missing link in literature. With respect to plate vibrations only plates of rectangular planform have been treated by means of this method. It is the scope of the present paper to give a unifying result for a whole class of complex structures, namely the class of plates with arbitrary polygonal planform with simply supported straight edges. The unifying approach is a result of non-dimensionalizing the governing equations by proper scaling, taking into account Berger´s approximation. That nonlinear approach makes use of results of linear vibrations of plates. Especially, the linear modes of simply supported plates of polygonal shape and hydrostatic in-plane forces are identical to those of effectively prestressed membranes of the same shape and mass distribution.

Berger's equations for large forced flexural vibrations of thermally loaded plates

Neglecting the influence of the second invariant of the midsurface strain tensor, Berger /6/ derived a decomposition of the von Karman plate equations /11/, that yields excellent numerical results in all the isothermal as well as thermal problems if the plate edges are assumed to be prevented from in-plane motions, see, eg /12/. The dynamic extension of Berger's static equations is due to Wah /13/, see Nayfeh and Mook /14/ for a review.

The strain energy of a Berger plate may be decomposed into the membrane energy

$$U_m = \frac{1}{2} D \int_A e^2 \, dA \tag{1}$$

with the midsurface dilatation

$$e = u_{,x} + v_{,y} + \frac{1}{2}\left(w_{,x}^2 + w_{,y}^2\right) \quad , \tag{2}$$

and into the bending energy

$$U_b = \frac{1}{2} K \int_A \left[w_{,xx}^2 + w_{,yy}^2 + 2\nu w_{,xx} w_{,yy} + 2(1-\nu) w_{,xy}^2\right] dA \quad , \tag{3}$$

where $K = Eh^3/12(1-\nu^2)$ and $D = Eh/(1-\nu^2)$ denote the bending and stretching stiffness, respectively. w is the plate deflection, and u, v denote the in-plane deformations. Integration is extended over the plate area A.
In the case of an in-plane heating of the plate, see Irschik /15/, the strain energy is increased by the term,

$$U_\theta = -(1+\nu) D \int_A e \, n_\theta^* dA \quad , \; n_\theta^* = \frac{1}{h} \int_{-h/2}^{h/2} \alpha\theta \, dz \quad , \tag{4}$$

where $\alpha\theta$ is the imposed thermal strain. It is assumed to be evenly distributed over the thickness of the plate. The potential energy of a pressure loading $f(\vec{x};t)$ becomes

$$W = -\int_A f w \, dA \quad , \tag{5}$$

and the kinetic energy due to flexural vibrations is defined by

$$T = \frac{1}{2} \int_A \dot{w}^2 \rho h \, dA \quad . \tag{6}$$

A dot indicates differentiation with respect to time and ρh is the mass density per unit plate area. By means of D'Alembert's principle the equations of motion of the nonlinear plate problem are obtained in the form

$$K\Delta\Delta w - n\Delta w + \rho h \ddot{w} = f \quad , \tag{7}$$

$$n = n|_{\bar{n}_\theta^* = 0} - K\bar{n}_\theta^*/h^2 \quad . \tag{8}$$

Berger's hydrostatic tensile in-plane force n is time dependent but it is constant within the plate domain. The geometric nonlinearity enters equ. (7) through

$$n|_{\overline{n}_\theta^*=0} = -\frac{6K}{Ah^2}\int_A w\Delta w dA \quad , \tag{9}$$

and the influence of heating is expressed by the integral over the plate domain

$$\overline{n}_\theta^* = \frac{12}{A}(1+\nu)\int_A n_\theta^* dA \quad . \tag{10}$$

Multimodal approach for simply supported plates of polygonal planform

Following /13/ and /7/, solutions of the problem are sought in the separated form

$$w(\vec{x},t) = \sum_{j=1}^{N} c_j\, q_j^*(t)\, w_j^*(\vec{x}) \quad , \tag{11}$$

where the multipliers c_j carry the dimension of length, and the spatial distributions w_j^* are characterized by their norm

$$\beta_j^* = \frac{1}{A}\int_A w_j^{*2}\, dA \quad . \tag{12}$$

A star denotes a non-dimensional quantity. Because this norm can be freely chosen, the c_j´s are not necessarily equal to the maximal amplitudes of the j-th mode of deflection.
In order to apply equ.(11) to the differential operator (7) in the form of a direct Galerkin procedure, the spatial distributions w_j^* must satisfy all the natural and dynamic boundary conditions of the plate. Considering plates that are rigid in shear, these boundary conditions in the case of straight and simply supported edges reduce to the Navier conditions

$$\Gamma: \; w = 0, \; \Delta w = 0 \quad , \tag{13}$$

see, eg /17/. In the following the plate is assumed to be of polygonal planform and simply supported. Therefore, the second-order, orthogonal set of Helmholtz-eigenfunctions

$$\Delta w_j^* + \alpha_j w_j^* = 0 \; , \; j = 1, 2 \ldots N \tag{14}$$

$$\Gamma: \; w_j^* = 0 \tag{15}$$

is used for the spatial distribution. Because of Dirichlet`s boundary condition (15), the original Navier conditions (13) as well as the field equation (7) with $f = 0$ are exactly satisfied by those Ansatz-functions. Recall that α_j is the j-th eigenvalue of an effectively prestressed membrane of the same planform and is related to the j-th linear natural frequency of the plate under consideration through (the thermal loading is absent)

$$\omega_{jL_{\overline{n}_\theta^*=0}} = \left[\frac{K\alpha_j^2}{\rho h}\right]^{1/2} , \tag{16}$$

see Leissa /18, p. 237/. Note that w_j^* also corresponds to the j-th eigenmode as well as to the j-th buckling mode of the linear plate with n= const., again see /18/. The force loading is also expanded in terms of the eigenfunctions of the linear problem,

$$f(\vec{x},t) = \sum_{j=1}^{N} c_{pj}\, f_j^*(t)\, w_j^*(\vec{x}) , \tag{17}$$

and Galerkin`s procedure finally leads to the coupled set of generalized nonlinear oscillator equations

$$\ddot{q}_j^* + \omega_{jL_{\overline{n}_\theta^*=0}}^2 q_j^* + n\frac{\alpha_j}{\rho h} q_j^* = \frac{c_{pj}}{c_j}\frac{1}{\rho h} f_j^* . \tag{18}$$

The influence of large amplitudes as well as that of the multimode approximation are combined in the time-variant in-plane force

$$n = \frac{K}{h^2}\left[6\sum_{k=1}^{N} \alpha_k \beta_k^* c_k^2 q_k^{*2} - \overline{n}_\theta^*\right] , \tag{19}$$

see equ. (8). Substituting equ.(19) into (18) yields the set of coupled generalized Duffing equations,

$$\ddot{q}_j^* + \gamma_{\theta j}^*\, \omega_{jL_{\overline{n}_\theta^*=0}}^2 q_j^* + \frac{6K\alpha_j}{\rho h^3} q_j^* \sum_{k=1}^{N} \alpha_k\, \beta_k^*\, c_k^2\, q_k^{*2} = \frac{c_{pj}}{c_j}\frac{1}{\rho h} f_j^* , \tag{20}$$

where

$$\gamma_{\theta j}^* = \left[1 - \frac{\overline{n}_\theta^*}{\alpha_j h^2}\right] \tag{21}$$

characterizes the influence of the thermal prestress. It is assumed that $\gamma_{\theta j}^* > 0$ is deterministic, likewise to the structural parameters K and ρh. Light modal damping of the viscous type is considered by adding

$$2\zeta_j\, \omega_{jL_{\overline{n}_\theta^*=0}}\, \dot{q}_j^* \equiv 2\lambda_j\, \dot{q}_j^* \tag{22}$$

to equ.(20).

When passing over to a nondimensional time scale

$$t^* = \left(\omega_{1L}\big|_{\bar{n}_\theta^*=0}\right) t \tag{23}$$

the equations of motion become

$$q_j^{*''} + 2\lambda_j^* q_j^{*'} + \gamma_{\theta j}^* \alpha_j^{*2} q_j^* + 6\alpha_j^* q_j^* \sum_k \alpha_k^* q_k^{*2} = f_j^* \ . \tag{24}$$

$(\ )'$ denotes the derivative with respect to t^* and

$$\lambda_j^* = \lambda_j / \left(\omega_{1L}\big|_{\bar{n}_\theta^*=0}\right) , \quad \alpha_j^* = \frac{\alpha_j}{\alpha_1} = \left(\frac{\omega_{jL}}{\omega_{1L}}\right)_{\bar{n}_\theta^*=0} \ . \tag{25}$$

In order to have the simple form of equ. (24), the multipliers of the generalized coordinates and forces are chosen to be

$$c_k^2 = h^2/\beta_k^* \ , \quad c_{pj} = c_j K \alpha_1^2 \ . \tag{26}$$

Using the potential function

$$V^* = \frac{1}{2} \sum_j \gamma_{\theta j}^* \alpha_j^{*2} q_j^{*2} + \frac{3}{2} \sum_j \sum_k \alpha_j^* \alpha_k^* q_j^{*2} q_k^{*2} \tag{27}$$

equ. (24) is put into the more general form

$$q_j^{*''} + 2\lambda_j^* q_j^{*'} + \frac{\partial V^*}{\partial q_j^*} = f_j^* \ , \tag{28}$$

that is characteristic for a large class of nonlinear problems.

Random vibrations

In the following analysis it is assumed that the generalized forces f_j^*'s are uncorrelated, zero-mean Gaussian white noise processes having the autocorrelation

$$\left\langle f_j^*(t^*) f_k^*(t^*+\tau^*) \right\rangle = 2D_j^* \delta(\tau^*) \delta_{jk} \ , \tag{29}$$

where δ_{jk} is the Kronecker delta and $\delta(\tau^*)$ the Dirac delta function, and D_j^* are constants. With these assumptions the FPK-equation for the stationary transition probability density $p(q_1^* \ldots q_N^*, q_1^{*'} \ldots q_N^{*'})$ can be derived, compare /1/ and /19/,

$$\sum_j \frac{\partial}{\partial q_j^{*'}}\left[\left(\frac{\partial V^*}{\partial q_j^*} + 2\lambda_j^* q_j^{*'}\right)p\right] - \sum_j \frac{\partial}{\partial q_j^*}\left(q_j^{*'}p\right) + \sum_j D_j^* \frac{\partial^2 p}{\partial q_j^{*'2}} = 0 \quad . \tag{30}$$

Equ. (30) has the following solution

$$p\left(q_1^* \ldots q_N^*,\, q_1^{*'} \ldots q_N^{*'}\right) = \overline{C} \exp\left\{-\frac{1}{\overline{K}}\left(\frac{1}{2}\sum_j q_j^{*'2} + V^*\right)\right\} \tag{31}$$

provided that

$$D_j^*/\left(2\lambda_j^*\right) = \overline{K} = \text{const. for all } j \quad , \tag{32}$$

and $\overline{C}$ is a normalizing constant.

Integration of the expression for p over the generalized velocities from $-\infty$ and $+\infty$ renders the probability density function of the modal amplitudes, see Ariaratnam /2/,

$$p\left(q_1^* \ldots q_N^*\right) = C \exp\left\{-V^*/\overline{K}\right\} \quad , \tag{33}$$

where C is evaluated from the following normalizing condition

$$\int_{-\infty}^{+\infty}{}_{\text{N-fold}} \cdots \int_{-\infty}^{+\infty} p\left(q_1^* \ldots q_n^*\right) \prod_j dq_j^* = 1 \quad . \tag{34}$$

Since the potential V^* is an even function for the case of symmetric vibrations considered here, the mean values of the dimensionless response vanishes. The second-order moments of the coupled modal deflection are determined by multifold integration

$$\left\langle q_j^* q_k^* \right\rangle = \int_{-\infty}^{+\infty}{}_{\text{N-fold}} \cdots \int_{-\infty}^{+\infty} q_j^* q_k^* \, p\left(q_1^* \ldots q_n^*\right) \prod_j dq_j^*$$

$$= C \int_{-\infty}^{+\infty}{}_{\text{N-fold}} \cdots \int_{-\infty}^{+\infty} q_j^* q_k^* \exp\left\{-\frac{1}{2\overline{K}}\left(\sum_j \gamma_{\theta j} \alpha_j^{*2} q_j^{*2} + 3\sum_j \sum_k \alpha_j^* \alpha_k^* q_j^{*2} q_k^{*2}\right)\right\} \prod_j dq_j^* \quad . \tag{35}$$

By substituting the results of equ.(35) the mean-square response of the nonlinear plate deflection becomes

$$\left\langle w^2(\vec{x}, t)\right\rangle = \sum_j \sum_k c_j c_k \left\langle q_j^*(t)\, q_k^*(t)\right\rangle w_j^*(\vec{x})\, w_k^*(\vec{x}) \quad . \tag{36}$$

In /2/ an integral of similar type of equ.(35) for the mean-square response $\left\langle q_j^{*2}\right\rangle$ has been solved approximately in the context of random vibrations of strings by replacing the exponentional function by the first two terms of its MacLaurin expansion. Analogously, Herbert /3/ gave the solution for a simply supported beam.

For $\gamma^*_{\theta j}=1$, it follows from a direct substitution of the results of Herbert /3/ that

$$\langle q_j^{*2}\rangle = \frac{\overline{K}}{\alpha_j^{*2}}\left\{\left[1-\frac{3}{2}\overline{K}\left\{\left(\sum_{k=1}^{N}\frac{1}{\alpha_k^*}\right)^2+2\sum_{k=1}^{N}\frac{1}{\alpha_k^{*2}}+\frac{4}{\alpha_j^*}\sum_{k=1}^{N}\frac{1}{\alpha_k^*}+\frac{8}{\alpha_j^{*2}}\right\}\right]\Big/\right.$$
$$\left.\left[1-\frac{3}{2}\overline{K}\left\{\left(\sum_{k=1}^{N}\frac{1}{\alpha_k^*}\right)^2+2\sum_{k=1}^{N}\frac{1}{\alpha_k^{*2}}\right\}\right]\right\} , \qquad (37)$$

is an approximation valid for small deviations of the nonlinear system behaviour from the linear solution, ie for $\overline{K}<<1$. For the single-mode approximation, N=1, there is $\overline{K}=D^*/2\zeta$, $\alpha_1^*=1$, and equ. (37) becomes independent from the special plate geometry:

$$\langle q_1^{*2}\rangle = \left\{1-9\left(D^*/\zeta\right)\Big/\left[1-\frac{9}{4}\left(D^*/\zeta\right)\right]\right\} D^*/2\zeta \quad . \qquad (38)$$

Fig. 1a gives this unifying single-mode representation as a function of $\overline{K}$.

A simple transformation from the nondimensional form of equ. (38) to the real plate under consideration brings the plate geometry subsequently into play, see equs.(11), (17), (26), where

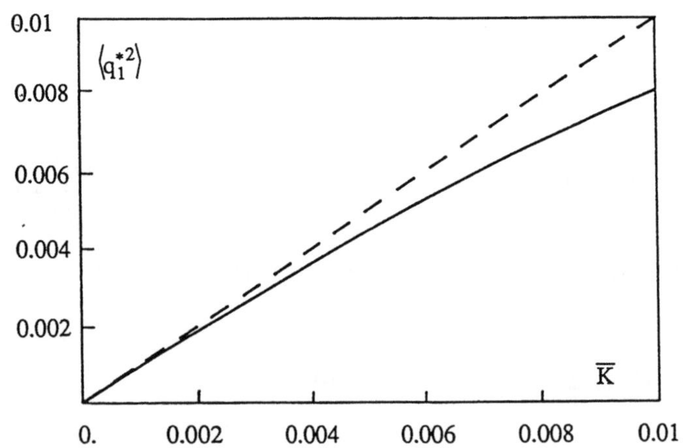

Fig. 1a. Linear (– – –) and nonlinear (——) nondimensional mean-square response $\langle q_1^{*2}\rangle$ as a function of $\overline{K}=D^*/2\zeta$ according to the approximation (37). Fig. 1a is independent of the special planform of the plate.

$$\langle q_1^2 \rangle = \langle q_1^{*2} \rangle \, h^2/\beta_1^* \quad , \tag{39}$$

$$D = \left(Kh/a^4\right)^2 D^* \tilde{\alpha}_1^4/\beta_1^* \ , \ \tilde{\alpha}_1 = \alpha_1 \, a^2 \ . \tag{40}$$

a is a characteristic length associated with the plate geometry.

Accordingly, Fig. 1b shows $\langle q_1^2 \rangle / h^2$ as a function of $D\left(a^4/Kh\right)^2/2\zeta$ for the plate strip ($\tilde{\alpha}_1 = \pi^2$, $\beta_1^* = 1/2$), rectangular plates of various aspect ratios ($\tilde{\alpha}_1 = \pi^2 \left[1+(a/b)^2\right]$, $\beta_1^* = 1/4$), and for a right-angled isosceles triangular plate ($\tilde{\alpha}_1 = 5\pi^2$, $\beta_1^* = 1/4$).

For plates with $\alpha_j^* = j^2$, eg for the plate strip, Fig. 2 visualizes the unifying representation of $\langle q_1^{*2} \rangle$, equ. (35) with j=k=1, for a wider range of $\overline{K}$, where the influence of the higher modes upon the first one is studied. A numerical integration of the N-fold integrals of equs. (34) and (35) is performed, where N=1, 2, 3, 4, 5, and $\gamma_{\theta j}^*=1$. It is seen that this influence becomes important for $\overline{K} > 0.1$.

Finally, Fig. 3 shows the influence of thermal prestress $\overline{n}_\theta^*$ on $\langle q_1^{*2} \rangle$ for plates with $\alpha_j^* = j^2$, where the integration is performed numerically for N=1 and N=3, respectively.

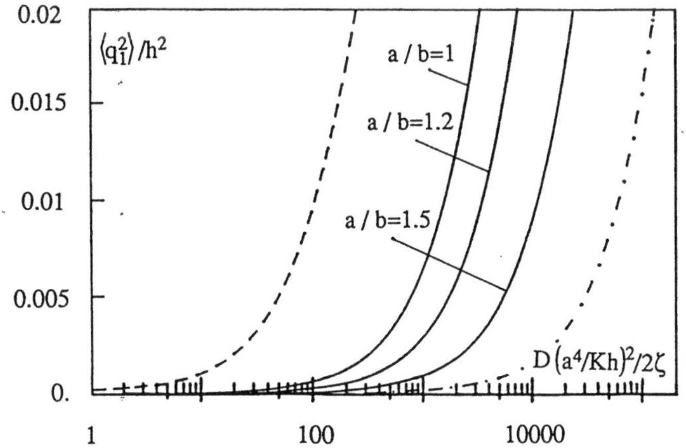

Fig. 1b. Nonlinear mean-square response $\langle q_1^2 \rangle / h^2$ as a function of $D\left(a^4/Kh\right)^2/2\zeta$ for various plate geometries: – – – plate strip, ——— rectangular plates, – · – · – right-angled isosceles triangular plate

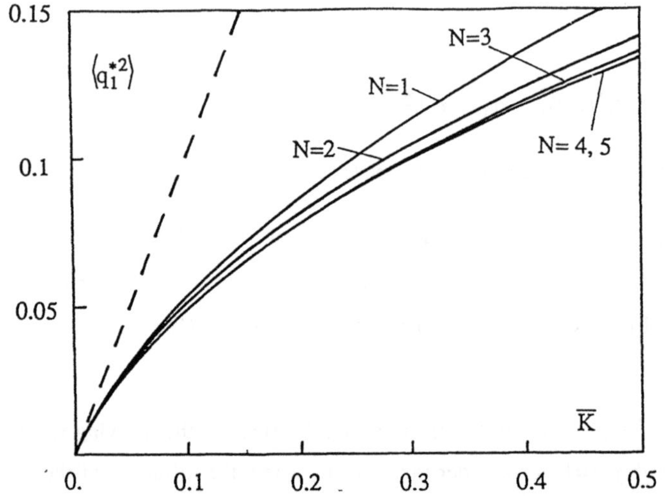

Fig. 2. Linear (– – –) and nonlinear (——) nondimensional mean-square response $\langle q_1^{*2} \rangle$ as a function of $\overline{K}$ according to equs.(34) and (35). Fig. 2 is valid for plates with $\alpha_j^* = j^2$, $j = 1, 2, \ldots N$, regardless of their special polygonal planform.

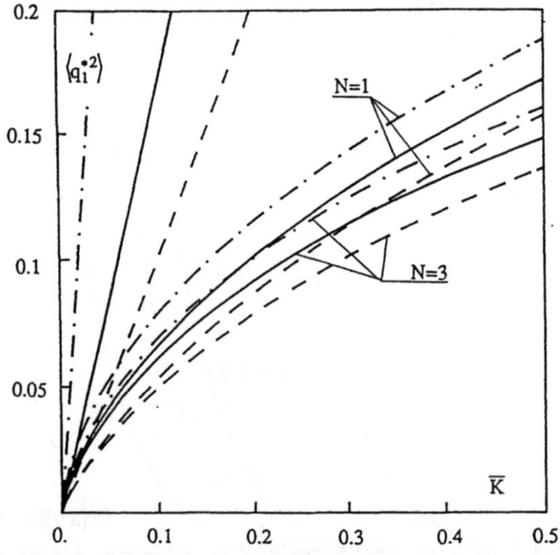

Fig. 3. Linear and nonlinear nondimensional mean-square response $\langle q_1^{*2} \rangle$ as a function of $\overline{K}$ and thermal prestress $\overline{n}_\theta^*$ according to equs.(21),(34) and (35). Fig. 3 is valid for plates with $\alpha_j^* = j^2$, $j = 1, 2, \ldots N$; $h/a = 0.1$.

– – – $\overline{n}_\theta^* = 0$, ——— $\overline{n}_\theta^* = 0.4$, – · – · – $\overline{n}_\theta^* = 0.8$

Conclusion:

A unifying nondimensional representation has been given in equs. (33)-(36) for the stationary transition probability function and the mean-square response of simply supported plates of arbitrary polygonal planform. Considering a multimodal expansion, the plate geometry enters only through the nondimensional frequency parameters α_j^*, while the single-mode expression (37) is completely independent of the geometry, and the corresponding result is presented in Fig. 1a. The influence of the higher modes is exemplarily studied in Fig. 2, where a numerical integration procedure has been applied to derive the variance over the full range of loading parameters.
The transformation of this nondimensional formulation to the real plate deflection under consideration is determined by the linear eigenvalues α_j and the norm β_j^* of the linear modes of the plate domain. Thus, the nonlinear plate problem is transferred to the solution of a deterministic linear boundary value problem, ie to the Helmholtz equation (14) with Dirichlet's boundary condition (15). The second-order statistics of bending moments can be determined within the nonlinear modal approximation.

Finally, it is noted that the nondimensional equations of motion (24) can be used as the starting point for the application of various approximate solution techniques of nonlinear probabilistic mechanics, see, eg /2/ and /20/. Since random vibrations in the postbuckling range are left to future studies a reference can only be made to the natural vibrations as examined in /21/.

References

1. Caughey, T.K.: Derivation and Application of the Fokker-Planck Equation to Discrete Nonlinear Dynamic Systems Subjected to White Random Excitation. J. Acoust. Soc. Am. 35 (1963), 1683-1692.

2. Ariaratnam, S.T.: Response of a Loaded Nonlinear String to Random Excitation. J. Appl. Mech. 29 (1962), 483-485.

3. Herbert , R.E.: Random Vibrations of a Nonlinear Elastic Beam. J. Acoust. Soc. Am. 36 (1964), 2090-2094.

4. Herbert , R.E.: Random Vibrations of Plates with Large Amplitudes. J. Appl. Mech. 32 (1965), 547-552.

5. Crandall, S.H.; Zhu, W.Q.: Random Vibration: A Survey of Recent Developments. J. Appl. Mech. 50 (1983), 953-962.

6. Berger, H.M.: A New Approach to the Analysis of Large Deflection of Plates. J. Appl. Mech. 22 (1955), 465-472.

7. Heuer, R.;Irschik, H.; Ziegler, F.: Multi-Modal Approach for Large Natural Flexural Vibrations of Thermally Stressed Plates. Nonlinear Dynamics 1 (1990), 449-458.

8. Heuer, R.;Irschik H.; Ziegler, F.: A BEM-Formulation for Nonlinear Plate Vibrations. In Proc. IUTAM / IACM-Symp. on Discretization Methods in Structural Mechanics, Vienna, Austria, 1989 (G. Kuhn, H. Mang, Eds.), Springer-Verlag, Berlin 1990, 341-351.

9. Heuer, R.;Irschik, H.; Fotiu, P.; Ziegler, F.: Nonlinear Flexural Vibrations of Layered Plates. Int. J. Solids Structures (1991), in press.

10. Lin, Y.K.: Probabilistic Theory of Structural Dynamics. McGraw-Hill, New York, 1967.

11. v. Karman, Th.: Festigkeitsprobleme im Maschinenbau. In: Encyklopädie der math. Wissenschaften, IV, Teubner, Leipzig, 1910, 350.

12. Nowinski, J.L.; Ohnabe, H.: On certain inconsistencies in Berger equations for large deflections of elastic plates. Int. J. Mech. Sci. 14 (1972), 165-170.

13. Wah, T.: Large amplitude flexural vibrations of rectangular plates. Int. J. Mech. Sci. 5 (1963), 425-438.

14. Nayfeh, A.; Mook, D.T.: Nonlinear Oscillations. Wiley, New York, 1979.

15. Irschik, H.: Large Thermoelastic Deflections and Stability of Simply Supported Polygonal Panels. Acta Mechanica 59 (1986), 31-46.

16. Irschik, H.: Influence of Large Amplitudes on Free Flexural Vibrations of Polygonal Shear-Deformable Plates - A Unifying Dimensionless Formulation. Int. J. Solids Structures 26 (1990), 675-681.

17. Timoshenko, S.; Woinowsky-Krieger, S.: Theory of Plates and Shells. McGraw-Hill, New York, 1959.

18. Leissa, A.W.: Vibrations of Plates. NASA-SP-160, Washington, 1969.

19. Parkus, H.: Random Processes in Mechanical Sciences. Proc. CISM-Course 9, Udine 1969, Springer-Verlag, Wien, 1969.

20. Hampl, N.C.; Schueller, G.I.: Probability Densities of the Response of Non-linear Structures under Stochastic Dynamic Excitation. In: Stochastic Structural Mechanics (Y.K. Lin, G.I. Schueller, Eds.), Lecture Notes in Engineering 31, Springer-Verlag, Berlin, 1987.

21. Heuer, R.; Irschik, H.; Ziegler, F.: Multi-Modal Formulation for Free Large Vibrations of Buckled Plates. In: Proc. 9th International Modal Analysis Conference, Florence, Italy, Union College, Schenectady, NY 12308, 1991, 96-100.

Structural Model Bifurcation under Random Excitation

R. A. Ibrahim, B. Lee and A. Afaneh

Wayne State University, Department of Mechanical Engineering
Detroit, Michigan 48202, U. S. A.

SUMMARY

Stochastic bifurcation of mean square response of a clamped-clamped beam under initial axial load excited by a wide band random excitation is investigated analytically, numerically, and experimentally. The nonlinear response is represented by the first three modes. The response statistics are examined in the neighborhood of a critical static axial load where the normal mode frequencies are commensurable. It is found that Gaussian closure fails to predict bifurcation of asymmetric modes. Both non-Gaussian closure and numerical simulation predicted bifurcation boundaries in terms of the axial load, excitation spectral density level, and damping ratios.

INTRODUCTION

The random response of a clamped-clamped beam has extensively been examined within the framework of one mode excitation. The early results go back to the work of Lyon, et a. [1] who considered narrow band random excitation of a hard-spring oscillator represented by Duffing equation. The purpose of their study was to find out if the well known jump phenomenon can occur under narrow band random excitation. The response was predicted by using the method of equivalent linearization and revealed multivalued response characteristics which have the same general appearance as those for sinusoidal forcing except that the peaks are much sharper. The Duffing oscillator under narrow band random excitation characteristics has received considerable attention by Lennox and Kuak [2], Rajan and Davies [3], Iyengar [4] and Roberts and Spanos [5]. The single mode modeling of continuous systems is less accurate due to the strong effect of the non-linear coupling [6,7]. The results of references [6,7] showed that the estimated mean square response of the first mode displacement is significantly affected by the nonlinear coupling with other modes. The present paper examines the stochastic bifurcation of the unexcited modes of a clamped-clamped beam when initial static axial load is applied at one end of the beam. Attention is focused to examine the response characteristics in the neighborhood of critical values of the axial load where the normal mode frequencies are nearly commensurable.

STOCHASTIC BIFURCATION ANALYSIS

In random structural dynamics, the meaning of "bifurcation under random perturbation" has not been clearly defined. In what follows, we will give a definition appropriate to the purpose of the

N. Bellomo · F. Casciati (Eds.)
Nonlinear Stochastic Mechanics
IUTAM Symposium Turin

present paper. To study the stochastic bifurcation problem, let α be a physical parameter in the differential equations of motion as represented by the state vector equations

$$\frac{d\mathbf{X}(t,\alpha)}{dt} = \mathbf{f}(\mathbf{X},t,\alpha) + \varepsilon\mathbf{g}(\mathbf{X},\xi(t),\alpha) \tag{1}$$

where $\mathbf{X}=\{X_1,X_2,...,X_n\}^T$ is an n-dimensional vector whose elements represent displacements and velocities, $\mathbf{f}=\{f_1,f_2,...,f_n\}^T$ is a vector-valued function, the vector $\mathbf{g}=\{g_1,g_2,...,g_n\}^T$ is a function of $\mathbf{X}$, time t, and the random vector process $\xi(t)=\{\xi_1(t),\xi_2(t),...,\xi_m(t)\}^T$ in m dimensions, both vectors $\mathbf{f}$ and $\mathbf{g}$ are also functions of the bifurcation parameter α, and ε is a small parameter depending on some physical constants of the system. Here the noise $\xi(t)$ may be additive and/or multiplicative. Under random excitation assume that the structure has an equilibrium position may be given by the trivial solution $\mathbf{X}=\mathbf{0}$ for any α and $\varepsilon\geq 0$. For $\varepsilon=0$, let α_c be the critical value above which ($\alpha>\alpha_c$) the equilibrium position $\mathbf{X}=\mathbf{0}$ loses its stability and bifurcates into another nearby equilibrium or steady-state solution. Since the motion is random, such a steady-state cannot exist in a deterministic sense. However, if $t\to\infty$, the solution $\mathbf{X}(t)$ approaches zero for $\alpha<\alpha_c(\varepsilon)$ and reaches a steady-state (statistically stationary) solution $\mathbf{X}(t)\approx\mathbf{Z}(t)\neq 0$ for $\alpha\geq\alpha_c(\varepsilon)$. Then we say that bifurcation takes place at the bifurcation point (critical value) $\alpha_c(\varepsilon)$. Here the steady-state solution $\mathbf{Z}(t)$ may be statistically stationary with a periodic component.

Suppose that the unperturbed system bifurcates from the equilibrium position at the origin to a limit cycle C. Then the state of the randomly perturbed system will fluctuate about the limit cycle. The stochastic stability of such motion arises naturally but has not been treated satisfactorily. For a white noise perturbation, the probability of escape from a periodic attractor (Limit Cycle C) is briefly discussed in Freidlin and Wentzell [8] who used asymptotic method for Ito's equation, and no explicit solutions were obtained. Recently, Namachchivaya [9] presented an excellent review pertaining to bifurcation problems in deterministic and stochastic systems. However, the available techniques are restricted to one or two codimension problems. For n coupled modes in continuous systems, it is very important to examine their interaction under random excitation in the neighborhood of the bifurcation of the non-excited mode. An example of a beam under axial static load P and subjected to a wide band lateral excitation Y(t) is considered. Figure 1 shows a clamped-clamped beam of mass m per unit length, length L, thickness h and width b. Applying Hamilton's principle and considering the elastic energy due to stretching of the midplane of the beam gives the equation of motion:

$$m\frac{\partial^2 U(x,t)}{\partial t^2} + c\frac{\partial U(x,t)}{\partial t} + EI\frac{\partial^4 U(x,t)}{\partial x^4} + P\frac{\partial^2 U(x,t)}{\partial x^2} - \frac{EA}{2L}\frac{\partial^2 U(x,t)}{\partial x^2}\int_0^L \left(\frac{\partial U(x,t)}{\partial x}\right)^2 dx = -m\frac{\partial^2 Y(t)}{\partial t^2} \tag{2}$$

and the boundary conditions

$$U(x,t)\big|_{x=0,L} = 0, \quad \frac{\partial U(x,t)}{\partial x}\Big|_{x=0,L} = 0 \tag{3}$$

The deflection of the beam is expanded to include n modes, i. e.,

$$U(x,t) = \sum_{i=1}^{n} q_i(t)\Phi_i(x) \tag{4}$$

where $q_i(t)$ are the generalized coordinates and $\Phi_i(x)$ are the corresponding shape functions. The shape functions which satisfy the boundary conditions are

$$\Phi_i(x) = A_i\left[\cos(\beta_i x) - \cosh(\beta_i x)\right] + \left[\sin(\beta_i x) - \sinh(\beta_i x)\right] \tag{5}$$

where $$A_i = -\left[\frac{\sin(\beta_i L) - \sinh(\beta_i L)}{\cos(\beta_i L) - \cosh(\beta_i L)}\right]$$

$\beta_i L = 4.730041$, 7.85305, and 10.995608 for i = 1, 2, and 3, respectively.

Substituting (4) into (2) and applying Galerkin's method results in a set of coupled nonlinear ordinary differential equations in n-modes. Under axial compression load, the first three modes will be considered and the corresponding equations in nondimensional form are:

$$\left.\begin{aligned}
&Q_1'' + 2\zeta_1 r_{11} Q_1' + r_{11}^2 Q_1 + r_{13}^2 Q_3 + \tilde{K}_{11} Q_1^3 + \tilde{K}_{12} Q_1 Q_2^2 \\
&\quad - \tilde{K}_{13} Q_1^2 Q_3 + \tilde{K}_{14} Q_1 Q_3^2 - \tilde{K}_{15} Q_2^2 Q_3 - \tilde{K}_{16} Q_3^3 = \tilde{f}_1 \tilde{Y}'' \\
&Q_2'' + 2\zeta_2 r_{22} Q_2' + r_{22}^2 Q_2 + \tilde{K}_{21} Q_1^2 Q_2 - \tilde{K}_{22} Q_1 Q_2 Q_3 + \tilde{K}_{23} Q_2 Q_3^2 + \tilde{K}_{24} Q_2^3 = 0 \\
&Q_3'' + 2\zeta_3 r_{33} Q_3' + r_{31}^2 Q_1 + r_{33}^2 Q_3 - \tilde{K}_{31} Q_1^3 - \tilde{K}_{32} Q_1 Q_2^2 \\
&\quad + \tilde{K}_{33} Q_1^2 Q_3 - \tilde{K}_{34} Q_1 Q_3^2 + \tilde{K}_{35} Q_2^2 Q_3 + \tilde{K}_{36} Q_3^3 = \tilde{f}_3 \tilde{Y}''
\end{aligned}\right\} \tag{6}$$

where a prime denotes differentiation with respect to $\tau = \omega_o t$,

$\omega_0^2 = 500.72 EI/mL^4$, $Q_i(\tau) = (L/h^2) q_i(t)$, $\tilde{Y}''(\tau) = (L/h^2\omega_0^2)\ddot{Y}(t)$,

$r_{11}^2 = 1-\alpha$, $r_{13}^2 = r_{31}^2 = 0.7676\alpha$, $r_{22}^2 = 7.595-3.674\alpha$,

$r_{33}^2 = 29.1-7.7979\alpha$, $\alpha = P/P_o$

and all other coefficients are fixed constants.

The linear coupling among symmetric modes arises from the axial static load. The nonlinear coupling exists due to the geometric nonlinearity. Let the normal mode frequencies of the linear coupled system (6) be ω_1, ω_2, and ω_3. These natural frequencies depend on the axial load ratio α. The system possesses internal resonance of combination type $\omega_3 = \omega_1 + 2\omega_2$ when $\alpha = 0.78$. Any deviation of $r = \omega_3/(\omega_1 + 2\omega_2)$ from unity measures the internal detuning which has a significant effect on the dynamic response characteristics of the system.

The external excitation is assumed Gaussian wide band process whose correlation time is much smaller than any characteristic period of oscillation of the system. It also assumed that the excitation has a flat spectral density of constant value 2D. In this case the response can be approximated by a Markov state vector which is established through the coordinates transformation:

$$\{Q_1, Q_1', Q_2, Q_2', Q_3, Q_3'\} = \{X_1, X_2, X_3, X_4, X_5, X_6\} = \mathbf{X} \tag{7}$$

If the random process $\tilde{Y}''(\tau)$ is replaced by the formal derivative of the Brownian process $B(\tau)$, i.e. $\tilde{Y}''(\tau) = \sigma dB(\tau)/d\tau$, where $\sigma^2 = 2D$, equations (6) may in turn be transformed into the Ito type equations:

$$\left.\begin{aligned}
&dX_1 = X_2 d\tau, \qquad dX_3 = X_4 d\tau, \qquad dX_5 = X_6 d\tau, \\
&dX_2 = (-2\zeta_1 r_{11} X_2 - r_{11}^2 X_1 - r_{13}^2 X_5 - \tilde{K}_{11} X_1^3 - \tilde{K}_{12} X_1 X_3^2 \\
&\quad + \tilde{K}_{13} X_1^2 X_5 - \tilde{K}_{14} X_1 X_5^2 + \tilde{K}_{15} X_3^2 X_5 + \tilde{K}_{16} X_5^3) d\tau + \tilde{f}_1 dB \\
&dX_4 = (-2\zeta_2 r_{22} X_4 - r_{22}^2 X_3 - \tilde{K}_{21} X_1^2 X_3 + \tilde{K}_{22} X_1 X_3 X_5 - \tilde{K}_{23} X_3 X_5^2 - \tilde{K}_{24} X_3^3) d\tau \\
&dX_6 = (-2\zeta_3 r_{33} X_6 - r_{31}^2 X_1 - r_{33}^2 X_5 + \tilde{K}_{31} X_1^3 + \tilde{K}_{32} X_1 X_3^2 \\
&\quad - \tilde{K}_{33} X_1^2 X_5 + \tilde{K}_{34} X_1 X_5^2 - \tilde{K}_{35} X_3^2 X_5 - \tilde{K}_{36} X_5^3) d\tau + \tilde{f}_3 dB
\end{aligned}\right\} \tag{8}$$

The joint probability density $p(\mathbf{X},\tau)$ of the response coordinates is described by the Fokker-Planck equation:

$$\frac{\partial p(\mathbf{X},\alpha,\tau)}{\partial \tau} = -\sum_{i=1}^{6} \frac{\partial}{\partial X_i}[a_i(\mathbf{X},\alpha,\tau) p(\mathbf{X},\alpha,\tau)] + \frac{1}{2}\sum_{i=1}^{6}\sum_{j=1}^{6} \frac{\partial^2}{\partial X_i \partial X_j}[b_{ij}(\mathbf{X},\alpha,\tau) p(\mathbf{X},\alpha,\tau)] \tag{9}$$

where $a_i(\mathbf{X},\alpha,\tau)$ and $b_{ij}(\mathbf{X},\alpha,\tau)$ are the first and second incremental moments.

Bifurcation for the system occurs at α_c if and only if equation (9) has a solution for $\alpha \geq \alpha_c(\varepsilon)$. It must be pointed out that the existence of a solution to equation (9) was used as a criterion for the bifurcation of nonlinear stochastic single-degree-of-freedom systems described by equation (1). It should not be confused with the bifurcation of the linear partial differential equation (9) itself, since, for the deterministic elliptic equation (Keller and Antman [10]), bifurcation is normally associated with a nonlinear boundary-value problem. Note that the trivial solution $\mathbf{X}=0$ may be regarded as having a Dirac delta probability density $\delta(\mathbf{X})$. Thus, if it exists, it bifurcates into a solution of $p(\mathbf{X},\alpha)$ of (9). Therefore, the definition given above may be called "bifurcation in (stationary) distribution." The solution of the Fokker-Plancke equation of the Markov vector (8) can be obtained since the stiffness nonlinearity is the gradient of the elastic potential function [11]. Alternatively, one can define "bifurcation in moments." To this end, let $m_K(\alpha)$ denote the K-th moment of the vector $\mathbf{X}$ as $t \to \infty$, i.e.

$$m_K(\alpha) = \lim_{t \to \infty} E|X_1(t,\alpha) X_2(t,\alpha) \ldots X_6(t,\alpha)|^K, \quad K \geq 1 \tag{10}$$

If the moment m_K diverges for $\alpha > \alpha_c$ but converges to a non-zero limit for $\alpha \leq \alpha_c$, we say that "the system bifurcates in K-th moment." We believe that these definitions clarify ambiguous statements in the literature.

Now, a general first order differential equation for the response moments will be generated for system (8). This equation is obtained by multiplying both sides of equation (9) by the scalar function $\Phi(\mathbf{X}) = X_1^{k1}X_2^{k2}X_3^{k3}X_4^{k4}X_5^{k5}X_6^{k6}$ and integrating by parts over the entire state space $-\infty < \mathbf{X} < \infty$. This process results in the general moment equation

$$\dot{m}_K = M_K(m_K, m_{K+1}, m_{K+2}, \alpha), \quad K = k_1 + k_2 + \ldots + k_6 \tag{11}$$

where $m_K = \int_{-\infty}^{\infty} \cdots \int_{-\infty}^{\infty} \Phi(X_1^{k1}X_2^{k2}X_3^{k3}X_4^{k4}X_5^{k5}X_6^{k6}) p(\mathbf{X},\alpha,\tau) dX_1 \cdots dX_6 = E[X_1^{k1} \ldots X_6^{k6}]$

Equation (11) constitutes an infinite set of coupled moment equations. In order to solve for the response statistics the generated moment equations must be truncated by an appropriate closure scheme [12]. Two truncation schemes are commonly used. These are the Gaussian and first-order non-Gaussian closures. In applying these closures precaution should be taken regarding properties of the response moments, such as the non-negativeness of even order moments and Schwartz's inequality. The Gaussian closure is carried out by setting third and fourth order cumulants to zero. According to the relationships between cumulants and moments third and fourth order joint moments are expressed by lower order moments as documented in reference [12]. In this case one can generate 27 coupled equations in the first (6 equations) and second (21 equations) order moments. The solution of the resulting closed 27 coupled equations is obtained numerically by using the DIVPAG subroutine (Adams-Moulton/Gear numerical integration method) of the IMSL Library (International Mathematical and Statistical Library). The numerical integration is carried out with double precision on SUN 3/260 computer using time increment $\Delta\tau = 0.5$ and integration accuracy 0.1D-03. The first-order non-Gaussian closure is carried out by setting fifth and sixth order cumulants to zero. Accordingly, one can express fifth and sixth order moments in terms of lower order moments [12]. For the present problem one needs to generate 209 coupled equations in the first through fourth order moments. The 209 differential equations are solved by numerical integration by using the DIVPAG subroutine.

Figs. (2a) and (2b) show the time history of the mean square responses as obtained by Gaussian and non-Gaussian closures for initial axial load ratios α=0.3 and 0.78 , respectively. These values of α correspond to internal detuning r= 0.8721 and 1, respectively. The two solutions are obtained for excitation spectral density 2D=0.5 and damping ratios ζ_1=0.005, ζ_2=0.006 and ζ_3=0.008. It is seen that the third mode response does not achieve a stationary value according to the Gaussian closure solution. The Gaussian closure fails to predict second mode bifurcation for all axial load levels. On the other hand, the non-Gaussian closure predicts second mode bifurcation which takes place due to energy flow from the symmetric excited modes at critical values of the axial load. The second mode response is also governed by the level of the axial load (or the internal detuning parameter r). Figure 2b also shows that the non-Gaussian solution is relatively higher than the Gaussian solution, which is particularly observed in the first mode. The critical value of α which separates two-mode response region I from three mode response region II on

σ^2-α plane is shown in Fig. (3). The bifurcation diagram given in Fig. 3 includes another region (III) in which the non-Gaussian solution is numerically unstable. In order to understand the effect of nonlinear interaction among the three modes, the normalized mean square responses are plotted against the axial load ratio α as shown in Fig. 4 for two excitation levels. The normalization is taken by dividing the mean square responses of the coupled equations by the corresponding non-Gaussian solution of the uncoupled nonlinear equations of the first and third mode. The second mode is normalized by the unimodal response of the first mode. It is clear, from Fig. 4, that energy is transferred from the externally excited (symmetric) modes to the non-excited modes only because of modal coupling. For higher excitation level the second mode bifurcation starts at smaller axial load ratio.

MONTE CARLO SIMULATION

The responsé statistics are estimated by numerically simulating the nonlinear coupled equations (6). The Monte Carlo simulation is carried out by generating 300 records of the random excitation $\widetilde{Y}''(\tau)$ with duration τ=1000 for each record. A sequence of 10,240 random numbers are sampled from every record. The IMSL DRNNOR subroutine (random number generation from a normal distribution using an inverse Cumulative Distribution Function (CDF) technique) is used. According to this subroutine, a uniform (0,1) random deviate is generated and the inverse of the normal distribution function is evaluated at each point. Equations (6) are then numerically solved by the IMSL DIVPAG subroutine. Response statistics such as mean squares, probability density, and power spectral density are evaluated by including the ILS (Interactive Laboratory System) software and other subroutines developed by the authors.

Figure 5 shows the time histories of mean square responses of the three modes as estimated by numerical simulation together with the non-Gaussian closure solution for comparison. These curves are obtained in the neighborhood of r=1, i.e., α=0.78. The numerical simulation exhibits severe nonstationarity in the second mode mean square response. The power spectra of both excitation and response processes are estimated by using the FFT order 9 and the rectangular window algorithm. The estimated spectra for the case of mixed mode interaction (α=0.78) are shown in Fig. 6. The marginal probability density functions for the two-mode (α=0.3) and three-mode (α=0.78) interactions are shown in Fig. 7. In both cases the effect of nonlinearity is drastically affecting the first mode which exhibit strong non-normality and multi-maxima. This feature is referred to as multifurcation [13].

EXPERIMENTAL RESULTS

The experimental model is made of a spring steel beam of length 0.25-m, width 0.013-m and thickness 0.0005-m. The beam is clamped from both ends and the base of its fixture is directly connected to the head of a VTS-400 electrodynamic shaker. The shaker thrust is 400 lb and provides up to 1 in. peak-to-peak stroke. The shaker receives a random signal from the HP 3565S computer. The excitation and response signals are read by the HP 3565 computer which can read signals from 8 channels simultaneously at 262kHz sampling rate. The HP VISTA software and the ILS software are used to estimate statistical parameters for the collected excitation and response signals. In addition to the subroutines available in both VISTA and ILS other programs are developed to determine mean squares and probability functions.

Based on preliminary tests it is decided to set the time duration of each test to 200 seconds which is found to be enough to exhibit all possible dynamic characteristics of the system and to establish the stationarity of the signal. Figure 8 shows time histories of mean square responses and excitation signals. It is seen that the shaker signal is always stationary while the beam response reaches a stationary state after transient period of 100 seconds. The power spectral density functions of the shaker and the beam are shown in Fig. 9. It is seen that there is a significant shift of the resonance peaks as the excitation level increases. Another important feature is that the response spectra become more wider as the excitation level increases. This feature implies that the stiffness nonlinearity enhances the degree of randomness of response signals. The probability density functions as estimated from the signals of the shaker and the clamped-clamped beam are plotted in Fig. 10. The response probability diagrams do not exhibit bifurcation as observed in the Monte Carlo simulation results. The main reason is due to the fact that the initial axial static load is very difficult to be adjusted to establish the internal resonance among the three modes. This difficulty arises mainly from the inevitable sliding of the model ends and the associated friction forces. At present the authors are designing a mechanism to adjust the critical axial load.

CONCLUSIONS

The stochastic bifurcation of a clamped-clamped beam under axial static load is investigated by three different approaches. The Gaussian closure fails to predict any nonzero motion for the second mode under all possible conditions of axial load and excitation levels. On the other hand, both non-Gaussian closure and Monte Carlo simulation predicted response bifurcation. The estimated probability density functions, from numerical simulation, displayed multifurcation. Depending on the value of axial static load the first three modes could be in internal resonance. In the neighborhood of this critical load the second mode is excited through nonlinear coupling. The experimental results reveal two interesting features as the excitation level increases. The first is a significant shift in the resonance peaks, while the second is a widening effect in the response spectra of each mode.

Acknowledgement: This research is supported by a grant from the National Science Foundation under grant No. 87-96342 and by additional funds from the Institute for Manufacturing Research of Wayne State University.

REFERENCES

1. R.H. Lyon, M. Heckl and C.B. Hazelgrove, Response of Hard-Spring Oscillator to Narrow-Band Excitation, *J. Acoustical Soc. America* **33**, 1404-1411, 1961.

2. W.C. Lennox and Y.C. Kuak, Narrow-Band Excitation of a Nonlinear Oscillator, *ASME J. Appl. Mech.* **43**, 340-344, 1976.

3. S. Rajan and G.G. Davies, Multiple Time Scaling of the Response of a Duffing Oscillator to Narrow-Band Random Excitation, *J. Sound and Vibration* **123**, 497-506, 1988.

4. R.N. Iyengar, Response of Nonlinear Systems to Narrow-Band Excitation, *Structural Safety* **6**, 177-185, 1989.

5. J.B. Roberts and P.D. Spanos, *Random Vibration and Statistical Linearization*, John Wiley & Sons, New York, 1990.

6. R.E. Herbert, Random Vibrations of a Nonlinear Elastic Beam, *J. Acoust. Society of Amer* **36**, 2090-2094, 1964.

7. H.R. Busby, and V.I. Weingarten, Response of Nonlinear Beam to Random Excitation, *ASCE J. Engineering Mech. Div.*, 55-68, 1973.

8. M.I. Freidlin and A.D. Wentzell, A. D., *Random Perturbations of Dynamical Systems*, Springer-Verlag, New York, 1984.

9. N. S. Namachchivaya, Stochastic Bifurcation, *Appl. Math. & Comp.* **38**, 101-159, 1990.

10. J.B. Keller and S. Antman, S., *Bifurcation Theory and Nonlinear Eigenvalue Problems*, Lecture Notes, Courant Institute of Math. Sci., New York University, N.Y., 1968.

11. S. T. Ariaratnam, Random Vibration of Nonlinear Suspensions, *J. Mechanical Engrg. Sci.* **2(3)**, 195-201, 1960.

12. R.A. Ibrahim, *Parametric Random Vibration*, John Wiley & Sons, New York, 1985.

13. T. Kapitaniak, *Chaos in Systems with Noise, World Scientific*, New Jersey, 1988.

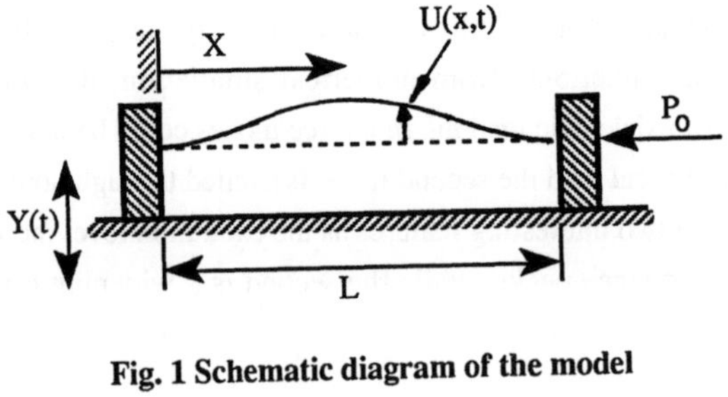

Fig. 1 Schematic diagram of the model

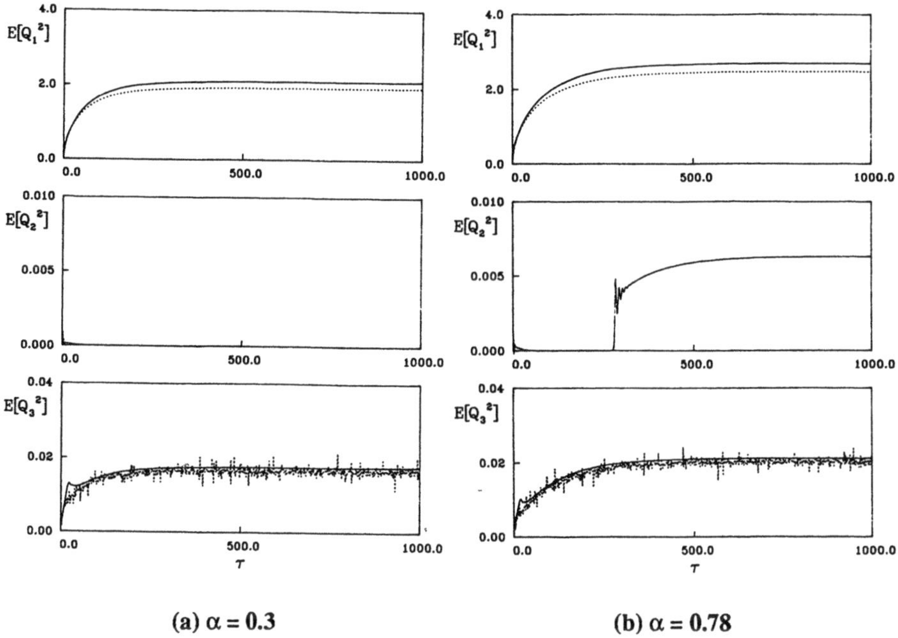

Fig. 2 Mean square responses for two axial load ratios

- - - - Gaussian solution ________ Non-Gaussian solution

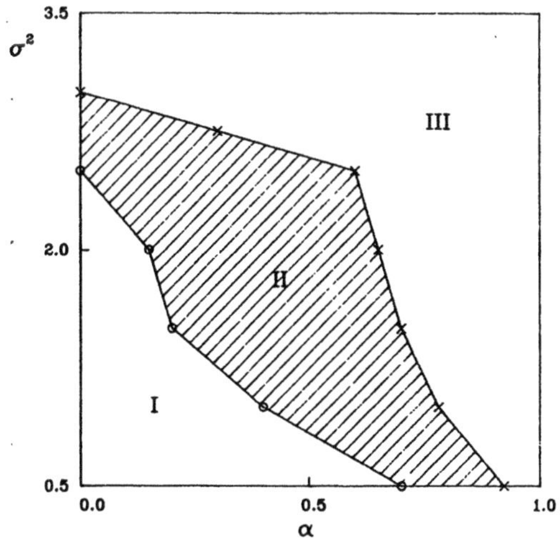

Fig. 3 Bifurcation diagram for two- and three- mode responses

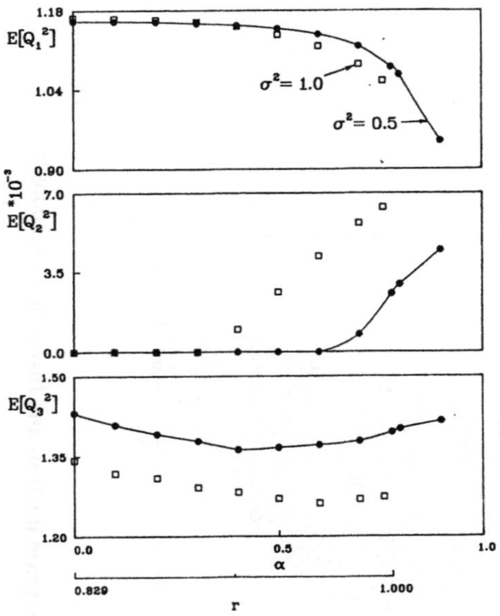

Fig. 4 Dependence of normalized mean square responses on axial load ratio

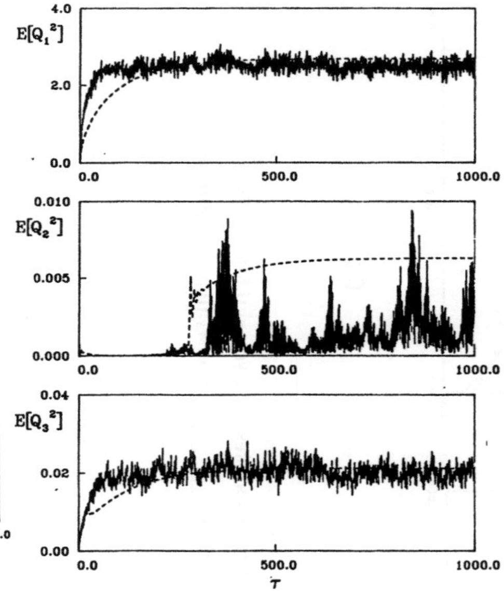

Fig. 5 Numerical simulation and non-Gaussian closure solutions

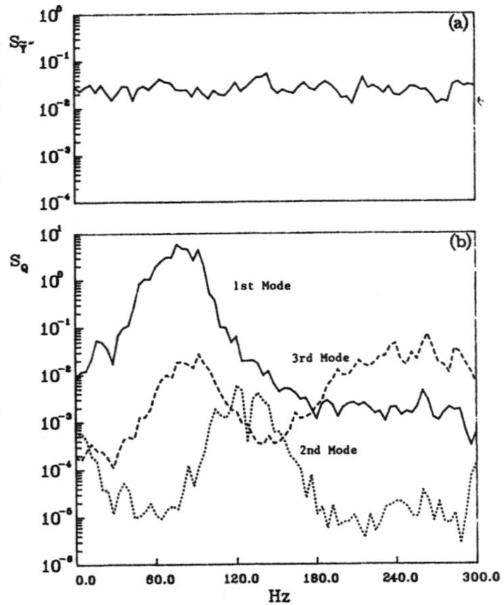

Fig. 6 (a) Excitation and (b) response spectra as estimated by Monte Carlo simulation

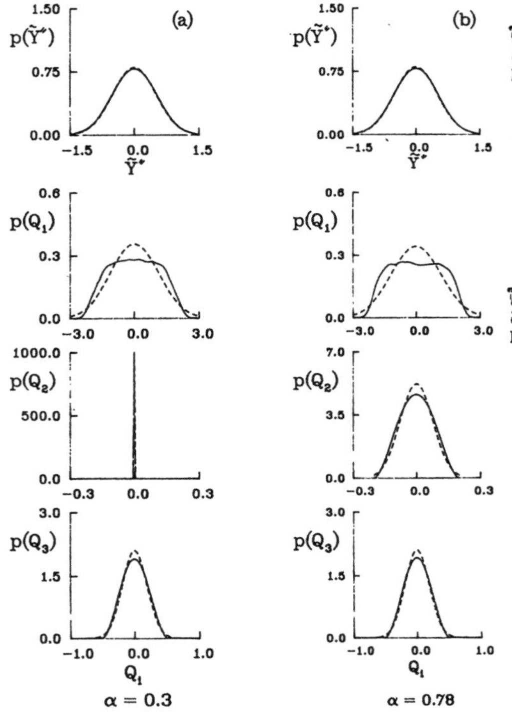

Fig. 7 Marginal probability densities of excitation and response coordinates as estimated from Monte Carlo simulation for two axial load parameters

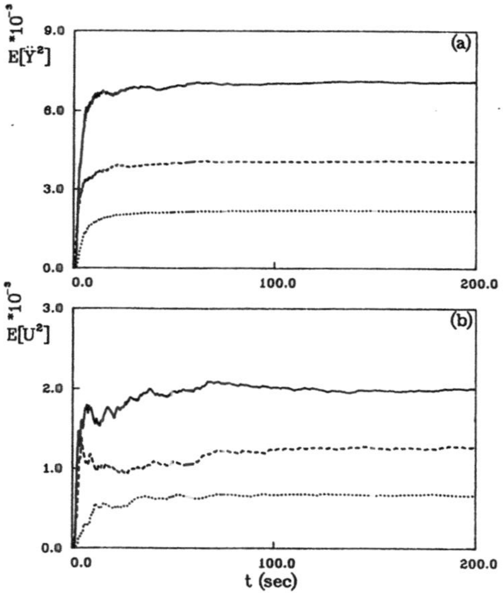

Fig. 8 Measured mean squares of (a) excitation and (b) response for three excitation levels

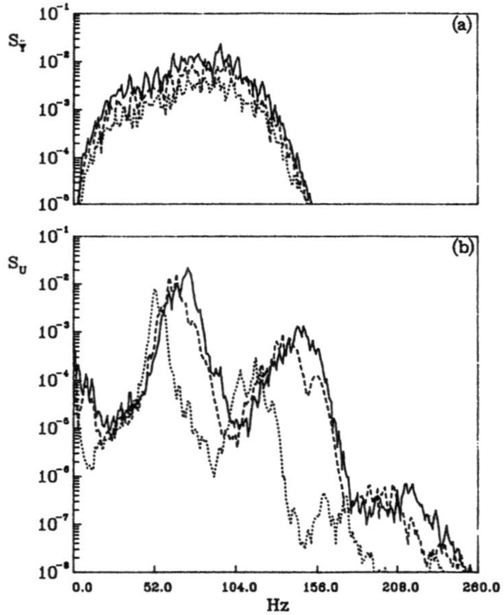

Fig. 9 Measured (a) excitation and response spectra for three excitation levels

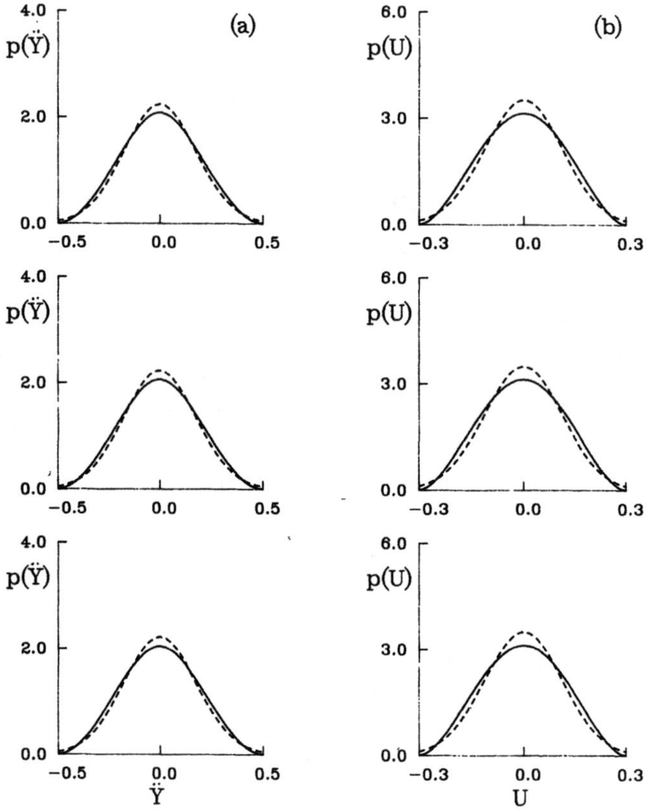

Fig. 10 Measured (a) excitation and (b) response probability density functions for three excitation levels

Approximate Analysis of Nonlinear Systems under Narrow Band Random Inputs

R.N. Iyengar
Department of Civil Engineering
Indian Institute of Science
Bangalore 560 012, INDIA

Introduction

Dynamic analysis of systems with nonlinear stiffness characteristics is important in many earthquake engineering problems. Eventhough the base seismic excitation could be modelled as a wideband random process, due to the filtering action of the primary structure, the excitation to the equipments at higher floor levels are narrowbanded. This problem has received considerable attention in recent years [1, 2, 3]. While, under a broadband excitation, atleast simple nonlinear oscillators can be solved, the corresponding solutions under a narrowband input are not available. Since a gaussian narrowband process is essentially a filtered whitenoise, the dimensionality of the system response vector would be atleast four. The Fokker-Planck (FP) equation for the transitional probability density can be easily written down, but obtaining a solution to this equation is a formidable task. The stochastic averaging technique reduces the number of variables to three [4], but the resulting FP equation is still difficult to handle. Thus equivalent linearization or closure methods have to be used to get approximate solutions. However, in the absence of solution error estimates these methods can only be evaluated with reference to extensive numerical simulations. The success of linearization methods in deterministic vibration theory may be in a large measure attributed to the selection of the nearly correct form of the starting solution. Even with stochastic inputs, the success of an approximation would much depend on whether the form of the assumed solution is good enough. Thus the usual equivalent linearization technique (ELT) fails with the Van der Pol oscillator unless, the form of the solution incorporates the existence of the limit cycle [5]. Under narrow band excitation ELT leads to another difficulty, namely that of multiple values for the moments. In such a situation the meansquare error of the linearized terms will have

N. Bellomo · F. Casciati (Eds.)
Nonlinear Stochastic Mechanics
IUTAM Symposium Turin

to be computed to decide which branch leads to the global minimum. An alternative to this would be a stochastic stability study of the first variational equation with the approximate solution appearing as a parametric term. This approach has been highlighted previously [3, 6] with reference to oscillators with cubic and quadratic nonlinearities. It may be noted here that the stability of an approximate solution for small variations is only a necessary condition for the solution to be realizable. In the previous studies by the author [3, 5, 6] it has been found that under narrowband inputs even when the ELT solution is singlevalued, it could be unstable in a wide parameter range. This observation indicates that under narrowband inputs ELT may not be very efficient unless improved further. In fact ELT does not incorporate information about the type of excitation but just minimizes a particular ensemble meansquare error. This may be contrasted with the deterministic case under periodic excitation, wherein linearization is achieved by minimizing the m.s. error over a period of oscillation [7]. Since with a narrowband processes a specific centre frequency can be identified, it would be interesting to study how the harmonic linearization technique (HLT) would approximate the nonlinear solution. Some preliminary results for the hardening Duffing oscillator are presented in this paper.

Hardening System

The cubic hardening system excited by a narrowband process can be described by

$$\ddot{x} + 2\eta\omega\dot{x} + \omega^2 x + \alpha\omega^2 x^3 = f(t) \tag{1}$$

$$\ddot{f} + 2\xi\lambda\dot{f} + \lambda^2 f = w(t) \tag{2}$$

$$\langle w(t_1)\, w(t_2)\rangle = I\delta(t_2 - t_1);\quad \sigma_f^2 = I/(4\xi\lambda^3) \tag{3}$$

In the case of a direct white noise excitation the presence of the nonlinear term decreases the response variance below its linear value. However, when f(t) is narrowbanded for some frequency ratios, the nonlinear system shows a variance several times higher than the variance of the linear system with $\alpha = 0$. With this in view, it is convenient to introduce the variables

$$y = x/\sigma_1\ ;\ \alpha\sigma_1^2 = \varepsilon < 1 \tag{4}$$

Here σ_1^2 is the linear system steady state variance x(t) when α = 0.

$$\sigma_1^2 = (\sigma_f^2/\omega^4)\ \frac{[1-\bar{\lambda}^2+(1+\xi\bar{\lambda})(\bar{\lambda}^2+4\xi\eta\bar{\lambda})]}{[(1-\bar{\lambda}^2)^2+4\eta\bar{\lambda}(1+\xi\bar{\lambda})(\xi+\eta\bar{\lambda})]} \tag{5}$$

$\bar{\lambda} = \lambda/\omega\ ;\ \bar{\xi} = \xi/\eta$

Thus the equation to be solved will be

$$\ddot{y} + 2\eta\omega\dot{y} + \omega^2 y + \omega^2 \varepsilon y^3 = f(t)/\sigma_1 \tag{6}$$

Probabilistic Linearization

When ξ is small, f(t) will have a strong centre frequency λ and hence as per the concepts of deterministic system theory, one could expect y(t) to respond mainly at the external frequency. Hence the solution form is taken as

$$y(t) = a(t)\sin(\lambda t-\theta) \tag{7}$$

where 'a' and θ are slowly varying random processes. Now replacing y^3 by γy and minimizing the m.s. error between the two terms over one period of oscillation, one gets

$$\gamma = [\int_0^{2\Pi} a^4 \sin^4 \psi d\psi \ / \int_0^{2\Pi} a^2 \sin^2 \psi d\psi\ ] = 0.75a^2 \tag{8}$$

The resulting linear equation is

$$\ddot{y} + 2\eta\omega\dot{y} + (\omega^2 + 0.75\ \varepsilon\ a^2)\ y = f(t)/\sigma_1 \tag{9}$$

Since a is slowly varying, it can be treated as a random variable. Hence, equation (8) can be easily solved conditioned on'a! (y|a) is a gaussian process and the steady state conditional variance is

$$\langle y^2|a\rangle = [\beta^2/(1+.75\varepsilon a^2)]\ \frac{[D+(1+\xi\bar{\lambda})(\bar{\lambda}^2+4\xi\eta\bar{\lambda})]}{[D^2+4\eta\bar{\lambda}(1+\xi\bar{\lambda})\{\xi(1+.75\varepsilon a^2)+\eta\bar{\lambda}\}]}$$

$$D = (1+0.75\ \varepsilon a^2 - \bar{\lambda}^2);\ \beta = \frac{\sigma_f}{\omega^2\sigma_1} \tag{10}$$

Thus the variance of y will be

$$\sigma_y^2 = \int_0^\infty \langle y^2|a\rangle\ p(a)\ da \tag{11}$$

This can be evaluated provided p(a) is known. Since this refers to the unknown amplitude density function of the original nonlinear equation, another assumption is required about its form. In the first approximation 'a' can be taken to be Rayleigh distributed with

$$p(a) = (a/\sigma_y^2)\ \text{ExP}(-0.5a^2/\sigma_y^2) \tag{12}$$

Substitution of this in eq. (11) provides a equation for finding σ_y^2. In Fig. 1, a typical result obtained by the above method is presented. A comparison is also provided with the ELT and simulation results previously obtained [3]. It is interesting to observe that the present probabilistic linearization technique (PLT) does not lead to multiple values for the variance. A structural stability analysis of the proposed solution to small variations is possible. The details have been previously presented [3]. The condition for structural stability will be

$$[-\eta/\bar{\lambda} + \text{Real} <\zeta\ (a)>] < 0$$

$$\zeta = 0.25\ [C_1^2 - 4(1-C_o)^2]^{1/2}$$

$$C_o = (1.5\ \varepsilon\ a^2 + 1-\eta^2)/\bar{\lambda}^2$$

$$C_1 = (1.5\ \varepsilon\ a^2/\bar{\lambda}^2) \tag{13}$$

The results of this condition are also marked in Fig. 1. It is observed that the assumed structure of the solution, namely a narrowband process with Rayleigh distributed amplitudes is not good enough for a range of values of $\bar{\lambda}$.

Improved Solution

The above solution represented by eq. (7) can be termed a small amplitude approximation to the nonlinear system of eq. (6). This follows, from the observation that eventhough ε is small, the solution becomes unstable as σ_y and hence the spread of the amplitude 'a' increases. This hints at the necessity of adding corrections to the basic approximation. Again, while a rigorous derivation of this correction is difficult, a reasonable estimation can be obtained as follows. Let the improved solution be

$$y = a\ \sin\ (\lambda t-\theta) + z = (y_1 + z) \tag{14}$$

where $y_1(t)$ is already known. Substitution of this eq. (6) leads to

$$\ddot{z} + 2\eta\omega\dot{z} + \omega^2 z + \varepsilon\omega^2 (3y_1^2 z + 3z^2 y_1 + z^3)$$
$$= 0.25 \ \varepsilon a^3 \sin 3 (\lambda t - \theta) \tag{15}$$

This is same as the first variational equation of eq. (6), if z^2, z^3 and a^3 are neglected as small. The excitation term in the above equation is nongaussian and hence, along with the nonlinear and time varying terms, exact determination of z(t) is not possible. However, the predominant part of z(t) is due to the excitation term. With this in view, the correction term for a given value of 'a' would be

$$z(t) = \frac{\varepsilon a^3}{4\sqrt{D_1}} \sin (3\lambda t - 3\theta - \delta)$$

$$D_1 = [(1+1.5\varepsilon a^2 - 9\bar{\lambda}^2)^2 + (6\eta\bar{\lambda})^2]$$
$$\cos\delta = (1+1.5\varepsilon a^2 - 9\bar{\lambda}^2)/\sqrt{D_1} \tag{16}$$
$$\sin\delta = 6\eta\bar{\lambda}/\sqrt{D_1}$$

Thus, the suggested improved approximation to the solution of eq. (6) would be

$$y = a \sin\psi + (0.25 \ \varepsilon a^3/\sqrt{D_1}) \sin (\psi+\phi) \tag{17}$$

where $\psi = (\lambda t-\theta)$ and $\phi = (2\lambda t - 2\theta-\delta)$

Now, the amplitude of y is given by

$$r^2 = \left(a + \frac{0.25\,\varepsilon a^3}{\sqrt{D_1}} \cos\phi\right)^2 + \left(\frac{0.25\,\varepsilon a^3}{\sqrt{D_1}} \sin\phi\right)^2 \tag{18}$$

This form immediately helps one in finding the non-Rayleigh distribution of the amplitude 'r'. Since p(a) is known from eq. (12), and θ is uniformly distributed in (0, 2Π) the above is a memoryless transformation to find p(r), at any time t.

<u>Stability of the Improved Approximation</u>

Whether eq. (17) represents an improvement or not can be checked by verifying its structural stability. This is carried out by considering the first variational equation

$$\ddot{z} + 2\eta\omega\dot{z} + \omega^2 (1+3\varepsilon y^2) z = 0 \tag{19}$$

where y is given by eq. (17). With the transformations

$$\lambda t = \tau, \ \omega_e = 1/\bar{\lambda}; \ z = v\exp(-\eta\omega_e\tau) \tag{20}$$

it can be shown that

$$\frac{d^2v}{d\tau^2} + [C_o - C_1 \cos 2(\tau-\theta) + C_2 \cos(2\tau-2\theta-\delta)]v = 0$$

$$C_o = 1.5 \ \varepsilon\omega_e^2 \ (a^2 + b^2) + \omega_3^2 \ (1-\eta^2)$$

$$C_1 = 1.5 \ \varepsilon\omega_e^2 \ a^2 \ ; \ C_2 = 3\varepsilon\omega_e^2 \ ab \qquad (21)$$

$$b = 0.25\,\varepsilon\, a^3 / \sqrt{D_1}$$

Now, since τ and θ are slowly varying, following the quasistatic averaging approach of Stratonovich [8] it can be shown that for asymptotic stability of the null solution of eq. (19), the condition will be

$$[-\eta\omega_e + \text{Real} <\zeta>] < 0$$

$$\zeta^2 = [0.0625 \ (C_1^2 + C_2^2) - 0.25(1-C_o)^2 - .125 \ C_1 \ C_2 \cos\delta] \qquad (22)$$

The response variance σ_y^2 as given by eq. (17) is shown in Fig. (1). It is found that the stability condition of eq. (22) is satisfied for all values of $\bar{\lambda}$.

Envelop distribution

The function r(t) is the envelop of the response process y(t). The density function of r in the first approximation will be p(r) = p(a) which is a Rayleigh density as per eq. (12). Since this approximation is unstable from $\bar{\lambda}$ = 1.625 to $\bar{\lambda}$= 2.025 it would be interesting to find p(r) as given by eq. (18) and compare it with numerically simulated results. In Figures 2, 3 and 4, typical response samples simulated on a digital computer are shown for $\bar{\lambda}$= 1, $\bar{\lambda}$= 2 and $\bar{\lambda}$= 3 respectively. It is noted that for $\bar{\lambda}$ = 1 the sample is typically narrowbanded. For $\bar{\lambda}$= 2, it is interesting to observe that the sample exhibits predominantly two levels of amplitudes. This feature has been previously observed by Lennox and Kuak [1] also. Further for $\bar{\lambda}$ = 3, higher frequencies are present but the amplitude is essentially at a single level. These observations can be quantified by estimating the probability density function of the envelop. In Figures 5, 6 and 7 the simulated results for the pdf of the local extreme values is compared with the theoretical prediction of eq. (18). For $\bar{\lambda}$ = 1 and $\bar{\lambda}$= 3, the response spread is smaller than the linear case (i.e. $\sigma_y < 1$) and hence the envelop is expected to be unimodal with a nearly Rayleigh distribution. This feature is reflected well in the comparisons of Fig. 5 and Fig. 7. In the intermediate range of $\bar{\lambda}$ when $\sigma_y > 1$, the response seeks higher values and thus, the contribution from second term in eq. (17) becomes very important. Without this term the envelop density function for $\bar{\lambda} = 2$ would still be unimodal. As seen from Fig. 6, for $\bar{\lambda} = 2$, the theoretical pdf correctly predicts a bimodal density function which compares qualitatively well with the simulated results.

Summary and Conclusion

An approximate analysis of the hardening cubic oscillator under narrowband excitation is presented in this study. A new linearization procedure called here the probabilistic linearization technique is adopted for this purpose. In this the given nonlinear system is approximated by a linear system with a random parameter. Further for the cubic oscillator it is shown how the first approximation may be further improved by including higher harmonic terms. The major limitation of the present solution is the assumption involved in obtaining the solution for z(t) in eq. (16). This solution ignores the fact that the excitation in eq. (15) is a random process. Also, the effect of the time varying parametric term y_1^2 is replaced by an average value. These limitations can be partly removed if nongaussian approximations of the type $y = au+bu^3$ are used where u(t) is already a good gaussian approximation to the solution. The unknowns a and b have to be found by appropriate Galerkin-type procedures.

References

1. W.C. Lennox and Y.C. Kuak (1976), "Narrow-band excitation of a nonlinear oscillator", J. of App. Mech., 43, 340-344.
2. H.G. Davies and D. Nand Lall (1981), "Phase plane for narrow-band random excitation of a Duffing oscillator", J. of Sound and Vib., 104, 277-283.
3. R.N. Iyengar (1988), "Stochastic response and stability of the Duffing oscillator under narrowband excitation", J. of Sound and Vib., 126, 255-263.

4. R.N. Iyengar (1989), "Response of nonlinear systems to narrow-band excitation", Structural Safety, 6, 177-185.
5. C.S. Manohar and R.N. Iyengar (1991), "Narrowband random excitation of a limit cycle system", Arch. of App. Mech., 61, 133-141.
6. G.V. Rao and R.N. Iyengar (1990), "Nonlinear planar response of a cable under random excitation", Prob. Engg. Mech., 5, 182-191.
7. P.T.D. Spanos and W.D. Iwan (1979), "Harmonic analysis of dynamic systems with nonsymmetric nonlinearities", J. Dyn. Syst., 101, 31-36.
8. R.L. Stratonovich (1963), Topics in the Theory of Random Noise, Vol. I. Gordon and Breach, New York.

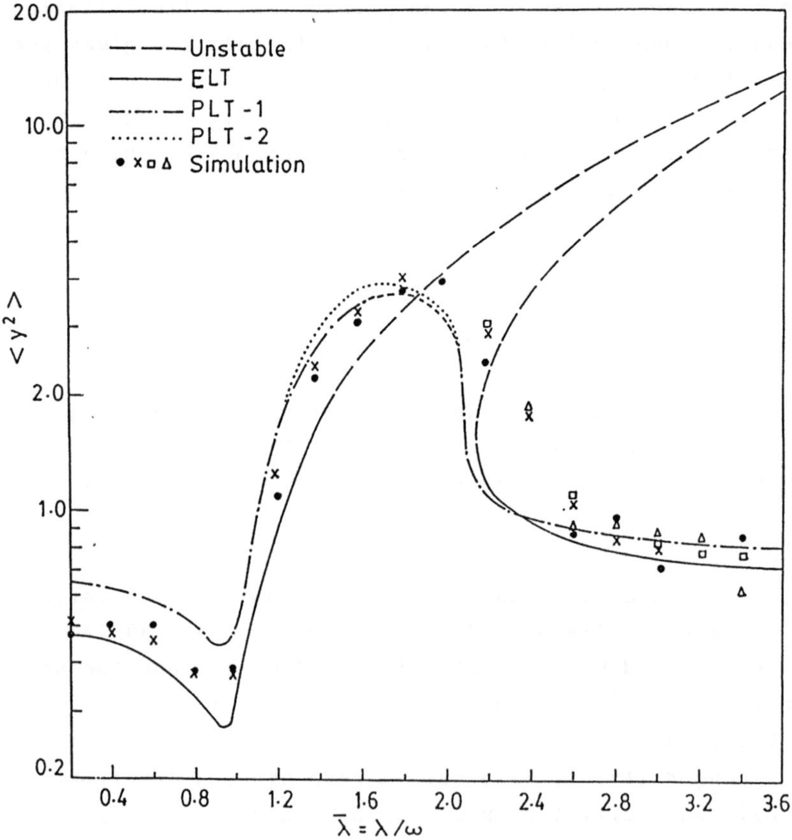

Fig.1. Steady state variance, $\varepsilon = 0{\cdot}3$, $\xi = 0{\cdot}02$, $\eta = 0{\cdot}08$

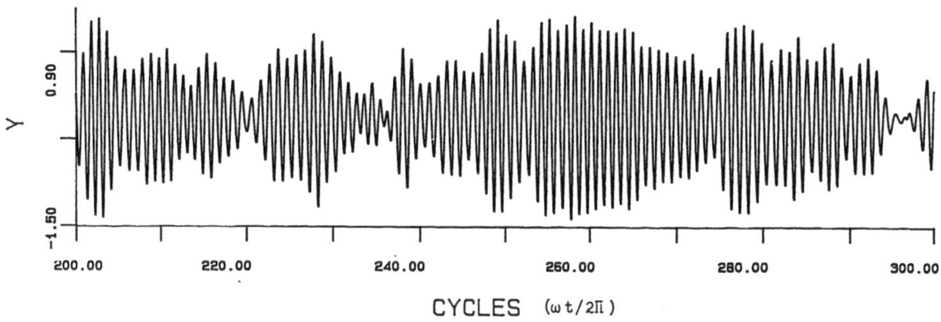

Fig.2. Sample Response, $\lambda/\omega = 1$, $\varepsilon = 0.3$, $\eta = .08$, $\xi = .02$

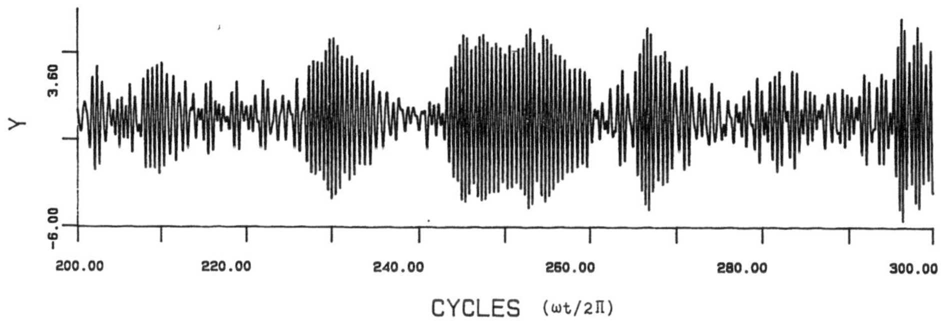

Fig. 3. Sample Response, $\lambda/\omega = 2$, $\varepsilon = 0.3$, $\eta = .08$, $\xi = .02$

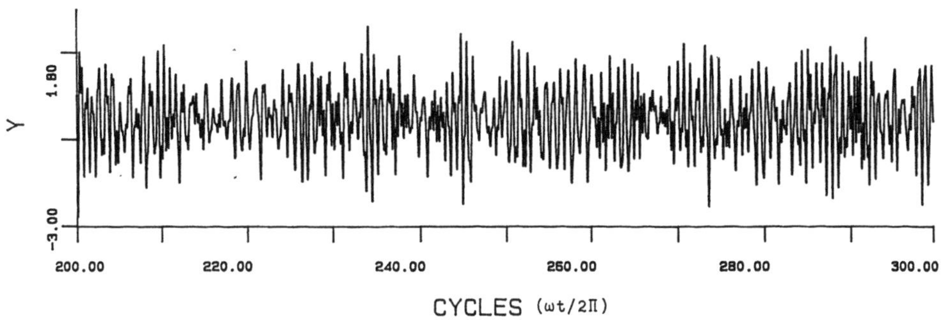

Fig. 4. Sample Response, $\lambda/\omega = 3$, $\varepsilon = 0.3$, $\eta = .08$, $\xi = .02$

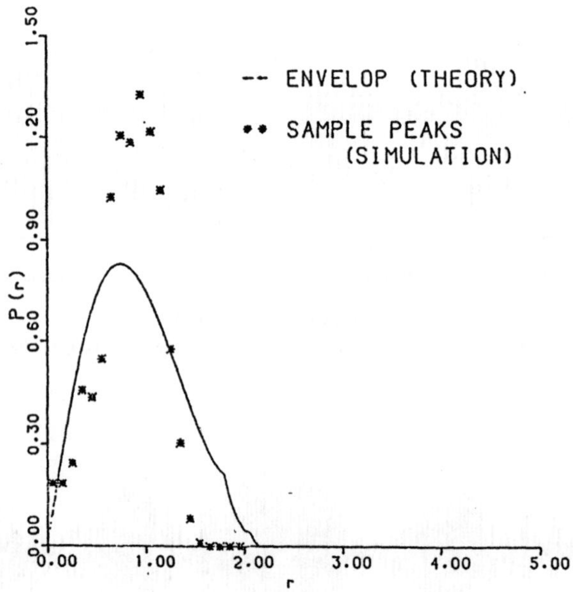

Fig.5. Probability density function of maximum $\bar{\lambda} = 1$

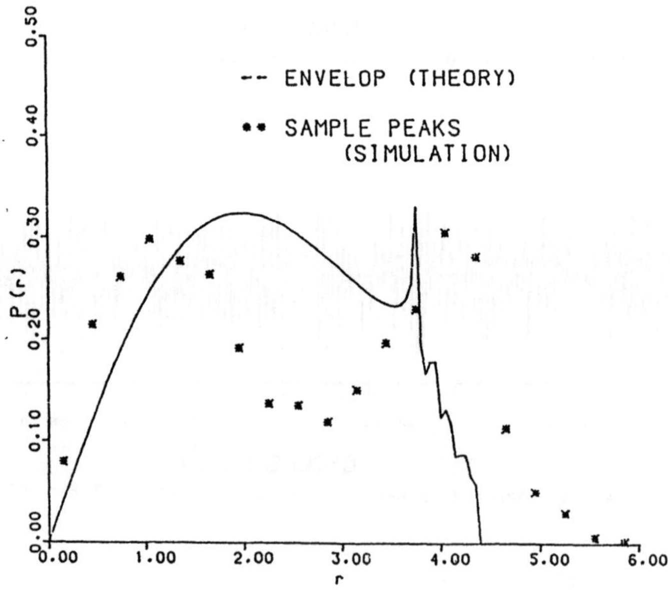

Fig.6. Probability density function of maximum $\bar{\lambda} = 2$

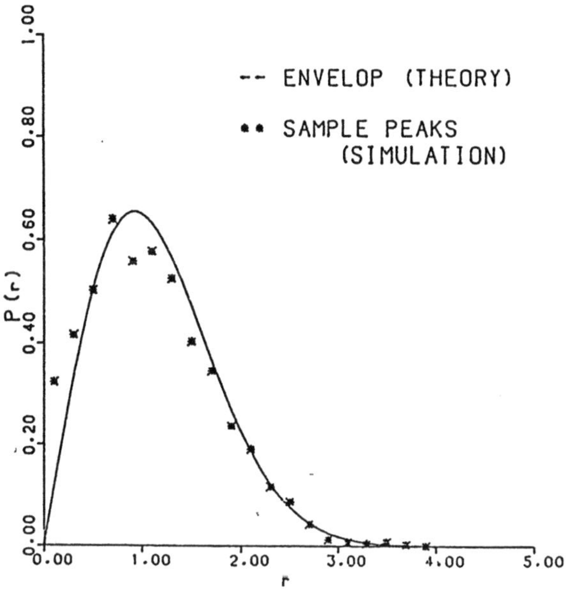

Fig.7. Probability density function of maximum $\bar{\lambda} = 3$

Maximum-Likelihood Parametric Identification Technique for Objects of Randomly Varying Structure

PIOTR KAZIMIERCZYK
Institut für Mechanik TH Darmstadt, Hochschulstr. 1, 6100 Darmstadt, B.R.D.*

Abstract:

Nonlinearities of many systems require descriptions, whose analytical forms depend on the actual state of the system or even on its past. Under random conditions both the coefficient functionals and the set of parameters have to be changed in random instances of time. Thus, not one model but many consecutive models have to be used in a random sequence. A maximum-likelihood (ML) method of on-line parametric identification of a wide class of such systems is concerned in this paper. The class consists of dynamical objects whose consecutive models take the form of a stochastic (ev. multidimensional) differential equations, which are affine in parameters but may be nonlinear with respect to the state (or state history). The answers to three major questions are given. Firstly, it is shown that, despite switching between models, the likelihood functional for each of the individual models can be defined similarly as in the case of one-model-systems and that the temporary change of a model does not affect the properties of ML estimator. Secondly, it is established that, if certain parameters appear in two (or more) models then the information gained while one of the models was being identified can be utilized during the identification of another model. This leads to the improvement of estimation in both models. The global likelihood function (for a group of models) is introduced to treat such cases. Thirdly, it is shown that also when the choice of the model, which should actually be considered depends on the (unknown) parameters being estimated then, in some cases, the consistent estimators can be established. The above is exemplified by results of estimation of parameters of two nonlinear oscillators. In one example, the restoring force is linear around the equilibrium and is strengthened apart from it. First, the identification is performed in each of these regions independently and next, simultaneous identification is applied. Another example concerns the hysteretic system with dry friction. Identification is performed in the state space (the data are the displacement and velocity only and the values of restoring force are not measured).

1. INTRODUCTION.

Let the behaviour of the object be described by the family of models (M_d), $d \in D$. Let each of the models be the (multidimensional) Itô stochastic differential equation of the form

$$dX_t^d = \mathbf{m}^d(X_\bullet{}^d, t; \theta^d)\,dt + \mathbf{b}^d{}_t(X_\bullet{}^d)\,dW_t^d \qquad (M_d)$$

where $\mathbf{m}^d(X_\bullet{}^d, t; \theta^d) = \mathbf{m}_0^d(X_\bullet{}^d, t) + \theta_1^d\,\mathbf{m}_1^d(X_\bullet{}^d, t) + \ldots + \theta_{k^d}^d\,\mathbf{m}_{k^d}^d(X_\bullet{}^d, t)$, $\theta^d = (\theta_j^d)$ are the parameters to be identified, $X_t^d = X_t^d(\omega)$, $W_t^d = W_t^d(\omega)$, $\omega \in \Omega$, $\{\Omega, \mathcal{F}, \mathcal{P}\}$ denotes a complete probability space, $X_t^d \in R^n$ —is a state vector of a system (n is the highest of the dimensions of models M_d – it is assumed that it is finite), $W_t^d \in R^l$ —is a family of mutually independent standard Wiener processes and m_j^d and b^d are assumed to be the known nonanticipating functionals of the state history.

* permanent address: Institute of Fundamental Technological Research Polish Academy of Sciences, ZTOC, ul. Świętokrzyska 21, 00-049 Warsaw, Poland

N. Bellomo · F. Casciati (Eds.)
Nonlinear Stochastic Mechanics
IUTAM Symposium Turin

It is assumed that at any given instant of time, t, the model describing evolution of the state has an index $d(t)$. In this case we shall say that the system is in *mode* $d(t)$. The actual mode may, in general, depend on chance: $d(t) = d(t, \omega)$. This means that the instants of time τ, when one of the models becomes the actual governing model substituting the model used previously, may also be random ($\tau = \tau(\omega)$).

Such a qualitative change of the system behaviour and consequently the necessity of changing the model occurs for instance, whenever a certain condition is turned on/off or when the regime of work is being adopted to varying external demands or because of the previously scheduled order of operations. Typical examples of such systems of varying structure are complicated robots and multifunction machinery. Reactions of some simple engineering elements however also show qualitative changes under varying conditions.

Several different models (instead of a single model) may also be desirable even if no abrupt changes occur in functioning of the system. For example, this is the case when the system is strongly nonlinear but, for some reasons (eg., for real time steering), linear approximations are to be used. Then, the linearization has to be performed at several points of the state space and each of different linear models so obtained is to be used within a defined region of the state space.

One of the examples considered in the following is of similar nature. Suppose one can trace the displacement y and the velocity $\dot{y}$ of a nonlinear oscillator

$$\ddot{y} + \theta_1 \dot{y} + f(y) = A \sin(\nu t + \phi) + \xi \tag{1}$$

where ξ is a white noise with intensity S and the stiffness characteristic $f(y)$ takes one of the two forms:

$$f(y) = f_1(y) = \theta_2 y, \qquad \text{for } |y| \le r \tag{2-1}$$

$$f(y) = f_2(y) = \theta_2 y + \theta_3 \frac{(|y| - r)^3}{1 + (|y| - r)^2} \operatorname{sgn}(y), \qquad \text{for } |y| \ge r \tag{2-2}$$

The system can be rewritten in the form (M_d), with $X^1 = X^2 = [y, \dot{y}]^T$ and with $\theta^1 = [\theta_1, \theta_2]^T$, $\theta^2 = [\theta_1, \theta_2, \theta_3]^T$:

$$dx_1^d = x_2^d \, dt$$

$$dx_2^d = [A \sin(\nu t + \phi) - \theta_1 x_2^d - f_d(x_1^d)]dt + S \, dW_t$$

When the system is in the first mode ($|y| \le r$) model M_1 is used and only parameters θ_1, θ_2 can be estimated. When y upcrosses r or downcrosses $-r$ the model switches to M_2

and all three parameters can be estimated. If only one of the models was in use one could construct likelihood function and ML estimators accordingly with the existing theory (see section 2). This theory was however developed for one model with smooth coefficients and for one continuous observation. Thus, the following questions arise. How to continue the estimation after consecutive changes of modes? How to use the information collected in the region $|y| \leq r$ while estimating parameter θ_3 ? The general answers to such questions are given in sections 3 and 5. They are illustrated by results of identification of system (1) (sections 4 and 6).

Another example of a system with randomly varying structure is an oscillator with the hysteretic restoring force (see Figure 1). Between mass m and a car a dry friction appears. The relative movement of these bodies is bounded by bumpers. The mass of the car and the friction in car axles are neglected. Springs k_1 are linear around the equilibrium and have cubic hardening characteristics apart from it; springs k_2 have logarithmic soft characteristics. The assumed global restoring force characteristic is shown in Figure 2.

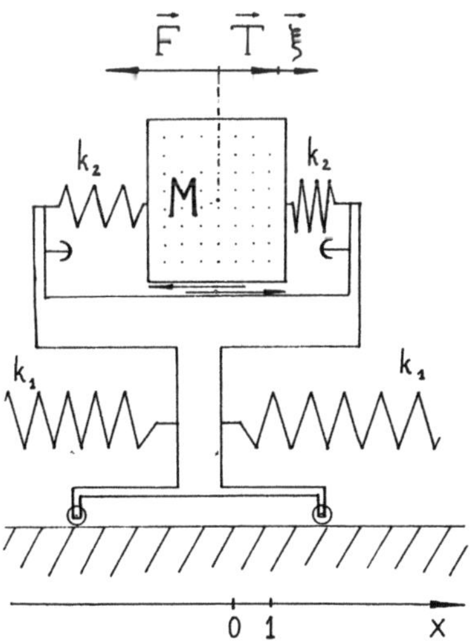

Figure 1

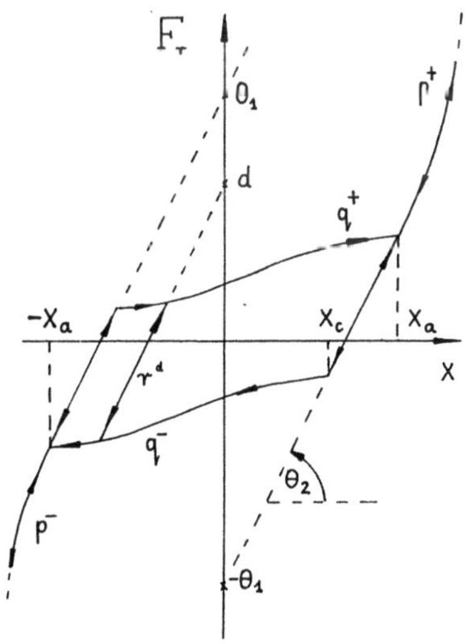

Figure 2

The system is symmetric. At point x_a mass m arrives at the right bumper; at point x_c spring k_2 is strong enough to push it apart from the bumper. Since the displacement is being observed (displacement and velocity of the mass m are the data for identification), the points x_a and x_c are known. The parameters of the functions $p^{\pm}(x) = \mp\theta_1 + \theta_2 x \pm 0.2(|x| - x_a)^3 \mathcal{H}(|x| - x_a)$ ($\mathcal{H}$ is the unit step function) and $q^{\pm}(x) = \theta_3 \ln(|x \mp \frac{x_a - x_c}{2}| + 1)\mathrm{sgn}(x \mp \frac{x_a - x_c}{2}) \pm \theta_2 \frac{x_a - x_c}{2}$ however, are not known without measuring the forces and are to be estimated (the functional dependence of the three parameters is not used during the identification). Note that at some instants of time the restoring force evolution may be described by one of the family of functions $r^d(x) = d + \theta_2 x$, $d \in (-\theta_1; \theta_1)$, rather than by function p or q. Parameter d depends functionally on the past of the system and on the values of parameters θ_i. Moreover, it is assumed that the relative displacement between mass and the car is not known (excluding the instants when the mass sticks one of the bumpers). In this case, in order to decide at which curve of the hysteresis lies the actual restoring force of the system, one has to use the state history as well as the estimates of parameters θ_i. Thus, the mode of the system depends on the parameters that are being identified. Nevertheless, as it is shown in section 7, the consistent ML estimators of these parameters can be constructed.

2. ML technique for Ito systems.

The method of ML estimation of parameters of Ito equations was developed already in the seventies (in the case where a datum is one continuous sample-path of a solution to a unique governing equation; cf. references in [1]). The generalization of this theory, enabling applications to a wide class of nonlinear engineering systems can be found in [1]. Under general conditions (cf. [1]) the log-likelihood function (the logarithm of mutual derivative of distributions) connected with the solutions of two Itô equations

$$dX = m\,dt + b\,dW \qquad (1\text{-x})$$

$$dZ = M\,dt + b\,dW \qquad (1\text{-z})$$

can be expressed as

$$L^{\tau}_{\sigma}(X) = \int_0^{\tau} (m - M)^T (bb^T)^+ \, dx_t - \frac{1}{2}\int_0^{\tau} (m^T(bb^T)^+ m - M^T(bb^T)^+ M)\, dt \qquad (3)$$

where $^+$ denotes the pseudoinversion operation and τ, σ are the Markov instances of starting observation and of estimation, respectively. Thus, for a single model of the form M_d (see sec.1) the ML estimator $\hat{\theta}$ maximizing the likelihood function (3) takes the form

$$\hat{\theta}(\omega) = F(X_\bullet) = \mathbf{G}^{-1}(X_\bullet)\,\mathbf{H}(X_\bullet). \qquad (4)$$

Here:

$$\mathbf{H}(X_\bullet) = \left[h_1 - g_{10}, \ldots, h_k - g_{k0}\right]^T,$$
$$\mathbf{G}(X_\bullet) = \left\{g_{ij}\right\}_{i,j=1,\ldots,k},$$

and (for $i = 1, \ldots, k$ and $j = 0, 1, \ldots, k$)

$$h_i = h_i(X_\bullet(\omega), \tau) = \int_\sigma^\tau \left[\mathbf{m}_i^T \left(\mathbf{b}\,\mathbf{b}^T\right)^+\right](X_s(\omega))\, dX_s(\omega),$$

$$g_{ij} = g_{ij}(X_\bullet(\omega), \tau) = \int_\sigma^\tau \left[\mathbf{m}_i^T \left(\mathbf{b}\,\mathbf{b}^T\right)^+ \mathbf{m}_j\right](X_s(\omega))\, ds,$$

If the eigenvalues of matrix $\mathbf{G}$ satisfy certain conditions of growth (as the time of observation grows; see [1] and references therein) then the above ML estimator is strongly consistent.

Both the actual recorded processes and the paths obtained through a simulation can be at best only approximated by a diffusion process resulting from an Itô equation model (1). For instance, during simulation experiments reported in the following, the polygonal approximation of the Wiener process was used. The sample paths of solution X were approximated by functions ϕ obtained through one-step forward difference integration (Euler method). Such approximations require the use of correction terms to estimate the value of MLE which would be obtained for a diffusion process. Even in the case where the form of a particular model does not require the well known Stratonovič (or Wong-Zakai, say) corrections in the governing equation, another correction is necessary to modify the formulae for estimator (for proper approximation of stochastic integrals by finite Cauchy integral sums). The problem of corrections required for the sake of ML estimation was studied in detail in [2].

As far as the example systems considered in this paper are concerned, the corrections prove to be nonvanishing in the case of system (1). Results reported in sections 4 and 6 were obtained with the use of corrections.

3. Identification of an individual model

Let for any of the models M_d the assumptions enabling the use of the theory recalled in section 2 hold true. Under these assumptions there exists a strong solution X^d of the equation M_d and, if the coefficient functionals depend on the state only, it possesses the strong Markov property. In this case, when the normal vectors of any of the switch surfaces in the state space are orthogonal to the field of operator $\mathbf{b}$ then the random instants of time τ_j^d at which the model M_d is being replaced by another model and the instants σ_j^d at which it becomes the governing model again form the increasing family of Markov times (see [3]). The case, when the above orthogonality assumption is replaced by the assumption that the switch surface is orthogonal to $\mathbf{b}(R^l)$ was considered in [4]. In the general case, where the change of models does not necessarily take place on surfaces bordering prescribed regions of the state space or,. where the coefficient functionals $\mathbf{m}$ and $\mathbf{b}$ depend on the past of the solution, the property that the instants τ_j^d and σ_j^d form the increasing family of Markov times will be treated as the assumption.

Under this assumption the likelihood functional for a model M_d can be defined as a sum of integrals of the form (3), that is

$$L_{\sigma\tau}^d = \sum_j L_{\sigma_{j-1}\tau_j}^d .$$

This simple definition is correct due to the fact that the likelihood function (3) as a sum of finite integrals of nonanticipating functionals of the Itô process is an additive function with respect to Markov bounds of integration. In consequence of the above definition the formulae for matrices G and H (see (4) and definitions below (4)) remain valid with the sole modification that the integration should be performed only over those periods of time, during which the system was in mode d. In other words, with each model an individual time can be associated:

$$t^d = \sum_{j \in \{i|\sigma_i < t\}} (\min(t, \tau_{j+1}) - \sigma_j)$$

Using this individual time the ML identification procedure recalled in the previous section can be applied to each of the individual models. If the eigenvalues of matrix G^d constructed in the above way satisfy the growth condition mentioned in section 2 (see [1]), then the ML estimator of parameter θ^d is strongly consistent independently on whether the estimators of parameters of other models converge or not.

4. Example 1; identification in individual modes.

The above technique has been applied to system (1). The values of parameters $\theta_1, \theta_2, \theta_3, r, A, D, \nu$ were equal to 0.1, 1.0, -1.0, 1.0, 1.0, 0.1, 1.0 respectively. The phase shift and the initial conditions were generated in a random way.

The pair θ_1, θ_2 was estimated on the basis of observations collected when the displacement was smaller than 1. Independently, using the observations collected when the displacement was greater than 1, the estimator of the triple $\theta_1, \theta_2, \theta_3$ was calculated. The mean square errors of these estimators were estimated by averaging square errors of 200 independent samples of estimators. The results are shown in Figures 3, 4, 5. In Figures 3 and 4 typical samples of estimators trajectories are represented. Lines 1, 2, and 3 are the values of estimators of parameters θ_1, θ_2 and θ_3, respectively.

In Figure 5 the estimates of mean square errors of estimators are drawn. Lines 1-1 and 1-2 concern the estimator of parameter θ_1 in first and second mode, respectively; lines 2-1 and 2-2 concern parameter θ_2, line 3 — parameter θ_3.

It is apparent from Figures 3 and 4 that the quality of estimation is very good. It should be noted however, that, as parameters θ_1 and θ_2 appear in both modes, two separate estimates for each of them were obtained. To use the collected information more efficiently one has to know how to combine these results. With better estimates of these parameters one should also be able to estimate the remaining parameter, θ_3, in a better way.

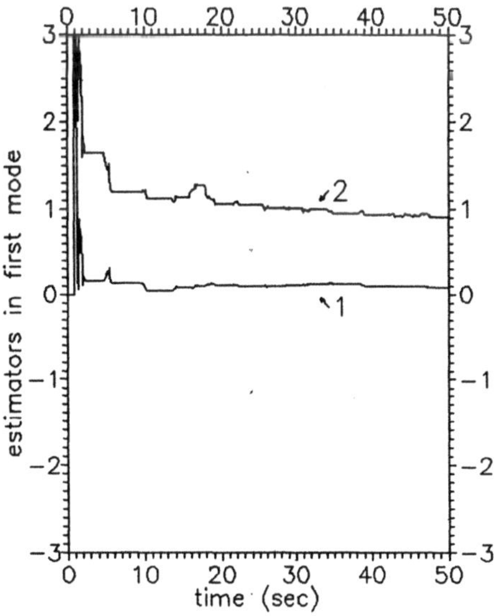

Figure 3

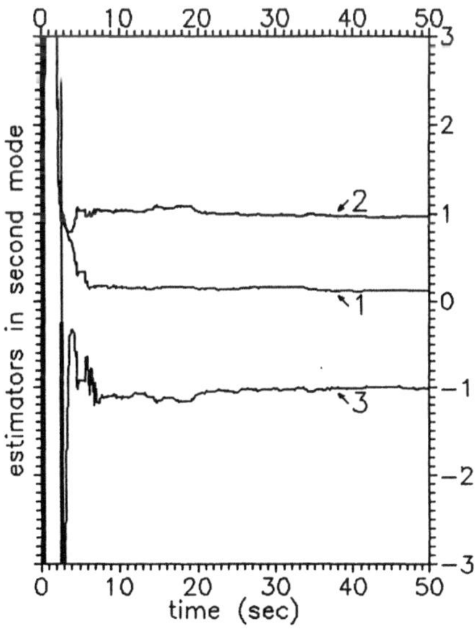

Figure 4

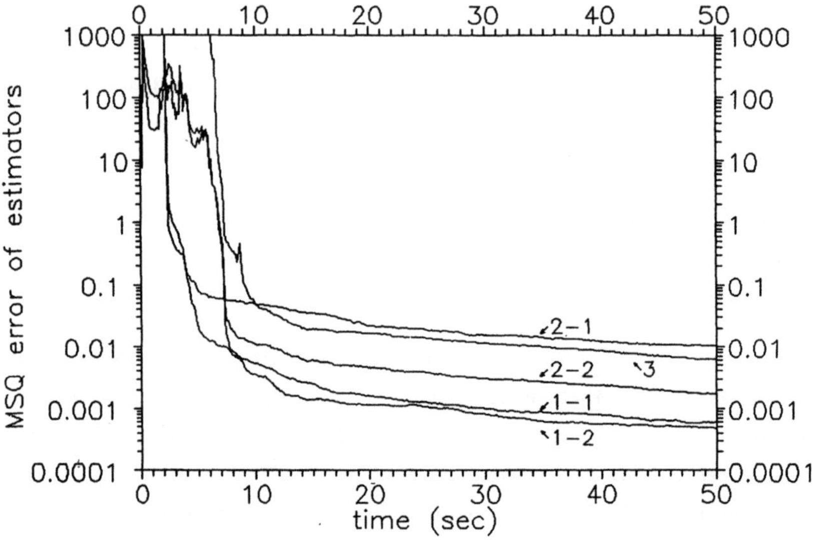

Figure 5

5. Simultaneous identification of a group of models.

If none of the parameters of one model appears in the second model then the behaviour of the system in the second mode cannot bring any information about the parameters of first model. If one (or more) parameters appear in two (or more) models however, then the information gained while the system is in one mode can be used to improve the estimation of all parameters entering this group of models.

The global equation describing evolution of the state of the system in a group of N models can be written as follows

$$dX_t = \tilde{\mathbf{m}}(X,t)\,dt + \tilde{\mathbf{b}}(X,t)dW_t + dD_t + dC_t \tag{5}$$

$$\tilde{\mathbf{m}}(x,t) = \sum_{i=1}^{N} \mathbf{m}^i_{t\omega}(x,t)\chi_{\Omega^i_t}(\omega)$$

$$\tilde{\mathbf{b}}(x,t) = \sum_{i=1}^{N} \mathbf{b}^i(x,t)\chi_{\Omega^i_t}(\omega).$$

The terms dD^i_t and dC^i_t describe annihilation and creation of a sample-path in the i-th model (tied with the change of the model in use). They shift the state of the system at instants τ_j^d and σ_j^d. $\Omega_t^i \in \mathcal{F} \cap \mathcal{F}_t^{W^i}$ is a time-dependent decomposition of a probability space. The Wiener process W consists of segments of individual Wiener processes. Construction of this process is straightforward as the instants τ_j^d and σ_j^d are Markov times with respect to the same filtration as all of the individual Wiener processes, which possess the strong Markov property. All parameters entering individual models form one multidimensional parameter (of course that requires a simple renumeration)

Now, one can use the same method for constructing the likelihood functional and forming of matrices $\mathbf{G}$ and $\mathbf{H}$ as in the case of the individual model (see sec. 3). As the random shift operators D and C act exclusively at instants of time τ_j^d and σ_j^d which are the bounds of integration for integrals defining elements of matrices $\mathbf{G}$ and $\mathbf{H}$ then they do not contribute to those integrals and are not important for the estimation.

Because of the properties of χ functions the elements of matrices $\mathbf{G}$ and $\mathbf{H}$ become the sums of those values, which would be obtained in individual models. Note that, if this procedure would be applied to two models having no parameters in common then the only result would be the increase of dimension of matrices. Due to properties of χ functions matrix $\mathbf{G}$ would take a diagonal matrix form:

$$\mathbf{G} = \begin{pmatrix} \mathbf{G}^1 & 0 \\ 0 & \mathbf{G}^2 \end{pmatrix}$$

and the identification would be again reduced to the independent calculations for each individual model. If, however, a group of parameters is present in two models, eg., θ_1, θ_2 in model M_1 and θ_2, θ_3 in model M_2 then, the matrix $\mathbf{G}$ takes the form

$$\mathbf{G} = \begin{pmatrix} g_{11}^1 & g_{12}^1 & 0 \\ g_{12}^1 & g_{22}^1 + g_{22}^2 & g_{23}^2 \\ 0 & g_{23}^2 & g_{33}^2 \end{pmatrix}$$

and the estimates are no longer independent: even while the system is remaining in mode 2, the estimate of parameter θ_1 is continuously being statistically improved although this parameter does not enter model 2.

6. Simultaneous identification of system (1).

The above technique was applied to system (1). The results obtained with the same set of parameters as described in section 4 are represented in Figure 6. The mean square error of estimates was, as previously, estimated from 200 independent estimation histories. It is seen that the quality of estimation was improved in all cases. The estimate of parameter θ_3

is better. This exemplifies the fact that the increased precision of estimation of a group of parameters entering many models enables improvement of estimation of remaining parameters.

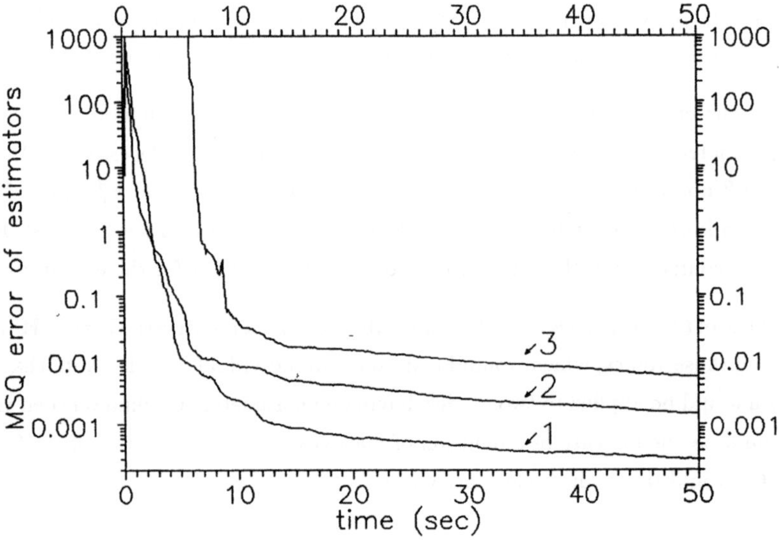

Figure 6

7. The hysteretic system

By considering this case we indicate another possibility of interchanging information about the parameters entering different models. The parameters θ_1, θ_2 fully describe the restoring force where $|x| > x_a$ as well as where $x_c < |x| < x_a$ and the last change of sign of velocity took place when the displacement was greater than x_a. If the oscillations have a stationary character and the time spent in this region (say in mode 1 for this system) tends to infinity, then the ML estimators of parameters θ_1, θ_2 calculated in the individual way described in sec. 3 converge to the true values with probability one. This information can be used to construct also a consistent estimate of parameter θ_3. In mode 2 the parameters θ_1, θ_2 were treated as known and their estimates resulting from mode 1 were used as the true values. Thus, in mode 2 the only parameter to be identified was the parameter θ_3. This trick not only simplifies the formulae (decreases the dimension of matrices) but also assures the convergence of the procedure (recall that all three estimates have to be used to recognize when the restoring force is given by one of functions q^+, q^- connected with parameter θ_3 – see section 1).

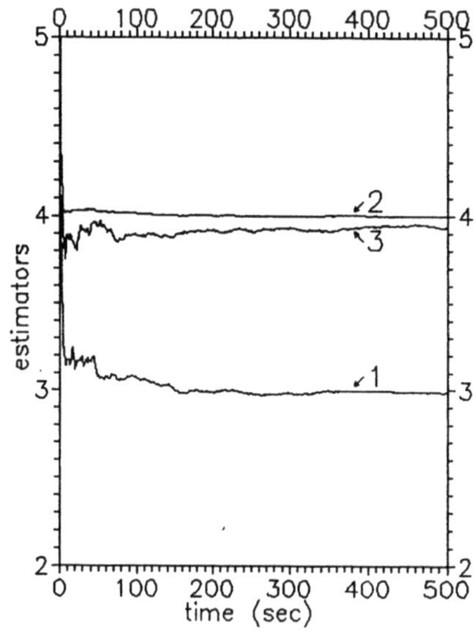

Figure 7

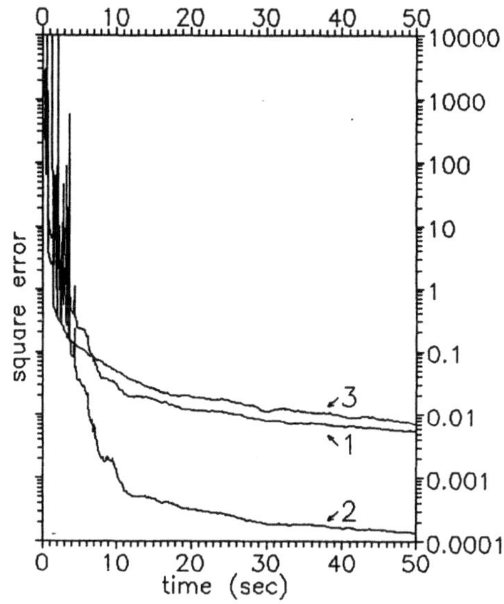

Figure 8

Results of estimation are represented in Figures 7 and 8. Lines 1, 2, 3 correspond with the estimators of parameters $\theta_1 = 3, \theta_2 = 4, \theta_3 = 3.954$, respectively. The intensity of external noise was 0.1, the mean external loading was a harmonic function with amplitude 4 and frequency 2. Figure 8 shows the estimates of mean square errors of estimators. The quality of estimation is apparently dependent on the magnitude of the diagonal element g_{ii} corresponding to the parameter θ_i. In the case considered the best are the estimates of parameter θ_2 since its cubic coefficient function is the most significant one. However, the estimates of parameters θ_1 and θ_3 are satisfactory. Even the worst estimate (for parameter θ_3) proved to be unbiased and its standard deviation from the true value, after 50 seconds of estimation, was less than two per cent of its true value.

8. Closing remarks.

The simultaneous use of several models requires modifications of the identification techniques developed for single model systems. The situation becomes especially involved if the evolution of the system depends on chance and the moments of change of the actual model become random. Not many results on modelling of such systems are available in the literature. Some classes of stochastic differential systems with variable structure were treated in [4], [5] (see references there). In [3] a method based upon the solution of an inverse problem for conditional moments of a response process was proposed for parametric identification of a class of such systems. In contrast to the case of linear stochastic differential systems, the moments of the response of systems with variable structure are involved in a complicated system of integro-partial-differential equations. Thus, in most practical situations, the identification procedure requires additional assumptions and approximations holding for almost linear quasistationary systems.

The maximum-likelihood method whose applicability to linear and nonlinear stochastic differential models is already well established can also be adopted to the systems with variable structure. In this paper the necessary mathematical prerequisites, formulae and analyses of properties of maximum-likelihood estimators for an important class of systems of interest are provided. The example cases studied show that the method can be successfully applied to strongly nonlinear engineering systems requiring models whose form as well as whose set of parameters vary in time in a random way.

ACKNOWLEDGEMENT. I thank Professor Kazimierz Sobczyk for turning my attention to the problems of varying structure and for his helpful bibliographical indications. I am also grateful to Alexander von Humboldt Foundation and to Professor Peter Hagedorn for assuring facilities enabling extensive computational experiments.

REFERENCES

1. Kazimierczyk P. *Maximum likelihood approach in parametric identification of stochastic differential models of engineering systems*, Structural Safety, 8 (1990) pp. 29-44.

2. Kazimierczyk P. *Correction formulas for estimators of drift parameters of an Itô equation*, IFTR Reports, 39/88, 37 pp., 14 Figs., 1988.

3. Kazimierczyk P. *Parametric identification of diffusions with constraints; applications in fracture mechanics and vehicle dynamics*, in: Proc. COSMEX Conf., Szklarska Poręba, Sept. 1989, pp. 249-264, World Scientific Publ. Inc., 1990.

4. Watanabe S. *On stochastic differential equations for multidimensional diffusion processes with boundary conditions*, J. of the Math. Kyoto Univ., vol. 11, pp. 169-180, 1971.

5. Kazakov I.E., Artem'ev V.M. *Optimizac'ja dinamičeskich sistem slučajnoj struktury*, Mir, 1980.

Controlling the Dynamics of a Random System

WOLFGANG KLIEMANN AND FRITZ COLONIUS

Department of Mathematics
Iowa State University
U.S.A.

Institut für Mathematik
Universität Augsburg
Germany

ABSTRACT. Random systems, dynamical systems and control systems can all be described as flows on (finite or infinite dimensional) spaces, which allows for the use of unified concepts in the analysis of their qualitative long term behavior. In particular there is a close connection between the attractors of an undisturbed system, the stationary and ergodic solutions of the system under Markovian parameter noise, the invariant control sets of an associated control system, and the (chaotic) attractors of the corresponding control flow. This correspondence is used to analyze the possible changes in the ergodic behavior of controlled random nonlinear systems. Discussing four examples in some detail reveals a link to stochastic bifurcation theory.

1. Introduction.

Random vibrations occur in physical systems (often modelled by ordinary differential equations), which are subjected to noise. As long as the resulting system is Gaussian and Markov (in particular the solution of a linear stochastic differential equation with additive noise), there is a well-known range of techniques for the analysis of such systems. Parameter excited and/or nonlinear systems are in general non-Gaussian, and a wide range of approaches have been developed to deal with these situations analytically, approximately or numerically, among them in particular stochastic linearization, averaging for systems on different time scales, computation of solutions of the corresponding Kolmogorov (or Fokker-Planck) equations. The textbooks of Dimentberg [17] and Sobczyk [26], as well as many papers in this volume describe the current state of art in these areas. They all have in common that they deal with the statistical properties of random systems, i.e. in the case of solutions of stochastic differential equations these approaches are based on the generator of the diffusion process.

However, for certain properties of the trajectories, like their exponential convergence, it is not appropriate to consider the generator: Baxendale [5] has an example, where for processes with the same generator the Lyapunov exponents, indicating exponential convergence or divergence with probability one of the trajectories, can be negative or positive. These exponents depend on the dynamics, i.e. the vectorfields of the random system. In recent years, the dynamical analysis of stability and bifurcations in stochastic systems based on Lyapunov exponents has made considerable progress, we just mention

Research supported in part by NSF grant DMS 8813976 and by DFG grants Co 124/6-1 and Co 124/8-1 .

N. Bellomo · F. Casciati (Eds.)
Nonlinear Stochastic Mechanics
IUTAM Symposium Turin

the papers of Arnold and Boxler [2], Boxler [7], and of Ariaratnam, Sri, and Wedig in these Proceedings.

In this paper we present another aspect of the dynamical analysis of random systems: we will gain insight by using concepts from the theory of dynamical systems and of control systems. We concentrate on the qualitative long term behavior, i.e. on stationarity and ergodicity, and analyze its change via deterministic control. (Note that this problem of controlling the dynamics of a random system is different from stochastic optimal control with adapted control processes.) After the introduction of some concepts and notations in Section 2, we will present an outline of the theory, linking stationarity to chaos and controllability. The examples in Section 4. show, how to apply this theory in some obvious and not so obvious situations. Besides for the control of random dynamics, these ideas also prove valuable for stability and stabilization of stochastic systems via Lyapunov exponents. We refer to [20], [9], and [10], where the last two papers also show some connections with stabilization theory for uncertain (deterministic) systems. Some of the common background material, which is needed for the following sections, can be found in the survey paper [20].

2. Random Vibrations, Stochastic Flows and Control Flows.

As a mathematical model of random vibrations we will use the following set-up: Given is a (deterministic) ordinary differential equation on a smooth manifold M, $\dim M = d$,

$$\dot{x} = X_0(x), \tag{2.1}$$

where X_0 is a smooth vectorfield, describing a physical system. The random system is

$$\dot{x} = X_0(x) + \sum_{i=1}^{m} \xi_t^i X_i(x) \qquad \text{on } M, \tag{2.2}$$

where $\left(\xi_t^i\right)_{i=1,\dots,m} = \xi_t$ is an m-dimensional noise process, and the vectorfields $X_1 \dots X_m$ describe, how the noise enters into the system. This model covers parameter and/or additive noise (in the latter case the vectorfields $X_1 \dots X_m$ are constant). In this paper we consider the case, where ξ_t is bounded (in particular non-Gaussian), i.e. we assume that ξ_t has values in a compact, convex set $V \subset \mathbb{R}^m$, which contains 0 in its interior. Furthermore, since we are interested in stationary behavior, we assume that ξ_t is (strict sense) stationary, i.e. all finite dimensional distributions are invariant under time shift.

In order to link the random system (2.2) with dynamical and control systems, we describe it as a stochastic flow. This means first of all that we need an appropriate probability space for ξ_t: Let $\mathcal{V} = \{v\colon \mathbb{R} \to V \subset \mathbb{R}^m$, locally integrable$\}$ be the trajectory space of ξ_t. Then the standard Kolmogorov construction (see e.g. Rozanov [24]) yields a σ- algebra $\mathcal{B}$ and a probability measure P on $(\mathcal{V}, \mathcal{B})$ such that: If $\theta_t\colon \mathcal{V} \to \mathcal{V}$, defined by $\theta_t v(\cdot) = v(t + \cdot)$, denotes the time shift on $\mathcal{V}$, then $\theta_t P = P$, i.e. shift invariance of P on $\mathcal{V}$ corresponds to shift invariance of the finite dimensional distributions of ξ_t, indicating stationarity. Now take $(\mathcal{V}, \mathcal{B}, P)$ as the probability space, then $\xi_t(v) = v(t)$ for $v \in \mathcal{V}$ is the equivalent noise model, which we will use from now on.

To capture the random dynamics of (2.2) we consider the family of differential equations

$$\dot{x} = X_0(x) + \sum_{i=1}^{m} v^i(t) X_i(x) \tag{2.3}$$

with $(v^i)_{i=1,\dots,m} = v \in \mathcal{V}$. The solutions are denoted by $\varphi(t,x,v)$ for $(v,x) \in \mathcal{V} \times M$, with initial value $\varphi(0,x,v) = x$ and using $v \in \mathcal{V}$ on the r.h.s. of (2.3). Putting things together, we arrive at the random flow description of (2.2):

$$\begin{aligned} \phi \colon \mathbb{R} \times \mathcal{V} \times M &\to \mathcal{V} \times M \\ (t, v, x) &\mapsto (\theta_t v, \varphi(t,x,v)) \end{aligned} \tag{2.4}$$

where the first component (i.e. the shift θ with invariant measure P) describes the stationary noise, and the second component describes the dynamics of the system under the noise. (2.4) is called a flow on $\mathcal{V} \times M$, because it has the group property $\phi_t \circ \phi_s = \phi_{t+s}$ for all times $t, s \in \mathbb{R}$ (we have used the notation $\phi_t = \phi(t,\cdot,\cdot)$). The shift component θ_t is also a flow (on $\mathcal{V}$), but the second component by itself does not satisfy this property, because of the time varying noise. It is called the skew component of ϕ, and the notion 'skew product flow' is often used for a flow ϕ on a product space with skew component. (If ξ_t is not a bounded, stationary process but white noise, one can also construct a corresponding stochastic flow for the system, see e.g. Kunita [21], Arnold and Crauel [3], or the survey paper [20].)

For the topological analysis of (2.4) as a dynamical system or a control system we need a topology on the trajectory space $\mathcal{V}$ (and we disregard the measure P of the stochastic flow). For control theoretic purposes the appropriate one is the weak* topology of $L^\infty(\mathbb{R}, \mathbb{R}^m) = \left(L^1(\mathbb{R}, \mathbb{R}^m)\right)^*$ (see e.g. [12]), since convergence of $v_n \to v$ in $\mathcal{V}$ then implies uniform convergence of the solutions $\varphi(\cdot, x, v_n) \to \varphi(\cdot, x, v)$ on finite time intervals. With this topology ϕ is continuous in all its components.

The next section is devoted to the study of ϕ from stochastic, dynamical systems, and control systems points of view. We will do this under an assumption which guarantees that we have the "right" state space for (2.2), namely

$$\dim \mathcal{LA}\{X_0, X_1, \dots, X_m\}(x) = d \quad \text{for all } x \in M. \tag{H}$$

This Lie algebra rank condition, which is discussed in detail e.g. in [20], means that the noise dynamics is sufficiently rich, i.e. that the set of points reached by some solution of (2.2) up to time $t > 0$ has nonvoid interior, or, in the language of Markov processes, that the supports of the transition probabilities up to time $t > 0$ have nonvoid interior. In fact, it is this assumption, on which in the white noise case the famous support theorem of Stroock and Varadhan is based, saying intuitively that the trajectories of (2.2) as a random system "agree" with the trajectories of (2.2) as a control system, see Stroock and Varadhan [27], or the survey [20] for details. In what follows, we develop a theory for real (bounded) noise systems, which reflects some of the consequences of the support theorem.

3. Stationary Random Systems, Controllability, and Chaotic Attractors.

In this section we will analyze the problem, under which conditions the random systems (2.2) has a bounded stationary solution. This is the basis for the subsequent discussion about changing the dynamical behavior via control.

As indicated in Section 2, stationarity means finding a flow invariant probability measure, i.e. for the stationarity of (2.2) under stationary noise ξ_t: Find an invariant measure μ on $\mathcal{V} \times M$, such that $\phi_t \mu = \mu$ for all $t \in \mathbb{R}$. Furthermore, since ξ_t and hence P on $\mathcal{V}$ are given, we need that the marginal of μ on $\mathcal{V}$ is P. This amounts to the following construction: μ on $\mathcal{V} \times M$ can be desintegrated as $\mu = \mu_v P$, where for each $v \in \mathcal{V}$ μ_v is a measure on M with the invariance property $\varphi(t, \cdot, v)\mu_v = \mu_{\theta_t v}$, see e.g. [11]. In order to find these measures, one has to study the ergodic properties of the flow ϕ on $\mathcal{V} \times M$. This can be done using the limit structure of ϕ. Carrying out this program, however, is quite a technical task, if done in full generality, and we refer to [11] and the references therein for several results in this direction.

Fortunately, the solution is much simpler, if we work in the context of Markov processes, i.e. stochastic processes, for which past and future are independent, given the presence. Since physical systems, described by ordinary differential equations, do not foresee their future, this seems to be a natural class of noises to consider for random vibrations. In particular, it includes the following models, which are quite common in random mechanics (colored noise):

$$d\eta_t = Z_0(\eta_t)dt + \sum_{i=1}^{k} Z_i(\eta_t) \circ dw_t^i \tag{3.1}$$

is a (nondegenerate) stationary diffusion process on a compact manifold N and $f\colon N \to V$ a smooth map with $f[N] = V$, and with coordinate functions $f_1, \ldots, f_m$. η is the 'background' noise, which enters the system $\dot{x} = X_0(x) + \sum_{i=1}^{m} f_i(\eta_t) X_i(x)$ via the transformation f. In this set-up we have: η_t is a Markov diffusion process, the pair (η_t, x_t) is a Markov diffusion process, and supp $P = \mathcal{N}$, where $\mathcal{N}$ is the trajectory space of η_t. (Strictly spaeking we also need that the initial values of η_t and of x_t are independent of the driving Wiener process W_t^i.) From now on we assume that these properties are satisfied.

In the Markov context stationarity of (2.2) can be described through the control structure of (2.3), which is quite surprising, because controllability is a finite time result, while stationarity and invariant measures refer to the long term behavior as $t \to \infty$. We need the following definitions and notions for the system (2.3):

Denote by $\mathcal{O}^+_{\leq T}(x) = \{y \in M;$ there is $v \in \mathcal{V}$ and $0 \leq t \leq T$ with $\varphi(t, x, v) = y\}$ the so called positive orbit of a point $x \in M$ up to time $T > 0$, and let $\mathcal{O}^+(x) = \bigcup_{T>0} \mathcal{O}^+_{\leq T}(x)$.

3.1. Definition. A set $D \subset M$ is a *control set* of (2.3), if $\overline{\mathcal{O}^+(x)} \supset D$ for all $x \in D$, and D is maximal with this property. $C \subset M$ is an *invariant control set*, if furthermore $\overline{\mathcal{O}^+(x)} = \overline{C}$ for all $x \in C$. (For a set A we denote by $\bar{A}$ its closure.)

In the context of this paper we are interested only in those control sets that have nonvoid interior, denoted by int D. Under our standing assumption (H), this is always true for invariant control sets, see e.g. [20].

In order to obtain concepts for the flow ϕ in (2.4), we have to lift the control sets D of (2.3), which are subsets of M, to the product space $\mathcal{V} \times M$. We define

$$\mathcal{D} = c\ell\{(v, x) \in \mathcal{V} \times M;\ \varphi(t, x, v) \in \text{int } D \text{ for all } t \in \mathbb{R}\}, \tag{3.2}$$

where "$c\ell$" is the closure in $\mathcal{V} \times M$ with respect to the described topologies. The interplay between the control structure of (2.2) and the dynamical systems structure of (2.3) yields the following result:

3.2. Theorem. (control and chaos)

(i) *Let $D \subset M$ be a control set of (2.3) with int $D \neq \phi$, and $\mathcal{D}$ its lift to $\mathcal{V} \times M$. Then the flow $(\mathcal{D}, \phi)$ is chaotic, i.e. (a) there is a dense set of periodic points, (b) $(\mathcal{D}, \phi)$ is maximal topologically transitive (and mixing), (c) $(\mathcal{D}, \phi)$ has sensitive dependence on initial values.*

(ii) *Vice versa, if $\mathcal{D} \subset \mathcal{V} \times M$ is maximal topologically transitive (or mixing) with int $\Pi_M \mathcal{D} \neq \phi$, then there is a control set $D \subset M$ of (2.3) with int D = int $\Pi_M \mathcal{D}$ and $c\ell D = c\ell \Pi_M \mathcal{D}$. Here $\Pi_M : \mathcal{V} \times M \to M$ denotes the projection.*

Theorem 3.2 says that the sets, in which the system (2.3) can be controlled, correspond to the chaotic components $\mathcal{D}$ of the flow ϕ. We will briefly recall the concepts used in Theorem 3.2: Let $\psi: \mathbb{R} \times S \to S$ be a continuous flow on a complete metric space. For a subset $A \subset S$ the limit set $\omega(A)$ is defined as $\omega(A) = \bigcap_{t \geq 0} c\ell\{\psi(\tau, A);\ \tau \in [t, \infty)\}$ and similarly for $\omega^*(A)$ using $t \leq 0$.

(S, ψ) is called topologically transitive, if there is $s \in S$ with $\omega(s) = S$, and topologically mixing, if for any two open sets $V_1, V_2 \subset S$ there are times $T_0 \in \mathbb{R}$, $T_1 > 0$ such that $\psi(-nT_1 + T_0, V_1) \cap V_2 \neq \phi$ for all $n \in \mathbb{N}$. (S, ψ) has sensitive dependence on initial values, if there is a uniform $\delta > 0$ such that for all $s \in S$ and all neighborhoods N of s there are $p \in N$ and $t > 0$ with $d(\psi(t, s), \psi(t, p)) \geq \delta$. Here $d(\cdot, \cdot)$ is the metric (distance) on S. Finally $s \in S$ is a periodic point of ψ, if there is a time T with $\psi(t, s) = \psi(T + t, s)$ for all $t \in \mathbb{R}$, and denseness means that the closure of the set of periodic points is all of S. (See e.g. Mañé [22] or Devaney [16] for these definitions.) Setting $S = \mathcal{V} \times M$ and $\psi = \phi$ one obtains the corresponding statements in Theorem 3.2, where "maximal" of course means a maximal (with respect to set inclusion) set with these properties.

A flow with the properties (a)–(c) from Theorem 3.2(i) is 'chaotic' (compare e.g. Devaney [16]), because it has the three ingredients: a dense set of regular (periodic) points, mixing, and nearby trajectories drift apart everywhere. (In differentiable dynamics, which we do not have because of the shift component θ, a concept similar to (c) can be based on hyperbolicity and Lyapunov exponents, see e.g. Mañé [22].)

Because of the maximal mixing in Theorem 3.2(ib), we expect the following 'meta theorem' to hold: Control sets D of (2.3) correspond to maximal mixing components

$\mathcal{D}$ of ϕ, which correspond to the recurrent components of the stochastic flow ϕ, since recurrence is a mixing property, namely that the solution process visits every open set infinitely often. For general stationary noises this meta theorem is 'almost' true, compare Section 4. in [11]. In the Markov context we arrive at an even stronger result, for which we need the following concepts.

Consider the control system (2.3) and its control sets D_α, $\alpha \in I$, where I is some index set. We define a (partial) *order* on the control sets through

$$D_\alpha \prec D_\beta \text{ if there exists } x \in D_\alpha,\ v \in \mathcal{V} \text{ and } t > 0 \text{ such that } \varphi(t,x,v) \in D_\beta \tag{3.3}$$

in other words, if D_β can be reached from D_α (but not vice versa, since control sets are maximal sets with the reachability condition in Definition 3.1). A control set D_γ is said to be maximal with respect to the order '$\prec$', if, whenever $D_\gamma \prec D_\delta$, then $D_\gamma = D_\delta$. It is easy to see that invariant control sets are maximal w.r.t. $\prec$.

Concerning the existence of invariant control sets for a control system (2.3) we mention that in each bounded, invariant subset of the state space M (in particular, if M itself is compact) there exists at least one (and at most finitely many) invariant control sets C; C is closed and has nonvoid interior. Furthermore, C is said to be ***isolated*** if there is an open neighborhood N of C such that $\Pi_M\omega(v,N) \subset C$ for all $v \in \mathcal{V}$. With these preparations we can state the following results on the generic behavior of the trajectories of (2.3);

3.3. Theorem. (genericity and attractors)

(i) *The set $\{(v,x) \in \mathcal{V} \times M;$ there is $t_0 \geq 0$ such that $\varphi(t,x,v) \in \operatorname{int} C$, some invariant control set for all $t \geq t_0$, or $\varphi(t,x,v) \to \infty$ for $t \to \infty\}$ is an open and dense set in $\mathcal{V} \times M$.*

(ii) *If $\mathcal{C} \subset \mathcal{V} \times M$ is the lift of an isolated invariant control set $C \subset M$, then $\mathcal{C}$ is a chaotic attractor of the flow $(\mathcal{V} \times M, \phi)$.*

Part (i), whose proof can be found in [13] , says that, except for a thin set of control functions $v \in \mathcal{V}$ and initial values $x \in M$, all trajectories will end up in an invariant control set or go towards the boundary of M. In particular, for all initial values in a bounded invariant subset of M, these trajectories will enter some invariant control set and stay there. Concerning part (ii), we first have to make precise the concept of an attractor: We are interested in the structure of the union of all ω-limit set $\cup\{\omega(v,x); (v,x) \in \mathcal{V} \times M\}$. Since this set is difficult to describe, it is replaced in dynamical systems theory by a (possibly somewhat larger) set CR of *chain recurrent points*, see e.g. Conley [15]: For a dynamical system (S,ψ) we define $x \in CR$ if for all $\varepsilon > 0$, $T > 0$ there are $n \in \mathbb{N}$, times $t_0,\dots,t_{n-1} \geq T$, and points $x = x_0,\dots,x_n = x$ in S with $d(\psi(t_i,x_i),x_{i+1}) < \varepsilon$ for $i = 0,\dots,n-1$. The connected components $\mathcal{M}_\alpha$, $\alpha \in \mathcal{J}$ ($\mathcal{J}$ is again some index set) of CR are the *Morse sets* of (S,ψ). Between the Morse sets there is a (partial) order defined by

$$\mathcal{M}_\alpha \prec \mathcal{M}_\beta \quad \text{if there is } s \in S \text{ with } \omega^*(s) \subset \mathcal{M}_\alpha \text{ and } \omega(s) \subset \mathcal{M}_\beta. \tag{3.4}$$

An *attractor* is a maximal Morse set w.r.t. the order $\prec$, see e.g. Ruelle [25]. Using these concepts for the flow $(\mathcal{V} \times M, \phi)$ and for the lifts $\mathcal{C}$ of isolated invariant control sets C one can now easily prove statement (ii).

In the Markov set-up described above, we obtain the following consequence:

3.4. Corollary. *$P\{v \in \mathcal{V}$; there is $t_0 \geq 0$ such that $\varphi(t,x,v) \in C$, some invariant control set, for all $t \geq t_0$, or $\varphi(t,x,v) \to \infty\} = 1$ for all $x \in M$.*

It is clear from this corollary that solutions of (2.2), such that (ξ_t, x_t) is a Markov process, can be stationary only in the invariant control sets C of (2.3). In order to show that there actually is a stationary solution in each C, one needs a tightness condition on the transition probabilities, see e.g. Hasminskii [19]. Since we are only interested in *bounded* stationary solutions, the situation is simple:

3.5. Corollary. *Let ξ_t be given by a background noise η_t as in (3.1). Then we have:*

(i) *In each bounded invariant control set C of (2.3) there exists a stationary solution (η_t^0, x_t^0) of (2.2). The corresponding invariant probability measure μ of (η_t^0, x_t^0) is a solution of the (stationary) Fokker-Planck equation for (η_t, x_t), whose marginal on the state space N of η_t is the invariant measure of η_t. μ has a smooth density with support $N \times C$, in particular the stationary Markov solution on C is unique.*

(ii) *All bounded stationary and ergodic Markov solutions of (2.2) are concentrated on $N \times C$ for some bounded invariant control set C of (2.3), i.e. they are of the form (i). All bounded stationary Markov solutions are finite convex combinations of these ergodic processes.*

(iii) *Any Markovian solution of (2.2), which enters some C with probability 1, will converge (in distribution) to the unique ergodic process in C, according to the strong law of large numbers.*

For remarks on the proofs of the Corollaries 3.4 and 3.5 see [20] and [11].

The results 3.3–3.5 allow the precise formulation of our meta theorem above in the Markovian context:

3.6. Theorem. *Let M be a bounded, invariant set and assume that the invariant control sets $C \subset M$ are isolated. Then the invariant control sets $C \subset M$ of (2.3) correspond exactly to the (chaotic) attractors $\mathcal{C}$ of $(\mathcal{V} \times M, \phi)$, which in turn correspond exactly to the stationary, ergodic Markov solutions (η_t^0, x_t^0) of (2.2) on $N \times M$.*

If the invariant control sets are not isolated, the statement above remains true with 'attractor' replaced by 'maximal topologically mixing component', where maximality is understood w.r.t. the order defined in (3.3).

Theorem 3.6 connects several interpretations of (2.4) as a control, a dynamical, and a random system. It does, however, not relate these interpretations to the behavior of the undisturbed physical system (2.1). We will now explain such a relationship for small, Markovian noise.

Scale the noise range V through $V^\varepsilon = \varepsilon \cdot V \subset \mathbb{R}^m$ for $\varepsilon > 0$, then $\bigcap_{\varepsilon>0} V^\varepsilon = \{0\}$, which corresponds to the undisturbed system (2.1). Denote by M_α, $\alpha \in \mathcal{J}$ the Morse sets of (2.1), defined as in the explanations after Theorem 3.3, but now for the flow $\chi: \mathbb{R} \times M \to M$ induced by (2.1) on the state space M. Assume that χ has finitely many attractors $A_1 \dots A_n$ in M. A point x in one of these attractors is called in inner point, if for all $\varepsilon > 0$ there exist times $T, S > 0$ such that $\varphi(T, x, 0) \in \operatorname{int} \mathcal{O}^{\varepsilon+}_{\leq S}(x)$. Here $\varphi(t, x, 0)$ is the trajectory of the undisturbed system, and $\mathcal{O}^{\varepsilon+}$ is the positive orbit of the control system (2.3) with range V^ε, compare the notation introduced before Definition 3.1. The inner point property is discussed in [13] (see also [20]) and can be verified through a Lie algebraic criterion similar to (H).

3.7. Theorem. *Consider the control system (2.3^ε) with control range V^ε. Assume that for some $\hat{\varepsilon} > 0$ M is a bounded, invariant set of ($2.3^{\hat{\varepsilon}}$) with finitely many attractors $A_1 \dots A_k$ of the undisturbed system (2.1), such that all points $x \in \bigcup_{i=1}^{k} A_i$ are inner points. Then there exists $\varepsilon^0 > 0$ such that*

(i) *for all $0 < \varepsilon < \varepsilon^0$ there exist invariant control sets C_i^ε of (2.3^ε) such that $A_i \subset \operatorname{int} C_i^\varepsilon$ for all $i \in \{1, \dots, k\}$;*
(ii) *vice versa, if for a family C^ε, $0 < \varepsilon < \varepsilon^0$, of invariant control sets we have $\bigcap_{\varepsilon>0} C^\varepsilon \neq \phi$, then $\bigcap_{\varepsilon>0} C^\varepsilon$ is one of the attractors $A_1, \dots, A_k$.*

This theorem says that for all ε the attractors of (2.1) "blow up" to corresponding invariant control sets of (2.3^ε). The result is a special case of Theorem 4.12 in [13]. Thus, we obtain under the assumptions of Theorems 3.6 and 3.7 an extension of our meta theorem:

3.8. Corollary. *For $\varepsilon > 0$ small enough, the following objects are in one-to-one correspondence:*

(a) *attractors of the undisturbed system (2.1) on M,*
(b) *invariant control sets of the control system (2.3^ε) on M,*
(c) *(chaotic) attractors of the flow ϕ in (2.4) on $\mathcal{V} \times M$,*

(or maximal topologically mixing components of ϕ, if the invariant control sets are not isolated),

(d) *stationary, ergodic solutions (η_t^0, x_t^0) of (2.2) on $N \times M$.*

Note that this correspondence also holds, if the attractors of (2.1) are not chaotic, even for linear systems with additive noise, if the origin is the unique attractor of $\dot{x} = Ax$. The chaotic nature of the attractor of the flow ϕ on $\mathcal{V} \times M$ comes from its infinite dimensional component, described by the shift on the space $\mathcal{V}$ of bounded trajectories. Corollary 3.8 applies in particular to stable fixed points and stable limit cycles of $\dot{x} = X_0(x)$, around which one finds, for small noises, unique stationary, ergodic solutions of the random system (2.2).

4. Control of Long Term Random Dynamics.

Section 3. was devoted to the study of the long term behavior of the random system (2.2). Now we introduce an additional control component and analyze the possible changes of the qualitative behavior via deterministic control functions. Let us consider

$$\dot{x} = X_0(x) + \sum_{i=1}^{m} \xi_t^i X_i(x) + \sum_{j=1}^{n} u_j(t) Y_j(x), \tag{4.1}$$

where the $Y_1 \dots Y_n$ are again smooth vectorfields on the state space M, and the admissible control functions $(u_j(t))_{j=1,\dots,n} = u(t)$ have values in a bounded set $U \subset \mathbb{R}^n$, again with $0 \in \text{int } U$, so that for $u \equiv 0$ the uncontrolled system corresponds to (2.2). We address the following problem: Find a control function $u(t)$ such that (4.1) has a stationary solution (or converges towards a stationary solution) with prescribed behavior. By prescribed behavior we mean here: possesses an invariant measure in a certain, given area. (We are not interested in the shape of the resulting invariant density, only in its support, since in this very general set-up analytical statements about the shape are not possible.) Recall that by the law of large numbers almost surely the trajectories of x_t will visit any open set in the support of the invariant measure infinitely often. Our results will therefore describe the area, in which the trajectories of the controlled system fluctuate.

We will continue to consider the Markov set up with noise processes described by (3.1), and we will use the correspondence between invariant control sets and ergodic solutions from Section 3. If the controlled equation is to have stationary solutions for t sufficiently large, then the vector fields in (4.1) have to be time independent for $t \geq t_0$ (see Hasminskii [19]), i.e. the controls $u(t)$ have to be constant for $t \geq t_0$. As we will see, this relates the problem studied here closely to the Markovian bifurcation theory of random systems.

Four examples will be studied according to the following scheme: Given a random system of the form (2.2) and control vectorfields Y_j with control range $U \subset \mathbb{R}^n$, and given certain required properties concerning the stationary behavior of the controlled system, is it possible to achieve this behavior. All examples will be in dimension 1 or 2, although the results work for any finite dimension, because the principles can be explained in an intuitive way for low dimensional state spaces. A software package is under development to compute the necessary invariant control sets ("CS" by Gerhard Häckl, University of Augsburg), which was used for Example 3.

4.1. Example. Consider the Verhulst equation in $\mathbb{R}^1$

$$\dot{x} = X_0(x) + \xi_t X_1(x) + u(t)Y_1(x) = \alpha x - x^2 + \xi_t x + ux, \tag{4.2}$$

$\alpha \in \mathbb{R}$, $\xi_t \in V = [-\delta, \delta] \subset \mathbb{R}$, $u(t) \in \mathbb{R}$. The goal is to obtain a stationary solution with support bounded away from 0.

For one dimensional systems it is convenient to represent the dynamics in the (u, x)-plane as a 'bifurcation diagram', see Figure 1. The arrows indicate the sign of the right hand side of (4.2), and the lines the zeros, i.e. the fixed points. Using the techniques described in [4] or [11], it is easy to find the control sets for each u and a given noise range V.

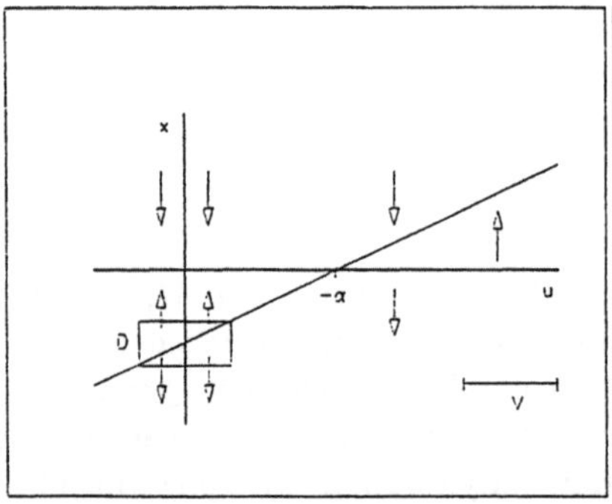

Figure 1: Uncontrolled system

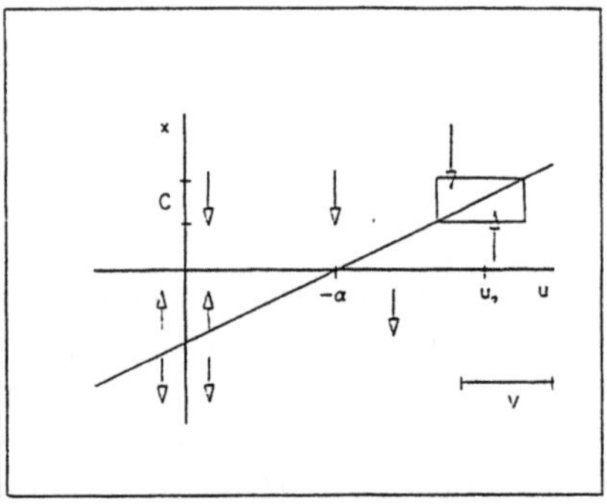

Figure 2: Controlled system

For the uncontrolled system, i.e. $u \equiv 0$, we obtain (see Figure 1): $x = 0$ is a fixed point (independent of the noise trajectory), there is one variant control set D, and no invariant control set (other than $\{x = 0\}$). Hence there is no stationary solution, bounded away from 0. Actually we obtain for the limit behavior of this system, with $D = (a, b)$: $x_t(y) \to 0$ almost surely for $t \to \infty$ for all initial values $y \geq b$, $x_t(y) \to -\infty$ almost surely for $t \to \infty$ for $y \leq a$, and for $y \in D$ there are positive probabilities p^0, p^∞ such that $P\{x_t(y) \to 0\} = p^0$ and $P\{x_t(y) \to -\infty\} = p^\infty$.

For the controlled system we have: (4.2) has a stationary solution, bounded away from 0, iff $u_0 > -\alpha + \delta$, see Figure 2. Here C is a unique invariant control set in $\mathbb{R}_+$. Increasing the control u means moving C, and hence the stationary solution, further away from 0. Note that for this system $x_t(y)$ converges towards the stationary solution iff the initial value $y > 0$. For nonpositive initial values the goal cannot be accomplished. Looking at Figures 1 and 2, these results are not surprising, but rather intuitive. We next consider a 2-dimensional system, which is also basically determined by its deterministic structure.

4.2. Example. Consider the Lotka-Volterra (or predator-prey) system in $\mathbb{R}^2_+$

$$\begin{pmatrix} \dot{x}_1 \\ \dot{x}_2 \end{pmatrix} = X_0(x) + \xi_t X_1(x) + u(t) Y_1(x) = \begin{pmatrix} (\alpha - \beta x_2) x_1 \\ -\gamma x_2 \end{pmatrix} + \xi_t \begin{pmatrix} 0 \\ x_1 x_2 \end{pmatrix} + u(t) \begin{pmatrix} -x_1^2 \\ 0 \end{pmatrix}, \tag{4.3}$$

where α, β, γ are positive constants, $\xi_t \in [a, b] \subset \mathbb{R}_+$ and $u(t) \in U \subset \mathbb{R}_+$. The goals are to (a) obtain a stationary solution in $\mathbb{R}_+$, bounded away from the x_1-and the x_2-axis, (b) obtain a stationary solution with $x_2 = 0$. Note that $x_1 = 0$ or $x_2 = 0$ indicates the extinction of the corresponding species.

The uncontrolled system with $u \equiv 0$: For $\xi_t \equiv v \in [a, b]$, the solutions are closed curves in $\mathbb{R}^2_+$, covering all of $\mathbb{R}^2_+$. Thus, even with the smallest noise in $[a, b]$, $a \neq b$, the system cannot have a bounded, stationary solution. Hence for all thresholds $L > 1$ and all initial values $(y_1, y_2) \in \mathbb{R}^2_+$ there exists a positive probability that the solution $x_t(y_1, y_2)$ will grow above L, and fall below $\frac{1}{L}$.

Controlled system with u small, i.e. $0 < u < \frac{\alpha a}{\gamma}$: For $\xi_t \equiv v \in [a,b]$, the solutions have a unique stable fixed point $\left(\frac{\gamma}{v}, \frac{\alpha v - u\gamma}{\beta v}\right)$ in $\mathbb{R}^2_+$. Thus, by Proposition 2.4 in [8], for each $u \in \left(0, \frac{\alpha a}{\gamma}\right)$ there exists a unique compact invariant control set $C_u \subset \mathbb{R}^2_+$, on which the system (4.3) has a unique stationary ergodic solution. If u increases within this interval, the control set C_u moves closer to the x_1-axis, i.e. the x_2-population becomes smaller. For all initial values in $\mathbb{R}^2_+$, the solutions of (4.3) convergence towards the stationary one.

Controlled system with u large, i.e. $u > \frac{\alpha b}{\gamma}$: There exists a unique stationary solution with support $\left[\frac{\alpha}{b}, \frac{\alpha}{a}\right] \times \{0\}$ on the positive x_1-axis, and all solutions of (4.3) with initial value in $\mathbb{R}^2_+$ converge towards this solution, i.e. the x_2-population becomes extinct w.p.1.

For a deterministic analysis of this system see e.g. Amann [1], for some aspects of a white noise analysis see e.g. Dimentberg [17].

The next example shows a 2-dimensional system, whose deterministic part can be controlled to stay in a certain area of the state space, while even small noise will drive it into a different region.

4.3. Example. Consider the model of a well-stirred chemical reactor in $M = (0,\infty) \times (0,1)$

$$\begin{aligned}\begin{pmatrix}\dot{x}_1\\ \dot{x}_2\end{pmatrix} &= X_0(x) + \xi_t X_1(x) + u(t) Y_1(x)\\ &= \begin{pmatrix}-x_1 - 0.15(x_1 - x_c) + 0.35(1 - x_2)e^{x_1}\\ -x_2 + 0.05(1 - x_2)e^{x_1}\end{pmatrix} + \xi_t \begin{pmatrix}x_c - x_1\\ 0\end{pmatrix} + u(t)\begin{pmatrix}x_c - x_1\\ 0\end{pmatrix},\end{aligned} \tag{4.4}$$

where x_1 denotes temperature, x_2 product concentration and noise and control affect the heat transfer coefficient. Noise and control affecting the same parameter means that one cannot steer the system precisely, but the control input into the reactor is disturbed by a (small) Markovian noise. For a discussion of the deterministic, uncontrolled system see e.g. Golubitsky and Schaeffer [18] or Poore [23]. We choose $u(t) \in U = [-0.15, 0.15]$ and $\xi_t \in (-\varepsilon, \varepsilon)$. The goal is to control the system such that the product concentration x_2 is as high as possible, under technological constraints.

The deterministic, controlled system (i.e. $\xi_t \equiv 0$) was analyzed in [14], and the results are shown in Figure 3. and 4. The control system has 3 control sets, 2 are invariant (the lower one, C_1, and the upper one, C_2), and one is variant (the middle one, D). Figure 3. shows these control sets, together with the phase portrait of the uncontrolled system. The problem is that the upper control set C_2, which is the most desirable region of operation because of the high product concentration, is technically not feasible, compare Bellman et al. [6]. However, the reactor with undisturbed controls can be steered from the region of attraction of the variant control set D, shown in Figure 4., into D and it can be kept in this set using for each initial value an appropriate control function $u(t)$, compare [14].

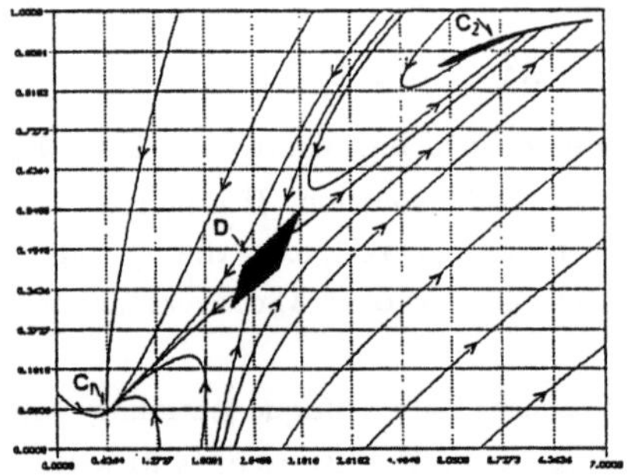

Figure 3: Control set of the undisturbed reactor model (4.4)

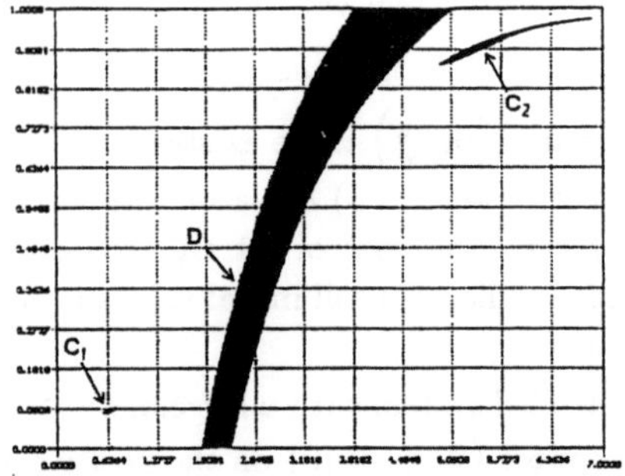

Figure 4: Region of attraction of the variant control set in (4.4)

Now, if the chosen control input $u(t)$ is disturbed by noise $\xi_t \in [-\varepsilon, \varepsilon]$, then by Corollary 3.4, for all $\varepsilon > 0$ however small, the system will tend towards one of the invariant control sets, which means lower product concentration in case of C_1, and destruction in case of C_2. Each of these possibilities will result with positive probability, if the initial value is in the region of attraction of D. If one starts to the left of this region, one converges to C_1 w.p.1, and similarly to C_2 for starting values to the right. This example shows that in order to stabilize input-disturbed systems in variant control sets, one has to design the controls depending on the noise.

4.4. Example. This example shows that even in one dimensional systems some surprising effects can occur. We will start with a 'bifurcation diagram' as in Figure 5. The noise and the control have the same dynamics (see e.g. Example 4.1 or 4.3), although the same effects can also occur, if this is not the case — the simple graphical representation, however, would not be possible. The size V of the noise is indicated in Figure 5. The goal is to obtain a stationary solution x_t with values above the point x_2. Figure 6 shows, for each control value u, the invariant control sets, corresponding to the noise range V, as areas around parts of the stable bifurcation curves.

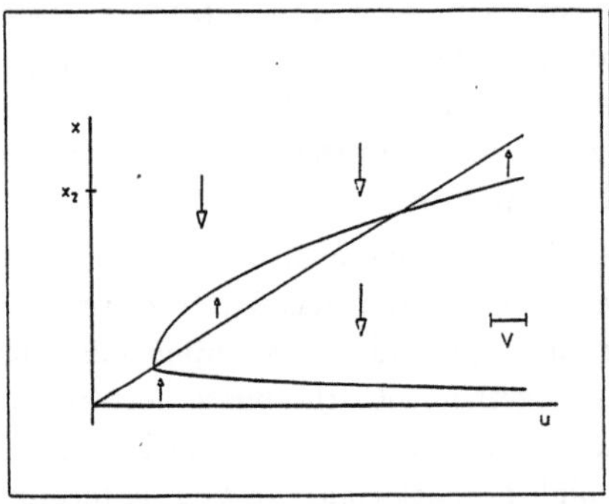

Figure 5: Bifurcation diagram of Example 4.4.

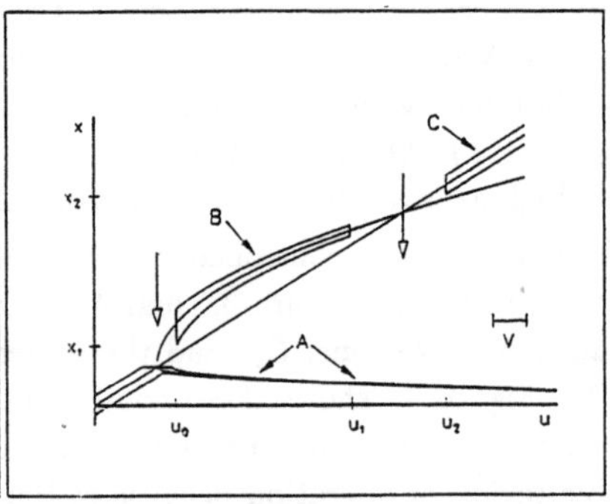

Figure 6: Invariant controls set of Example 4.4

For $u \equiv 0$ there is a unique invariant control set with low x-values. Increasing the control u leads for initial values $< x_1$ to unique invariant control sets in the area A around the lower bifurcation curve, i.e. the goal cannot be reached. For initial values in $[x_1, x_2)$ a control $\geq u_0$ leads to stationary solutions in the area B. However, increasing u above u_1 will cause the system to converge towards a stationary solution in area A. Finally, if the initial value is $\geq x_2$, a control $\geq u_2$ will result in a unique stationary solution in area C. Hence we see that, although the points 0 and x_2 are linked by a continuous line of stable bifurcation branches, it is not possible to steer the system to a stationary solution around x_2 for initial values $< x_2$. This is prohibited by the random dynamics of the Markov system. Note that this effect is present even for small noises; increasing the range V will lead to smaller areas B, and C will be shifted to larger x-values. For V large enough, the area B will disappear completely.

The separation between areas A and B is an example of 'noise induced symmetry breaking', while the separation between B and C (and the subsequent convergence towards A) has not been discussed in the literature. These effects, and many others, can be analyzed in the context of stochastic bifurcation theory for systems with bounded Markov noise.

Conclusions. We summarize our findings in the examples above in a few rules:

(a) Control to stationary Markov solutions in prescribed areas of the state space needs invariant control sets in these areas.

(b) The control to a stationary solution consisting of one point s_0 (i.e. the invariant measure is the Dirac measure at x_0) requires the existence of a control $u^0 \in U$ such that x_0 is a fixed point of $\dot{x} = X_0(x) + \sum \xi_t^i X_i(x) + \sum u_j^0 Y_j(x)$ for all $X_1 \ldots X_m$.

(c) The bifurcation behavior of the random systems with u as a (multidimensional) bifurcation parameter, and its control structure determine the the possible supports of stationary, ergodic solutions.

(d) A deterministically controlled system may drastically change its behavior in the presence of noise, if the control is designed independent of the noise.

References.

[1] Amann, H. (1983), 'Gewöhnliche Differentialgleichungen,' De Gruyter.

[2] Arnold, L., P. Boxler (1991),'Stochastic bifurcation: Instructive examples in dimension one,' to appear in: 'Stochastic Flows,' Pinsky, M., V. Wihstutz (eds.), Birkhäuser.

[3] Arnold, L., H. Crauel (1991), Introduction to 'Lyapunov Exponents, Proceedings of the Oberwolfach Conference 1990,' Arnold, L., H. Crauel, J.-P. Eckmann (eds.), Springer.

[4] Arnold, L., W. Kliemann (1983), 'Qualitative theory of stochastic systems,' In: Probabilistic Analysis and Related Topics Vol. 3,' Bharucha-Reid, A. T. (ed.), Academic Press, 1–79.

[5] Baxendale, P. (1986), 'Asymptotic behavior of stochastic flows of diffeomorphisms.' In: Proceedings of the 15th SPA, Nagoya 1985. LN Mathematics 1203, Springer, 1–19.

[6] Bellman, R., J. Bentsman, S. M. Meerkov (1983), 'Vibrational control of systems with arrhenius dynamics,' J. Math. Anal. Appl. **91**, 152–191.

[7] Boxler, P. (1989), 'A stochastic version of center manifold theory,' Probab. Th. Related Fields **83**, 509–545.
[8] Colonius, F., W. Kliemann (1989), 'Infinite time optimal control and periodicity,' Appl. Math. Opt. **20**, 113–130.
[9] ______________ (1990), 'Stability radii and Lyapunov exponents.' In: Control of Uncertain Systems, Hinrichsen D., B. Mårtensson (eds.) Birkhäuser, 19–55.
[10] ______________ (1991), 'Stabilization of uncertain linear systems,' to appear in: Modeling and Control of Uncertain Systems, DiMasi, G., A. Gombani, A. Kurzhanski (eds.), Birkhäuser.
[11] ______________ (1991), 'Remarks on ergodic theory of stochastic flows and control flows,' to appear in: 'Stochastic Flows,' Pinsky, M., V. Wihstutz (eds.), Birkhäuser.
[12] ______________ (1991), 'Some aspects of control systems as dynamical systems,' submitted.
[13] ______________ (1991), 'Limit behavior and genericity for nonlinear control systems,' submitted.
[14] ______________ (1991), 'Kontrolltheorie and Dynamische Systeme,' submitted.
[15] Conley, C. (1978), 'Isolated Invariant Sets and the Morse Index,' Regional Conference Series in Mathematics no. 38, American Mathematical Society.
[16] Devaney, R. L. (1986), 'An Introduction to Chaotic Dynamical Systems,' Benjamin-Cummings.
[17] Dimentberg, M. F. (1988), 'Statistical Dynamics of Nonlinear and Time-Varying Systems,' Research Studies Press, Wiley.
[18] Golubitsky, M., D. G. Schaeffer (1985), 'Singularities and Groups in Bifurcation Theory,' Springer.
[19] Hasminskii, R. Z. (1980), 'Stochastic Stability of Differential Equations,' Sijthoff and Nordhoff. (Russian edition 1969).
[20] Kliemann, W. (1988), 'Analysis of nonlinear stochastic systems,' in: Analysis and Estimation of Stochastic Mechanical Systems, W. Schiehlen, W. Wedig (eds.), Springer, 43–102.
[21] Kunita, H. (1990), 'Stochastic Flows and Stochastic Differential Equations,' Cambridge University Press.
[22] Mañé, R. (1987), 'Ergodic Theory and Differentiable Dynamics,' Springer.
[23] Poore, A. B. (1974), 'A model equation arising from chemical reactor theory,' Arch. Rational Mech. Anal. **52**, 358–388.
[24] Rozanov, Y. A. (1967), 'Stationary Random Processes,' Holden Day.
[25] Ruelle, D. (1989), 'Elements of Differentiable Dynamics and Bifurcation Theory,' Academic Press.
[26] Sobczyk, K. (1991), 'Stochastic Differential Equations,' Kluwer.
[27] Stroock, D. W., S. R. S. Varadhan (1972), 'On the support of diffusion processes with applications to the strong maximum principle,' Proc. 6th Berkeley Symp. Math. Stat. Probab., Vol. 3, 333–359.

Structural Safety Control and Stochastic Aspects

S.C. Liu, H.J. Lagorio, and K.P. Chong
Engineering Directorate
National Science Foundation
Washington, D.C. 20550
U.S.A.

ABSTRACT

Recently, increasing interest has been shown in structural safety design using active and hybrid control systems. This interest is driven by the rapid development of intelligent control systems, advanced sensor technology, signal processing, and successful applications in the control of large flexible space structures.

This paper focuses on the potential and application of advanced control technology in land-based civil structures for safety protection against natural hazard forces such as winds and earthquakes. An assessment of the state-of-the-art in structural control technology that involves multiple, diverse fields is made in this paper, and, on this basis, an integrated comprehensive research program is outlined. Treating input forces as random loads, one of the most important aspects of structural control problems, e.g., its stochastic nature and implications in the formulation, solution, and practical implementation process, are highlighted.

I. INTRODUCTION

Recently, considerable success has been achieved in applying control system theories in aeronautic and aerospace structures, particularly large, flexible systems. Meanwhile, rapid technical advances have also been made in many fundamental fronts; these include smart materials and systems, high-tech sensors, detecting and monitoring systems, innovative actuator dynamics, and development of advanced intelligent, dynamic control systems. Driven by such development and by the need to search for a new form of structures for future construction, civil engineers began to focus their attention to the prospect of designing future land-based structures such as buildings and bridges to possess their own intelligence for better performance, serviceability, services and maintenance, adaptation to local and global environmental conditions, and most importantly, for improved safety protection against hostile natural hazard loads as compared to their conventional counterpart structures existing today.

An active or hybrid controlled structure, which uses a combination of active and passive devices, can be viewed under the general context of "intelligent structures". The intelligence can be programmed or built into the structure by material composition and processing, real-time measurement and monitoring of dynamic response, detection of microfracture, yielding, or other failure, or conditioning through computerized servo system to adapt in a controlled manner to external forces or other stimulus. In a general sense, a "controlled" structure usually contains, through ingenious engineering design, its own sensors, data acquisition and processing, actuators, computational/control, and feedback capabilities.

It appears that such technologies, when sufficiently developed and broadly implemented, offer an excellent potential to impact the safety design and construction of new structures in the future. Furthermore, the dynamic control and intelligence technology also provides a viable, alternative solution to the nation's enormous problems in the rehabilitation of existing structures which are deficient in their seismic, wind, or other resistant capacities.

It is generally agreed that dynamic control and intelligence technology holds considerable promises in the design applications of civil engineering structures, new and existing. By its interdisciplinary nature, such technology also has the potential impact to the future development of a diverse field of scientific or technical subjects, such as material science, vibration, biotechnology, neural network, and new devices.

In order to forge ahead with an intensive research and development program focusing on seismic or wind safety of civil engineering structures through the use of modern control systems, a systematic assessment of the current state of technology in some of the major technical areas involved must be made. Through

N. Bellomo · F. Casciati (Eds.)
Nonlinear Stochastic Mechanics
IUTAM Symposium Turin

such an evaluation, a clear picture for the future research needs, which would lead us to the evental goal of achieving a broad, practical implementation of structural control technology, will begin to emerge.

II. TECHNOLOGY ASSESSMENT AND RESEARCH NEEDS

Over the last few years, intensive development efforts have been made for the purpose of charting a long-term national research agenda on structural control. An important capstone of such an agenda is to foster, on a coordinated basis, multidisciplinary research and development of passive, active and hybrid control technology for application to structures and nonstructural building systems under dynamic loads. Collaborative efforts among researchers in interdisciplinary areas involving academics, industrial experts and professional practitioners are considered extremely essential. Such program development efforts include

(1) In 1989, a U.S. Panel on Structural Control Research was established under the auspices of NSF. The Panel has the responsibility of:

(a) Facilitating the transmission of information concerning state-of-the-art developments in the field.
(b) Identifying and prioritizing needed research and development.
(c) Developing preliminary plans for analytical and experimental advancement in the field.
(d) Developing plans for the performance of full-scale testing and demonstration.

The Panel's organization consists of an Executive Committee with Professor George W. Housner as its Chairman, and seven (7) Working Groups in: (1) Analytical Methods, (2) Experimental Methods, (3) Building Applications, (4) Non-Building Applications, (5) Interdisciplinary Approaches, (6) International Coordination, and (7) Information Dissemination. In the meantime, a Japanese counterpart panel on Seismic Control Research was also established by the Science Council of Japan and Dr. T. Kobori is the Chairman of the Japanese Panel. Both the U.S. and japan panels have been working closely in the development of a joint US-Panel research agenda in the areas of active and hybrid control.

(2) On July 23 to 26, 1990, an International Workshop on Intelligent Structures was held in Taipei, Taiwan. The emphasis of the workshop was on the development, application, and modification of active/hybrid control theories, and through innovative materials processing and system concepts, to achieve designed intelligence of civil engineering structures. Three major theme areas for the technical program of the workshop were established: (a) Sensing and monitoring techniques, (b) Structural control, and (c) Intelligent systems [1].

(3) On October 15 and 16, 1990, a workshop was held at the Electrical Power Research Institute (EPRI) under the joint auspices of NSF and EPRI. The objective of this workshop is to evaluate the status of and research needs in the area of Intelligent Control Systems [2]. The technical program was structured on four technical areas: (a) General theoretical foundations assessment, (b) Knowledge-based systems, (c) Application related research, and (d) Software/hardware requirements.

(4) On October 25 and 25, 1990, the U.S. Panel on Structural Control Research organized and convened the "U.S. National Workshop on Structural Control Research" at the University of Southern California (USC), Los Angeles. The workshop identified four specific areas related to earthquake safety for technical discussions. They are: (a) Analytical research, (b) Experimental research, (c) Building applications, and (d) Non-building applications. Proceedings of the Workshop is now available from USC [3].

(5) On February 25 and 26, 1991, NSF held another Workshop on "Sensors and Signal Processing for Structural Control" in Washington, D.C. This Workshop focused on assessing the current state of knowledge and technology concerning sensors and measurement techniques, signal processing, actuator mechanics as they applied to structural control [4].

The essential conclusions and recommendations of these workshops as summarized below provide a comprehensive technical assessment of the state-of-the-art technology in intelligent control systems and structural control research.

II.1. Fundamental Research in Intelligent Control Systems

As mentioned before, an active or hybrid controlled structure might also be looked at from a broad prospective as an "intelligent" controlled system. Structures which undergo dynamic motions due to winds or ground shaking maybe considered as real-time systems whose complexity in control increases with (1) the number of functions to be controlled, (2) the rate at which functions to be controlled and (3) the amount of information to be considered for a control decision [2]. From the basic research and development viewpoint it is of increased interest to engineers in intelligent systems which possess improved flexibility in terms of application, operability (in terms of reliability and performance), and availability within the context of serviceability and maintenance.

At an NSF/EPRI Workshop [2] the following research commendations were made for the purpose of advancing the development of complex dynamical control systems and to introduce new technologies for improvement of quality, safety, flexibility, operability, and availability of such systems:

- Establishment of theoretical foundation.
- Techniques for technology transfer and implementation.
- Advanced hardware (e.g., sensors and actuators) and software (e.g., information processing and knowledge interpretation).
- Baseline technology for system design and implementation.
- Test beds for system verification and validation.

II.2. Sensors, Actuators, and Signal Processing Research

Sensors and actuators are analogous to the nerve and muscle systems, respectively, of a human body - which is itself an intelligent system. The signals that are "sensed" by these sensors, modified by the actuators at input and various response locations must be processed in real time under very restrictive conditions. These topics require special attention in terms of the unique requirements imposed upon them by structural control as compared with applications in other fields.

The sensor and actuator technologies and signal processing techniques as applied to structural control were discussed at a special workshop at NSF in February, 1991 [4]. The major considerations behind this workshop were that while these topics have been advanced quite dramatically in their respective fields in recent years, these developments--often of advanced nature--are not familiar to structural engineers. Structural engineers working the control areas need to know the baseline technologies in these areas, products and technique available, and about their utilities and limitations. On the otherhand, specialists in sensors, devices, control systems, and information theory need to know what structural/architectural measurements are needed and what kind of information processing are to be dealt with. The technical assessment can be briefly summarized as follows.

(a) Sensors - Through an examination of previous research programs in this topic area, sensor needs by structural engineers, required measurement parameters of civil structures, new sensor types, etc. the following specific recommendations are identified.

- Explore issues of integration and installation of sensors in/or intelligent civil structures.

- Pursue advanced techniques for in-service monitoring and control.

- Address the need for application-specific damage assessment and Non-Destructive Evaluation for critical structures.

- Investigate frontier technologies which will enable new types of measurements to be made.

- Advance basic research to the proof-of-concept level, integrating developing technologies into quantitative demonstrations of capability.

- Encourage and/or require interdisciplinary cooperation, a path to information dissemination, and a means for technology transfer.

- Establish Centers of Excellence to address the increasingly complex needs for sensing and control in intelligent civil structures.

(b) Actuators - Central to all actuator questions was the need to transform energy from one form to another. As a result, questions related to practical constraints, cost/effectiveness, control philosophy and system design are carefully examined. The research needs identified through such an exercise are:

- Develop standards for structural control systems. Specifically, test procedures and performance criteria, including control parameter levels.

- Undertake theoretical studies to contrast the uses of distributed versus discrete actuators.

- Research programs focused on actuators and systems for existing structures.

- Research on the use of actuators as structural elements.

- Research programs emphasized on cross-disciplinary and undertaken by multiple principal investigators.

(c) Signal Processing - For signal processing in structural control, the interface with sensors and actuators has to be addressed first so that the input and output of a signal processing system can be defined. This interface relates to the sensor type, number of sensors required and the manner they are deployed. From these questions the type and volume of signals to be processed, dynamic range of input signals, and the accuracy and speed requirements for processing can be determined. Similar analysis for actuators will provide guidelines for output signal processing.

The recommendations of the signal processing research needs are:

- Encourage interdisciplinary research team activities with emphasis on close work relationships among team members.

- Identify existing signal processing techniques suitable for structural control in order to avoid duplication of work.

- Formulate signal processing problems that cannot be solved using existing techniques and present the challenge to the signal processing community.

II.3. Control Research as Pertaining to Civil Engineering Structures

The technical evaluation was made through a special workshop on the subject area conducted at the University of Southern California in October, 1990 [3]. The essential findings are given below:

(a) Analytical Methods

Special attention is needed to study the nonlinear control theory, and interactive effects between "actuator dynamics" and "structural dynamics" warrant particular study. In connection with uncertainties in structural modeling and loading, the robustness of control schemes is important and probabilistic models may be required. Other important issues include: redundancy in sensors, development of "smart structures" such as embedding active elements in structural members, time delay effects caused by actuator dynamics or by computer processing, and the potential of adaptive control algorithms. Finally, the importance of integrating design of structures and control systems into a complete seamless process is emphasized.

(b) Experimental Approaches

Relatively little experimental research has been done in active, passive, or hybrid control systems, and a great deal of future effort is needed which must occur simultaneously with correlative analytical studies. The following major issues are emphasized.

(i) Component Tests - Standardized tests for each component of the control system need to be developed. These include the performance of different actuators, pumps, and servo-hydraulic systems, sensors, computer control systems, etc. The time-delay effects of these elements, their reliability and interactive effects should be evaluated.

(ii) Component Development - Sensors need to be tested and developed and what response quantities need to be measured and to what degree of accuracy the measurement must be made should be investigated. More compact and efficient actuators, development of hardware in conjunction with software algorithms, and the use of single component, serial components and paraller systems, should be developed.

(iii) Model Tests - Laboratory tests of small-scale and medium-scale control systems and controlled structural models should be conducted utilizing existing experimental facilities.

(iv) Full-Scale Tests - Control systems need to be tested on full-scale structures before practical implementation is possible. Such tests are particularly suitable for US-Japan cooperative efforts because of test facilities available in Japan.

In addition to above, the development of active structural elements such as active braces that induce changes in structural stiffness and passive devices which can function to distribute model kinetic energy or dissipate input energy are also important and could lead to significant benefits. Development and deployment of adaptive systems and hybrid systems are encouraged for systems which undergo nonlinear response and to help reduce the external power supply requirement.

III. STOCHASTIC CONSIDERATIONS IN STRUCTURAL CONTROL

Stochastic aspects of structural control originates primarily from the random characteristics of the input forcing functions, such as earthquakes and winds. Because such forcing functions as expected to be experienced by a structure during its service life cannot be predicted precisely in the deterministic sense, engineers have resorted to the random vibration approaches to structural analysis and design under such conditions. To illustrate such a situation, a schematic diagram of active control, as shown in Fig. 1 [5], is helpful.

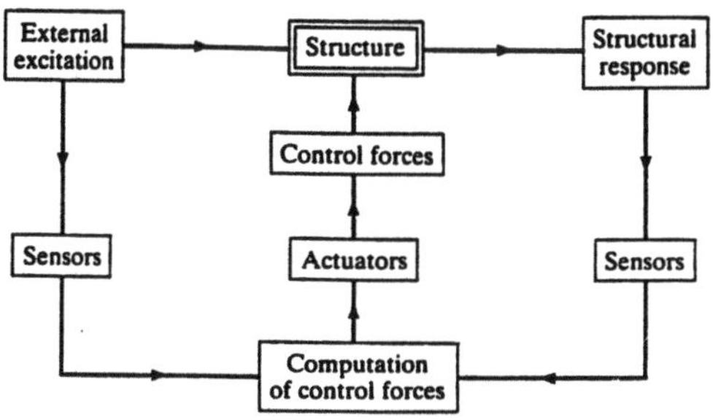

Figure 1. Active Control of Structures

When only the structural response variables are measured, the control configuration is referred to as closed-loop control since the structural response is continually monitored and this information is used to make continual corrections to the applied control forces. An open-loop control results when the control forces are regulated only by the measured excitation. In the case where the information on both the response quantities and excitation are utilized for control design, the term closed-open-loop control is often used.

Using the above closed-open-loop control configuration and idealizing the structure by an n-degree of freedom linear system subjected to an external earthquake excitation $\ddot{X}_0(t)$, the matrix equation of motion, in state-space representation, can be written as

$$\dot{\underline{Z}}(t) = \underline{A}\,\underline{Z}(t) + \underline{B}\,\underline{U}(t) + \underline{W}_1\,\ddot{X}_0(t) \qquad (1)$$

in which an under bar denotes a vector or matrix. In Eq. (1). $\underline{Z}(t)$ = a 2n state vector, $\underline{U}(t)$ = a r-dimensional control vector, $\underline{A}$ = a (2nx2n) system matrix, representing the structural characteristics of the building, $\underline{B}$ = a (2nxr) location matrix specifying the location of r controllers and $\underline{W}_1$ is an appropriate 2n vector denoting the effect of the earthquake ground acceleration $\ddot{X}_0(t)$. The state vector $\underline{Z}(t)$ consists of the displacement vector $\underline{Y}(t)$ and velocity vector $\dot{\underline{Y}}(t)$, all relative to the ground, as

$$\underline{Z}(t) = \left[\begin{array}{c} \underline{Y}(t) \\ \hline \dot{\underline{Y}}(t) \end{array}\right] \qquad (2)$$

with the initial condition $\underline{Z}(0) = 0$. Matrices $\underline{A}$, $\underline{B}$ and $\underline{W}_1$ are given in Ref 6.

Using a given optimal control law, a performance function or index J(t) can be defined, for example.

$$J^*(t) = \dot{\underline{Z}}'(t)\underline{Q}^*\,\dot{\underline{Z}}(t) + \underline{U}'(t)\,\underline{R}\,\underline{U}(t) \qquad (3)$$

is obtained using the instantaneous optimal control law [6] in which a prime denotes the transpose of a vector or matrix, $\underline{Q}^*$ is a (2nx2n) positive semi-definite weighting matrix and $\underline{R}$ is a (rxr) positive definite weighting matrix. It can be noticed that $J^*(t)$ involves both the responses and control forces.

The optimal control vector $\underline{U}(t)$ can be obtained by minimizing $J^*(t)$. Depending on the way the control vector is regulated, one obtains optimal closed-loop control (feedback), optimal open-loop control (feedforward) and optimal closed-open-loop control (feedforward and feedback).

It can be shown that [6]

$$\underline{U}(t) = -\underline{R}^{-1}\,\underline{B}'\,\underline{Q}^*\,\dot{\underline{Z}}(t) \qquad \text{Closed-loop control} \qquad (4a)$$

$$\underline{U}(t) = -\Delta t/2\,\underline{R}^{-1}\,\underline{B}'\,\underline{Q}\,\underline{Z}(t) \qquad \text{Instantaneous Optimal Control} \qquad (4b)$$

$$\underline{U}(t) = -\,1/2\,\underline{R}^{-1}\,\underline{B}'\,\underline{P}\,\underline{Z}(t) \qquad \text{Linear Quadratic Optimal Control} \qquad (4c)$$

where $\underline{Q}$ is a (2nx2n) positive semi-definite weighting matrix and $\underline{P}$ is a (2nx2n) Riccati matrix.

Given $\ddot{X}_o(t)$, the controlled structural response $\underline{Z}(t)$ can be solved numerically by using Eq. (1) and Eq. (4a), (4b), or (4c).

It can be seen that $\underline{Z}(t)$ and $\underline{U}(t)$ are stochastic when the forcing function $\ddot{X}_o(t)$ is stochastic. In this case, the established random vibration methods can be used to obtain the solutions. As such, the current state-of-the-art development in random vibration theory sets the limits to the solution approach. In general, when the input is a simple stationary process and the structure is a linear system, the stochastic solutions can be normally found.

The most commonly used stochastic models for $\ddot{X}_0(t)$, among others, include:

a) white noise
b) shot noise
c) linearly filtered white noise or shot noise
d) modulated stationary process
e) evolutionary nonstationary process

The degree of complexity increases, as expected, when the input becomes nonstationary or when the structure undergoes nonlinear inelastic motions. In such situations, only limited solutions can be found in special cases and in general, numerical approach has to be employed to find solutions on a case by case basis.

IV. SUMMARY AND CONCLUDING REMARKS

Over the last two decades, considerable progress has been made to improve structural safety through research in structural, earthquake, and wind engineering. The possibility of applying control systems to mitigate the structural damage against earthquakes and winds opens the door to a wide range of research challenges for multidisciplinary investigators in the field. Further, such challenges can be metr most effectively through international research cooperation. In this spirit, the mechanisms which might be provided by the International Decade for Natural Disaster Reduction (IDNDR) program can efficiently serve as a launching pad for the research projects as identified.

Based upon the technical assessment presented above, it is now possible to formulate a comprehensive program to facilitate and accelerate research and development in structural control. This program will be multi-disciplinary in nature, clearly needs close technical coordination among projects, and best suited for a team research team efforts involving academic, industrial, and governmental participation. The key research areas shall include, but not limited to, the following:

- New materials
- Performance evaluation, analytical and experimental verification.
- Standardized test procedures and performance criteria.
- Integrated structural and architectural systems design techniques and standards.
- Control algorithm. development, stochastic modeling and analysis, robustness, and reliability, adaptivity, and optimization.
- Sensors, actuators and monitoring system development and optimal deployment.
- Advanced computer technology incorporating artificial intelligence, neural network, fuzzy logic, etc. for on-line targeting, detection, monitoring, diagnosis, and control.
- Intelligent robots and devices.
- Modeling, signal processing, and computer simulations.
- Practical methods and techniques for design, manufacturing, fabrication, and implementation.
- Laboratory and field experiments of systems and devices under actual or simulated dynamic excitation.
- Design and demonstration projects of control systems as applied to new or existing structures.

- Innovations in base isolation, energy dissipation devices, and other passive techniques.

In some of the research areas described above, collaboration among researchers, practicing engineers, and manufacturers is necessary. Results obtained from such a coordinated research effort are expected to generate new knowledge, techniques, and a better understanding of the working principles as well as practical limitations inherent in these control systems so that conditions of applicability and design limits of each class of systems can be established and evaluated.

V. REFERENCES

1. Chong, K.P., Liu, S.C., and Li, J.C., Editors, Intelligent Structures (1990). Elsevier Applied Science, London and New York, 460 pp.

2. Shoueshi, R. and Wormley, D., Editors, Proceedings of Workshop on Research Needs in Intelligent Control Systems (1990). NSF/EPRI.

3. Housner, G.W. and Masri, S., Editors, Proceedings of U.S. National Workshop on Structural Control Research, October 1990, USC Publication No. CE-9013. 251 pp.

4. Roberts, R., and Li, W.P., Editors, Summary Report of Workshop on Sensors and Signal Processing for Structural Control, NSF, February 1991, Lehigh University.

5. Soong, T.T., Active Structural Control, Longmal Scientific & Technical Publications, London, 1990.

6. Yang, J.N., Li, A., and Liu, S.C., "Instantaneous Optimal Control with Acceleration and Velocity Feedback", to appear in Journal of Probabilistic Mechanics.

Stochastic Computational Mechanics in Brittle Fracture and Fatigue

W. K. Liu*, Y. J. Lua** and T. Belytschko*

Department of Mechanical Engineering
Northwestern University
Evanston, Illinois

ABSTRACT

The probabilistic finite element method (PFEM), which is based on a second-order perturbation is developed for non-linear problems. In order to incorporate the probabilistic distribution for the compatibility condition, constitutive law, equilibrium, domain, and boundary conditions into the PFEM, the probabilistic Hu-Washizu variational principal is applied. The first- and second- order moments of the response variable are determined from the statistics of the random parameters and loads. The fusion of the PFEM and the first order reliability analysis provides a means for determining fatigue life and its stochastic nature which results from the randomness in the load, crack length, crack location and orientation, and crack growth parameters. A stochastic damage model is employed to explore the brittle fracture of advanced materials. The model, based on the macrocrack-microcrack interaction, incorporates uncertainties in locations and orientations of microcracks. The statistical nature of the fracture toughness is compared with the Neville function, which gives correct fits to sets of experimental data from advanced materials. Finally, a stochastic approach to the curved fatigue crack growth reliability is outlined.

1. INTRODUCTION

Fracture and fatigue are an important factor in structural design and safety of structures. Due to the uncertainties in the environment (load and temperature) and the randomness of unavoidable defects, a stochastic approach employing a theory of stochastic processes and statistics is required to determine the behavior of the structure and provide a rational reliability analysis.

Stochastic computational mechanics, which is the basis of the reliability and risk analysis of structures and mechanical components and systems, has become a potentially useful tool for addressing problems in reliability engineering. The basic aspects of probabilistic mechanics and its application to the analysis of uncertain structural systems are summarized and recapitulated in the recent book by Liu and Belytschko [1]. The methods and applications of reliability theory have been described in many textbooks such as Ang and Tang [2], Augusti et al [3] and Madsen et al [4]. An important class of problems for which reliability methods are well suited is the prediction of the safety of flawed structures. Based on the failure mechanism in the material, the problems can be classified into two types:

---

* Professor, Department of Mechanical Engineering
** Post-Doctoral Fellow, Department of Mechanical Engineering

N. Bellomo · F. Casciati (Eds.)
Nonlinear Stochastic Mechanics
IUTAM Symposium Turin

1. ductile failure problems, where failure is expected to result from the growth of a critical flaw until it can lead to failure of the structure.

2. brittle failure problems, where the material contains flaws with the random location and orientation. The major question is the reliability of the structure in the presence of these flaws.

The first category of problems are of particular relevance to the safety of aging structures and aircraft. In addition to uncertainties in loads, material parameters and component geometry, the randomness in the sizes and locations of the initial defects and crack growth parameters plays an important role. The PFEM (see Liu et al [5-9]) coupled with the first order reliability analysis has been developed by Besterfield et al [10-11] to address this category of problems. The second category of problems are closely related to the heterogeneity of advanced materials such as ceramics, fiber composites and polycrystalline alloys. Due to the intrinsic microcracks, which are randomly distributed and oriented, a substantial dispersion in the value of fracture toughness has been observed from testing of advanced materials. Both indirect models (Hillerborg et al [12] and Horii and Nemat-Nasser [13]) and direct models (Rose [14] and Kachanov [15] have been proposed to account for the effects of the near tip microcrack zone. A stochastic damage model based on the macrocrack-microcrack interaction has been developed by Lua et al [16-17] to quantify the inherent statistical distribution of the fracture toughness of advanced materials. In this paper, attention is devoted to the stochastic approaches for both categories of problems.

In Section 2, the PFEM, which combines probability theory and statistics with the finite element method, is outlined for nonlinear problems. The accuracy and efficiency of the method is illustrated via the problem of a ten-bar truss with material and geometrical nonlinearities. The fusion of the PFEM and the first order reliability analysis is presented in Section 3 for determining the probability of fatigue failure under cyclic loading. The performance of the method is demonstrated by a classical Mode I fatigue crack growth problem. In Section 4, a stochastic damage model is proposed to quantify the inherent statistical distribution of the fracture toughness of an advanced material. The mixed boundary integral equation formulation and its application to the curved fatigue crack growth reliability is discussed in Section 5.

2. PFEM FOR NON-LINEAR PROBLEMS

The formulation of PFEM for non-linear problems based on the Hu-Washizu Variational Principle is outlined. The Probabilistic Hu-Washizu Variational Principle (PHWVP) allows incorporation of probabilistic distribution for the compatibility condition, constitutive law, equilibrium, domain, and boundary conditions into the PFEM. An additional advantage of using the Hu-Washizu variational principle involves the elimination of the locking phenomena (Belytschko and Bachrach [18]) and suppression of hourglass modes (Belytschko et al [19]). Using matrix notation, the Hu-Washizu Variational Principle (HWVP) for nonlinear problems is given by

$$\int_{\Omega} \delta\varepsilon^{T} (\psi - \sigma)\, d\Omega \; - \int_{\Omega} \delta\sigma^{T} (\varepsilon - \nabla \mathbf{u})\, d\Omega$$

$$+ \int_{\Omega} \delta(\nabla \mathbf{u})^{T} \sigma\, d\Omega \; - \int_{\Omega} \delta \mathbf{u}^{T} \mathbf{F}\, d\Omega \; - \int_{\partial\Omega_h} \delta \mathbf{u}^{T} \mathbf{h}\, d\Gamma = 0 \tag{1}$$

where ε, σ, and $\mathbf{u}$ are independent random field variables representing the nonsymmetric measure of the strain, first Piola-Kirchhoff stress, and displacement, respectively. In Eq. (1): Ω, $\partial\Omega_h$, $\mathbf{F}$, $\mathbf{h}$, and $\nabla \mathbf{u}$ represent the domain, traction boundary, body force vector, prescribed traction vector, and the nonsymmetric part of the displacement gradient, respectively; δ represents a virtual quantity. The surface and volume integrals in Eq. (1) can be expressed via parametric representation:

$$d\Gamma = J_s\, dA \quad \text{and} \quad d\Omega = J_v\, dR \tag{2a, b}$$

respectively, where J_s and J_v represent the surface and volume Jacobians, respectively; and R and A represent the reference domain and boundary, respectively. Random domains and boundaries are incorporated into the formulation through randomness in the gradient operator and Jacobians. The application of second-order perturbation techniques in the HWVP involves the expansion of all random functions about the mean value of the random field $b(\mathbf{x})$ denoted by $\bar{b}(\mathbf{x})$ and retaining only up to second-order terms, i.e., for a given small parameter [ξ = the scale of randomness in $b(\mathbf{x})$], the random function ψ is expanded about $\bar{b}$ at a given point $\mathbf{x}$ in the reference domain as follows:

$$\psi \approx \psi^{0} + \xi\,(\psi' + C^{0}\varepsilon') + \xi^{2}\,(\psi'' + C'\varepsilon' + C^{0}\varepsilon'') \tag{3}$$

where the superscripts nought, prime, and double prime represent the random functions evaluated at $\bar{b}$, the first-order variation due to variations in $\bar{b}$, and the second-order variation, respectively. The first elasticity tensor, $\mathbf{C}$ in Eq. (3) is given by

$$C_{ijkm} = \frac{\partial^2 W}{\partial G_{ij}\, \partial G_{km}} \tag{4}$$

where W is the strain energy density function; and $\mathbf{G}$ is the deformation gradient. After substituting the second order perturbations of all these random functions (ψ, ε, σ, $\mathbf{F}$, $\mathbf{h}$, J_S, J_V, $\nabla\mathbf{u}$, $\delta(\nabla\mathbf{u})^{T}$) into Eq. (1), the zeroth-, first-, and second- order nonlinear PHWVP can be obtained as shown by Liu et al [9].

In addition to the usual finite element approximation of the displacement field, the random fields are also discretized with q shape functions (see Liu et al [9]). Substituting the given approximation of all random functions into the zero-order, first-order, and second-order PHWVPs and using the three stationary conditions (strain-displacement, stress-strain, and equilibrium), the zeroth, first and second order equations can be obtained. After determining the zeroth-, first-, and second- order solutions, the expectations and autocovariance matrices for the displacements, strains, and stresses can be obtained (see Liu et al [9]).

The applicability and effectiveness of the PFEM for nonlinear problems is demonstrated by the following ten-bar truss problem (see Liu et al [5]). The problem statement is depicted in Fig. 1.

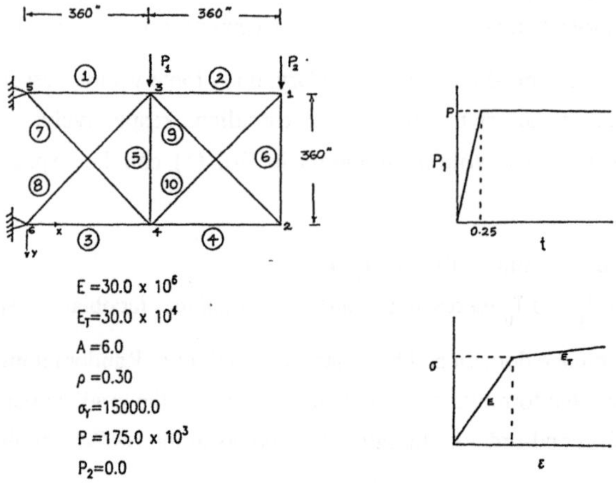

Fig.1. Problem Statement : Ten-Bar Nonlinear Structure.

The load time function shown in Fig. 1 is applied at node 3. Under this particular load-time history, four of the ten bars (elements 1, 3, 7, and 8) will yield. The yield stresses of these four elements are assumed as normal random variables with the coefficient of variation δ=0.05. The remaining six elements, which do not yield, are considered as deterministic variables. The computed results are compared with those obtained using (1) Monte Carlo Simulation (MCS) and

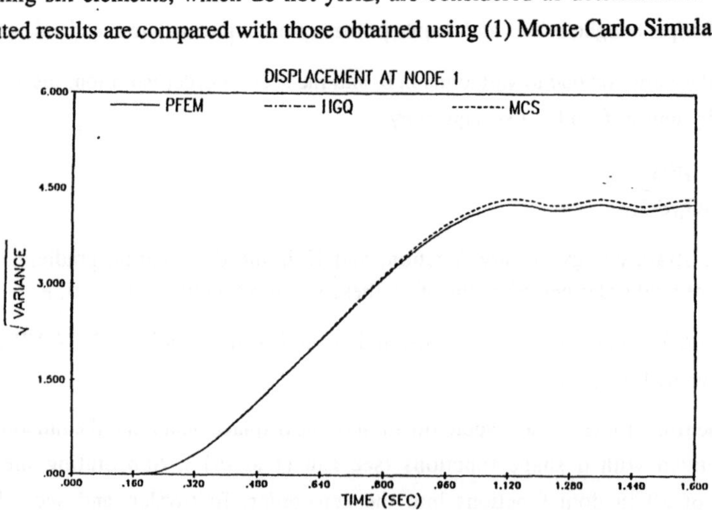

Fig. 2. Comparison of the Variance of the y-Displacement of Node 1.

(2) Hermite-Gauss Quadrature (HGQ) schemes. The variance of the y-displacement of node 1 is depicted in Fig. 2 for all the three numerical methods-PFEM, HGQ and MCS. The maximum coefficient of variation of the displacement at node 1 is found to be 0.13. The variance of stress in member 1 obtained from PFEM, HGQ and MCS is plotted in Fig. 3. The maximum coefficient

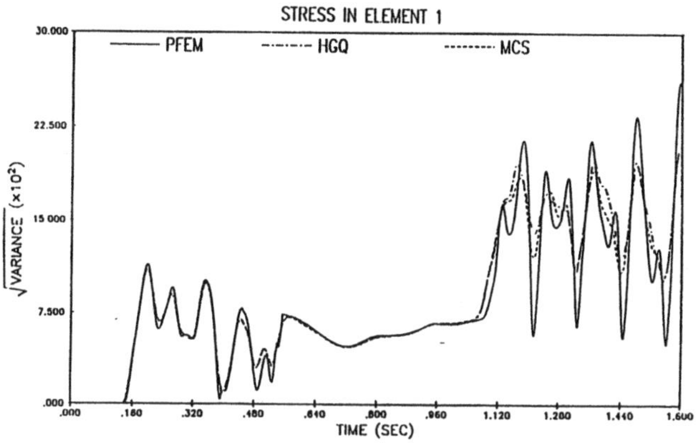

Fig. 3. Comparison of the Variance of Stress in Member 1.

of variation of the stress in member 1 is 0.11. From Figs. 2 and 3, the results obtained from these three methods (PFEM, HGQ and MCS) compare quite well.

3. PFEM FOR FATIGUE CRACK RELIABILITY ANALYSIS

In this section, the application of PFEM to the reliability analysis under fatigue crack growth is described. The uncertainties -in load, material properties, component geometry and crack geometry- are represented by a q-dimensional vector of random variables denoted by $\mathbf{b} = [b_1, \ldots, b_q]^T$. The performance function g(**b**) is introduced to describe the following three states of a system:

$$g(\mathbf{b}) = 0 \text{ (limit state)} \qquad g(\mathbf{b}) < 0 \text{ (failure state)} \qquad g(\mathbf{b}) > 0 \text{ (safe state)} \tag{5a, b, c}$$

Using the Rosenblatt transformation (Rosenblatt [20]), the general correlated random variables **b** can be transformed into the uncorrelated standard normal vector **r**. The reliability index β is defined by the minimum distance from the origin to the limit state surface in the **r**- space, i.e.

$$\beta = \min (\mathbf{r}^T\mathbf{r})^{1/2} \text{ , subject to } g(\mathbf{r}) = 0 \tag{6}$$

For the present fatigue crack growth problem, the performance function is given by

$$g = T - T_s \tag{7}$$

where T_s is the service life of the component, and T is given by the Paris law (Paris and Erdogan [21]), i.e.,

$$T = \int_{a_1}^{a_f} \frac{da}{D\,(\Delta k_{eq})^n} \qquad (8)$$

where a_1 and a_f are the initial and final crack lengths, respectively, da is the random crack path; D and n are primarily material parameters but can also depend on the loading and environmental effects; and $\Delta k_{eq}(a)$ is the range of equivalent Mode I stress intensity factors, i.e.,

$$\Delta k_{eq} = k_{eq}^{max} - k_{eq}^{min} \qquad (9)$$

where k_{eq}^{min} and k_{eq}^{max} are the minimum and maximum equivalent Mode I stress intensity factors, respectively, associated with the minimum and maximum cyclic applied stresses, respectively. If the minimum equivalent Mode I stress intensity factor is assumed to be zero, then

$$\Delta k_{eq} = k_{eq}^{max} \equiv k_{eq} \qquad (10)$$

The calculation of the reliability index β by the first-order probability theory is performed by solving a constrained optimization problem. The method of Lagrange multipliers has been employed by Besterfield et al [11] to incorporate all constraints such as equation of equilibrium, limit state condition, minimum distance condition and crack direction law in the formulation. The PFEM described in Sec. 2 can be applied to perform the sensitivity analysis, which is required in the determination of the most probable point on the failure surface. The final optimization problem for fatigue crack growth along with solution procedures has been given by Besterfield et al [11]. In order to demonstrate the performance of the method for reliability analysis against failure due to fatigue crack growth, a classical Mode I fatigue problem is presented (see Besterfield et al [11]). Figure 4 shows a finite rectangular plate with a single edge crack of length

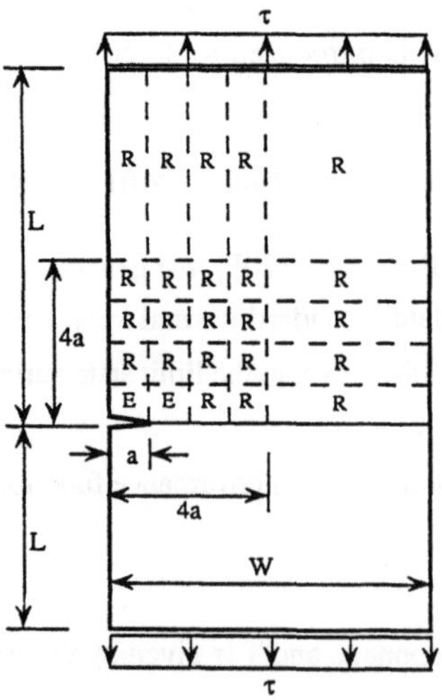

Fig. 4. Model for Single Edge-Cracked Beam with an Applied Load.

a subjected to a distributed load. The problem parameters along with their second-moment statistics are listed in Table 1. Due to symmetry, two enriched 9-node elements and twenty-three

Parameter	Mean	Standard Deviation	Percent
Length (L)	10.0 in	0.0	0.0
Width (W)	4.0 in	0.0	0.0
Thickness (t)	1.0 in	0.0	0.0
Young's Modulus (E)	30,000.0 ksi	0.0	0.0
Poisson's Ratio (ν)	0.3	0.0	0.0
Applied Stress (τ)	12.0 ksi	3.0 ksi	25.0
Initial Crack Length (a_1)	0.01 in	0.01 in	100.0
Final crack Length (a_f)	0.1 in	0.01 in	10.0
Fatigue Parameter (D)	1.0×10^{-10}	3.0×10^{-11}	30.0
Fatigue Parameter (n)	3.25	0.08	2.5

Table 1. Problem Constants: Single Edge-Cracked Plate with a Distributed Load.

regular 9-node elements are depicted on the upper half of the plate. In Fig. 5, the reliability index obtained by PFEM is plotted versus the service life for various types of uncertainties. As shown in Fig. 5, when all five of the parameters are treated as random, the resulting effect is much

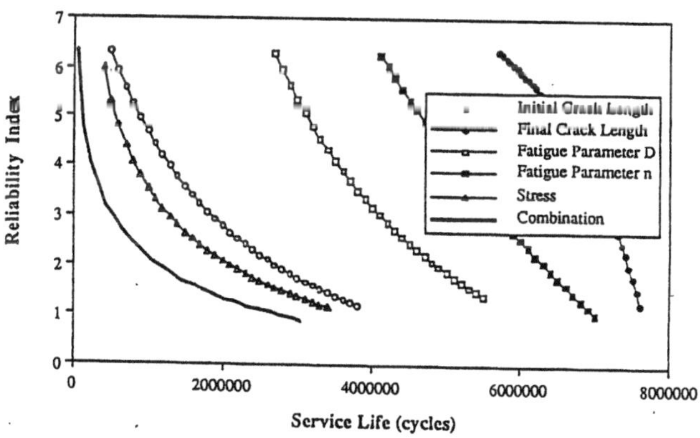

Fig. 5. Reliability Index for the FEM Solution Comparing the Effects of Uncertainty in the Individual Variables and Their Combined Effect.

greater than any one individual effect.

4. BRITTLE FRACTURE OF AN ADVANCED MATERIAL

Most of advanced materials such as ceramics, composite and polycrystalline alloys are multi-phase solids. Due to the heterogeneity in these multi-phase materials, a substantial dispersion in the value of fracture toughness has been observed from experimental tests. A large scatter in observed values of material properties is mainly attributed to intrinsic microcracks, which are randomly distributed as a result of the applied load and residual stress. In order to quantify the inherent statistical distribution of the fracture toughness, a stochastic damage model was proposed recently by Lua et al [16-17]. The effects of uncertainties in locations, orientations and numbers of microcracks at a macro-tip have been studied quantatively by Lua et al [17].

The progressive failure process of an advanced material under external loading is characterized by the following phases: 1) pre-existing random microcracks along the weakest matrix-inclusion interface; 2) nucleation and propagation of microcracks due to the increase of the load; 3) formation of an extensive array of microcrack surrounding a macro-tip resulting in damage saturation; 4) unstable growth of the main crack through coalescence of near tip microcracks. Since the fracture toughness determined at the onset of macrocrack extension is of particular interest from the point of view of safety and reliability, it is imperative to develop a model to quantify its inherent statistical distribution. As the onset of macrocrack growth occurs before an appreciable wake zone is developed, a frontal microcrack zone confined to a region ahead of the main crack is considered here. A central macrocrack of length $2a_0$ with a near tip frontal damage zone situated at (RST,0) (or (-RST,0)) in an infinite body subjected to remote loading is depicted in Fig. 6. It can be expected that the concentration of microcracks in a near tip zone is extremely

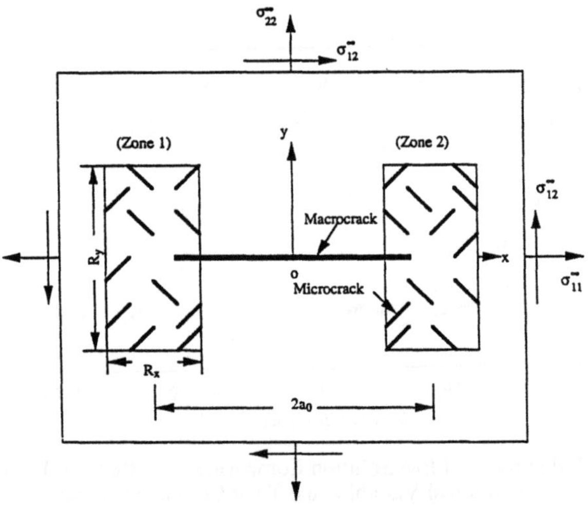

Fig. 6. A Damage Saturation Model for a Brittle Multi-Phase Material.

high at the onset of macrocrack extension. Owing to large uncertainties in the number of microcracks and the geometrical parameters of microcracks, the traction boundary integral equation method in conjunction with computer simulation has been developed to perform the analysis (see Lua et al [16-17]).

The microcrack density parameter W, which has been used extensively as a damage parameter in continuum damage mechanics can be defined as

$$W = \frac{A_c}{A_d} \tag{11}$$

where A_c ($=4M\,a^2$) is the total area of the clouds of microcracks in a damage zone and A_d ($=R_x * R_y$) is the area of the frontal damage zone (see Fig. 6). The mixed- mode crack growth criterion based on the minimum strain energy density proposed by Sih [22] is given by

$$\frac{dS}{d\theta} = 0 \;\; ; \;\; \frac{d^2S}{d^2\theta} > 0 \quad \text{at } \theta = \theta^0 \, ; \; S(\theta^0) = S_{cr} \tag{12}$$

where S_{cr} is the critical value of the intensity of strain energy density. Since the value of the applied intensity, S^{app}, at which $S=S_{cr}$ is the measured fracture toughness (S_c^{app}), the relation between S_c^{app} and S is given by

$$S_c^{app} = \frac{S_0^{app}}{S} S_{cr} \tag{13}$$

where S_{cr} is the base toughness and S_0^{app} is the intensity of the strain energy density associated with the applied load t_0. By applying the reciprocal relation between the measured fracture toughness S_c^{app} and the local intensity of strain energy density S (13), the inherent distribution of the measured fracture toughness (S_c^{app}) can be predicted from the statistical distribution of S, which is the output of the present computer-simulation model.

The Neville statistical function has been used successfully to fit statistically distributed fracture toughness data of an advanced material (see Neville [23]). Fig. 7 shows both the Neville and

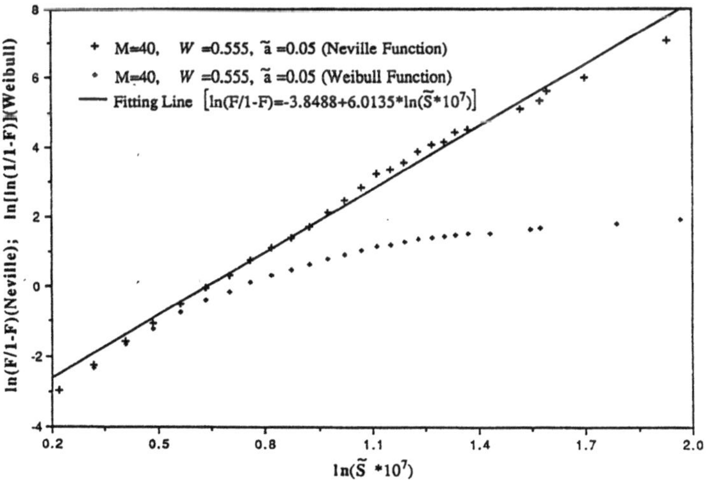

Fig. 7. Neville and Weibull Plots of the Intensity of Strain Energy Density $\widetilde{S}$ for the High Microcrack Concentration W=0.555

two-parameter Weibull (see Lewis [24]) probability plots for the set of data $\widetilde{S}$ associated with the high microcrack concentration W = 0.5555 on probability papers. The Neville function

$(\ln\left(\frac{F}{1-F}\right) \sim \ln(\widetilde{S}*10^7))$ provides a very good fits for $W = 0.5555$ indicated by the coincidence of points and the straight line over the whole range of $\widetilde{S}$ (see Fig. 7). On the contrary, the Weibull plot $(\ln\left[\ln\left(\frac{1}{1-F}\right)\right] \sim \ln(\widetilde{S}*10^7))$ shown in Fig. 7 curves from data with a distinct curvature. As indicated by Neville [25], the Weibull distribution does not provide a good fit for heterogeneous materials due to the assumption of equal probability of failure in the material. This assumption is rarely true for the material with cracks because the presence of near tip singular fields cannot provide equally likely failure. Due to the existence of the reciprocal relation between S and $\widetilde{S}_c^{app}$, it can be concluded directly that the statistical nature of the measured fracture toughness $\widetilde{S}_c^{app}$ is also described by the analytical model based on the Neville distribution function.

5. FATIGUE FAILURE OF AN ADVANCED MATERIAL

Two experimental observations have been obtained from fatigue tests of an advanced material, namely, the non-unique crack path and a substantial scatter in the value of fatigue life. The non-unique crack trajectories for multiple experiments under identical loading conditions is mainly attributed to the existence of random microdefects, such as voids, inclusions, and microcracks. Due to the uncertainty in microdefect, crack geometry, external load and fatigue parameters, the fatigue life of a brittle material can no longer be treated as a single-valued property.
The existence of random microdefects perturbs the near tip stress field of a crack resulting in a random curvilinear crack path. For curvilinear crack growth problems, near tip singular elements coupled with continuous remeshings are required in the finite element methods. An alternative approach based on the boundary integral equation formulation can be employed to avoid the remeshing process. In order to study the effects of various of microdefects such as voids, rigid inclusions and transformation inclusions on the crack path and fatigue life, the mixed boundary integral equation has been formulated by Lua et al [26] for a multi-connected region.

The applicability and versatility of this mixed boundary integral equation has been demonstrated in [26] through the study of elastic interaction between a fatigue crack and a microdefect, such as a void, a rigid inclusion or a transformation inclusion. The effects of uncertainty in microdefect, crack geometry, external load and fatigue parameters on the fatigue life have been studied by Lua et al [27-28] using computer simulations and structural reliability methods. As shown in [27], the statistical distribution of fatigue life of an advanced material agrees well with the Weibull distribution. The response surface methodology developed by Faravelli [29-30] along with an iteration scheme is employed in [28] to perform the reliability and sensitivity analysis.

6. CONCLUDING REMARKS

Stochastic computational mechanics based on Finite Element Method (FEM) or Boundary Integral Equation Method (BIEM) has been shown to be a promising and practical approach to problems with fatigue and fracture uncertainties. For problems with small uncertainty, the PFEM, which is based on a second-order perturbation, is more efficient than simulation techniques. The ability of PFEM to determine the probability of fatigue failure due to uncertainties in the component geometry, applied loads, material properties and crack geometry has been shown through a numerical example. Associated studies in which PFEM has been compared to Monte Carlo methods show that even in these highly nonlinear problems, reliability methods are quite accurate. For the problem of the brittle fracture of an advanced material, a stochastic damage model has been proposed to quantify its statistical distribution of fracture toughness. The mixed boundary integral equation coupled with the response surface approach has been employed to the reliability and sensitivity analysis of a curvilinear fatigue crack growth problem. The BIEs in conjunction with the computer simulation has been used to justify the reliability methods.

ACKNOWLEDGEMENTS

The support of NASA Lewis Grant No. NAG 3-822 for this research and the encouragement of Dr. Christos Chamis are gratefully acknowledged.

REFERENCES

[1] Liu, W.K. and T. Belytschko (1989), *Computational Mechanics of Probabilistic and Reliability Analysis*, Elme Press International.

[2] Ang, A.H.S. and Tang, W.H. (1984), *Probability Concepts in Engineering Planning and Design: Volume II - Decision, Risk, and Reliability*, John Wiley & Sons, New York .

[3] Augusti, G., Baratta, A. and Casciati, F. (1984), *Probabilistic Methods in Structural Engineering*, Chapman and Hall, London .

[4] Madsen, H.O., Krenk, S. and Lind, N.C. (1986), *Methods of Structural Safety*, Prentice-Hall, Inc., Englewood Cliffs, N.J.

[5] Liu, W.K., Belytschko, T. and Mani, A. (1986), " Probabilistic Finite Elements for Nonlinear Structural Dynamics," *Comput. Methods Appl. Mech. Eng.*, Vol. 56.

[6] Liu, W.K., Belytschko, T. and Mani, A. (1986), " Random Field Finite Elements," *Int. J. Numer. Methods Eng.*, Vol. 23, pp. 1831-1845.

[7] Liu, W.K., Belytschko, T. and Mani (1987), " Applications of Probabilistic Finite Element Methods in Elastic/Plastic Dynamics," *J. Eng. Ind.*, ASME, Vol. 109, pp.2-8.

[8] Liu, W. K., Mani, A. and Belytschko, T. (1988). " Finite Element Methods in Probabilistic Mechanics," *Probabilistic Engng Mech.*, Vol. 2, 201-213.

[9] Liu, W.K., Besterfield, G.H. and Belytschko, T. (1988), " Variational Approach to Probabilistic finite Elements," *J. Eng. Mech.* ASCE, Vol. 114, pp. 2115-2133.

[10] Besterfield, G.H., Liu, W.K., Lawrence, M.A., and Belytschko, T.B. (1990), " Brittle Fracture Reliability by Probabilistic Finite Elements," *J. Eng. Mech.*, ASCE, Vol. 116, pp. 642-659.

[11] Besterfield, G.H., Liu, W.K., Lawrence, M. and Belytschko, T. (1991), " Fatigue Crack Growth Reliability by Probabilistic Finite Elements," *Comput. Methods Appl. Mech. Eng.*, Vol. 86, pp.297-320.

[12] Hillerborg, A., Modeer, M. and Peterson, P.E. (1976), " Analysis of Crack Formation and Crack Growth in Concrete by Means of Fracture Mechanics and Finite Elements," *Cement Coner. Res.*, Vol. 6, pp. 773-782.

[13] Horri, H. and Nemat-Nasser, S. (1983), " Overall Moduli of Solids with Microcracks: Load-Induced Anisotropy, " *J. Mech. Phys. Solids*, Vol. 31, 155-171.

[14] Rose, L.R.F. (1986), " Microcrack Interaction with a Main Crack," *Int. J. Fracture Mech.*, Vol. 31, pp. 233-242.

[15] Kachanov, M. (1986), " On Crack-Microcrack Interactions," *Int. J. Fracture Mech.*, Vol. 30, pp. R65-R72.

[16] Lua, Y.J, Liu, W.K. and Belytschko, T. (1990), " A Stochastic Damage Model for the Rupture Prediction of a Multi-Phase Solid: Part I: Parametric Studies," accepted for publication in *Int. J. Fracture Mech.*

[17] Lua, Y.J, Liu, W.K. and Belytschko, T. (1990), " A Stochastic Damage Model for the Rupture Prediction of a Multi-Phase Solid: Part II: Statistical Approach," accepted for publication in to *Int. J. Fracture Mech.*

[18] Belytschko, T. and Bachrach, W.E. (1986), " Simple Quadrilateral with High-Course Mesh Accuracy," *Comput. Methods Appl. Mech. Eng.*, Vol. 54, pp.279-301.

[19] Belytschko, T. et al. (1984), " Hourglass Control in Linear and Nonlinear Problems," *Comput. Methods Appl. Mech. Eng.*, Vol. 43, pp. 251-276.

[20] Rosenblatt, M. (1952), " Remarks on a Multivariate Transformation," *Annals of Mathematical Statistics*, Vol. 23, pp. 470-472.

[21] Paris, P.C. and Erdogan, F. (1963), " A Critical Analysis of Crack Propagation Laws," *J. Basic Eng.*, ASME, Vol. 85, pp. 528-534.

[22] Sih, G.C. (1974), " Strain Energy Density Factor Applied to Mixed Mode Crack Problems," *Int. J. Fracture Mech.*, Vol. 10, pp.305-322.

[23] Neville, D.J. (1990), " Application of a New Statistical Function for Fracture Toughness to Failures at Microcracks in Brittle Materials," *Int. J. Fracture Mech.*, Vol. 44, pp. 79-96.

[24] Lewis, E.E. (1987), *Introduction to Reliability Engineering*,John Wiley & Sons,New York.

[25] Neville, D.J. (1987), " A New Statistical Distribution Function for Fracture Tiughness," *Proc. R. Soc. Lond.*, A, Vol. 410, pp. 421-442.

[26] Lua, Y. J., Liu, W. K. and Belytschko, T. (1991), " Mixed Boundary Integral Equations Approach to the Study of the Effect of Microdefects on the Fatigue Life and Crack Trajectory," in preparation.

[27] Lua, Y. J., Liu, W. K. and Belytschko, T. (1991), " Application of Mixed Boundary Integral Equations to the Statistic Analysis of Curvilinear Fatigue Crack Growth," in preparation.

[28] Lua, Y. J., Liu, W. K. and Belytschko, T. (1991), " Stochastic Boundary Element Method to the Reliability Analysis of Curvilinear Fatigue Crack Growth," in preparation.

[29] Faravelli, L. (1989), " Response Surface Approach for Reliability Analysis," *J. of Eng. Mech.* ASCE, Vol. 115, pp. 2763-2781.

[30] Faravelli, L. and Bigi, D. (1990), " Stochastic Finite Element for Crash Problems," *Structural Safety*, Vol. 8, pp. 113-130.

Flow of a Thin Liquid Film over a Rough Rotating Disk

F. Ma

Associate Professor of Applied Mechanics
Department of Mechanical Engineering
University of California
Berkeley, California 94720, USA

ABSTRACT

The depletion of thin liquid films due to the combined effect of centrifugation, surface roughness, and air-shear is addressed. A stochastic model of surface roughness is devised, and the equations describing the flow of a thin viscous film over a rough rotating disk are solved by Monte Carlo techniques. Depletion histories of the thin film are given for several cases involving both deterministic and random surface asperities. It is found that surface roughness of a rotating disk markedly enhances the retention of the thin liquid film. The thinner the liquid layer, the more significant is the effect of surface roughness.

1. INTRODUCTION

Magnetic disk systems are commonly used for information storage and retrieval. In rigid disk drives, the slider carrying the read-write circuitry flies within a fraction of a micron from the disk over an air bearing during operation. When the unit is first started or turned off, however, the slider lacks the air bearing and slides in contact with the disk. In order to protect the disk as well as the slider from excessive wear, a thin layer of lubricant is applied to the surface of the disk during the manufacturing process. This lubricant exists in the form of a thin film of viscous liquid, which has 10-30 nm thickness for a particulate disk and 0.5-4.0 nm for a thin-film disk. The lubricant tends to drift outward because of the rotation of the disk drive, which may spin at thousands of revolutions per minute. Since it is important that a sufficient amount of lubricant should always remain on the disk, the study of the detailed mechanism that holds the lubricant onto the surface becomes an important issue. This is the motivation of the present article.

N. Bellomo · F. Casciati (Eds.)
Nonlinear Stochastic Mechanics
IUTAM Symposium Turin

The flow of a thin viscous layer over a rotating disk has been considered in the literature [1-8], with particular relevance to spin coating. Spin coating has been extensively used to produce thin liquid layers; common examples are the phosphor coatings on television screens and the photoresist films on silicon wafers. In a well-known investigation, Emslie et al. [1] developed a model of Newtonian flow on a horizontal surface to obtain the distribution of the liquid as a function of time. Their study involves balancing the viscous forces with inertial forces, and the influence of air-shear and surface roughness of the rotating disk has been neglected. Middleman [2] and Rehg and Higgins [6] recently studied the effect of air-shear on spin coating. They observed significant increase in depletion due to the induced air-flow on the free liquid surface. McConnell [3] has conducted experiments to demonstrate the validity of the classical model of Emslie et al. when air-shear can be neglected. For conditions in which air circulation takes place, McConnell has obtained data that would qualitatively support the observations of Middleman. Thus, the models developed by Emslie et al. and Middleman may be employed in applications with reasonable confidence. For the case of very thin liquid layers, however, it is known that the theoretical predictions of both models have appreciably exaggerated the thinning of the liquid film at large values of time. Based upon extensive experimental data, it is thought that the discrepancy arises mainly from the omission of the surface roughness in the models [9]. Surface asperities of the disk may play a dominant role in retaining the liquid layer on the disk, but an attempt to evaluate the effect of surface roughness on the retention of a liquid layer has not been reported in the open literature.

This article sets out to explore the effect of surface roughness on the retention of thin liquid films in the presence of centrifugation and air-shear. Surface roughness is represented by either deterministic curves or by spatial stochastic processes. Monte Carlo simulations are performed whenever statistical models of surface asperities are used. Although a limited set of data from simulations is presented, extensive numerical calculations have been performed to support the qualitative results reported herein. This paper generalizes an earlier investigation conducted by the author [7,8], in which the effect of air-shear is neglected. Incorporation of air-shear requires substantial restructuring of the theoretical and numerical algorithms; one fundamental objective is to highlight the interplay of surface roughness, centrifugation, and air-shear on the depletion of a liquid film. The organization of this paper is as follows. In

Section 2, the equations governing the flow of a thin layer over a rough rotating surface are derived. Stochastic modelling of the surface roughness is briefly described in Section 3. The important results on liquid retention are highlighted in Section 4, in which a comparison is made with data obtained by Middleman [2] and Yanagisawa [5]. A summary of the findings is given in Section 5. The terms, thin liquid layer and lubricant, will be used interchangeably throughout the paper.

2. MATHEMATICAL MODEL OF VISCOUS FLOW

Consider the flow of a thin viscous layer over a rough rotating disk, for which the flow is accelerated by the induced air-shear at the liquid-air interface. It is assumed that (i) the plane of the rotating disk is horizontal, (ii) the liquid layer is radially symmetric, and so thin that gravitation has negligible effect in distributing the layer, (iii) the flow is Newtonian and nonvolatile, with a radial velocity so small that Coriolis forces may be ignored, and (iv) the radial velocity of the induced air-flow is substantially greater than that of the liquid layer. These assumptions are valid in most engineering applications [1-8]. A system of cylindrical coordinates (r, θ, z) which rotates with the disk at an angular velocity ω about the z-axis is used. The situation is clarified in Fig. 1, in which $z = 0$ is a reference elevation which may conveniently

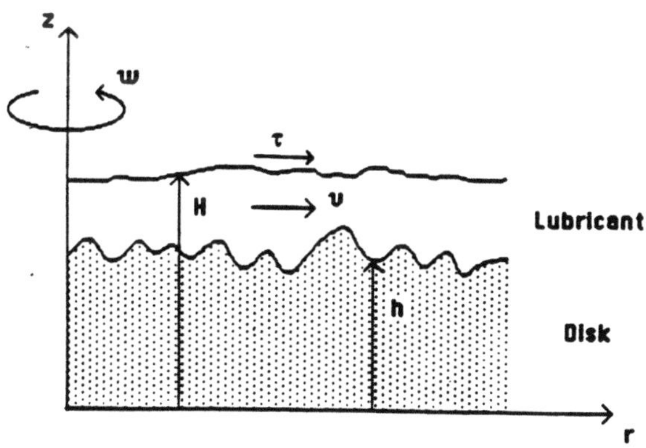

Fig. 1 Viscous flow of a thin layer over a rough rotating disk.

be taken as the center of the thickness of the disk, $v(r, z, t)$ is the radial velocity of the liquid, $h(r)$ is the height of the rough disk surface, and $H(r, t)$ is the height of the free surface of the viscous liquid. When the first three assumptions are incorporated in the Navier-Stokes equation, we obtain the quasi-steady equation of flow, in the form

$$-\eta\frac{\partial^2 v}{\partial z^2} = \rho\omega^2 r - \frac{\partial}{\partial r}\left[\frac{\gamma}{R}\right], \tag{1}$$

where η is the viscosity, γ is the coefficient of the surface tension, and ρ is the density of the liquid. In the above equation, R is the curvature of the free liquid surface in the radial direction given by

$$\frac{1}{R} = -\left[\frac{\partial^2 H}{\partial r^2} + \frac{1}{r}\frac{\partial H}{\partial r}\left[1 + \left[\frac{\partial H}{\partial r}\right]^2\right]\right]\left[1 + \left[\frac{\partial H}{\partial r}\right]^2\right]^{-\frac{3}{2}}. \tag{2}$$

Since there is no flow on the disk surface, the boundary condition at the liquid-solid interface is

$$v(r, h, t) = 0. \tag{3}$$

As the disk spins, ambient air is drawn in towards the disk and acquires rotational motion in a thin layer adjacent to the liquid surface. This circulating air also moves outwards in the radial direction due to centrifugation, thereby imposing a shearing force at the liquid-air interface. The induced air-flow enhances the depletion of the lubricant. Let τ be the shear stress directed radially on the free surface of the lubricant. The boundary condition at the liquid-air interface can be written as

$$\eta\frac{\partial v}{\partial z}(r, H, t) = \tau. \tag{4}$$

If assumption (iv) is satisfied, Middleman [2] points out that a formula by Cochran [10] can be exploited to give the approximation

$$\tau = \left[\frac{1}{2}\omega^{\frac{3}{2}}\nu_{air}^{-\frac{1}{2}}\eta_{air}\right]r, \tag{5}$$

where ν_{air} is the kinematic viscosity of the air, and $\eta_{air} = \rho_{air}\nu_{air}$ is the viscosity of the air. Defining the radial flow per unit length by

$$q = \int_h^H v dz, \tag{6}$$

the continuity equation can be written as

$$r\frac{\partial H}{\partial t} = -\frac{\partial}{\partial r}(rq). \tag{7}$$

Equations (1) and (7), together with the appropriate initial and boundary conditions, provide a description of the flow of a thin viscous film over a rough rotating disk. They represent a generalization of the mathematical models employed by Emslie et al. [1] and Middleman [2]. Any rough disk surface can be considered by prescribing the function $h(r)$, and the remaining task is to determine the free liquid surface $H(r,t)$. The volume of lubricant retained on the disk at instant $\overline{T}$, $Q(\overline{T})$, can then be obtained by integration.

A Crank-Nicolson finite difference scheme is used to solve the equations of viscous flow. This numerical scheme has been found to be quite stable, with global error of the order A more detailed description of the numerical method is provided in an earlier paper by Ma and Hwang [8]. On the average, each calculation to $\overline{T}$=10 with a prescribed disk surface consumes about 110s in CPU time with the IBM 3090 computer.

3. SURFACE ROUGHNESS AND STOCHASTICITY

It has been realized that models of surface asperities could be so complex that it would be necessary to draw upon statistical techniques for an adequate description. Thus Greenwood and Williamson [11] developed a model of a rough but nominally flat surface, consisting of a series of spherical caps each with the same radius of curvature, and having a probabilistic distribution of heights. Later, Whitehouse and Archard [12] considered the representation of a surface profile as a random signal, the waveform being characterized by the autocorrelation and probability density of the heights. Although these earlier works are concerned with surface asperities of a much larger scale than those existing on magnetic disk drives, it appears that the conceptualization expounded therein may be readily modified and exploited for the current study.

To develop a model of the surface roughness for a magnetic storage disk, the rough surface is treated as a random field with an exponential-type correlation structure and a log-normal distribution of heights [8]. Now, it has been observed by experiments that there are isolated pores scattered on the surface of a particulate magnetic disk, the dimensions of these pores being one order of magnitude larger than the minute log-normal fluctuations of the heights. It means that two length scales would be required to characterize the surface asperities. To this end a sequence of random locations on the disk surface is generated, such that the distances between the locations follow an exponential density function. Around each random location a pore is constructed, the shape of which is determined by controlled Brownian motion. When the small-scale and large-scale surface asperities are superimposed upon each other, the resulting surface possesses two length scales in the surface asperities, and such a model appears to provide an adequate characterization of the surface topography in three dimensions with radial symmetry. A more detailed description of the statistical model of surface asperities is provided in a recent paper by Ma and Hwang [8]. Robustness of the stochastic description of surface roughness has been examined by Kim et al. [13]. Figure 2 displays one computer-generated realization of the random disk surface, which closely resembles the surface roughness shown by micrographs of one class of particulate disks. For a thin-film magnetic disk, a single length scale is adequate for the description of surface asperities. Since the qualitative results for both types of disks are identical, we shall dwell upon the case of particulate disks only.

4. RESULTS AND DISCUSSION

The results of several simulations will be reported in this section, primarily to highlight the interplay of surface roughness, centrifugation, and air-shear on the depletion of lubricant. To make the presentation more focused, the initial height of the lubricant has been assumed constant. One reason for this choice is obvious: uniform distribution of the initial height of lubricant is frequently used in the study of spin coating [2,5]. Most of the literature in spin coating addresses the change in the height of the lubricant as a function of time. When the surface of the rotating disk is rough, however, it is more appropriate to examine the change in the volume of the lubricant as time progresses. If Q_o is the total volume of lubricant initially on the rough disk, then the relative volume retained at time $\bar{T}$ is $Q(\bar{T})/Q_o$, where $Q(\bar{T})$ is the volume of lubricant that remains on the rotating disk at $\bar{T}$. We shall use the mean of the rela-

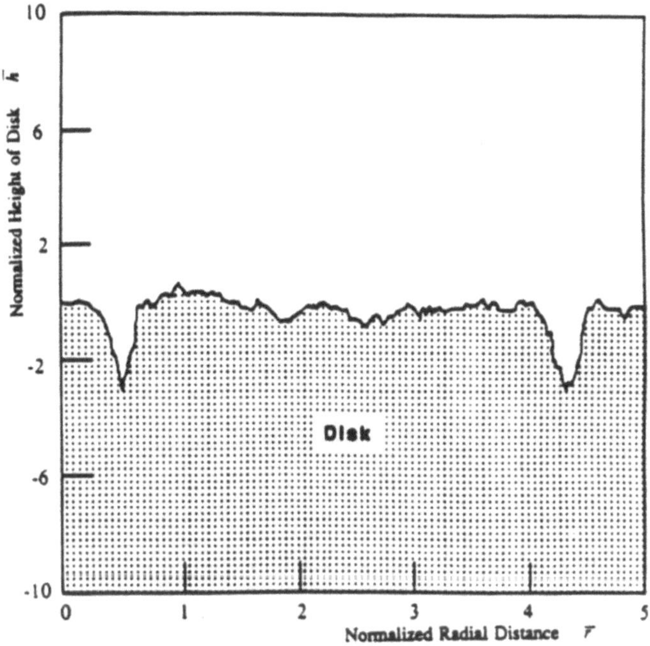

Fig. 2 Typical surface roughness of a magnetic storage disk.

tive volume retained at $\bar{T}$ whenever surface roughness is modeled as a spatial stochastic process.

A typical simulation of depletion is exhibited in Fig. 3. The dashed line represents the classical retention curve derivable from the solution of Emslie et al. [1], for which a smooth disk is assumed and air-shear is neglected. Suppose a randomly rough disk surface, of which the surface asperities are fairly slight, is used. The retention curve of the same lubricant flowing on such a rough surface in the absence of air-shear is given by the dotted line. This dotted line follows the classical retention curve quite closely at an initial stage, which is not surprising in the light of the experimental observations made by McConnell [3]. As the lubricant layer gradually decreases in thickness, surface roughness becomes imposing and causes the dotted line to depart from classical theory. In other words, surface roughness eventually

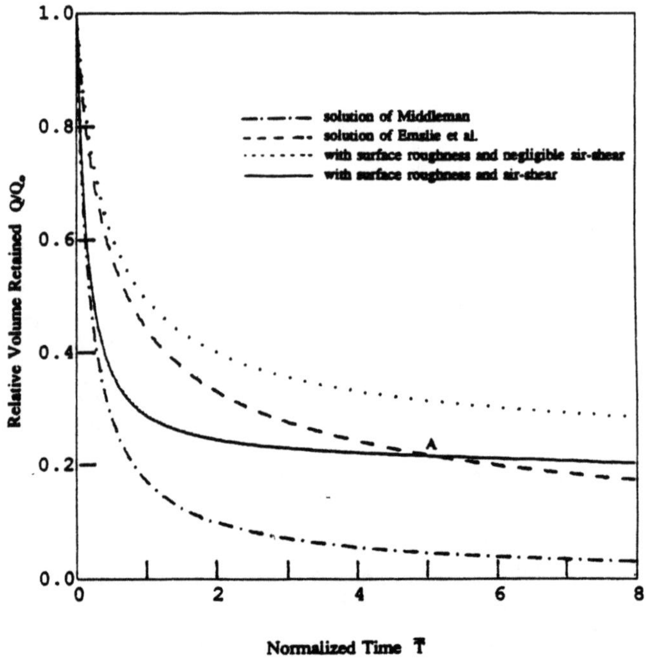

Fig. 3 Lubricant retention under various conditions.

overcomes centrifugation and the dotted line approaches a nonzero asymptotic limit. This implies that some lubricant will always remain on the rough disk. Define

$$N_{af} = \frac{3\left(\frac{1}{2}\omega^{\frac{3}{2}}\nu_{air}^{-\frac{1}{2}}\eta_{air}\right)}{2\rho\omega^2 H_o}, \tag{8}$$

where H_o is the initial height of the free surface of the liquid layer. The dimensionless parameter N_{af} represents the ratio of the effect of air-shear to the effect of centrifugation. Returning to a smooth disk and incorporating fairly moderate air-shear on the free surface of the lubricant, a retention curve represented by the half-dotted line is obtained, for which $N_{af} = 2.0$. This retention curve is derivable from the analysis of Middleman [2], and demonstrates that air-shear greatly increases the depletion of lubricant. If we now take into account both surface roughness and air-shear, a retention curve given by the solid line is obtained. The solid line intersects the classical retention curve at a point A. In the period prior to A, air-shear dominates and there is increased depletion. Therefore, the solid line lies below the classical retention curve prior to time A. When the instant A is reached, the lubricant becomes thin enough, so that surface roughness takes over and the solid line lies above the classical retention curve. Surface roughness is in full play after A, and the solid line eventually approaches a nonzero asymptotic limit. Although the magnitude of this new asymptotic limit has been compromised by air-shear, the existence of a nonzero asymptotic limit implies that surface roughness eventually dominates over air-shear. The results of the simulation associated with Fig. 3 have been reported in dimensionless variables. A set of realistic dimensional quantities is to have ω, γ, ρ, ρ_{air}, η, η_{air}, and H_o equal to 100π rad/s, 20.0 dyn/cm, 1.9 g/cm^3, 1.18×10^{-3} g/cm^3, 1.0 g / $(cm{\cdot}s)$, $1.81\times10^{-4} g/(cm{\cdot}s)$, and 50.0 nm, respectively. The large-scale roughness is then separated by a mean distance of 5.0 μm apart, and the mean depth of the large-scale roughness is about 0.9 μm.

Finally, comparison of theory with experiment is sought. Yanagisawa [5] has conducted an experiment on spin coating of perfluoro-polyether of molecular weight of 4000 with negligible vapor pressure and a viscosity value of 30 cS at 20° C. The mean initial thickness of the liquid layer is 25 nm, which is thinned to less than 10 nm after 200 hrs at 3400 rpm. In this case, $N_{af} = 6.67$. Middleman has used a parameter β in interpreting the experiment of Yanagisawa. It turns out that $N_{af} = 1/\beta$, so that $\beta = 0.15$. The data of Yanagisawa, the classical retention curve, as well as the retention curve obtained by Middleman, are all plotted in Fig. 4. As we can see, the classical retention curve overestimates the experimental data, and the retention curve by Middleman underestimates the data. Both retention curves have neglected surface roughness. Yanagisawa indicates that the disk has been polished to obtain 5 nm surface roughness, which is 20% to 60% of his measured film thickness. If we assume that the roughness pattern is sinusoidal, then a new retention curve can be obtained by simulation, as shown in Fig. 4. This new retention curve lies between the classical curve and the curve by Middleman within the time period shown, and it matches the experimental data more closely. Moreover, the new retention curve displays a tendency to approach a nonzero asymptotic limit. The new retention curve still underestimates the data of Yanagisawa. This may be due to the various approximations used in the theoretical model, as well as possible limitations in the accuracy of the experimental data. In addition, the disk used by Yanagisawa must be randomly rough to some degree. But information of this kind has not been given by Yanagisawa. Nevertheless, the new retention curve reported in this article provides a much closer representation of reality, particularly in displaying the tendency towards asymptotic retention.

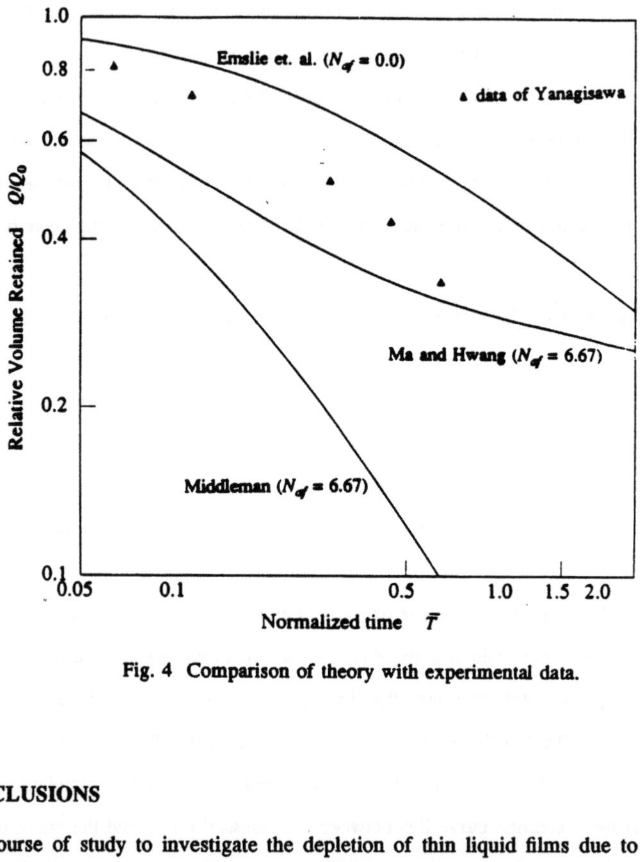

Fig. 4 Comparison of theory with experimental data.

5. CONCLUSIONS

A course of study to investigate the depletion of thin liquid films due to the combined effect of centrifugation, surface roughness, and air-shear has been reported in this paper. The equations describing the flow of a thin viscous film over a rough rotating disk have been developed and subsequently solved by numerical techniques. Surface roughness has been represented by either deterministic curves or by spatial stochastic processes. Monte Carlo simulations have been performed whenever statistical models of surface roughness are used. Depletion histories of the thin film have been given for several cases involving both deterministic and random surface asperities. Although a limited set of data is presented, extensive numerical calculations have been performed by the author, and all numerical simulations have yielded identical qualitative results on liquid retention. The major results, summarized in the following, are applicable to the flow of a thin viscous film over a rough rotating disk.

(1) Surface roughness of a rotating disk plays an important role in lubricant retention; the thinner the lubricant layer the more dominant is the effect of surface roughness.

(2) The influence of surface roughness is more significant than air-shear or inertial forces in the depletion of very thin films.

(3) The asymptotic retention of a liquid layer represents, to first order, a compromise between surface roughness, centrifugation, and air-shear. A theoretical model that incorporates all three factors can provide a close representation of reality.

The results stated in this paper, we believe, are both reliable and important. However, these results have been obtained from a radially symmetric model of flow, and as such they are only as credible as the assumptions underlying the model. Among other things, it is hoped that this paper would serve to point to directions along which further research efforts should be made. The model developed in this paper may be improved upon by incorporating disjoining pressure terms in equation (1). Another extension is also obvious. Surface roughness used so far is radially symmetric. Three-dimensional random roughness can be generated if the flow is extended to depend on the angular orientation. The study of the influence of other factors, such as vertical flow of the lubricant, disturbances induced by the slider, and topographic structures of surface roughness, is also worthwhile in a subsequent course of investigation.

ACKNOWLEDGMENTS

This research has been supported in part by the National Science Foundation under Grant No. MSS-8657619, and by the Computer Mechanics Laboratory of the University of California at Berkeley. Opinions, findings, and conclusions expressed in this paper are those of the author and do not necessarily reflect the views of the National Science Foundation.

REFERENCES

1. Emslie, A. G., Bonner, F. T., and Peck, L. G. (1958) "Flow of a Viscous Liquid on a Rotating Disk," *J. Appl. Phys.*, Vol. 29, pp. 858-862.

2. Middleman, S. (1987) "The Effect of Induced Air-Flow on the Spin Coating of Viscous Liquids," *J. Appl. Phys.*, Vol. 62, pp. 2530-2532.

3. McConnell, W. H. (1988) "On the Rate of Thinning of Thin Liquid Films on a Rotating Disk," *J. Appl. Phys.*, Vol. 64, pp. 2232-2233.

4. Langlois, W. E. (1984) "Incorporation of Nonclassical Effects into the Hydrodynamics of Thin Liquid Films," IBM Research Report No. RJ 4413, San Jose, CA.

5. Yanagisawa, M. (1987) "Slip Effect for Thin Liquid Film on a Rotating Disk," *J. Appl. Phys.*, Vol. 61, pp. 1034-1037.

6. Rehg, T. J., and Higgins, B. G. (1988) "The Effects of Inertia and Interfacial Shear on Film Flow on a Rotating Disk," *Phys. Fluids*, Vol. 31, pp. 1360-1371.

7. Hwang, J. H., and Ma, F. (1989) "On the Flow of a Thin Liquid Film over a Rough Rotating Disk," *J. Appl. Phys.*, Vol. 66, pp. 388-394.

8. Ma, F., and Hwang, J. H. (1989) "Stochastic Simulation of the Flow of a Thin Liquid Film over a Rough Rotating Disk," *J. Appl. Phys.*, Vol. 66, pp. 5026-5033.

9. Ko, M., Oxsen, M. E., and Prime, R. B. (1987) "Polymeric Additives for Magnetic Coating Materials," US Patent No. 4,546,035.

10. Cochran, W. G. (1934) "The Flow due to a Rotating Disc," *Proc. Cambridge Philos. Soc.*, Vol. 30, pp. 365-375.

11. Greenwood, J. A., and Williamson, J. B. P. (1966) "Contact of Nominally Flat Surfaces," *Proc. R. Soc. London*, Ser. A, Vol. 295, pp. 300-319.

12. Whitehouse, D. J., and Archard, J. F. (1970) "The Properties of Random Surfaces of Significance in their Contact," *Proc. R. Soc. London*, Ser. A, Vol. 316, pp. 97-121.

13. Kim, S., Kim, J. S., and Ma, F. (1991) "On the Flow of a Thin Liquid Film over a Rotating Disk," *J. Appl. Phys.*, Vol. 69, pp. 2593-2601.

Evaluation of a Stochastic Approach for Reliability Centered Maintenance Airframes

S.D. MANNING* and J.N. YANG**

* Structures Technology Staff, General Dynamics, Fort Worth Division, Fort Worth, Texas 76101
** Professor, Department of Civil Engineering, University of California, Irvine, California 92717

Summary

A reliability centered maintenance analysis (RCMA) method for metallic airframes is described, demonstrated and evaluated. This method is useful for evaluating aircraft structural maintenance and supportability requirements and options in terms of risk. It can be used to estimate the average maintenance manhour per flight hour (MMH/FH) and life-cycle-costs. The RCMA method, based on a stochastic crack growth approach, can be implemented using deterministic crack growth analysis (DCGA) results, an estimate of the crack growth life dispersion (e.g., material and service usage), σ_z, allowable risk rate, R_a, limit and conditional probability of a class A mishap, P_a. The sensitivity of RCMA inspection predictions (initial and reinspection intervals) to variations in analysis variables is investigated using a cutout in an aluminum-lithium cheek frame. Results are presented in useful formats for evaluating options in terms of risk. For comparable percentage variations in the baseline RCMA variables, the stress level, followed by σ_z, affects the RCMA inspection predictions more than either R_a or P_a.

Introduction

Aircraft structural inspection and repair maintenance is essential for ensuring aircraft structural integrity and operational readiness against the consequences of fatigue cracking in service. Effective analysis tools are needed to determine structural inspection requirements (i.e., initial and reinspection intervals) due to the risks and economic impact of crack growth accumulations in multiple structural details.

Reliability centered maintenance (RCM) [1] is a systematic approach for establishing aircraft preventive maintenance (PM) requirements. RCM is concerned with various failure modes for the airframe, mechanical and electrical equipment. This paper concerns the fatigue cracking failure mode in metallic structures.

A stochastic crack growth analysis (SCGA) method developed by Yang et al. [e.g. 2-5] has recently been adapted for RCMA of metallic airframes [6,7] (see Fig. 1). SCGA is an enhancement of the DCGA method commonly used to ensure damage tolerance, durability and aircraft structural integrity program (ASIP) requirements [8]. A stochastic crack growth model is the nucleus of the RCMA method for determining inspection intervals for metallic structures.

N. Bellomo · F. Casciati (Eds.)
Nonlinear Stochastic Mechanics
IUTAM Symposium Turin
© Springer-Verlag Berlin Heidelberg 1992

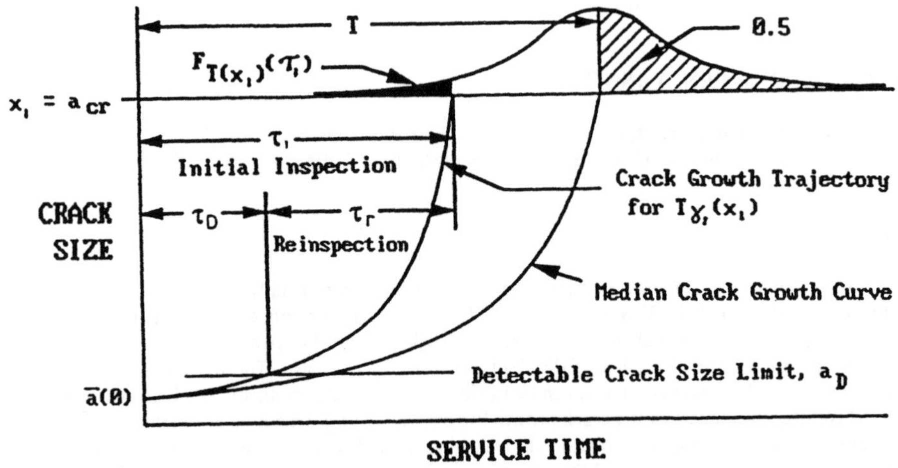

Fig. 1 Reliability Centered Maintenance Analysis Approach

The ASIP method [8] uses a DCGA and a fixed confidence factor of 2 to establish initial and reinspection interval requirements. Whereas, the RCMA method tailors the crack growth life dispersion from various sources (e.g., material, manufacturing and service usage) to the particular application. The research described in this paper is part of a continuing evaluation of the SCGA methods, including the RCMA method.

The objectives of this paper are to: (1) describe and demonstrate the RCMA method and procedures for determining structural inspection requirements (i.e., initial and reinspection intervals) for metallic airframes, (2) evaluate the sensitivity of RCMA inspection predictions to variations in analysis variables (i.e., stress level, σ_Z, P_a and R_a) and (3) show useful formats for presenting RCMA predictions. This paper is based on a more comprehensive investigation reported elsewhere [9]. Due to space limitations, this paper summarizes essential details, results and conclusions.

Technical Approach

Technical approaches for RCMA are described briefly in this section. Details are given elsewhere [6-9].

RCMA Method

The risk rate (or probability of a class A mishap per flight hour) of an aircraft is expressed as follows

$$R_a = P_a P_f(\tau)/\tau \tag{1}$$

in which R_a is the risk rate of an aircraft in $(0,\tau)$, $P_f(\tau)$ is the probability of failure of a control point in $(0,\tau)$ which is equal to the probability that the crack size $a(\tau)$ will exceed the critical crack size $x_1 = a_{cr}$. This probability is identical to the distribution function of service time, $T(x_1)$, to reach the critical crack size $x_1 = a_{cr}$, evaluated at time τ, i.e., $P_f(\tau) = P[T(x_1) \leq \tau] = F_{T(x_1)}(\tau)$. Hence, the service time, τ, can be obtained from Eq. (1) as follows

$$\tau = [P_a/R_a] F_{T(x_1)}(\tau) \tag{2}$$

Equation (2) is used to determine initial and reinspection intervals and to evaluate inspection technique options for selected control points. The conditional probability of a class A mishap, P_a, is estimated using an event tree logic to predict the consequence probability should a control point fail [e.g. 7]. Similarly, the allowable risk rate, R_a, is based on the applicable air vehicle specification, historical data for structural maintenance, and judgement. The probability of failure of a control point, $P_f(\tau)$, in $(0,\tau)$ or the distribution function of the service time $T(x_1)$ to reach the critical crack size $x_1 = a_{cr}$ in Eq. (2) is determined using SCGA methodologies developed by Yang and Manning [2-5] in the following.

A stochastic crack growth rate model suggested by Yang and Manning [2-5] for practical applications is represented by

$$da(t)/dt = Xg(\Delta K, R, ...) \tag{3}$$

in which da(t)/dt is the crack growth rate (per flight hour), t is the flight hour, $g(\Delta K, R, ...)$ is a general deterministic function of crack growth variables, such as ΔK (stress intensity range), R (stress ratio), etc. In Eq. (3), X is a lognormal random variable with a median value of 1.0 and a log standard deviation σ_z (also called the "crack growth life dispersion parameter"). Thus, the deterministic crack growth equation used for damage tolerance and durability analyses given by

$$da(t)/dt = g(\Delta K, R, ...) \tag{4}$$

represents the median crack growth rate in Eq. (1); whereas, the lognormal random variable X takes care of the statistical variability of the crack growth accumulation. The distribution function of the lognormal random variable X is given by

$$F_x(x) = P[X \leq x] = \Phi[lnx/\sigma_z] \tag{5}$$

in which $\Phi[\ ]$ is the standard normal distribution function and σ_z is the standard deviation of $Z = \ln X$ or log standard deviation of X. The stochastic crack growth model, Eq. (1), is referred to as the "lognormal random variable model"; see Yang et al. [2-5].

The cumulative distribution function of the service time, $T(x_1)$, to reach any crack size x_1 from a given median crack size $\bar{a}(0)$, denoted by $F_{T(x_1)}(\tau) = P[T(x_1) \leq \tau]$, can be derived from Eq. (3)-(5) as [2-5]

$$F_{T(x_1)}(\tau) = \Phi\{[\ln(\tau) - \ln(\bar{t}(x_1))]/\sigma_z\} \tag{6}$$

in which $\bar{t}(x_1)$ is the median service time to reach any crack size x_1 from the median initial crack size $\bar{a}(0)$. $\bar{t}(x_1)$, a function of the crack size x_1, is referred to as the "median crack growth curve".

The median crack growth curve $\bar{t}(x_1)$ can be established using either DCGA predictions, Eq. (4), i.e., based on a suitable analytical crack growth computer program [e.g. 10], or fractographic results. When fractographic data or sample service crack growth trajectories are available, the method of maximum likelihood can be used to estimate the median crack growth curve, $\bar{t}(x_1)$ [e.g., 2-5]. The log standard deviation, σ_z in Eq. (5), represents the crack growth life dispersion. σ_z can be estimated from a set of sample service crack growth trajectories that include variabilities contributed by material crack growth resistance, service loading spectra, etc., using the method of maximum likelihood [e.g., 2-5].

The RCMA initial inspection, τ_1, for the control point can be determined using Eqs. (2) and (6) and the following iterative procedure: (i) with the applicable risk rate R_a, e.g., $R_{a(max)}$ or $R_{a(min)}$, and the conditional probability of a class A mishap, P_a, the probability of failure of a control point per flight hour

$$F_{T(x_1)}(\tau)/\tau = R_a/P_a \tag{7}$$

is computed from Eq. (2), and (ii) given the median initial flaw size $\bar{a}(0)$, the critical flaw size $x_1 = a_{cr}$, the median crack growth curve $\bar{t}(x_1)$ and σ_z, $F_{T(x_1)}(\tau)$ can be computed from Eq. (6) for a given service time τ. By trial and error, the required service time τ_1 can be determined which satisfies Eq. (7).

The RCMA reinspection interval τ_r, can be determined using the same iterative procedure described above in which the median initial flaw size, $\bar{a}(0)$, is replaced by a_D. An alternate approach for determining τ_r, based on probabilistic crack growth trajectories for γ percentiles, denoted by $T_{\gamma_1}(x_1)$ in Fig. 1, is described elsewhere [2-4,9].

Maintenance manhours per flight hour (MMH/FH) is an important criterion for evaluating airframe supportability and maintainability for a component, an aircraft or a fleet of aircraft. Strictly speaking, the MMH/FH is a statistical variable. While the exact statistical distribution of MMH/FH may not be needed, the expected (mean) value and standard deviation can be estimated as follows.

Divide the critical component into m stress regions. In each stress region, the stress level at each structural detail (control point) is approximately the same. The probability that the crack size at a structural detail in the ith stress region will exceed x_1 in the service time $(0,\tau)$ is referred to as the "probability of crack exceedance", denoted by $p(i,\tau)$ [e.g.,2], i.e., $p(i,\tau) = P[a(\tau)>x_1]$. The probability of crack exceedance is identical to the probability that the service time, $T(x_1)$, for a crack to reach a given crack size x_1 is smaller than τ, which is the distribution function of $T(x_1)$ evaluated at τ, i.e., $p(i,\tau) = F_{T(x_1)}(\tau)$. It follows from Eq. (6) that the crack exceedance probability is given by

$$p(i,\tau) = \Phi\left\{[\ln(\tau) - \ln(\bar{t}(x_1))]/\sigma_z\right\} \tag{8}$$

in which $\bar{t}(x_1)$ is the median crack growth curve evaluated at x_1 for details in the ith stress region. In the ith stress region, let N_i be the total number of details and R_i be the maintenance manhours per structural detail required for the inspection and repair when the crack size exceeds x_1. Then, with the assumption that the crack growth accumulation in one detail is statistically independent of the growth in neighboring details, the expected (average) MMH/FH for the entire critical component consisting of m stress regions, denoted by E[MMH/FH], is obtained as [9]

$$E[MMH/FH] = \sum_{i=1}^{m} N_i p(i,\tau) R_i / \tau \tag{9}$$

and the corresponding standard deviation, $\sigma_{MMH/FH}$, is

$$\sigma_{MMH/FH} = \left\{\sum_{i=1}^{m} N_i p(i,\tau)\,[1 - p(i,\tau)]\,R_i^2/\tau^2\right\}^{1/2} \tag{10}$$

Equations (9) and (10) can be used to quantify the required MMH/FH for a critical component or an aircraft.

Evaluation and Demonstration of RCMA Method

RCMA initial interval (τ_1) and reinspection interval (τ_r) predictions (see Fig. 1) were determined for a cutout in an aluminum-lithium cheek frame from a fighter fuselage component. The following RCMA variable values were considered: stress level (18.5 ksi, 20.6 ksi and 22.7 ksi), σ_z = 0.131(M), 0.249(U), 0.281(M+U) and 0.337, R_a (1E - 08 and 5E - 08), and P_a = 0.9. Crack growth dispersion contributions are noted as follows: M = material, U = service usage, and M+U = material and service usage combined. Other variable values were also considered for investigating the sensitivity of RCMA τ_1 and τ_r predictions to variable variations. RCMA predictions were based on the methods described in Section 2.0, Ref. 2 and the following. The stress levels considered included the effect of stress gradient at the edge of the hole. Deterministic crack growth curve predictions for three stress levels (i.e., 18.5 ksi, 20.6 ksi and 22.7 ksi) were determined using the ADAMSystem [10]. These crack grwoth predictions were based on a "durability analysis" initial flaw size of a(0) = 0.005" (corner flaw) and a fighter load spectrum. For the sensitivity study, crack growth curves for other stress levels were approximated using the method recommended in Ref. 11. This involved scaling the median crack growth curved for the baseline stress level (20.6 ksi) to the selected stress level. A detectable crack size limit of a_D = 0.03" (eddy current inspection technique) was assumed. The median service time to reach the critical crack size and a_D for three stress levels are summarized in Table 1.

Table 1. Median Service Time to Reach Selected Crack Size

Stress Level (ksi)	a_{cr} (In.)	Median Service Time (Flight Hours)	
		$t(a_{cr})$	$t(a_D=0.03")$
18.5	0.555	36565	8043
20.6	0.45	28500	6383
22.7	0.372	22772	5170

$\sigma_z = 0.131(M)$ was determined using da/dN versus delta-k data, the Forman crack growth model and the method described in Ref. 9. $\sigma_z = 0.249(U)$ is based on crack growth analysis predictions for fleet tracking results for 32 fighter aircraft tracking control points and the method recommended by Yang and Manning [2]. Material and usage variabilities were combined into a single $\sigma_z = 0.281(M+U)$ using the square root of the sum of the squares [12,13]. Material and usage variabilities were assumed to be statistically independent [12,13]. Other σ_z values were considered (see Figs. 5 and 6).

Historical data for a fighter aircraft were used to establish the following maximum and minimum allowable risk rate limits [6,7]: $R_{a(max)}$ = (1E-05)/200 = 5E-08; $R_{a(min)}$ = (1E-05)/1000 = 1E-08. Other R_a values were also considered.

The conditional probability of a class A mishap was assumed to be $P_a = 0.9$; i.e., there is a 90% probability of losing the aircraft if the crack in the cutout grows to the critical crack size in-flight. Other P_a values were also considered.

Selected RCMA results and applicable parameter values are shown in Figs. 2-6 and Table 1. The results shown in Figs. 2 and 3 reflect a baseline stress level = 20.6 ksi, $x_1 = a_{cr} = 0.45"$ and $P_a = 0.9$. In Figs. 3 and 4 the straight lines between the horizontal boxes represent lines of constant risk rate levels. Average MMH/FH versus initial inspection predictions are shown in Fig. 4 for three different stress levels. These results are based on $x_1 = 0.10"$, $\sigma_z = 0.281(M+U)$, Eqs. (8)-(9), and the following assumptions: 1000 details in the population and 10 manhours/detail is required for inspection and repair maintenance. RCMA sensitivity analysis results for τ_1 and τ_r are shown in Figs. 5 and 6, respectively. These results reflect the following applicable baseline values: stress level = 20.6 ksi, $\sigma_z = 0.281(M+U)$, $P_a = 0.9$, and R_a = 5E-08. The maximum allowable value of P_a is 1.0. To cover the same range of ratios for each of the four variables considered in Figs. 5 and 6, τ_1 and τ_r values for $P_a = 1.0$ were used.

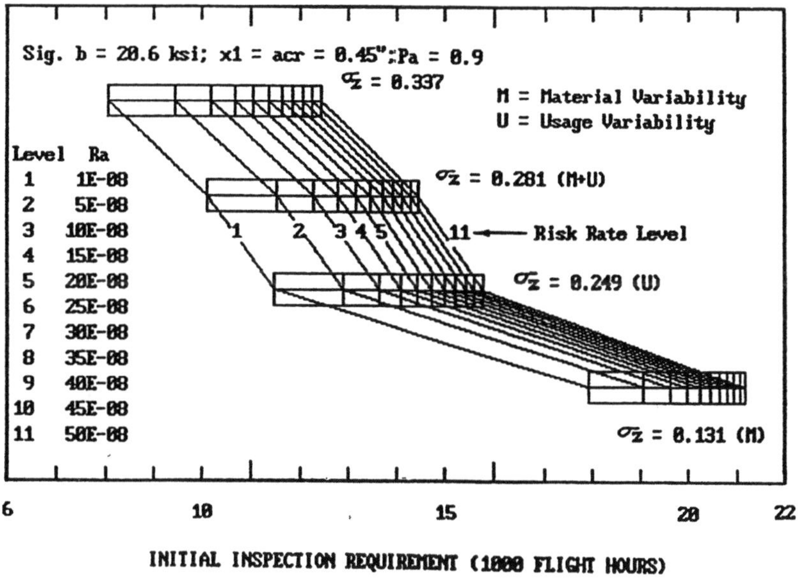

Fig. 2 Effect of σ_z and Risk Rate on RCMA Initial Inspection Predictions

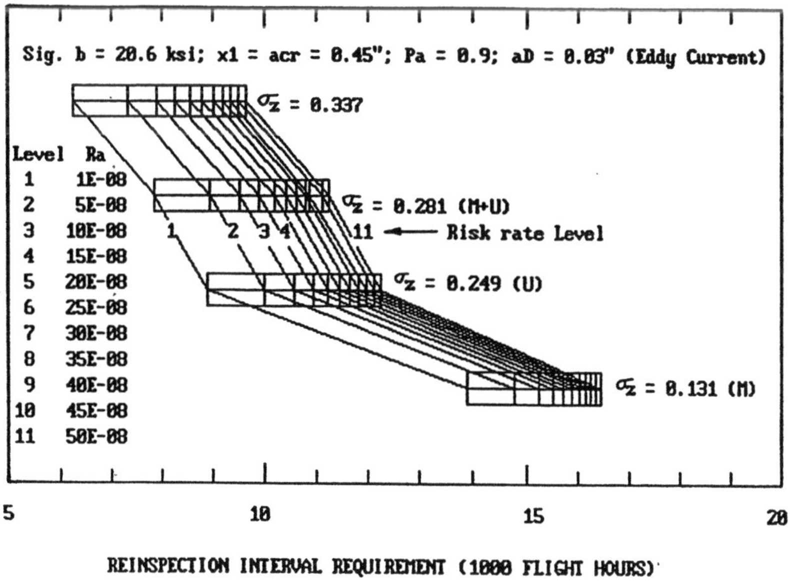

Fig. 3 Effect of σ_z, Risk rate and Eddy Current Inspection (a_D = 0.03") on RCMA Reinspection Interval Predictions

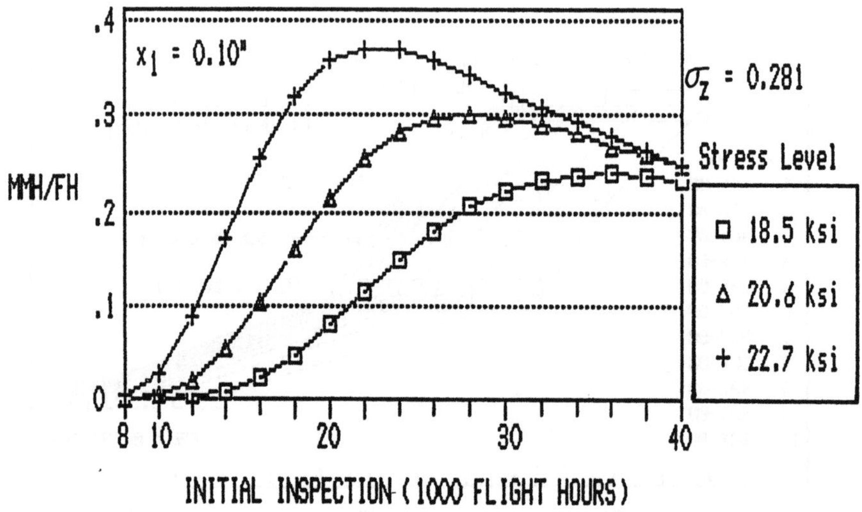

Fig. 4 Effect of Stress Level on Maintenance Manhour Per Flight Hour Predictions

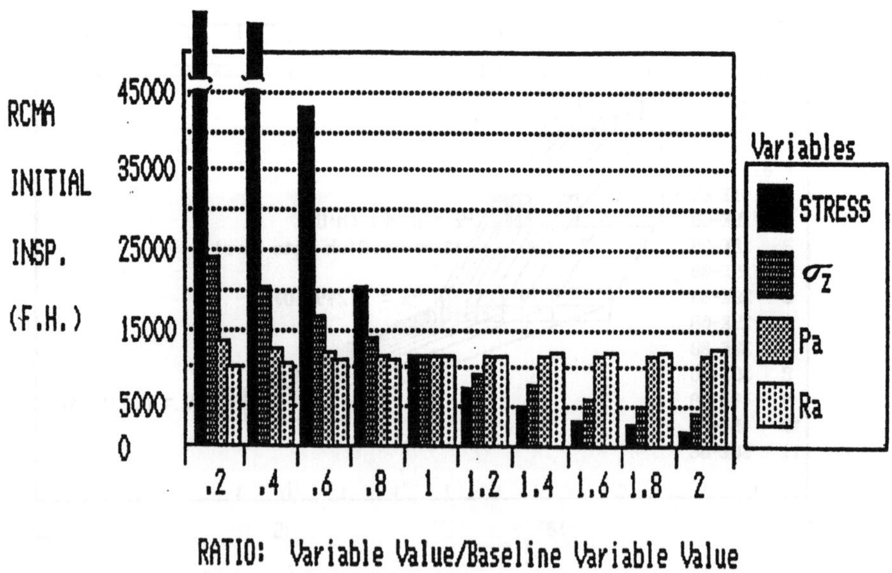

Fig. 5 Sensitivity of RCMA Initial Inspection Predictions to Selected Variables

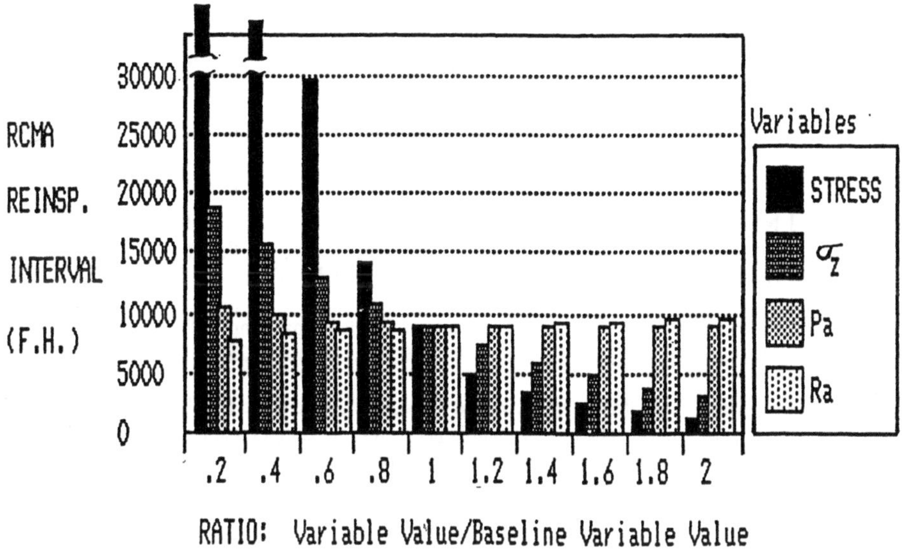

Fig. 6 Sensitivity of RCMA Reinspection Interval Predictions to Selected Variables

Conclusions

A RCMA method has been described, demonstrated and evaluated using a cutout in a metallic cheek frame. Useful formats for presenting RCMA results have been illustrated. Also, the sensitivity of RCMA inspection predictions to stress level, σ_z, R_a and P_a has been investigated. For the same percentage variation in the baseline variable values, it was found that the stress level, followed by σ_z, affects the RCMA results more than either R_a and P_a. Ideally, the median crack growth curve should reflect the expected normal fleet usage and the median initial flaw size. Shorter inspection predictions will result if initial flaw sizes larger than the median are used.

Further research is needed to: (1) compare predictions based on the RCMA and ASIP methods and to better understand the implications, (2) establish a suitable range of σ_z values for different materials, structural detail types (e.g., fastener holes, cutouts, fillets, lugs), (3) validate the methodology for an aircraft component with multiple structural control points, (4) determine a suitable range of median initial flaw sizes for different materials, types of structural details, and manufacturing quality, and (5) evaluate the methodology for structural design or life extension applications. The authors are investigating these aspects.

Acknowledgement

This research was partially supported by the Air Force Wright Aeronautical Laboratories, Wright-Patterson Air Force Base, Ohio, under contract No. F33615-87-C-3207.

References

1. Matteson, T.B., "The Origin of Reliability Centered Maintenance", Institute of Industrial Engineers (IIE), Proc. 6th Inter. Maintainability Conf., Atlanta, GA, Oct. 11-13, 1989.

2. Yang, J.N. and Manning, S.D., "Stochastic Crack Growth Methodologies for Metallic Airframes", Engineering Fracture Mechanics, Vol. 37, No. 5, pp. 1105-1124, 1990.

3. Yang, J.N. and Manning, S.D., "Application of Probabilistic Approach for Crack Growth Damage Accumulation in Metallic Structures", Structural Safety and Reliability, Vol. II., Ed. by A. H-S. Ang, M. Shinozuka and G.I. Schueller, Pub. ASCE, NY, pp. 1475-1482, 1990.

4. Yang, J.N. and Donath, R.C., "Statistical Fatigue Crack Propagation in Fastener Holes Under Spectrum Loading", J. of Aircraft, AIAA, Vol. 20, No. 12, pp. 1028-1032, Dec. 1983.

5. Yang, J.N., Hsi, W.H., Manning, S.D. and Rudd, J.L., "Stochastic Crack Growth Models for Application to Aircraft Structures", Probabilistic Fracture Mechanics and Reliability, Chapter IV, Ed. by J. W. Provan, Pub. Martinus Nijhoff, Dordrecht, pp. 171-212, 1987.

6. Manning, S.D., Yang, J.N., Pretzer, F.L. and Marler, J.E., "Reliability Centered Maintenance for Metallic Airframes Based on a Stochastic Crack Growth Approach", Advances in Fatigue Lifetime Predictive Techniques, ASTM STP 1122, Ed. by M.R. Mitchell and R.W. Landgraf, ASTM, Philadelphia, 1991.

7. Pretzer, F.L. et al., "F-16 Reliability Centered Maintenance Analyses", General Dynamics/Fort Worth Division, Final Report 16PR8829, July 1, 1989.

8. "Aircraft Structural Integrity Program, Airplane Requirements", Military Standard MIL-STD-1530A, Air Force Aeronautical Systems Division, Wright-Patterson Air Force Base, OH, Dec. 1975.

9. Manning, S.D. and Yang, J.N., "Evaluation of Two Stochastic Crack Growth Analysis Methods for Aircraft Structural Maintenance", paper to be submitted to J. of Eng. Fracture Mech., 1991.

10. Roach, G.R., McComb, T.H. and Chung, J.H., "ADAMSys User's Manual", Structural and Design Dept., General Dynamics/Fort Worth Division, July 1987.

11. Gallagher, J.P., Giessler, F.J., Berens, A.P. and Engles, Jr., R.M., "USAF Damage Tolerant Design Handbook: Guidelines for the Analysis and Design of Damage Tolerant Aircraft Structures", AFWAL-TR-82-3073, Air Force Wright Aeronautical Laboratories, Wright-Patterson Air Force Base, OH, pp. 10.4.1-10.4.13, May 1984.

12. Yang, J.N. and Chen, S., "Fatigue Reliability of Gas Turbine Engine Components Under Scheduled Inspection Maintenance," J. of Aircraft, AIAA, Vol. 22, No. 5, May 1985.

13. Yang, J.N. and Chen, S., "An Exploratory Study of Retirement-For-Cause for Gas Turbine Engine Components", J. of Propulsion and Power, AIAA, Vol. 2, No. 1, Jan. 1986.

Random Flow-Induced Oscillations with Impacts of a Set of Coupled Pendulums

A.I.MENYAILOV
USSR Academy of Sciences
Institute for Problems in Mechanics
117526 Moscow, USSR

Summary

Oscillations of a set of coupled hinged spherical pendulums, a model of many internal elements of power plant , are situated within a channel of circular or almost circular cross-section with small gaps and excited by an upward flow of one-phase and two-phase viscous fluid, are studied both analytically and numerically. Various regimes of motion are identified: plane oscillations with alternating impacts; rotation along channel wall, without impacts; polarised oscillations with impacts and with slowly rotating plane.For plane motion maximum impact impulse and maximum frequency were found analytically.Random excitations due to two-phase flow with small intensities change the situation,leading to transitions between various types of motion. Mean times beetween transitions were established by numerical simulations.

Introduction

Axial flows of viscous one-phase or two-phase fluid around elastic cylinders in channels with narrow gaps may lead to flutter-type or divergence-type instabilities. Numerous studies of these phenomena [1] were concerned with small oscillations, without impacts, and with one-phase fluid. However, in many structures with small gaps between cylindrical elements and channel walls the impacts become possible even for small lateral dynamic displacements [2]. Impacts may lead to an intensive wear and fretting of the contact surfaces with possible loss of sealing of channel tube. The plane impact motions of a simple pendulum with rigid barriers excited by a parallel flow were considered in [3]. Later studies for a single and double spherical pendulums revealed certain new qualitatively different types of motion, especially for two-phase flow of fluid, which are presented in this report.

Equations of motion

Consider a two-mass sphercal physical pendulum, which is con-

N. Bellomo · F. Casciati (Eds.)
Nonlinear Stochastic Mechanics
IUTAM Symposium Turin

sisted of two cylinders, installed with small gaps=d within narrow tubes or channel with axial fluid flow(one-phase or two-phase) which has velocity V and fluid density ρ_f for one-phase flow and $V=V_o(1+\zeta(t))$ and $\rho=\rho_{fo}(1+\zeta(t))$ for two-phase flow respectively. We shall use spherical system of coordinates (φ_1, $\varphi_2, \theta_1, \theta_2$). The pendulums impact the wall of channel when the conditions $\varphi_i=r_i$ where r_i- dimensionless gaps is satisfied, $r_i=d/l_i$, d-gap, l_i-length of i-cylinder. We assume that the velocities $\varphi_1', \theta_1', \varphi_2', \theta_2'$ will chande during the impact accoding to finite relations:

$$\varphi'_{1+}=-R_n\varphi'_{1-}, \quad \varphi'_{2+}=-R_n\varphi'_{2-}$$

$$\theta'_{1+}=(1-R_\tau)\theta'_{1-}, \theta'_{2+}=(1-R_\tau)\theta'_{2-}$$

where φ'_{i-}, θ'_{i-}, φ'_{i+}, φ'_{i+} are preimpact and afterimpact pendulum velocities and R_n, R_τ are the impact factors of normal part and braking of tangential part of impact velocity. The equations of motion of system were derived taking into account forces:

$$P_1= 2\rho_f AC_m|V(\xi,t)|\partial^2 r(\xi,t)/\partial t\partial\xi$$

where A - cross-section of cylinder, r and ξ - radius-vector and flowing coordinate of cylinder respectively, C_m- coefficient of added mass;

$$P_2= 0.5\rho_f DC_n|V(\xi,t)|[\dot{r}(\xi,t)-V(\xi,t)]$$

where D is the diameter of the cylinder, C_n is the frictional coefficient;

$$P_3= C_d D|\dot{r}(\xi,t)|\dot{r}(\xi,t)$$

where C_d is a drag coefficient in still fluid;

$$P_4= C_k|r(\xi,t)|^2 r(\xi,t)$$

where C_k is the frictional coefficient in hinge-joint. Forces P_1-inviscid hydrodynamic force, P_2-frictional force, P_3-quadratic force at zero flow velocity, P_4-frictional force in joints. The model was developed on the basis the method of Lagrange. The forces associated with the structure itself were taken into account in the kinetic and potential energines of the system consists of two sherical pendulums. The fluid forces were be incorporated party in the kinetic energy party left as generalized forces. The fluid forces on each cylinder must be calculated by integrating the hydrodynamic forces acting on each element of cylinder per unit length. As a result of these calculations we obtain the following equations of motion of system (Fig.1).

$$M\ddot{q} + B\dot{q} + SV\dot{q} + (K + K_1V^2)SIN(q) - K_2V^2COS(q) + (C_dD_1 + C_kD_2)\dot{q} = 0 \quad (1)$$

$$q=(\varphi_1, \theta_1, \varphi_2, \theta_2)^T, \quad SIN(q)= (Sin\varphi_1, Sin\theta_1, Sin\varphi_2, Sin\theta_2)^T,$$

$$COS(q)= (Cos\varphi_1, Cos\theta_1, Cos\varphi_2, Cos\theta_2)^T$$

All matrices in eqs. (1) have dimention 4x4 and too complicated to be presented here. Derivation of eqs.(1) was verified by computer program REDUCE.

Analytical solution for partial case of common model-plane model

Consider a plane two-body pendulum. In this case $\theta_1=\theta_2=const$ and we have $\theta_1'=\theta_2'=\theta_1''=\theta_2''=0$, and system (1) in dimensionless parameteres reduced to:

for $|\psi_1|<\pi/2$, $|\psi_2|<\pi/2$

$$\begin{aligned}\ddot{\psi}_1 &= \alpha_1\dot{\psi}_1 + \beta_1\dot{\psi}_2 + \gamma_1\psi_1 + \delta_1\psi_2\\ \ddot{\psi}_2 &= \alpha_2\dot{\psi}_1 + \beta_2\dot{\psi}_2 + \gamma_2\psi_1 + \delta_2\psi_2\end{aligned} \quad (2)$$

with $\alpha_i, \beta_i, \gamma_i, \delta_i$ (i=1,2) being much smaller than unity; and for $|\psi_1|=\pi/2$, $|\psi_2|=\pi/2$

$$\dot{\psi}_{1+} = -R_n\dot{\psi}_{1-}, \quad \dot{\psi}_{2+} = -R_n\dot{\psi}_{2-} \quad (3)$$

Coefficients $\alpha_i, \beta_i, \gamma_i, \delta_i$ are proportional:

$$\alpha_i \sim U, \ \beta_i \sim U, \ \gamma_i \sim U^2, \ \delta_i \sim U^2$$

where U is dimensionless velocity of flow, $\psi_i = \pi\varphi_i^*/d$, $\varphi_1^*=\varphi_1$, $\varphi_2^*= \varphi_1 + \varphi_2$, i=1,2. This system may be studied analytically for one-fase flow by using piecewise-linear trasformation [4] and average method [5]. Following the piecewise-linear transformation involve the functions:

$$\psi_1 = \Pi(x) + \lambda S(x), \quad \psi_2 = \Pi(y)+\lambda S(y),$$

$$\lambda = \frac{2(1-R_n)}{(1+R_n)}, \quad \Pi'(x)= M(x), \quad S(x)= \int\Pi(x)dx, \quad (')=d/dx$$

$$\Pi(x) = \begin{cases} x\ , & -/2 \quad x \quad /2 \\ -x\ , & /2 \quad x\ 3\ /2 \\ -x\ , & 3\ /2 \quad x - /2 \end{cases}$$

$$\Pi(x) = \Pi(x + 2\pi)$$

Which reduce eqs. (2), (3) to the "nonimpact" form. Introducing variables $W_1=\dot{x}$ and $W_2=\dot{y}$ we get

$$\dot{x} = W_1,\ \dot{y} = W_2,\ \dot{W}_1 = -\lambda W_1^2 + \alpha_1 W_2 + \beta_1 M(y)M(x)W_2 + \gamma_1 M(x)\Pi(x) + \delta_1 \Pi(y)M(x),\quad \dot{W}_2 = -\lambda W_2^2 + \alpha_2 M(x)M(y) + \beta_2 W_2 + \gamma_2 \Pi(x)M(y) + \delta_2 M(y)\Pi(x) \qquad (4)$$

Following the everage method we shall consider resonant motion with $W_2-W_1 \sim \varepsilon$, $\varepsilon \ll 1$, corresponding to pendulums with close natural frequencies. Applying averaging over the period yields the following system of equations:

$$\dot{\bar{\theta}} = \bar{W}_1 - \bar{W}_2,\ \dot{\bar{W}}_1 = \bar{W}_1^2 + {}_1\bar{W}_1 + 2/\pi\beta_1\ \Pi(\bar{\theta}+\pi/2)\bar{W}_2 + 2/\pi\delta_1 S(\bar{\theta}- /2),\quad \dot{\bar{W}}_2 = -\lambda\bar{W}_2^2 + 2/\pi\alpha_2\bar{W}_1\Pi(\bar{\theta}+\pi/2) + \beta_2\bar{W}_2 + 2/\pi\gamma_2 S(\bar{\theta}- /2)$$

with its steady-state solution providing impact frequency $W^* = W_1 = W_2$ and velocity $V^* = W^*(M(\pi/2 + \Pi(\pi/2)$ of self-oscillations as the functions of flow velocity; these functions V*, W* are illustrated in Figs. 2 and 3 respectively.

Numerical analysis of the equatins of motions

Consider one-body spherical pendulum. Equations of motion (1) between impacts can be written in dimensionless parameters:

$$\tau = \Omega t,\ \Omega^2 = 3(1-\rho_f/\rho)g(l^2-l_o^2)/[2(1-C_m\rho_f/\rho)(l^3-l_o^3)],$$
$$U=V/l\Omega,\ \beta = 3(l^5-l_o^5)C_k/[5A\rho(l^3-l_o^3)],\ \alpha = 3(l^4-l_o^4)C_d/[4\rho A(l^3-l_o^3)],\ \gamma = \rho_f Dl/2A\rho,\ \delta = 3\rho_f Dl^2(l^2-l_o^2)/[4A\rho\cdot(l^3-l_o^3)],\ \varepsilon_1 = l\rho_f(l_2^2-l_o^2)/[\rho(l^3-l_o^3)],\ d/d \equiv '$$

as for $\varphi < r$

$$\varphi'' - \varphi\theta'^2 + [\alpha\sqrt{\varphi'^2 + \varphi^2\theta'^2} + \beta - (\gamma C_n + \varepsilon_1 C_m)U] \cdot$$

$$\varphi'/(1 + C_m \varrho_f/\varrho) + [1 + \delta C_n U^2/(1+C_m \varrho_f/\varrho]\varphi = 0,$$

$$\theta'' + 2\theta'\varphi'/\varphi + [\alpha\sqrt{\varphi'^2 + \varphi^2\theta'^2} + \beta - (\gamma C_n + \varepsilon_1 C_m)U]\theta'/$$

$$(1 + C_m \varrho_f/\varrho) = 0$$

and $\varphi = r$

$$\varphi'_+ = -R_n\varphi'_- \quad , \quad \theta'_+ = (1 - R_\tau)\theta'_-$$

where ϱ is a cylinder density. For one-phase flow the trajectories of pendulum in the circular channel are shown in Fig.4. The self-oscillation characteristics will change in dependence on breaking coefficient R_τ according to diagrams in Fig.5. For small R_τ the self-oscillations approach the circling along the wall. For high values of breaking coefficient $R_\tau(R_\tau \to 1)$ the polarized self-oscillation arises and the pendulum impacts the wall in the opposite points of the channel. The polarized impact self-oscillations is the most danger case of self-oscillation from point of view of the intensity of impact loading - impact velocity has maximum, Fig.6. In case of slightly eccentric channel the following trens are found. the motion of double pendulum is almost always polarized along a small axis of the elleptic channel. Increasing the tangential restitution factor leads to bifurcations of the impact motion: first the regime with 4 impacts per one rotation around the channel appears, then 6 impacts, etc.. Hysteretic phenomena are observed in these transitions, i.e. critical values of the restitution factor are found to be different when it is decreasing. Furthermore, chaotic motion are observed within small ranges of this factor $R_\tau = 0.418 \div 0.421$ and $R_\tau = 0.4 \div 0.41$ Fig.7.For numerical treatment we used values of parameters:

$C_m = 6$, $\varrho_f = 1000 kg/m^3$, $\varrho = 11800 kg/m^3$, $l = 1.54 m$,

$l_o = 0.4 m$, $A = 0.000354 m^2$, $D = 0.02153 m$, $C_d = 0.127 kg/m^2$,

$C_k = 0.2 kg/m^3 sec$, $d = 0.0045 m$

For two-phase flow, when $V = V_o(1 + \zeta(t))$ and $\varrho_f = \varrho_{fo}(1 + \zeta(t))$, where $\zeta(t)$ is a zero-mean random process,there are transitins from one regime to another regime in depend of rms of $\zeta(t)$ - σ. These results are summarized in Fig.8 for mean time of transition from regime with 2 impacts per one rotation to regime with 4 impacts per one rotation pendulum, $\sigma = 0.01 \div 0.5$.

Conclusions

1. Dynamics of model of fuel bundles of channel type nuclear reactor was studied analytically and numerically for cases one -phase and two-phase coolant flow.
2. Three types of motion of fuel bundle were identified: i)plane motion with impactsii)plane motion with slow rotation, with impacts iii) rotational motion alongthe wall, without impacts.
3. For plane motion maximum impact momentum and maximum frequency were found analytically.
4. Random excitations due to two-phase flow with small intesities change the situation leading to transitions between various types of motion. Mean times between transitions were established by numerical simulations for more important cases, the eccentrical channel.

References

1. Paidoussis,M.P. A review of flow-induced vibrations in reactors and reactor components. Nuclear Engineering and Design, 1983, v.74, 1
2. Menyailov,A.I.,Peterka,F. Simulation oscillations internal elements of reactors. Transaction of Symposium on Modelling, Ostrava, 1987.
3. Menyailov,A.I.,Peterka,F.,Chipera,S. Vibroimpact self-oscillations of spherical pendulum in axial fluid flow. Transaction of 1-st Conference of Socialist Countries, Prague, 1987.
4. Zhuravlev,V.PH. Method of special function for vibroimpact systems analysis. Izvestia Akad.Nauk SSSR. Mechanika tverdogo tela (Mech.Solids), 1976, 2, 13-17.
5. Bogoliubov,N.N.,Mitropolsky,JU.A. Asymptotic methods in theory of nonlinear oscillations (in Russian), Moscow, Nauka, 1974.

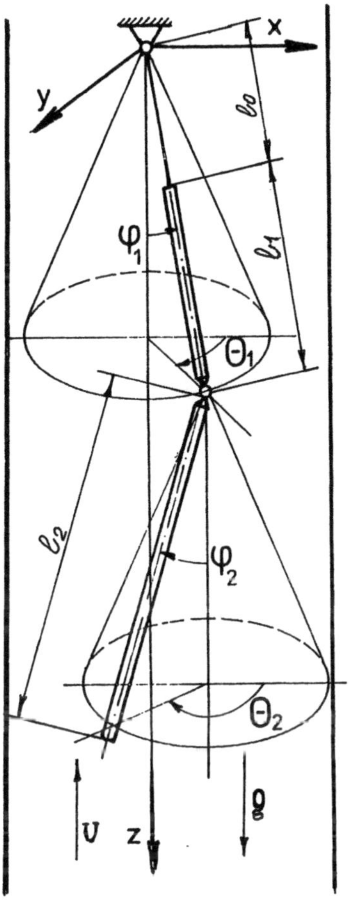

Fig.1 Double pendulum in flow channel. Arrangment, dimensiom and coordinates

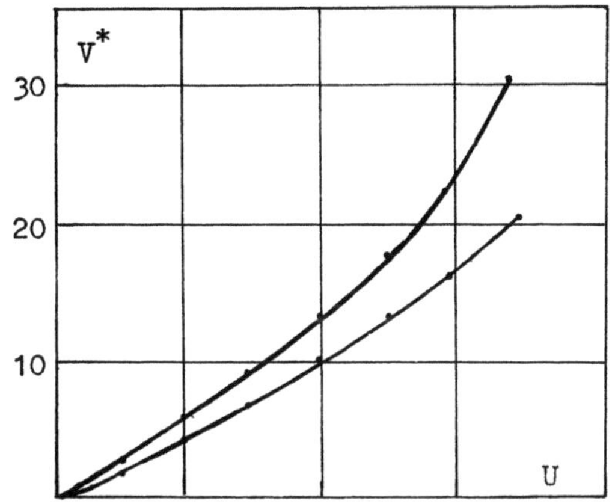

Fig.2 Impact velocity of pendulum as function of flow-rate

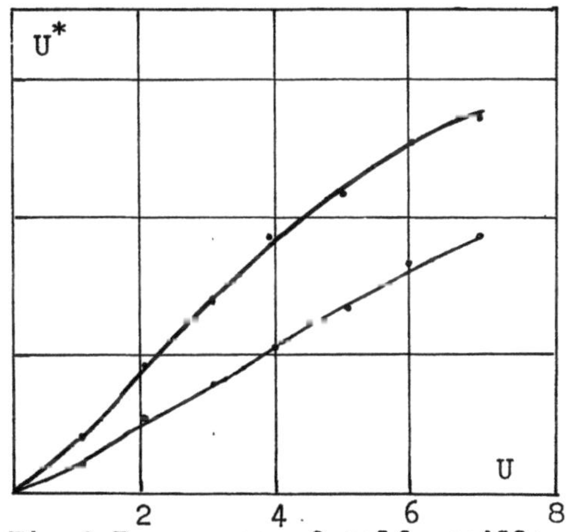

Fig.3 Frequency of self-oscillations of the pendulums as a function of flow-rate

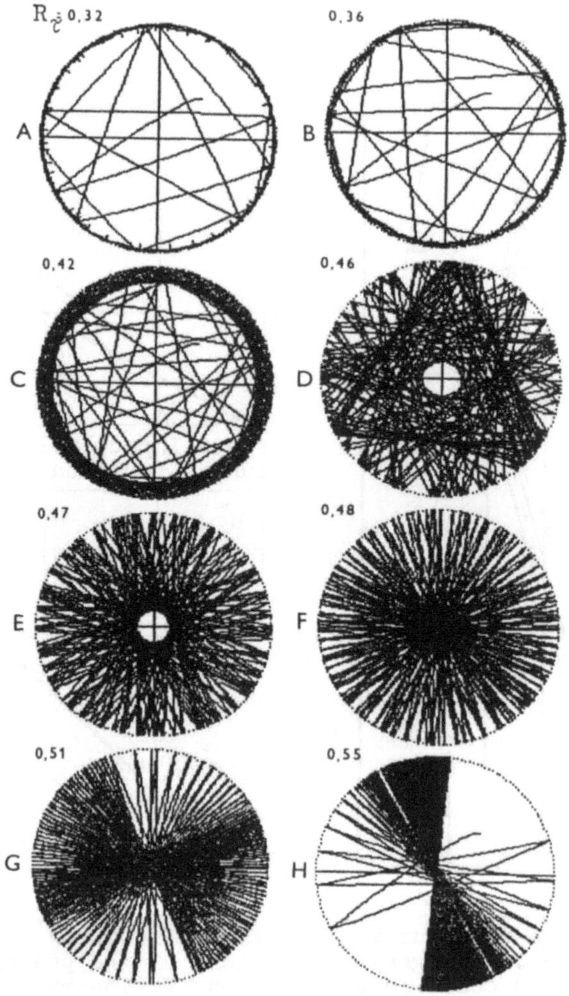

Fig.4 Typical trajectory of the system of the pendulums movement within channel

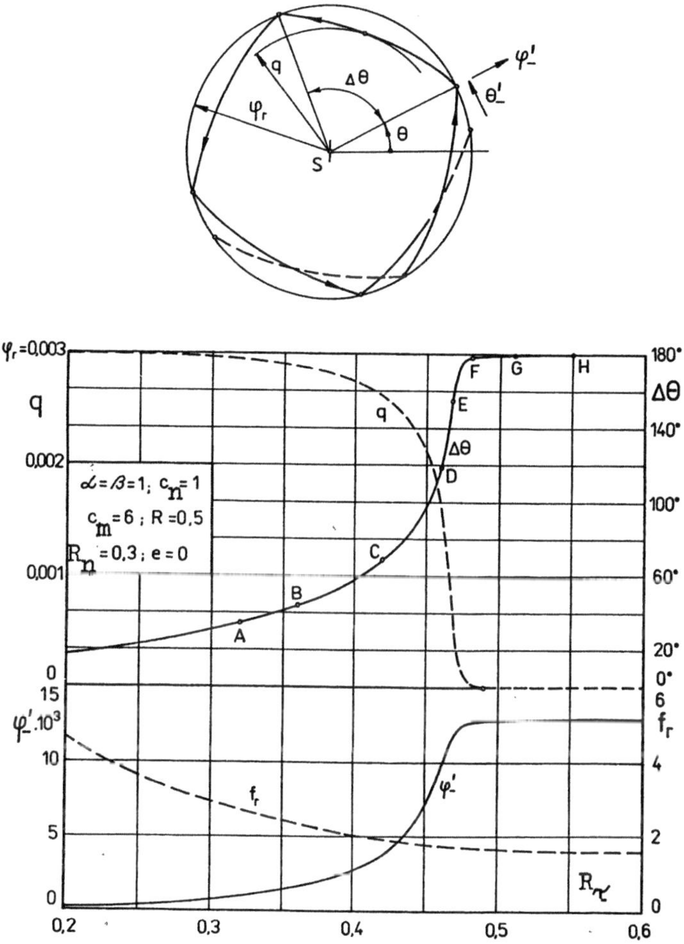

Fig.5 Tangential velocity of the pendulum and angle-difference between consequent impacts of the pendulum as function of impact factor

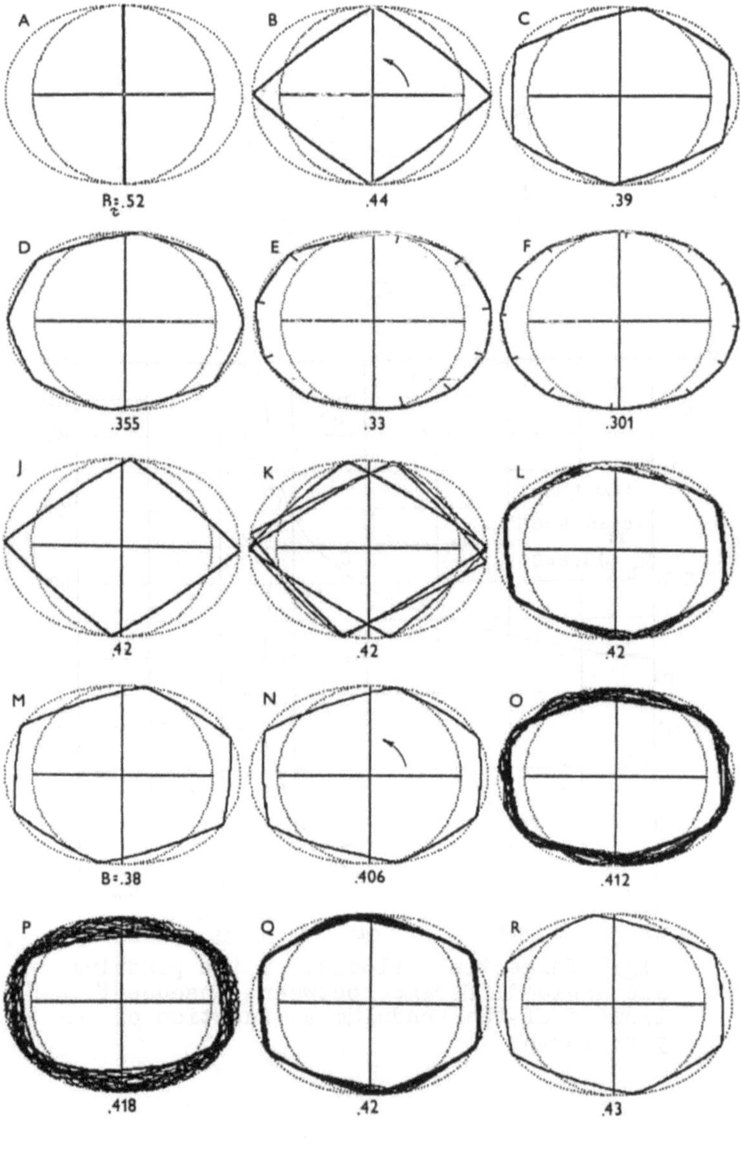
A
B
C
R=.52
.44
.39
D
E
F
.355
.33
.301
J
K
L
.42
.42
.42
M
N
O
B=.38
.406
.412
P
Q
R
.418
.42
.43

Fig.6

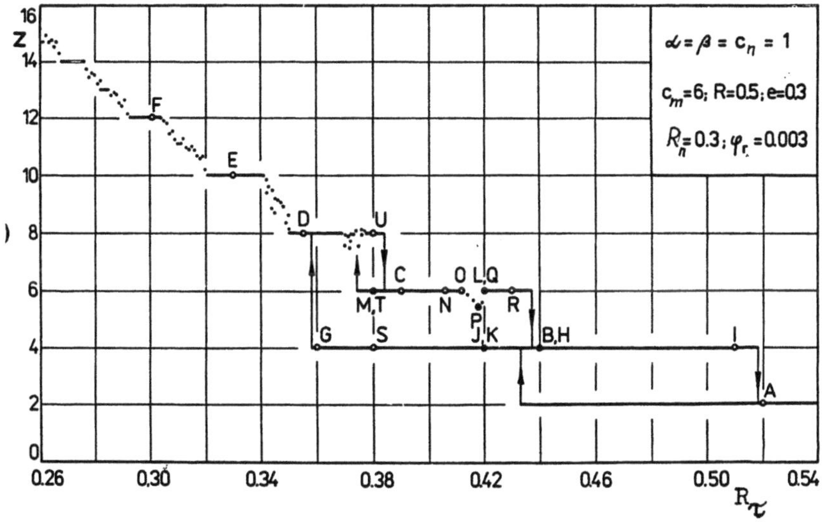

Fig.7 Number of impacts per one rotation around the channel as a function of the tangential part of the impact factor

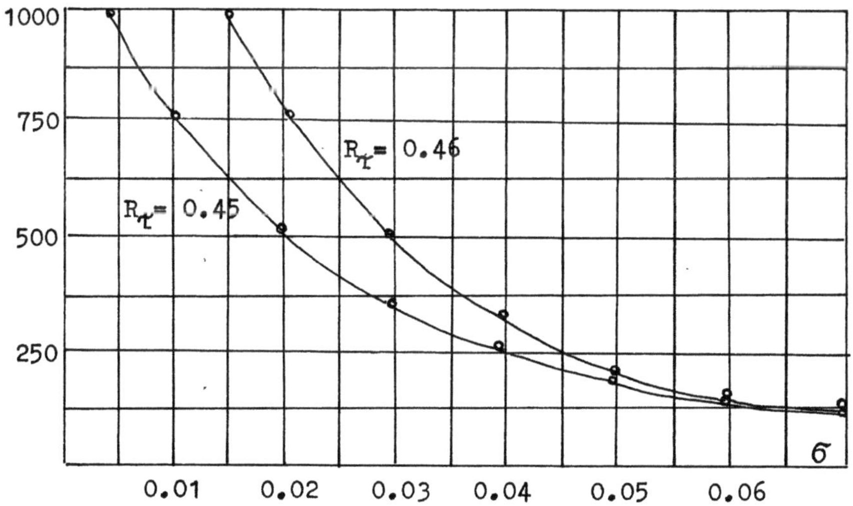

Fig.8 Mean times between trasitions from regim with two impacts to regim with four impacts per one rotation pendulum within eccentrical channel

Response Statistics of Nonlinear Dynamic Systems by Path Integration

Arvid Naess and **John M. Johnsen**

Department of Civil Engineering
Norwegian Institute of Technology, Trondheim, Norway

ABSTRACT

The paper presents a method for calculating the response statistics of nonlinear dynamic systems excited by a white noise or filtered white noise process. The method, which is based on the path integral solution technique, is still under development, but experience so far indicates that it is singularly well suited for numerical calculation of the response statistics of nonlinear systems to which can be associated a Markov vector process whose probability density satisfies a Fokker-Planck-Kolmogorov (FPK) equation. The method is a viable alternative to the direct numerical solution of the FPK equation. A key feature of the method is the possibility of obtaining highly accurate solutions at very low probability levels. Also, there seems to be almost no limitation on the type of nonlinearity that can be accomodated. On the negative side there are clear limitations of the method concerning required computer resources.

INTRODUCTION

The theory of Markov diffusion processes has been used for a number of years to investigate the response statistics of nonlinear dynamic oscillators subjected to broad band random excitation. An important property of Markov diffusion processes is that their transition probability densities (TPD) and probability density functions (PDF) satisfy a partial differential equation known as the Fokker-Planck-Kolmogorov (FPK) equation, or forward Kolmogorov equation. Considerable effort has been directed toward a numerical solution of the FPK equation associated with a particular dynamic model. This has frequently proved to be difficult, and instead of trying to solve the FPK equation itself, one often has to be content with studying the corresponding, simpler Pontryagin-Vitt (PV) equations for the statistical moments. For reviews and further references to the literature, the reader may consult ref. [1-3].

Quite recently there has been some progress in the direct numerical solution of FPK equations. Of recent contributions it is worth noting those by Langtangen [4] and Enneking et al. [5]. However, it is really with the first-passage problem that it has been possible to establish the most satisfactory results. This is due to the fact that when Markov diffusion processes are used for this problem, the probability distribution of the time to first passage satisfies a backward Kolmogorov equation. Compared to the FPK equation, it has turned out that the backward Kolmogorov equation can be solved by more efficient numerical tools. For a review of results and methods for first-passage problems, see ref. [2,3,6].

An alternative to the direct numerical solution of the FPK equation is provided by the path integral solution (PIS) [7]. The PIS is a formal solution of the FPK equation which exploits

N. Bellomo · F. Casciati (Eds.)
Nonlinear Stochastic Mechanics
IUTAM Symposium Turin

the fact that that the associated Markov vector process locally looks like a Brownian motion. This makes it possible to express the TPD of the Markov process in closed form for a short time step. The Markov character can then be invoked to build a global (in time) solution from local (in time) TPDs. It seems that very few researchers have taken advantage of the PIS as a basis for developing numerical procedures to solve the FPK equation. The first systematic effort known to the present authors, which is similar to the method described here, is reported in [8]. Later the method was applied to a nonlinear random vibration problem by Kapitaniak [9]. Recent contributions are reported in [10-12].

The main efforts in the early stages of our own work with the development of numerical PIS procedures, were invested in establishing efficient numerical interpolation and smoothing methods. This was done to be able to fully exploit what we discovered to be one of the most important characteristics of the path integral solution, viz. the possibility to obtain accurate solutions down to very low probability levels, which is important for extreme value prediction. Another important feature of the path integral solution is the numerical stability that can be attained in the solution procedure. These properties are intimately related to the way the path integral solution is established, viz. by following the evolution in time of the statistics of the system. By this, the numerical calculation of the statistical distribution of the state space vector is closely connected to the physics of the dynamic model. This is in contrast to a direct numerical solution of the FPK equation, where such a connection is only implicit. Another important conclusion from our experience with the PIS method, is that there is practically no limitation on the nonlinearities that can be accommodated.

The accuracy and efficiency of the method has been tested by application to several nonlinear oscillators for which closed-form solutions are available, specifically white noise excited Duffing and Caughey oscillators, which lead to two-dimensional (2D) state spaces [10]. Some results for these type of oscillators will be given below. The method has also been applied to specific problems leading to 3D state spaces, and very satisfactory results have been obtained [12]. Our experience so far indicates that two-dimensional problems can be solved routinely with high accuracy, requiring a few minutes CPU time on a work station (DEC 3100). However, the solution of 3D problems requires more care in the sense that computer capacity starts to become an issue of importance. In such cases the CPU time easily runs into many hours. For problems with dimensions larger than 3, supercomputers would be needed.

In the present paper, the numerical PIS method will also be used to study the response statistics of parametrically excited nonlinear oscillators.

THE PATH INTEGRAL SOLUTION

To describe the basic ideas behind the path integral solution technique in a simple manner, it is expedient to explain the method for a scalar Markov process. Hence, let $X(t)$ denote a scalar Markov process with a TPD $p(x,t|x',t')$ satisfying the FPK equation

$$\frac{\partial p(x,t|x',t')}{\partial t} = -\frac{\partial}{\partial x}\{\, \alpha(x,t)\, p(x,t|x',t')\,\} + \frac{1}{2}\frac{\partial^2}{\partial x^2}\{\, \beta(x,t)\, p(x,t|x',t')\,\} \tag{1}$$

The functions $\alpha(x,t)$ and $\beta(x,t) \geq 0$ are determined by the coefficients in the SDE of the dynamic model, which in this case will assume the form, see e.g. ref. [7,13].

$$dX(t) = \alpha(X(t),t)dt + \sqrt{\beta(X(t),t)}\,dW(t) \qquad (t \geq 0) \tag{2}$$

Here $W(t)$ denotes a standard Brownian motion [7,13]. The key to the solution technique we have applied, lies in the observation that an approximate solution to equation (1) can always be obtained if the time-step $t - t'$ is small enough and the fact that every TPD can be expressed in terms of a product of short-time TPD's. Specifically, it can be shown that up to corrections of order τ^2, the following relation obtains [7]

$$p(x,t+\tau|x',t) = \frac{1}{\sqrt{2\pi\beta(x',t)\tau}} \exp\left\{ - \frac{(x-x' - \alpha(x',t)\tau)^2}{2\beta(x',t)\tau}\right\} \tag{3}$$

Also, by dividing the time interval (t',t) into N small time intervals of length $\tau = (t-t')/N$, it is obtained that ($t_j = t' + j\tau$, $t = t_N$, $x = x^{(N)}$, $t' = t_0$, $x' = x^{(0)}$)

$$p(x,t|x',t') = \int_{-\infty}^{\infty}\cdots\cdots\int_{-\infty}^{\infty} \prod_{j=1}^{N} p(x^{(j)},t_j|x^{(j-1)},t_{j-1})dx^{(1)}...dx^{(N-1)} \tag{4}$$

Hence, by combining equations (3) and (4), a formal (approximate) solution to the FPK-equation can be written down. Equation (4), which is often referred to as a path integral solution (PIS), is the gist of our numerical solution procedure. It is realized that a numerical solution according to this method, implies that one follows the evolution in time of the conditional probability density function (PDF) of the Markov process $X(t)$ after the start condition $X(t') = x'$.

NUMERICAL IMPLEMENTATION

The numerical implementation of the solution technique involves the following two elements: The stepping-stone for the time evolution of the conditional PDF is given by the Chapman-Kolmogorov equation

$$p(x^{(j)},t_j|x',t') = \int_{-\infty}^{\infty} p(x^{(j)},t_j|x^{(j-1)},t_{j-1})\; p(x^{(j-1)},t_{j-1}|x',t')dx^{(j-1)} \tag{5}$$

In the numerical solution, the discretization of the state space makes it appropriate to employ an interpolation procedure to increase the numerical efficiency. We have found that interpolation by cubic B-splines [14] offers the desired flexibility and accuracy for the type of problem considered here. So, at each time step $t_{j-1} \rightarrow t_j$, $p(x^{(j-1)},t_{j-1}|x',t')$ is represented as a cubic B-spline series

$$p(x^{(j-1)},t_{j-1}|x',t') = \sum_{k=1}^{n} \Gamma_{j-1,k}\, B_k(x^{(j-1)}) \tag{6}$$

where n = number of grid points in state space, $\{B_k(\cdot)\}_{k=1}^{n}$ is a basis of B-splines [14] and

$\Gamma_{j-1,k}$ are the interpolation coefficients. This representation makes it possible to retain high numerical accuracy even with a fairly coarse basic grid. By substituting from equation (6) into equation (5) it is obtained that

$$p(x^{(j)},t_j|x',t') = \sum_{k=1}^{n} \Gamma_{j-1,k} \int_{-\infty}^{\infty} p(x^{(j)},t_j|x^{(j-1)},t_{j-1})\, B_k(x^{(j-1)})\, dx^{(j-1)} \tag{7}$$

In this paper we shall consider only homogeneous Markov processes, i.e. processes with TPDs that do not depend on absolute time, but only on the time increment τ. From equation (3), this is seen to imply that $\alpha(\cdot)$ and $\beta(\cdot)$ only depend on x'. Hence, for every j,

$$p(x^{(j)},t_j|x^{(j-1)},t_{j-1}) = p(x^{(j)},\tau|x^{(j-1)},0) \tag{8}$$

From equation (7) it now follows that for a fixed value of the time increment τ, each of the integrals on the right hand side of the equation can be calculated once and for all and stored for repeated use. That is, the following parameters are calculated initially and stored

$$B_{k,l} = \int_{-\infty}^{\infty} p(x_l^{(j)},\tau|x^{(j-1)},0)\, B_k(x^{(j-1)})\, dx^{(j-1)}, \quad l = 1,2,\dots,n \tag{9}$$

Here, the index l refers to the points of the grid. It may be noted here that due to the properties of the TPD for small τ, the matrix $(B_{k,l})$ have a strongly banded character with the elements decreasing rapidly away from the main diagonal. This has important implications for the efficiency of the computer program. Let $p_l^{(j)} = p(x_l^{(j)},t_j|x',t')$. Then equation (7) can be rewritten as

$$p_l^{(j)} = \sum_{k=1}^{n} \Gamma_{j-1,k}\, B_{k,l} \tag{10}$$

Applying standard spline interpolation procedure [14] to the calculated values of the TPD at the grid points, viz. $p_l^{(j)}$, a new set of interpolation coefficients $\{\Gamma_{j,k}\}_{k=1}^{n}$ are calculated, and an updated representation of the conditional PDF for time step j is obtained, cf. equation (6). For each time step, the normalization of the resulting PDF given by equation (6) is checked. That is, if

$$\int_{-\infty}^{\infty} p(x^{(j)},t_j|x',t')\, dx^{(j)} = \sum_{k=1}^{n} \Gamma_{j,k} \int_{-\infty}^{\infty} B_k(x)dx = q_j \tag{11}$$

and $q_j \neq 1.0$ within the specified accuracy, then the following replacement is made to restore the correct normalization

$$\Gamma_{j,k} \leftarrow \frac{\Gamma_{j,k}}{q_j} \tag{12}$$

This way of implementing the numerical calculation of the PIS, has so far proved to yield a very stable and accurate procedure.

EXAMPLES

The accuracy and power of the developed PIS procedure will be illustrated by application to specific case studies taken from two classes of dynamic models. The equation of motion for the state space vector $\boldsymbol{Z} = (Z_1,Z_2)^T = (X,\dot{X})^T$ can for both classes be represented as follows

$$d\dot{\boldsymbol{Z}}(t) = \boldsymbol{m}[\boldsymbol{Z}(t)]dt + \sigma[\boldsymbol{Z}(t)]\ d\boldsymbol{W}(t) \tag{13}$$

Here $\boldsymbol{X} = (X,\dot{X})^T$, $\boldsymbol{m}(\cdot) = (m_1(\cdot),m_2(\cdot))^T$, $\sigma(\cdot) = (\sigma_{ij}(\cdot))$, $i = 1,2$; $j = 1,2,3$ and $\boldsymbol{W}(t) = (W_1(t),W_2(t),W_3(t))^T$, where the $W_j(t)$, $j = 1,2,3$, are independent standard (real) Brownian motions. Further, $\boldsymbol{m}(\cdot)$ and $\sigma(\cdot)$ are such that $m_1(Z_1,Z_2) = Z_2$ and $\sigma_{1j}(\cdot) = 0$ for $j = 1,2,3$. With suitable restrictions on $\boldsymbol{m}(\cdot)$ and $\sigma(\cdot)$ [7,13], $\boldsymbol{Z}(t)$ becomes a Markov diffusion process. Proceeding as in Risken [7], it can be shown that, up to correction terms of order τ^2, the associated TPD assumes the form

$$p(z,\tau\,|\,z',0) = \delta(z_1 - z_1' - z_2'\tau) \cdot \tilde{p}(z_2,\tau\,|\,z',0) \tag{14}$$

$\tilde{p}(z_2,\tau\,|\,z',0)$ is given by the relation

$$\tilde{p}(z_2,\tau\,|\,z',0) = \frac{1}{\sqrt{2\pi\beta(z')\tau}} \exp\left\{-\frac{(\,z_2 - z_2' - m_2(z')\tau\,)^2}{2\beta(z')\tau}\right\} \tag{15}$$

where

$$\beta(z') = \sum_{j=1}^{3} \sigma_{2j}(z')^2 \tag{16}$$

By combining equations (14) and (15), and applying the solution technique described in the previous section, the TPD $p(z,t\,|\,z',t')$ for large $t-t'$ can be calculated. By this, the time evolution of the system when it starts from e.g. rest, can be studied. The stationary PDF is obtained in the limit as $t - t' \to \infty$.

Example 1 - Caughey oscillator

There is a class of dynamic models for which there exist an analytical solution for the stationary joint PDF of $\boldsymbol{Z}$. A member of this class may be called a Caughey oscillator [15]. The generic equation of motion for this oscillator can be written as

$$\ddot{X} + \dot{X}g(E) + h(X) = \Gamma\ N(t) \tag{17}$$

$N(t)$ denotes a stationary, zero-mean Gaussian white noise satisfying $E[N(t)\ N(t+\tau)] = \delta(\tau)$, where $\delta(\cdot)$ denotes Dirac's delta function, Γ is a positive constant and $g(E)$ is a function of the total energy $E = E(X,\dot{X})$ given as follows

$$E = \frac{1}{2}\dot{X}^2 + V(X) \tag{18}$$

where

$$V(x) = \int_0^x h(s)ds \tag{19}$$

For this example $m_2(x,\dot{x}) = -\dot{x}g[E(x,\dot{x})] - h(x)$, and we may choose $\sigma_{21} = \sigma_{22} = W_1 = W_2 = 0$. $\sigma_{23} = \Gamma$ and $dW_3(t) = N(t)dt$. The stationary, joint PDF, denoted by $p_s(\cdot)$, is then determined by the relation [15,16]

$$p_s(x,\dot{x}) = C\exp\{-\frac{2}{\Gamma^2}\int_0^{E'} g(s)ds\} \tag{20}$$

where $E' = \dot{x}^2/2 + V(x)$, and C is a normalization constant to ensure a total probability equal to 1.0.

For the illustration purposes in this paper, we have chosen the following special case of equation (17)

$$\ddot{X}(t) + 2\xi\dot{X}(t)\{1 + \varepsilon\sqrt{\frac{1}{2}\dot{X}^2(t) + \frac{1}{2}X^2(t) + \frac{1}{4}\lambda X^4(t)}\} + X(t) + \lambda X^3(t) = 2\sqrt{\xi}N(t) \tag{21}$$

with parameters ξ, ε and λ.

The stationary PDF only depends on the parameters ε and λ, and the numerical solution for the following set of parameter values has been calculated $(\varepsilon,\lambda) = (0,0)$ (Gaussian response), $(0, 0.2)$ (Duffing oscillator) and $(0.5, 0.1)$. The calculations were carried out with the same number of grid points on both axes in state space, viz. 45. Since the resulting PDFs are actually independent of ξ, the value $\xi = 0.1$ was chosen for the Gaussian and Duffing cases, while $\xi = 0.5$ was adopted for the last case. The time increments used were $\tau = 0.0025$ sec., 0.001 sec. and 0.02 sec., respectively. The total CPU time on a DEC station 3100 was about 5 minutes for each case. In Figs. 1 and 2 are shown the marginal PDFs of the displacement response for the three case studies considered, together with the corresponding analytical solutions. In Fig. 3 are given the corresponding analytical and numerical results for the mean upcrossing rate. It is seen that in all three cases the agreement between the numerical PIS and the analytical solution is very good over the whole range of probability levels given. In fact, the accuracy can be retained down to much lower probability levels ($\approx 10^{-10}$) at a moderate increase in computer time.

Example 2 - Parametric and external excitation

In this example, the response statistics of a nonlinear oscillator subjected to both external and parametric random excitation will be illustrated by applying the methodology of the paper to four specific case studies.

The equation of motion of the oscillator is the following [16]

$$\ddot{X} + 2\xi\,[1 + \tilde{N}_1(t)]\,\dot{X} + \gamma\,[X^2 + \frac{\dot{X}^2}{\omega_o^2}]\,\dot{X} + \omega_o^2\,[1 + \tilde{N}_2(t)]\,X = \tilde{N}_3(t) \qquad (22)$$

Here ξ, γ and ω_o are positive constants, $\tilde{N}_j(t)$, $j = 1,2,3$, are independent Gaussian white noises satisfying

$$E[\tilde{N}_j(t)\tilde{N}_j(t+\tau)] = \Gamma_j^2\,\delta(\tau), \quad j = 1,2,3 \qquad (23)$$

where Γ_j are positive constants. For this example it is found that $m_2(x,\dot{x}) = -2\xi\dot{x} - \gamma\,[x^2+\dot{x}^2/\omega_o^2] - \omega_o^2 x$, $\sigma_{21}(x,\dot{x}) = -2\xi\dot{x}\Gamma_1$, $\sigma_{22}(x,\dot{x}) = -\omega_o^2 x\Gamma_2$ and $\sigma_{23}(x,\dot{x}) = \Gamma_3$.

This model was studied by Dimentberg [17], who showed that when

$$\omega_o^2\,\Gamma_2^2 = 4\xi^2\,\Gamma_1^2 \qquad (24)$$

a closed-form expression for the stationary, joint PDF can be obtained. It is given as

$$p_s(x,\dot{x}) = C\,\frac{\exp\{-\mu\,(x^2 + \dot{x}^2/\omega_o^2)\}}{(\kappa + x^2 + \dot{x}^2/\omega_o^2)^{\delta-\kappa\mu}} \qquad (25)$$

where C is a normalization constant and

$$\kappa = \frac{\Gamma_3^2}{\omega_o^4\Gamma_2^2}, \qquad \delta = \frac{2\xi}{\omega_o^2\Gamma_2^2} + \frac{1}{2}, \qquad \mu = \frac{\gamma}{\omega_o^2\Gamma_2^2} \qquad (26)$$

By this, we have the opportunity to test the accuracy of the PIS method for this kind of dynamic model. The results of two particular cases will be presented.

Case 1 : Here the following parameter values were used. $\xi = 0.1$, $\gamma = 0.1$, $\omega_o = 1.0$, $\Gamma_1^2 = 2.5$, $\Gamma_2^2 = 0.1$, $\Gamma_3^2 = 0.3$. For the numerical calculations a grid size of 49x49 points and a time increment $\tau = 0.01$ sec. was used. The total CPU time on a DEC 3100 work station was 3 minutes for the PIS calculation. The results for the analytical and numerical solutions are given in Figs. 4 - 6. In Figs. 4 and 5 are shown the marginal PDF of the displacement response and in Fig. 6 is shown the corresponding mean upcrossing rate, see also ref. [11].

Case 2 : In this case the following set of parameters were used. $\xi = 0.1$, $\gamma = 0.4$, $\omega_o = 1.0$, $\Gamma_1^2 = 5.0$, $\Gamma_2^2 = 0.2$, $\Gamma_3^2 = 0.3$. A grid size of 51x51 points together with a time increment $\tau = 0.01$ sec. were chosen. The CPU time was about the same as in the previous case. The same results as for Case 1 are presented in Figs. 4 - 6.

For the next two case studies there is no analytical solution available.

Case 3 : The parameter set $\xi = -0.1$, $\gamma = 0.1$, $\omega_o = 1.0$, $\Gamma_1^2 = 0.1$, $\Gamma_2^2 = 0.1$, $\Gamma_3^2 = 0.3$ was adopted. The grid size was 71x71 points, with a time increment $\tau = 0.01$ sec. as before. The CPU time for this case was about six minutes. The results are given in Figs. 7 - 9. In Fig. 7 is shown the marginal PDF of the displacement response, while the conditional PDF of the displacement response for zero velocity is given in Fig. 8, cf. also ref. [11]. Fig. 9 shows the corresponding mean upcrossing rate.

Case 4 : This case is like the previous one, except that $\xi = -0.3$, $\gamma = 0.4$. The results are given in Figs. 7 - 9.

CONCLUSIONS

A general method for calculating the response statistics of nonlinear dynamic systems excited by white noise or filtered white noise has been described. Application of the method leads to estimates of the joint PDF of the state variables. Its practical application is at present limited to single-degree-of-freedom dynamic models. The example calculations presented shows that the method gives very accurate estimates of the PDF. In fact, for every example having analytical solution on which the method has been tested, complete agreement has been found with proper choice of grid size and time increment. Here, of course, only a few particular cases can be given.

Experience with the method indicates that two-dimensional problems can be solved routinely with high accuracy requiring a few minutes CPU time on a work station (DEC station 3100). The solution of three-dimensional problems requires more care in the sense that computer capacity starts to become an issue of importance. In such cases the CPU time easily runs into hours.

REFERENCES

1. Roberts, J. B. Response of Nonlinear Mechanical Systems to Random Excitations. Part 1: Markov Metods, *Shock and Vibration Digest*, Vol. 13, No. 5, pp.17-28, 1981.

2. Crandall, S. H. and Zhu, W. Q. Random Vibration: A Survey of Recent Developments, *ASME Journal of Applied Mechanics*, Vol. 50, pp. 953-962, 1983

3. Roberts, J. B. First-Passage Probabilities for Randomly Excited Systems: Diffusion Methods, *Probabilistic Engineering Mechanics*, Vol. 1, No. 2, pp. 66-81, 1986.

4. Langtangen, H. P. A general numerical solution method for Fokker-Planck equations with applications to structural reliability, *Probabilistic Engineering Mechanics*, Vol. 6, No. 1, pp. 33-48, 1991.

5. Enneking, T. J., Spencer Jr., B. F. and Kinnmark, I. P. E. Stationary Two-State Variable Problems in Stochastic Mechanics, *Journal of Engineering Mechanics*, Vol. 116, No. 2, pp. 343-358, 1991.

6. Spencer, Jr., B. F. *Reliability of Randomly Excited Hysteretic Structures*, Lecture Notes in Engineering, Springer Verlag Berlin, Heidelberg 1986.

7. Risken, H. *The Fokker Planck Equation*, Second Edition, Springer Verlag Berlin, Heidelberg 1989.

8. Wehner, M. F. and Wolfer, W. G. Numerical evaluation of path-integral solutions to Fokker-planck equations, *Physical Review A*, Vol. 27, No. 5, pp. 2663-2670, 1983.

9. Kapitaniak, T. Stochastic Response with Bifurcations to Non-Linear Duffing's Oscillator, *Journal of Sound and Vibration*, Vol. 102, No. 3, pp. 440-441, 1985.

10. Naess, A. and Johnsen, J. M. Direct Numerical Simulation of the Response Statistics of Nonlinear Dynamic Systems, *Proc. Scandinavian Forum for Stochastic Mechanics*, Lund Institute of Technology, Lund, Sweden, August 1990.

11. Sun, J. Q. and Hsu, C. S. The Generalized Cell Mapping Method in Nonlinear Random Vibration Based Upon Short-Time Gaussian Approximation, *Journal of Applied Mechanics*, Vol. 57, pp. 1018-1025, 1990.

12. Johnsen, J. M. and Naess, A. Time Variant Wave Drift Damping and its Effect on the Response Statistics of Moored Offshore Structures, *Proc. 1st Int Conference on Offshore and Polar Engineering* (ISOPE-91), Edinburgh, August 1991.

13. Wong, E. and Hajek, B. *Stochastic Processes in Engineering Systems*, Springer Verlag, New York, 1985.

14. de Boor, C. *A Practical Guide to Splines*, Springer Verlag, New York, 1978.

15. Caughey, T. K. Nonlinear theory of random vibrations, In: *Advances in Applied Mechanics*, Academic Press, New York, 1971.

16. Lin, Y. K. *Probabilistic Theory of Structural Dynamics*, McGraw-Hill, New York, 1967.

17. Dimentberg, M. F. An Exact Solution to a Certain Non-Linear Random Vibration Problem, *Int. Journal of Non-Linear Mechanics*, Vol. 17, No. 4, pp.231-236, 1982.

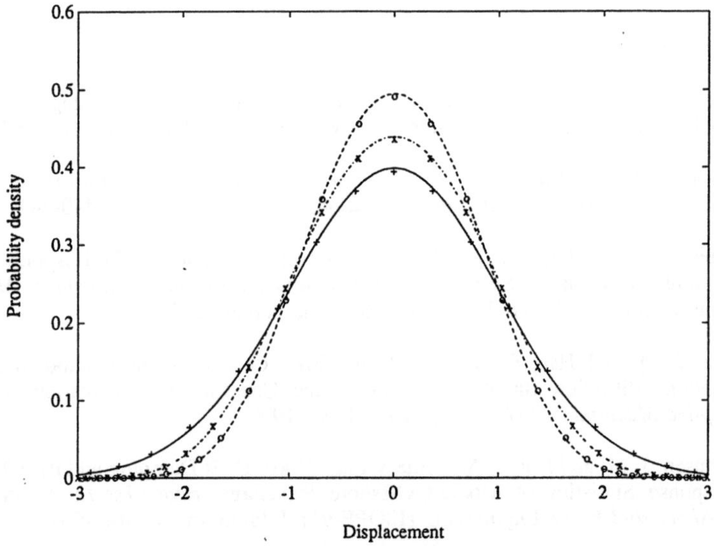

Figure 1 PDF of displacement response for the Caughey oscillator in example 1. Analytical solutions: ——, $\varepsilon = 0$, $\lambda = 0$; —·—·, $\varepsilon = 0$, $\lambda = 0.2$; ----, $\varepsilon = 0.5$, $\lambda = 0.1$. Numerical PIS: +, $\varepsilon = 0$, $\lambda = 0$; x, $\varepsilon = 0$, $\lambda = 0.2$; o, $\varepsilon = 0.5$, $\lambda = 0.1$.

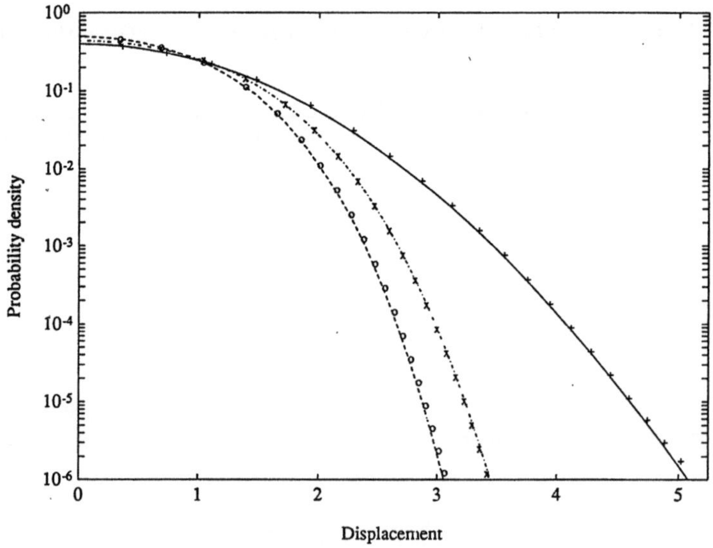

Figure 2 PDF of displacement response for the Caughey oscillator in example 1. Key as in Fig.1.

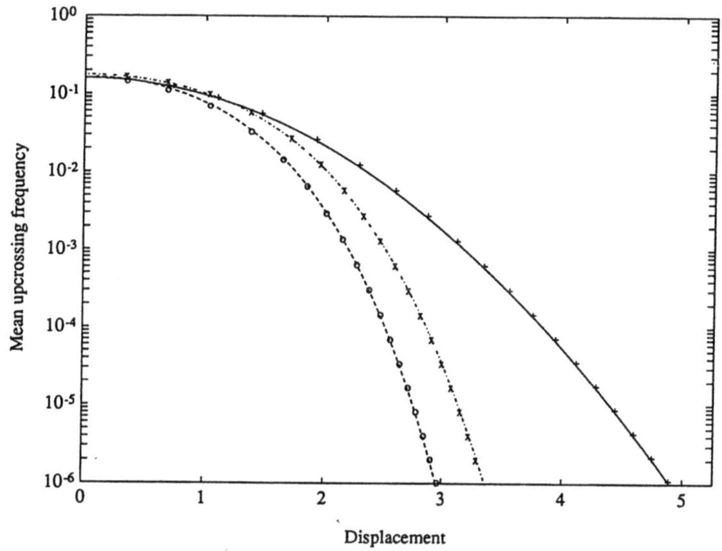

Figure 3 Mean upcrossing rate for the Caughey oscillator in example 1. Key as in Fig.1.

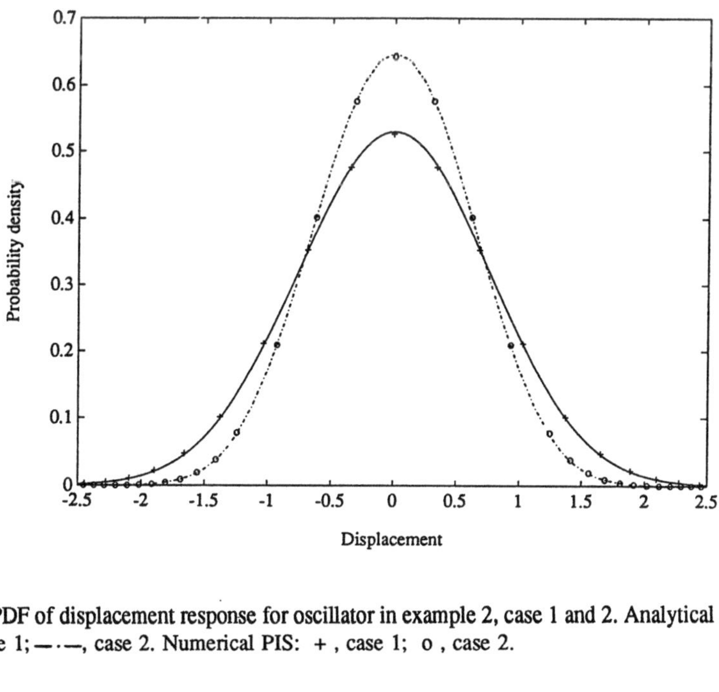

Figure 4 PDF of displacement response for oscillator in example 2, case 1 and 2. Analytical solutions: ——, case 1; —·—, case 2. Numerical PIS: +, case 1; o, case 2.

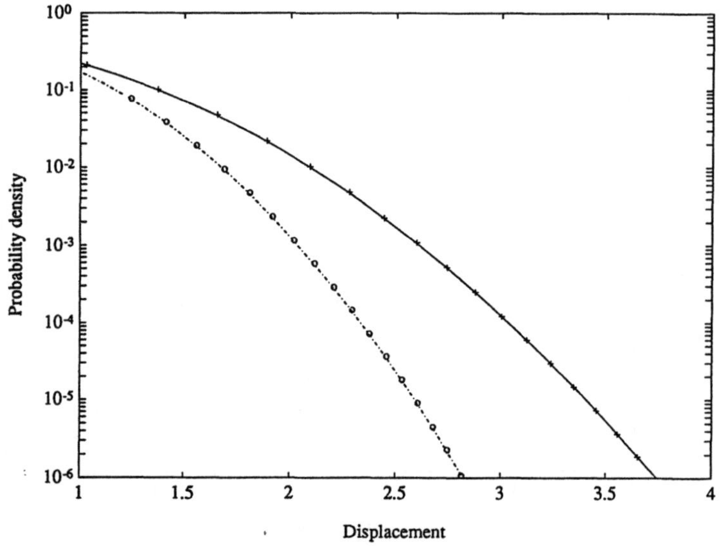

Figure 5 PDF of displacement response for oscillator in example 2, case 1 and 2. Key as in Fig. 4.

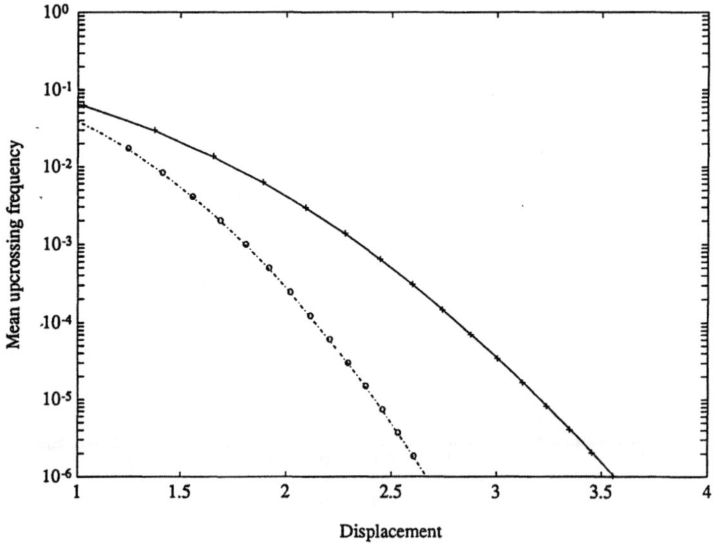

Figure 6 Mean upcrossing rate for the oscillator in example 2, case 1 and 2. Key as in Fig. 4.

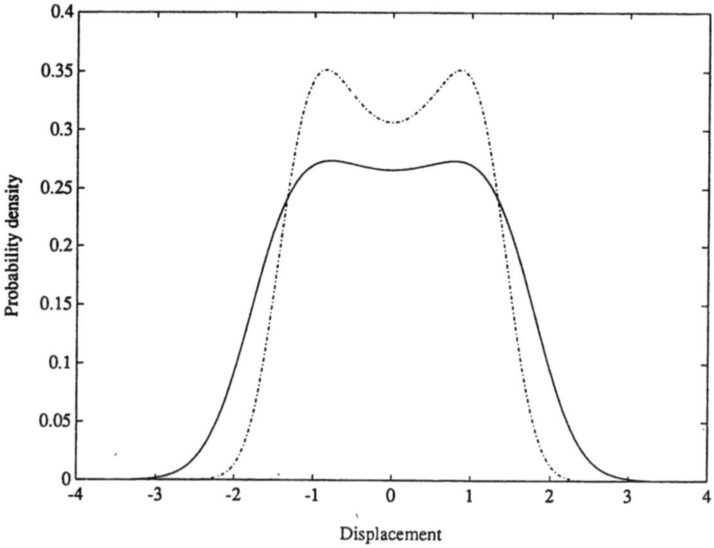

Figure 7 PDF of displacement response for oscillator in example 2, case 3 and 4. Numerical PIS: —— , case 3; —·— ,case 4.

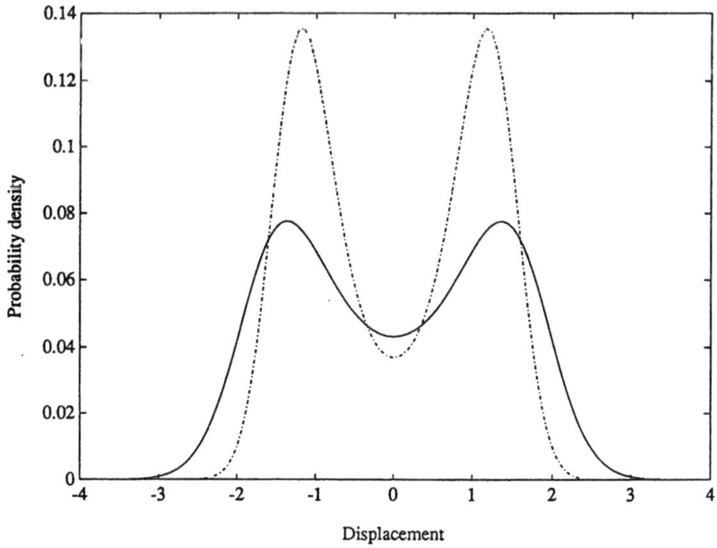

Figure 8 Conditional PDF of displacement response at zero velocity for oscillator in example 2, case 3 and 4. Key as in Fig. 7.

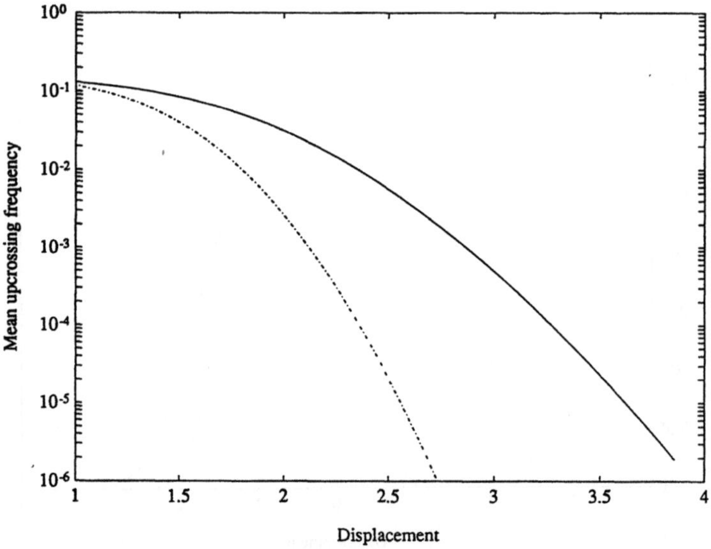

Figure 9 Mean upcrossing rate for oscillator in example 2, case 3 and 4. Key as in Fig. 7.

A Comprehensive Stationary Non-Gaussian Analysis of BWB Hysteresis

Mohammad Noori, Mechanical Engineering Department, Worcester Polytechnic Institute, Worcester, Massachusetts 01609, U.S.A.

Hamid Davoodi, Department of Mechanical Engineering, University of Puerto Rico, Mayaguez, Puerto Rico, 00708

Thomas T. Baber, Civil Engineering Department, University of Virginia, Charlottesville, Virginia, 22901, U.S.A.

SUMMARY

A thorough non-Gaussian stationary analysis of a nondimensionalized form of a Bouc-Wen-Baber smooth hysteresis model is presented. Stationary moments of up to 6th order are generated via 3000 sample Monte Carlo simulation in order to assess the convergence of the non-Gaussian response statistics. The degree of deviation from Gaussian response is assessed by comparison of the simulated probability density functions of various response variables with a corresponding unit normal density function. The effect of variation of hysteresis shape control parameters and other important system and excitation parameters, such as the postyield to preyield stiffness ratio, and power spectral density level on response coordinate moments of various orders is studied. The effect of these parameters on correlation coefficients, probability density of the displacement, and other statistical indices are thoroughly studied. This study provides considerable insight about the range of important system and excitation parameters influencing system response. These information on the characteristic behaviors of BWB hysteresis model is useful for non-Gaussian analysis of MDOF hysteresis models and comparative studies with approximation techniques.

INTRODUCTION

In recent years, the problem of random vibration analysis of non-linear systems in general, and hysteretic systems in particular, has attracted considerable attention. This has been due to the significance of this problem in response analysis of structural or mechanical systems subjected to earthquakes, winds, acoustic and sea wave loadings. Extensive analyses of systems with various types of piecewise linear or smoothly varying hysteresis models have been conducted by researchers interested in the response of hysteretic systems to random or complex deterministic excitations. Comprehensive reviews of these works can be found in references [1-17]. Rate type hysteresis models, as typified by the model

N. Bellomo · F. Casciati (Eds.)
Nonlinear Stochastic Mechanics
IUTAM Symposium Turin

originated by Bouc [18], and extended by Wen [16], Baber and Wen [1], and others [4,5,10] (and referred to herein as the BWB model), have proven particularly useful. Applications and extensions of the BWB model have included stochastic computational mechanics [19,20], modelling base isolation for primary structures [21-23], hysteretically yielding frames [1,3,6,7], systems identification of inelastically behaving structures [24], hysteretic columns [25], stochastic damage analysis of inelastic structures [26], stochastic control [27], and other studies [28].

Since the exact analysis of hysteretic systems under random excitation is difficult, a variety of approximation methods have been developed and utilized for random vibration of hysteretic systems. The method of stochastic equivalent linearization has been used most extensively in the analysis of systems with general hysteretic behavior [1-7,10,14,16,29-34]. However, due to highly nonlinear behavior of BWB and other general hysteresis models, particularly at high levels of excitation, the Gaussian response assumption for these types of hysteretically degrading systems is not valid, and this has been addressed in the literature [4,5,8,11-13,26,35-37]. Nevertheless, information on higher order response statistics for hysteretically yielding structures is very important in the reliability and random vibration analysis of these systems. Moreover, the method of equivalent linearization is not capable of predicting spectral properties of the response for nonlinear systems since the spectra of the equivalent linear system response span only the frequency range of the input spectrum [14,15]. This limitation, for example, precludes drift of a linearized model of a hysteretic system under realistic (i.e. high pass filtered) models of noisy excitations.

Researchers have recently focused their efforts in developing non-Gaussian solution techniques for the response analysis of these systems [8,36,37]. Noori et al [11-13] and Davoodi and Noori [8] extended the cumulant-neglect approach proposed by Wu and Lin [26] and Ibrahim [38,39] to the Gaussian and non-Gaussian analysis of BWB and three other general pinching hystereses [8,11,12]. In addition, recently Liu [37], presented another non-Gaussian response analysis of BWB hysteresis based on an extension of Crandall's approach [40]. These works have been the first attempt for the stationary or non-stationary non-Gaussian response analysis of systems with general hysteresis. In all these non-Gaussian analyses, emphasis has been placed on the nonstationary response of systems, with the goal of incorporating system degradation. The aforementioned work have shown that non-Gaussian analysis of general hysteretic systems is a practical task and information on higher order moments for hysteretically yielding systems can be obtained. These information, which have not been available in the past, are valuable for reliability analysis of these systems and are also particularly important in random vibration analysis of MDOF hysteretically yielding structures.

Although these studies have revealed new information about the non-Gaussian nature of BWB system response, the presented probability density functions of response variables are not realizable, i.e. are negative at the tails. This problem is due to "cumulant-neglect" type formulation and the inherent problems with orthogonal series expansions utilized for non-Gaussian density function. Other problems have also been reported on the use of "cumulant-neglect" type analyses [8,12,35,41,42]. One suggestion is to seek alternative approaches for developing non-Gaussian probability density functions [35]. These problems should be addressed before this approximation method can be utilized for non-Gaussian analysis of MDOF systems.

In addition, the response characteristic behavior of hysteresis models such as BWB, depends on the system and excitation parameters. Thus, a thorough and comprehensive parameter study of a general hysteresis, e.g. BWB, under simulated stationary non-Gaussian conditions can reveal information which can be useful for development of a new approximation technique. To the best of the authors' knowledge, only one comprehensive stationary analysis of general hysteresis [36] has been reported in the literature which explores some important characteristics of BWB model. However, there are numerous other characteristics that need to be explored. Hence, in this paper, a thorough study of stationary non-Gaussian response statistics of a BWB general hysteresis model is undertaken. Non-Gaussian stationary moments of up to 6th order are generated using 3000 sample Monte Carlo simulation. The degree of deviation from Gaussian response is assessed by comparison of the probability densities of response statistics with a corresponding unit Gaussian probability density function. The effect of variation of parameters controlling the shape of hysteresis on moments of various orders of response coordinates is studied. Also the variation of important system and excitation parameters on statistical indices are thoroughly studied. This investigation provides new information about the range of parameters and other unexplored characteristic behaviors of this hysteresis model as they are needed for non-Gaussian analysis of hysteresis models. These data will be valuable for extension of non-Gaussian analysis to MDOF and reliability analysis of hysteretically yielding structures.

STATIONARY NON-GAUSSIAN ANALYSIS OF BWB HYSTERESIS MODEL

The nonlinear system to be studied is a single degree of freedom (SDOF)system with the nondimensionalized governing differential equations presented for the first time in this study

$$v'' + 2\zeta v' + \alpha v + (1-\alpha)z = f - (v_g)'' \tag{1-a}$$

$$(v_g)'' = (v_o)'' - W = 2\zeta_g.\Omega_g + \Omega_g{}^2.v_o \tag{1-b}$$

$$(v_o)'' + 2\zeta_g.\Omega_g.(v_o)' + \Omega_g{}^2 = W(\tau) \tag{1-c}$$

$$z' = v' - \beta v'|z|^n - \gamma|v'||z|^{(n-1)}z \tag{1-d}$$

in which $v = u/(u_y)$ = nondimensional displacement; $u_y = Z_{ult}/k$; (k = initial tangent stiffness); Z_{ult} = elasto-plastic yield restoring force; Z/Z_{ult} = nondimensional hysteretic restoring force; $v_g = u_g/u_y$ = nondimensional ground displacement; v_o = nondimensional base filter variable; W = nondimensional white noise input to base filter; f = nondimensional white noise forcing; $\tau = \omega t$ = nondimensional time; $\omega^2 = k/m$ = linear natural frequency; and $\Omega_g = \omega_g/\omega$ = base frequency/structure frequency ratio. Differentiation (' and ") is with respect to τ. Hence, v' and v" are nondimensional velocity and accelerations respectively. In the case of a simple white noise input, to be considered herein, the equations simplify to equations (1-a) and (1-d), and the $(v_g)''$ term disappears. In equation (1-d), β, γ and n are the hysteresis shape parameters. In the most general form, if deterioration is excluded, the response is a function of α, β, γ, n, ζ, ζ_g, Ω_g and the level of forcing. For the cases considered in detail here, the response varies with α, β, γ, n, ζ and the forcing level. β, γ, and η determine the hysteresis shape; and A determines the tangent stiffness. Parameter n controls the smoothness of this model, the general slopes of the curve is controlled by $\beta+\gamma$, and γ controls the slimness of the hysteresis loop. The BWB model can represent a variety of hysteretic, deteriorating type of behavior with a considerable range of energy dissipation [1,16,17].

NUMERICAL ANALYSES

The nondimensionalization significantly reduces the number of distinct cases that need explicit consideration. Even so, it is not feasible to address all cases within this paper. Hence, representative statistics are presented for the cases $\alpha = 1/21$, $\zeta = 0$, and n = 1, with $\beta = 0.1$, 0.5, and 0.9, and with $\gamma = 1 - \beta$. Moment statistics are presented for v, v', and z for these cases, for varying power spectral density function levels. Figures 1, 2, and 3 show the second order moment estimates at various PSD levels for displacement, velocity and restoring force respectively and for the cases $\beta = 0.1$, 0.5, and 0.9. Figures 4, 5, and 6 show 4th order moments for the same response coordinates, and Figures 7,8, and 9 show 6th order moments corresponding to these cases. Detailed probability density function estimates are presented in Figures 10, 11, and 12 for the case $\beta = \gamma = 0.5$, and the power spectral density function levels 0.0016, 0.0159, and 0.3183. In particular, it is important to notice the significant deviation from Gaussian behavior displayed by the density function of z, even at relatively low forcing levels. This suggests that z may not be well modeled by any density function that assumes unbounded behavior, even if a series expansion is permitted. As can be seen, all response moment plots are in log-log scale since the logarithmic form may tend to obscure differences that might be significant when simulation results are compared with other approximation analyses. These plots, Figures 1-9, reflect the continued growth of all moments as predicted by the simulation. The rates of growth of the higher order moments can be compared with the rates of growth obtained if the displacement, velocity, and restoring force were Gaussian, as a measure of the deviation of the distributions from Gaussian behavior. The density function plots are of particular interest. The data was reduced to standard form by subtracting the mean and dividing by the standard deviation. The density function estimates were then constructed from a normalized histogram of the 3000 data points in the ensemble, and smoothed by using AXUM algorithm. The density functions for displacement and velocity appear to agree well with the Gaussian density function within the range roughly +/- 2.5 standard deviations. Beyond that range, the tail behavior begins to deviate from the Gaussian significantly. The density

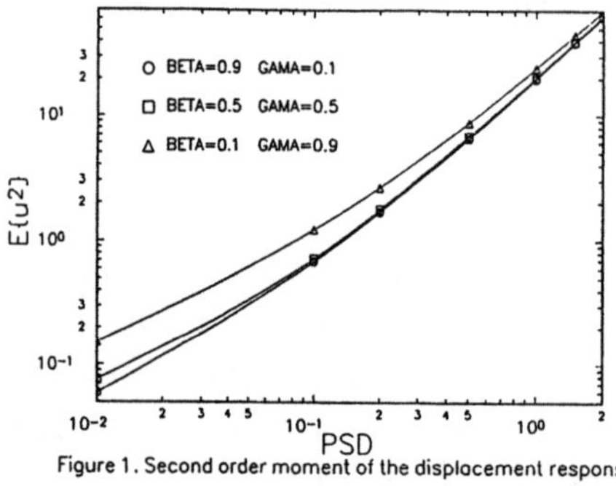

Figure 1. Second order moment of the displacement response.

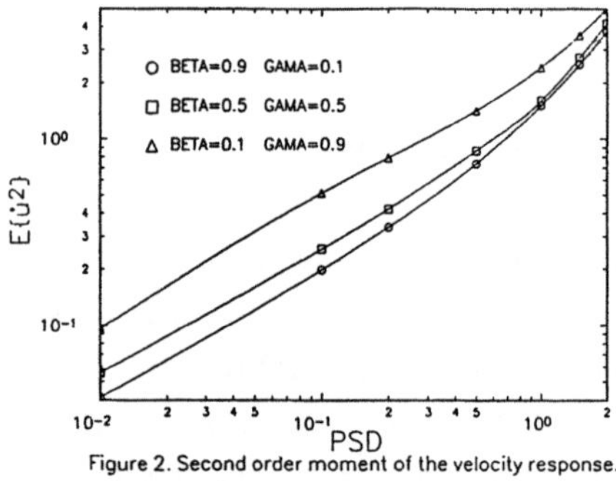

Figure 2. Second order moment of the velocity response.

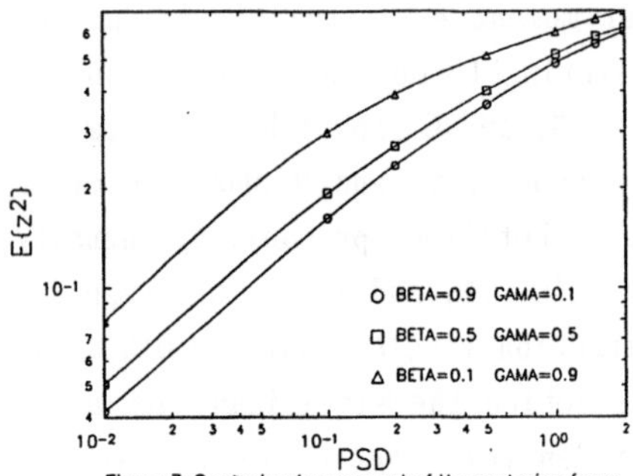

Figure 3. Second order moment of the restoring force.

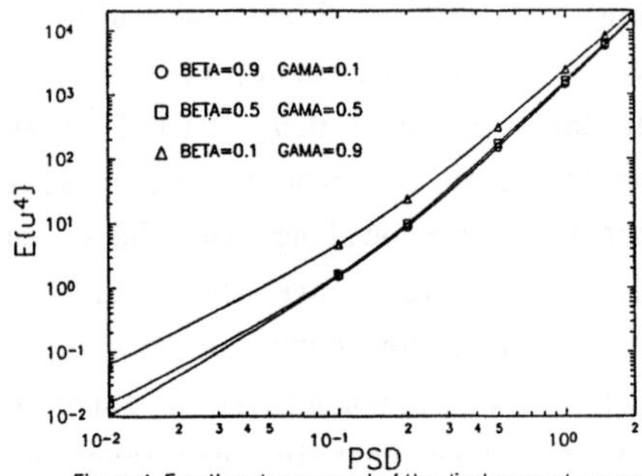

Figure 4. Fourth order moment of the displacement response.

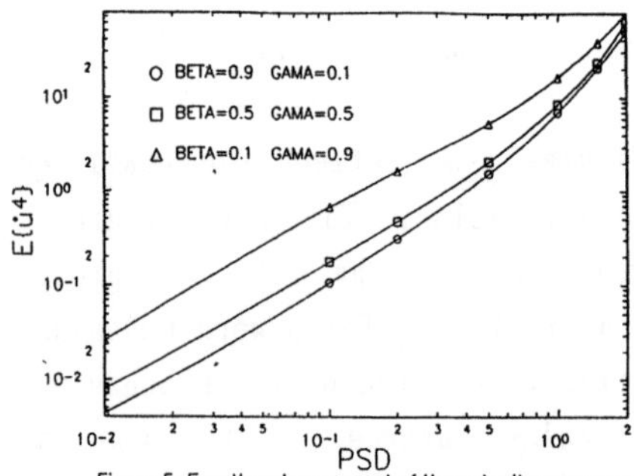

Figure 5. Fourth order moment of the velocity response.

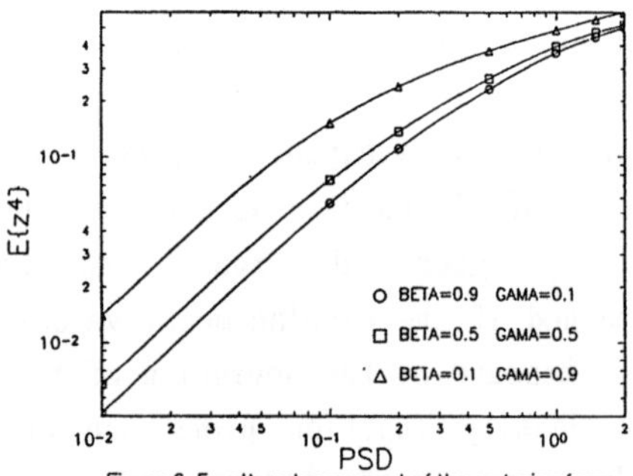

Figure 6. Fourth order moment of the restoring force.

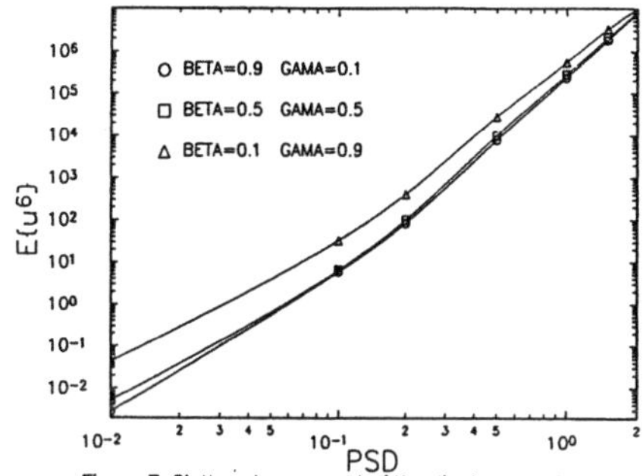

Figure 7. Sixth order moment of the displacement response.

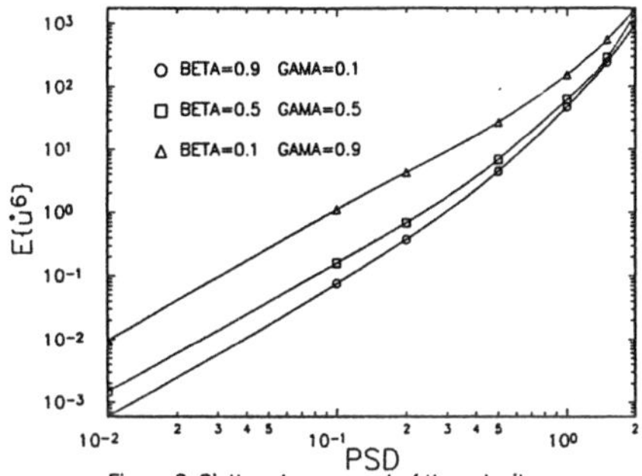

Figure 8. Sixth order moment of the velocity response.

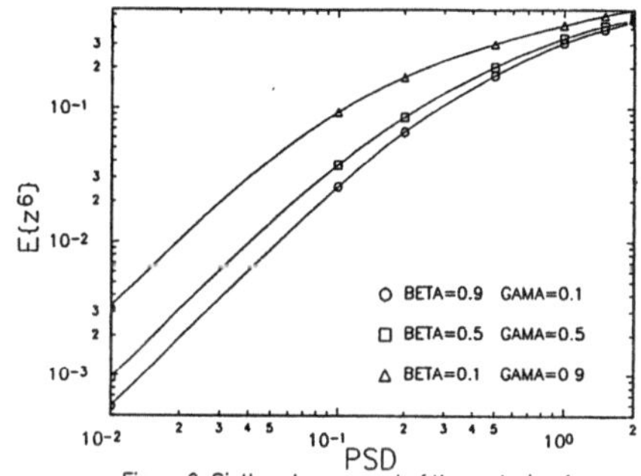

Figure 9. Sixth order moment of the restoring force.

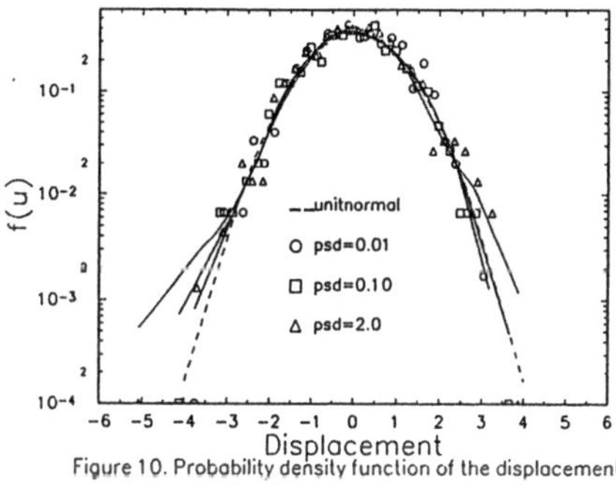

Figure 10. Probability density function of the displacement.

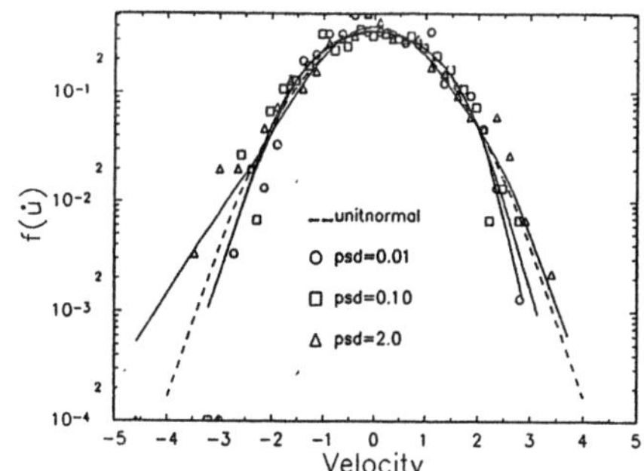

Figure 11. Probability density function of the velocity.

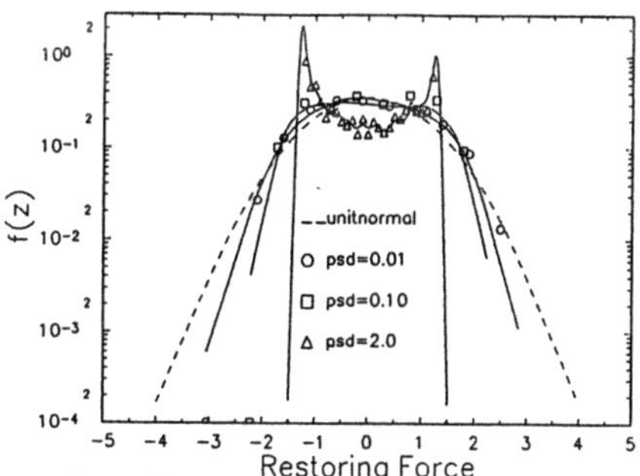

Figure 12. Probability density function of the restoring force.

functions for restoring force differ from the Gaussian density even for small levels of response. This difference becomes more pronounced as the response level increases. At the highest level of response, the density for restoring force does not resemble the Gaussian distribution at all.

CONCLUSION

A thorough non-Gaussian stationary analysis of a nondimensionalized BWB smooth hysteresis model was studied. The behavior observed in the rate of growth of higher order moments as compared with the rates obtained if the response moments were Gaussian, and the corresponding density functions indicate that the restoring force may not be modelled by any density function that assumes unbounded behavior. This indicates that series expansions utilized in cumulant-neglect type approximations which are based on such assumption are not suitable for hysteretic restoring force of this type. On the other hand, displacement and velocity responses reflect a Gaussian type behavior at all excitation levels considered. This study provides a valuable information about the range of important system and excitation parameters influencing the system response.

REFERENCES

1. Baber, T.T. and Wen, Y.K.: Stochastic Response of Multistory Yielding Frames. J. of Earthq. Eng. and Struc. Dyn., 1982, (10), 403-416.
2. Baber, T.T.: Nonzero Mean Random Vibration of Hysteretic Systems. ASCE, J. of Eng. Mech. 1984, (110), 1036-1049.
3. Baber, T.T.: Nonzero Mean Random Vibration of Hysteretic Frames. Computers and Structures, 1986, (23), 265-277.
4. Baber, T.T. and Noori, M.: Random Vibration of Degrading, Pinching Systems. ASCE, J. of Eng. Mech., 1985, (111), 1010-1027.
5. Baber, T.T. and Noori, M.: Modeling General Hysteresis Behavior and Random Vibration Application. Trans. of ASME, J. of Vib., Acous., Str., and Rel.in Des., 1986, (108), 411-420.
6. Casciati, F. and Faravelli, L.: Stochastic equivalent linearization in 3-D hysteretic frames. Trans. of 9th Int. Conf. on Struc. Mech. in Reactor Technology, 1987, (M), pp 453-458.
7. Casciati, F. and Faravelli, L.: Non-linear seismic analysis of three-dimensional frames. Proc. of Int. Conf. on Des., Const. and Repair of Building Structures in Earthq. Zone, 1987, 104-108.

8. Davoodi, H. and Noori, M.N.: Extension of An Ito-Based Approximation Technique for Random Vibration of A BBW General Hysteresis Model, Part II: Non-Gaussian Analysis. J. of Sound and Vib., 1990, **(139)**, No. 3.
9. Minami, T. and Osawa, Y.: Elastic-Plastic Response Spectra for Different Hysteretic Rules. Earthq. Eng. and Struc. Dyn., 1988 **(16)**, 555-568.
10. Noori, M., Davoodi, H. and Choi, J.D.: Zero and Nonzero Mean Random Vibration Analysis of a New General Hysteresis Model. J. of Prob. Eng. Mech., 1986, **(4)**, 192-201.
11. Noori, M. and Padula, M.: Application of A New Approximation Method To Random Vibration of A General Hysteresis. Journal of Nonlinear Dynamics (to appear).
12. Noori, M.N., Davoodi, H. and Saffar, A.: An Ito-Based General Approximation Method for Random Vibration of Hysteretic Systems, Part I: Gaussian Analysis. J. of Sound and Vib., 1988, **(127)**, No. 2, 331-342.
13. Noori, M.N., Davoodi, H.: Comparison Between Equivalent Linearization and Gaussian Closure for Random Vibration of Several Nonlinear Systems. Int. J. of Eng. Science, 1990, **(28)**, No. 9, 897-905.
14. Roberts, J.B. and Spanos, P.D.: Random Vibration and Statistical Linearization, 1990, John Wiley and Sons, New York.
15. Thyagarajan, R.S.: Modeling and Analysis of Hysteretic Structural Behavior. Report No. EERL 89-03, Caltech Pasadena, California, 1989.
16. Wen, Y. K.: Method for Random Vibration of Hysteretic Systems. J. of the Eng. Mech. Div., ASCE, 1976, **(102)**, EM2, 249-263.
17. Wen, Y.K.: Methods of Random Vibration for Inelastic Structures. Appl. Mech. Rev., 1989, **(42)**, No. 2, 39-52
18. Bouc, R.: Modele Mathematique d'Hysteresis. Acustica, 1961, **(24)**, 16-25, Marseille, France.
19. Mohammadi, J. and Amin, M.: Nonlinear Stochastic Finite Element Analysis of Pipes on Hysteretic Supports Under Seismic Excitation. Comp. Prob. Meth., Proc. of The Joint ASME/SES Conf., 1988, Berkeley, California, **AMD 93**, 123-133.
20. Simulescu, I., Mochio, T. and Shinozuka, M.: Equivalent Linearization Method in Nonlinear FEM. ASCE, J. of Eng. Mech., 1989, pp 475-492.
21. Constantinou, M.C. and Tadjbakhsh, I.G.: Hysteretic Dampers in Base Isolation: Random Approach. ASCE, J. of Struc. Eng., 1985, **(4)**, 705-721.

22. Igusa, T.: Response Characteristics of an Inelastic Two-Degree-of-Freedom Primary-Secondary System. Report No. 87-7/TI-01, Department of Civil Eng., Northwestern University, 1987.
23. Su, L., Ahmadi, G. and Tadjabaksh I.G.: A Comparative Study of Different Base Isolators. Proc. of ASCE, Struc. Congress 87, 1987, Orlando, Florida. 15-26.
24. Mohammad Yar, A. and Hammond, T.: Modelling and Response of Bilinear Hysteretic Systems. ASCE, J. of Eng. Mech., 1987, **(113)**, 1000-1013.
25. Orabi, I., Ahmadi, G. and Su, L.: Hysteretic Column under Earthquake Excitations. ASCE, J. of Eng. Mech., 1989, **(1)**, 33-51.
26. Wu, W.F. and Lin, Y.K.: Cumulant-Neglect Closure for Nonlinear Oscillators Under Random Parametric and 8External Excitations. Int. J. of Nonl. Mech., 1984, **(19)**, 349-362.
27. Minai, R. and Suzuki, Y.: Stochastic Estimate and Control of Hysteretic Structural Systems. Presented at the Second International Conference on Stochastic Structural Dynamics, Boca Raton, FL, May 9-11, 1990.
28. Yang, C.Y., Cheng, A.H. and Roy, V.: Chaotic and Stochastic Dynamics for a Nonlinear Structure with Hysteresis and Degradation. J. of Prob. Eng. Mech. (To Appear).
29. Park, Y.J., Wen Y.K and Ang, A.H-S.: Random Vibration of Hysteretic Systems Under Bi-Directional Ground Motions. Earth. Eng. and Struc. Dyn., 1986, **(14)**, 543-557.
30. Spanos, P-T.: Stochastic Analysis of An Osci llator with Non-Linear Damping, Int. J. of Nonl. Mech., 1978, **(13)**, 249-259.
31. Spanos, P-T.: Formulation of Stochastic Linearization for Symmetric or Assymetric MDOF Nonlinear Systems. Trans. of ASME, J. of Appl. Mech., 1980, **(1)**, 209-211.
32. Spanos, P-T.: Stochastic Linearization in Structural Dynamics. Applied Mech. Rev., 1981, **(1)**, 1-8.
33. Sues, R.H., Wen Y.K. and Ang, A.H-S.: Safety Evaluation of Structures to Earthquakes. ASCE, Proc. of the Symp. on Prob. Meth. in Struc. Eng., 1981, St. Louis, Missouri., 358-377.
34. Wen, Y.K.: Stochastic Response and Damage Analysis of Inelastic Structures. J. of Prob. Eng. Mech., 1986, **(1)**, No. 1, pp 49-57.
35. Hampl, N.C., and Schuller, G.I.: Probability Densities of the Response of Nonlinear Structures Under Stochastic Dynamic Excitation. J. of Probabilistic Engineering Mechanics, 1989, **(4)**, No. 1, 2-9.

36. Lin, Y.K.: A New Solution Technique For Randomly Excited Hysteretic Structures. Technical Report NCEER-88-0012, Center for Applied Stochastic Research, Boca Raton, Florida, 1988.
37. Liu, Q.: Response Analysis of a Hysteretic System Under Random Excitation. Department of Mechanical Engineering, University of New Brunswick, Canada, 1990.
38. Ibrahim, R.A.: Parametric Random Vibration, Letchworth, Hertfordshire, England: Research Studies Press Ltd, 1985.
39. Ibrahim, R.A., Soundararajan, A. and Heo, H.: Stochastic Response of Nonlinear Dynamic Systems Based on Non-Gaussian Closure. Trans. of the ASME, J. of Appl. Mech., 1985, (52), 965-970.
40. Crandall, S.H.: Non-Gaussian Closure for Random Vibration of a Non-Linear Oscillator. Int. J. of Nonlinear Mechanics, 1980, (20), 303-313.
41. Fan, F.G., and Ahmadi, G.: Loss of Accuracy and Nonuniqueness of Sound and Vibration of Solutions Generated by Equivalent Linearization and Cumulant-Neglect Methods. (To Appear).
42. Sun, J.Q., and Hsu, C.S.: Cumulant-Neglect Closure Method for Nonlinear Systems Under Random Excitations. ASME Journal of Applied Mechanics, 1987, (54), 649-655.

A Practical Approach to Predict the Stochastic Response on Many-DOF-Systems Modeled by Finite Elements

H.J. PRADLWARTER and G.I. SCHUËLLER

Institute of Engineering Mechanics
University of Innsbruck, Austria

Summary
The stochastic response of hysteretic MDOF systems is most conveniently determined by statistical equivalent linearization (EQL). Traditionally, EQL assumes the nonlinear hysteretic stochastic response to be normally distributed, although it is well known that this assumption is not justified and may lead to considerable errors in the prediction of the second moment properties. In the present paper, the consideration of non-Gaussian response properties is shown. The suggested method utilizes nonlinear transformations between the nonlinear and linearized state vector and Nataf's representation of the n-dimensional joint distribution by its marginal distributions. A method to evaluate the required nonlinear transformation is proposed. The applicability of the introduced concept is shown in a numerical example.

1. Introduction

At the present state-of-the-art in nonlinear structural dynamics, the deterministic as well as the stochastic response analysis reveal quite significant deficiencies. The use of Finite Element (FE) procedures is already more than one decade common practice in deterministic structural analysis. Continuous developments in numerical FE-procedures together with the constantly decreasing computational costs allow fairly accurate and detailed nonlinear structural models. In many cases, however, the loads and some structural parameters are known only approximately, i.e. the loading and the structural parameters reveal random properties. The stochastic approach, on the contrary, is capable to include random loading and uncertainties of the structural parameters. Stochastic methods, however, are often applicable only for quite simplified and strongly idealized structural models. Most of the available approaches are quite sophisticated and can be applied successfully only to systems with very few degrees-of-freedom. They provide useful and very important insight on the characteristic behavior of the nonlinear stochastic response. Many approaches in stochastic analysis, however, cannot be generalized to determine the stochastic response of systems with many degrees of freedom (say >10) or these methods cannot compete with straightforward Monte Carlo simulation with respect to computational efficiency, accuracy and simplicity. For a wider applicability of stochastic nonlinear dynamics in engineering practice, the procedure has to meet the following requirements: (a) It should be applicable to any type of structures descretized by a MDOF-system and to general Finite Element models, (b) the effort and computational costs required to

N. Bellomo · F. Casciati (Eds.)
Nonlinear Stochastic Mechanics
IUTAM Symposium Turin

obtain the desired stochastic response should be considerably less than for straightforward Monte Carlo simulation, (c) the procedure should apply to general nonstationary excitation characteristics described by evolutionary spectra as well as parametric excitation, (d) the approach should allow to account for uncertain structural parameters. None of the available stochastic methods can meet all requirements listed above. The most severe discrepancy between requirement and capabilities hold for item (a) and (d). In fact, almost all procedures to evaluate the nonlinear stochastic response, such as those based on the Fokker-Planck equation [1,2] and closure techniques [3] etc., are only applicable if the structure can be idealized by one, or at best, by very few degrees of freedom. Alternative approaches, such as the perturbation method meet the requirement (d) but are restricted to weakly nonlinear structures [4].

In the following the method of statistical equivalent linearization (EQL) is discussed. The method has the potential to meet the requirement (a) and (b), but does not apply for parametric excitation and for uncertain structural parameters. EQL replaces the set of coupled nonlinear differential equations by a "statistically equivalent" linear set for which well developed and efficient procedures exist to determine the stochastic response. One basic drawback of EQL is that only second moment properties of the response quantities are obtained. Since the nonlinear response is known to be non-Gaussian, the first two moments are insufficient for a complete characterization of the response. The method provides an overall description of the stochastic response, but cannot be used for reliability analysis where the probability density function at the tails must be utilized. For evaluating probabilities of failure in the range $< 10^{-4}$, advanced Monte Carlo procedures [5] might be applied for a credible reliability analysis. However, the second moments of the stochastic response remain of paramount importance to identify all critical stochastic response quantities. Investigations on the accuracy of statistical equivalent linearization [6] showed that errors in the prediction of the variance of the response may exceed 25% standard deviation for a special type of nonlinearity (Bouc's model [7]). The reasons for the deviation from the true answer arise from differences between the assumed and true shape of the distribution of the stochastic response. The conventional equivalent linearization assumes a jointly normal distribution (e.g. [7,8]). Generally this assumption is not justified for nonlinear systems even if the excitation is normally distributed. The accuracy of EQL can only be improved essentially if the assumed shape fits the true shape of the distribution well. The word "shape" refers to the *type* of distribution (e.g. normal, lognormal, exponential) in that sense, that the shape determines all higher central moments for a fixed standard deviation or covariance matrix. Hence, the stochastic response is determined uniquely by its first two moments and its shape. In the following sections, the consideration of non-Gaussian response properties by means of nonlinear transformations is shown.

2. Gaussian Statistical Equivalent Linearization

Structural models with nonlinear hysteretic restoring forces follow a set of nonlinear first order differential equations

$$\dot{\mathbf{y}}(t) = \mathbf{g}(\mathbf{y}(t)) + \mathbf{b}(t) + \mathbf{w}(t) \tag{1}$$

where $\mathbf{y}(t)$ represents the state vector of the response which comprises displacements, velocities, restoring force parameter and response quantitites of a linear or nonlinear filter. The vector $\mathbf{b}(t)$ denotes the mean of the excitation, $\mathbf{w}(t)$ a vector of components with zero mean white noise and $\mathbf{g}(\mathbf{y}(t))$ a nonlinear vector function of components of $\mathbf{y}(t)$. Since there are no exact methods available to analyze the stochastic response of a nonlinear system without resorting to simulation, approximate solution techniques are employed. The most general and efficient method to treat nonlinear stochastic differential equations is the method of statistical linearization initially developed by Booton [9], Kazakov [10] and Caughey [11]. Following Booton, the nonlinear vector function $\mathbf{g}(\mathbf{y}(t))$ is linearized,

$$\mathbf{g}(\mathbf{y}(t)) \cong \hat{\mathbf{g}}(t) + \mathbf{A}(t) \cdot (\mathbf{y}(t) - E\{\mathbf{y}(t)\}) \tag{2}$$

by minimizing the difference $\mathbf{e}(t)$ between the model and the linearized nonlinear system respectively in a mean square sense, i.e.:

$$\mathbf{e}(t) = \mathbf{g}(\mathbf{y}(t)) - \hat{\mathbf{g}}(t) - \mathbf{A}(t) \cdot (\mathbf{y}(t) - E\{\mathbf{y}(t)\}) \tag{3}$$

$$E\{\mathbf{e}^T(t)\mathbf{e}(t)\} \rightarrow \text{Minimum!} \tag{4}$$

The solution is,

$$\hat{\mathbf{g}}(t) = E\{\mathbf{g}(\mathbf{y}(t))\} \tag{5}$$

$$\mathbf{A}(t) = E\{\mathbf{g}(\mathbf{y}(t)) \cdot (\mathbf{y}(t) - E\{\mathbf{y}(t)\})^T\} \cdot \mathbf{S}^{-1}(t) \tag{6}$$

where $\mathbf{S}$ denotes the covariance matrix. Replacing in eq.(1) the nonlinear vector function $\mathbf{g}(\mathbf{y}(t))$ by the linearized version in eq.(2) leads to the following statistically equivalent linear system,

$$\dot{\mathbf{x}}(t) = \mathbf{A}(t) \cdot \mathbf{x}(t) + \mathbf{c}(t) + \mathbf{w}(t) \tag{7}$$

where $\mathbf{x}$ denotes the linearized state vector and $\mathbf{c}(t)$ reads:

$$\mathbf{c}(t) = E\{\mathbf{g}(\mathbf{y}(t))\} + \mathbf{b}(t) - \mathbf{A}(t) \cdot E\{\mathbf{y}(t)\} \tag{8}$$

It can be shown (see [12,13]) that the linearized system in eq.(7) leads *exactly* to the first two moments of the nonlinear stochastic response, i.e.

$$E\{\mathbf{y}(t)\} = E\{\mathbf{x}(t)\} \tag{9}$$

$$E\{\mathbf{y}(t) \cdot \mathbf{y}^T(t)\} = E\{\mathbf{x}(t) \cdot \mathbf{x}^T(t)\} \tag{10}$$

in case eq.(6) and (8) are satisfied. At first sight, it seems that it might be possible to obtain the exact first two moments of the nonlinear stochastic response, i.e. the mean $E\{\mathbf{y}(t)\}$ and the covariance matrix $\mathbf{S}(t)$, by means of statistical equivalent linearization. However, this is not the case, since eq.(6) and (8) require the joint distribution $f_{\mathbf{Y}(t)}(\mathbf{y})$ of the nonlinear response vector. Clearly, the joint distribution is not available, since if it were known, there would be no need to linearize. Instead of the true statistics being applied, the conventional practice is to assume the nonlinear response to be normally distributed, i.e. $f_{\mathbf{Y}(t)}(\mathbf{y}) = f_{\mathbf{X}(t)}(\mathbf{y})$. For this case in [14] an alternative approach to determine the linearization coefficients of the matrix $\mathbf{A}(t)$ is proposed:

$$A_{ij}(t) = E\left\{\frac{\partial g_i(\mathbf{x}(t))}{\partial x_j}\right\} \tag{11}$$

where $E\{\mathbf{x}(t)\} = 0$ has been assumed. Atalik and Utku proved eq.(11) to be identical to eq.(6) in case $\mathbf{x}(t)$ is normally distributed and the system satisfies quite general conditions. Since eq.(11) is a convenient tool to evaluate the linearization coefficients, the question arises whether or not this approach is valid for non-Gaussian distributions as well. Unfortunately, this is not the case [15]. Hence, eq.(11) is valid if and only if (iff) the nonlinear response $\mathbf{y}(t)$ respectively $\mathbf{x}(t)$ might be assumed to be normally distributed.

3. Non-Gaussian Statistical Equivalent Linearization

It has been mentioned in the previous section that an unbiased estimate of the second moments can be obtained only if the true joint distribution is utilized by evaluating the expectations of eq. (6) and (8). Clearly, this joint distribution is not known. Improvements in accuracy can be achieved, however, in case some additional information on the shape of the joint distribution can be estimated. In this paper, non-Gaussian response properties are taken into account by a suitable nonlinear transformation $\mathbf{y}(\mathbf{x})$ between the normally distributed state vector $\mathbf{x}$ and the non-Gaussian distributed state vector $\mathbf{y}$. It can be recognized from the formal relation

$$\mathbf{y}(\mathbf{x}) = F_{\mathbf{Y}}^{-1}(F_{\mathbf{X}}(\mathbf{x})) \tag{12}$$

that the nonlinear transformation $\mathbf{y}(\mathbf{x})$ is an equivalent notion for the non-Gaussian joint probability density function $f_{\mathbf{Y}}(\mathbf{y})$. In case such a transformation is established, all expectations required in eq. (6) and (8) can be determined, e.g.

$$E\{\mathbf{g}(\mathbf{y}(t))\mathbf{y}^T(t)\} = \int_{-\infty}^{+\infty} \mathbf{g}(\mathbf{y}(\mathbf{x}))\mathbf{y}^T(\mathbf{x})\, f_{\mathbf{X}(t)}(\mathbf{x})\, d\mathbf{x} \tag{13}$$

It can be recognized further from the eqs. (9) and (10), that irrespective of any specific transformation for a particular problem, the transformation must be invariant for the first two moments, i.e.:

$$E\{y_i(\mathbf{x})\} = E\{x_i\} \tag{14}$$

$$E\{y_i(\mathbf{x})y_j(\mathbf{x})\} = E\{x_i x_j\} \tag{15}$$

In other words, the transformation must not change the properties of the first two moments. Hence, the transformation contains only information on the shape of the distribution, but not on the first two moments of the response. A basic problem by determining the required nonlinear transformation $\mathbf{y}(\mathbf{x})$ is the lack of a general but simple way to represent multidimensional non-Gaussian joint distributions. The Gram Charlier series of Hermite polynomials have been utilized in most applications when non-Gaussian response properties are evaluated by deriving higher order moment equations and a closure technique (e.g. [16]). For practical reasons, however, this type of representation is applicable for systems with very few degrees of freedom only. For multidimensional non-Gaussian joint distributions, the required number of terms increases very fast to an amount which is no longer tractable. In the area of reliability analysis, a most useful and practical representation of non-Gaussian joint distributions is attributed to Nataf [17]. Nataf's representation is capable of satisfying pre-described marginal distributions and correlations of a given covariance matrix [18]. The main advantage of Nataf's model is its applicability to n-dimensional non-Gaussian joint distributions, where n is large. Since Nataf's model is an approximation, it has certainly also specific disadvantages. For example, it is not capable of representing higher order moments (higher than two) between two random variables correctly. This implies, in case two normally distributed random variables are uncorrelated but not independent, the correct distribution is not jointly normal, whereas Nataf's representation leads to a jointly normal distribution, i.e. independent random variables. Since Nataf's representation of the n-dimensional joint distribution requires just n marginal distributions or densities $f_{Y_j}(y_j)$, the general transformation $y_j(\mathbf{x})$ simplifies considerably to $y_j(x_j)$. The model by Nataf leads to the following representation of the joint PDF:

$$f_{\mathbf{Y}}(\mathbf{y}) = f_{\mathbf{X}}(\mathbf{x},\mathbf{R}') \prod_{k=1}^{N} \left|\frac{\partial y_k}{\partial x_k}\right|^{-1} \tag{16}$$

where $f_{\mathbf{X}}(\mathbf{x},\mathbf{R}')$ is a jointly normal PDF with a matrix of correlation coefficients $\mathbf{R}'$ whose components ρ'_{ij} are defined by the integral equation

$$\rho_{Y_iY_j} = \int_{-\infty}^{+\infty}\int_{-\infty}^{+\infty} \left(\frac{y_i(x_i)-E\{x_i\}}{\sigma_{X_i}}\right)\left(\frac{y_j(x_j)-E\{x_j\}}{\sigma_{X_j}}\right) f_{X_iX_j}(x_i,x_j;\rho'_{ij})\, dx_i dx_j \tag{17}$$

The accuracy of the above outlined procedure depends mainly on two ingredients. First, on the appropriateness of the representation of the joint distribution by marginal distributions in eq.(16). Second, on the accuracy of the utilized marginal distributions, defined by normalized nonlinear relations $\hat{y}_i(\hat{x}_i)$, in which $\hat{x}_i$ and $\hat{y}_i$ are defined as follows:

$$\hat{x}_i = \frac{x_i - E\{x_i\}}{\sigma_{x_i}} \quad \text{and} \quad \hat{y}_i = \frac{y_i - E\{x_i\}}{\sigma_{x_i}} \tag{18}$$

The evaluation of $\hat{y}_i(\hat{x}_i)$ is shown in the following section.

4. Nonlinear Hysteretic Restoring Force and Associated Nonlinear Transformation

For illustration of the non-Gaussian equivalent linearization procedure, consider first a SDOF-system with a hysteretic nonlinear restoring force

$$m \cdot \ddot{u} + c \cdot \dot{u} + r = f(t) \tag{19}$$

where u represents the displacement, f the random forcing function, m the mass, c the damping and a dot "·" indicates differentiation with respect to time. The restoring force

$$r = \alpha k u + (1-\alpha)\, kz \tag{20}$$

is assumed to consist of a linear part αku and a hysteretic part $(1-\alpha)\, kz$, where k represents the stiffness, z an auxiliary variable and α a non-dimensional constant $1 > \alpha \geq 0$. The auxiliary variable (dimension of displacements) determines the hysteretic charcteristics defined by an incremental formulation,

$$\dot{z} = \dot{u} \cdot h\,[z, \mathrm{sgn}(\dot{u})] \tag{21}$$

where $h[\cdot]$ is a non-dimensional function given by

$$h\,[z, \mathrm{sgn}(\dot{u})] = \begin{cases} 1 & \text{for } \mathrm{sgn}(\dot{u})\ z \leq z_y \\ \dfrac{z_p - \mathrm{sign}(\dot{u})z}{z_p - z_y} & \text{for } \mathrm{sgn}(\dot{u})\ z > z_p \end{cases} \tag{22}$$

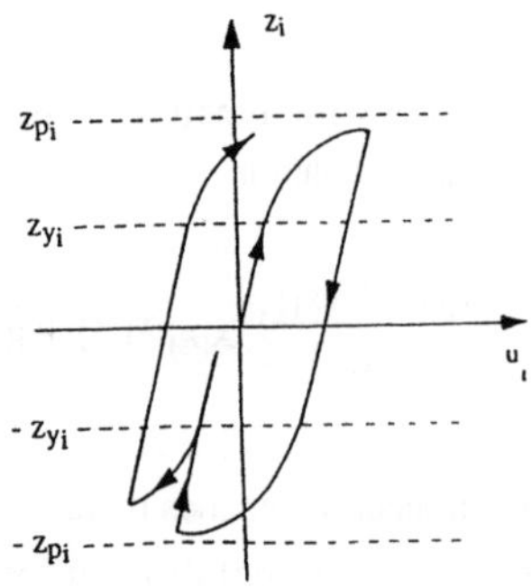

Fig. 1: Hysteretic restoring force

The characteristics of the hysteretic restoring force, defined by the auxiliary variable z, are shown in Fig. 1. It can be recognized that z remains linear for $|z| \le z_y$ as is observed in most structures. The variable z_p represents the ultimate restoring force reached by a smooth transition from the linear range. It should be noted that the restoring force law defined above can be reduced to the well known Bouc's model

$$\dot{z} = A\dot{u} - \beta\, |\dot{u}|\, |z|^{n-1}\, z - \gamma\, \dot{u}\, |z|^{n} \tag{23}$$

by taking $z_y = 0$, $z_p = A/(\beta + \gamma)$, $n = 1$ and $\beta = \gamma$.

Next, suitable transformations $y_k(x_k)$, required to establish the marginal distributions of Nataf's model for the joint PDF, are derived. Let **x** be the linearized normally distributed state vector and **y** the non-Gaussian distributed state vector, defined by

$$\mathbf{x}^T = (x_1, x_2, x_3) = (u, \dot{u}, z)_L \tag{24}$$

$$\mathbf{y}^T = (y_1, y_2, y_3) = (u, \dot{u}, z)_N \tag{25}$$

Obviously, the response quantities y_1 and y_3 cannot be normally distributed. First of all, $y_3 = z$ is bounded by $-z_p < y_3 < z_p$, whereas x_3 covers an infinite range $-\infty < x_3 < \infty$. Hence, the PDF $f_{Y_3}(y_3)$ is zero outside the range $[-z_p, z_p]$, but the density shows a considerable increase for $|y_3|$ near z_p. The quantity y_1 representing the non-Gaussian distributed displacement u is not bounded. It must be expected, however, that the PDF $f_{Y_1}(y_1)$ does not decay in the tails as fast as a normal distribution predicts. In terms of higher central moments, the inequalities $E\{y_1^4\} > 3\sigma_1^4$, $E\{y_1^6\} > 15\sigma_1^6$, $E\{y_3^4\} < 3\sigma_3^4$, $E\{y_3^6\} < 15\sigma_3^6$ are expected to hold. For a quantitative relation between linearized and nonlinear components of the state vector kinetic as well as potential energy relations are utilized. It is reasonable to require that the distribution of the energy of the linearized system and of the nonlinear system are approximately identical. It follows then readily from a consideration of the kinetic energy that the velocity y_2 is approximately normally distributed. For deriving the nonlinear transformation $y_1(x_1)$ and $y_3(x_3)$, the differential equation (21) is integrated leading to the relations

$$z(u) = \operatorname{sgn}(u - u_0) \begin{cases} |u - u_0| & \text{for } |u - u_0| \le z_y \\ z_p - (z_p - z_y) \exp\{ - \dfrac{|u - u_0| - z_y}{z_p - z_y} \} & \text{for } |u - u_0| > z_y \end{cases} \tag{26}$$

$$u(z) = u_0 + \operatorname{sgn}(z) \begin{cases} |z| & \text{for } |z| \le z_y \\ z_y + (z_p - z_y) \ln \dfrac{z_p - z_y}{z_p - |z|} & \text{for } |z| > z_y \end{cases} \tag{27}$$

The value u_0 is defined as displacement conditioned on $z = 0$ and might be interpreted as drift. From the viewpoint of energy, the drift u_0 can be associated with the energy dissipation of the linearized system due to hysteretic loops. It affects mainly the velocity y_2 which is known to be captured well by EQL. It is therefore justified to assume $u_0 = 0$ in the following expressions for the potential energy. Integration of the above nonlinear relations leads to the following potential energy associated with the displacemtens x_1 and y_1

$$\Pi(x_1) = \kappa_1 k \frac{x_1^2}{2} = \Pi(y_1) =$$

$$= k \cdot \begin{cases} \frac{y_1^2}{2} \quad \text{for } |y_1| \le z_y, \text{ else} \\ \alpha \frac{y_1^2}{2} + (1-\alpha) \cdot [\frac{z_y^2}{2} + z_p(|y_1|-z_y)-(z_p-z_y)^2\{1-\exp(-\frac{|y_1|-z_y}{z_p-z_y})\}] \end{cases} \tag{28}$$

and for the auxiliary variables x_3 and y_3 respectively

$$\Pi(x_3) = \kappa_3 k \frac{x_3^2}{2} = \Pi(y_3) =$$

$$= k \cdot \begin{cases} \frac{y_3^2}{2} \quad \text{for } |y_3| \le z_y, \text{ else} \\ \frac{z_y^2}{2} - (|y_3|-z_y)\,(z_p-z_y) + z_p(z_p-z_y) \ln \frac{z_p-z_y}{z_p-|y_3|} \end{cases} \tag{29}$$

The above two equations establish the desired nonlinear transformations $y_1(x_1)$ and $y_3(x_3)$, where the parameters κ_1 and κ_3 are determined by the constraints $E\{x_1^2\} = E\{y_1^2(x_1)\}$ and $E\{x_3^2\} = E\{y_3^2(x_3)\}$ in eq.(15). Finally, it should be mentioned that alternative procedures to estimate the required nonlinear transformations have been developed by either using Monte Carlo simulation [19], or the Fokker-Planck equation of the nonlinear and the linearized system respectively [20].

5. Numerical Example

The application of the introduced non-Gaussian equivalent linearization is demonstrated by analyzing a six storey (2-dim.) frame structure utilizing static condensation. The frame can be approximated accurately by a simply coupled six-degree of freedom shear beam model. In this model, the mass of the i-th floor is represented by the variable m_i and the restoring force r_i acting between the (i-1)th and the i-th floor

$$r_i = \alpha_i\, k_i \cdot u_i + (1-\alpha_i)\, k_i\, z_i \tag{30}$$

according to eq.(20), where u_i denotes now the relative displacement of the i-th floor to the (i-1)th floor. The structural parameters of the stiffness k_i and the constants z_{y_i} and z_{p_i} defining

floor i	1	2	3	4	5	6
k_i [MN/m]	305.8	309.2	260.5	264.0	216.4	220.0
z_{y_i} [mm]	6.4	7.9	7.9	9.5	10.2	12.5
z_{p_i} [mm]	10.7	13.3	12.2	14.2	15.7	17.3

Tab. 1: Stiffness k_i and the constants z_{y_i} and z_{p_i}; $1 \le i \le 6$

the hysteretic characteristics (see eq.(20) and (21)) are given in Tab. 1. All floors have equal masses of 81500 kg. The constants α_i have been asumed for all floors equal $\alpha_i = 0.05$. For the initial linear range, the eigenfrequencies are $\{f_i\} = \{0.643, 1.796, 2.877, 3.796, 4.485, 5.072$ [Hz] $\}$. In the linear range, the damping ratio $\xi = 0.03$ is assumed.

In the following the stationary stochastic response obtained by traditional Gaussian and introduced non-Gaussian equivalent linearization is compared. The frame is subjected to horizontal stationary ground acceleration $a_g(t)$ modelled by filtered white noise

$$a_g(t) = \xi_g(t)\,\Omega\,\dot{x}_g(t)$$
$$\ddot{x}_g + 2\xi_g\,\Omega\,\dot{x}_g + \Omega^2\,x_g \qquad (31)$$

where x_g is the displacement of the filter, the parameter ξ_g and Ω determine the spectral density of the ground acceleration, and w(t) denotes stationary white noise with a constant spectral density S_0. The value Ω 0.10 rad/sec and $\xi_g = 0.10$ are utilized. The stationary stochastic response has been determined for the range $0 \le S_0 \le 2.0$ m^3/sec^3, resulting in the standard deviation of the ground acceleration σ_{ag} in the range $0 \le \sigma_{ag} \le 2.50$ m/sec^2. In the first and fifth floor, the strongest nonlinear hysteretic effects are observed. The standard deviation of the auxiliary variable z_1 for the first floor is shown in Fig. 2 for three cases. The full line shows the standard deviation obtained by non-Gaussian EQL and the nonlinear transformations defined by eq.(28) and (29). The broken and dotted lines respectively show the standard deviations obtained by traditional Gaussian EQL. Assuming the auxiliary variable z_i normally distributed, one obtains a PDF $f_{Z_i}(z_i) > 0$ also for values $|z_i| > z_{p_i}$. Physically, however, the absolute value $|z_i|$ must be less than z_{p_i}. It is then not clear, how the increment $\dot{z}$ in eq.(21) should be chosen for $|z_i| > z_{p_i}$. The broken line shows the result obtained by assuming $h[z, \text{sign}(\dot{u}_i)] = 0$ for $|z_i| > z_{p_i}$, whereas the dotted line shows the results by assuming that eq.(22) holds also for $|z_i| > z_{p_i}$. It should be noted that a similar assumption is implied for the results in closed form by Wen [7] using Bouc's model. Fig. 2 shows clearly that the somewhat arbitrary assumption for $|z_i| > z_{p_i}$ strongly affects the result by traditional EQL. This problem does of course not arise in non-Gaussian EQL, since the PDF $f_{Z_i}(z_i) = 0$ for $|z_i| > z_{p_i}$. It is seen in Fig. 2 that Gaussian EQL may considerably over- or underestimate the standard deviation of the restoring force. As a consequence, the displacements may be under- or overestimated as shown in Fig. 3.

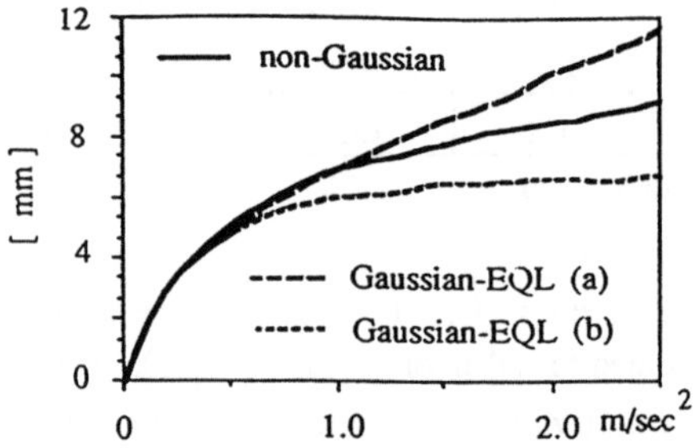

Fig.2: Standard deviation of auxiliary variable z_1

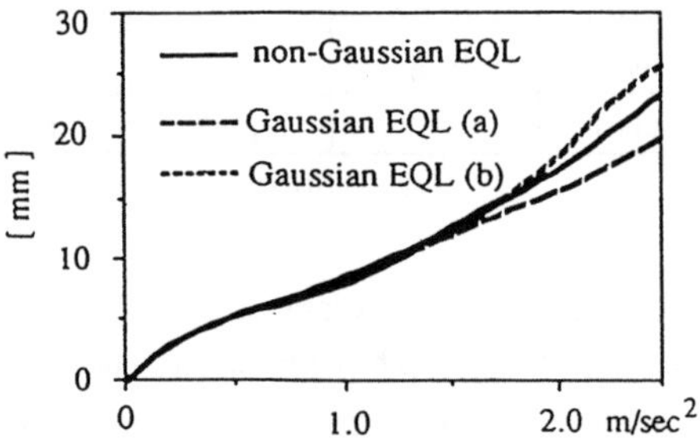

Fig.3: Standard deviation of displacement u_1

6. Conclusions

From these results the following conclusions can be drawn:

(1) Non-Gaussian stochastic response properties can be taken into account within the framework of statistical equivalent linearization by utilizing a nonlinear transformation $\mathbf{y}(\mathbf{x})$ between the normally distributed linearized state vector $\mathbf{x}$ and the nonlinear state vector $\mathbf{y}$.

(2) The utilization of Nataf's representation of the joint distribution is suggested. Since this model requires only information on the marginals of the non-Gaussian distribution, the representation of n-dimensional distribution does not cause any serious problem.

(3) A method to determine quantitatively the shape of the non-Gaussian mariginals is proposed considering the energy distribution.

(4) Traditional (Gaussian) statistical equivalent linearization may lead to ambiguous results as shown in the numerical example.

7. Acknowledgement

This research is partially supported by the Austrian Industrial Research Promotion Fund (FFF) under contract No. 6/546 which is gratefully acknowledged by the authors.

References

[1] Lin, Y.K. : "Probabilistic Theory of Structural Dynamics", McGraw-Hill, New York, 1967; Reprint Robert E. Krieger, Huntington, 1976.

[2] Cai, G.Q., Lin, Y.K.: "On exact stationary solutions of equivalent non-linear stochastic systems", *Int. J. Non-Linear Mech.*, Vol. 23, No.4, 1988, pp.315-325.

[3] Wu, W.F., Lin, Y.K.: "Cumulant-neglect Closure for Nonlinear Oscillators under Random Parametric and External Excitation", *International Journal of Non-Linear Mechanics*, 19 (1984) pp.349-362.

[4] Benaroya, H., Rehak, M: "Finite Element Methods in Probabilistic Structural Analysis: A selective review", *Applied Mechanics Review*, 41(5) (1988) pp.201-213.

[5] Schuëller, G.I., Pradlwarter, H.J., Bucher C.G.: "Efficient Computational Procedures For Reliability Estimates of MDOF Systems", to appear: *Int. J. Non-Linear Mechanics*, 1991

[6] Pradlwarter, H.J., Schuëller, G.I.: "Accuracy and Limitations of the Method of Equivalent Linearization for Hysteretic Multi-Storey Structures", F. Ziegler, G.I. Schuëller (Eds.). Nonlinear Stochastic Dynamic Engineering Systems, Proc. IUTAM Symp., Innsbruck, Springer Verlag Berlin Heidelberg, 1988, pp.3-21.

[7] Wen, Y.K.: "Method for Random Vibration of Hysteretic Systems", *Journal of the Engineering Mechanics Division*, ASCE, April 1976, Vol.102, pp.249-263.

[8] Spanos, P.-T.D.: "Stochastic Linearization in Structural Dynamics", *Applied Mechanics Review*, Vol.34 (1981) pp.1-8.

[9] Booton, R.C.: "The Analysis of Nonlinear Control System with Random Inputs", Proc. MRI Symposium on Nonlinear Circuits, Polytechnic Inst. of Brooklyn, 1953, pp.341-344.

[10] Kazakov, I.E.: "An Approximate Technique for the Statistical Investigations of Nonlinear Systems", Tr. VVIA im Prof. n.E. Zhukovskogo, No.394, 1952.

[11] Caughey T.K.: "Equivalent linearization techniques", *Journal of the Acoustical Society of America*, Vol.35, 1965, pp.1706-1711.

[12] Kozin, F.: " The Method of Statistical Linearization for Nonlinear Stochastic Vibrations", F. Ziegler, G.I. Schuëller (Eds): Nonlinear Stochastic Dynamic Engineering Systems, Proc., IUTAM Symposium Innsbruck/Igls, Austria, 1987, Springer Verlag Berlin Heidelberg, 1988, pp.44-56.

[13] Pradlwarter, H.J.: "On the Existence of "True" Equivalent Linear Systems for the Evaluation of the Nonlinear Stochastic Response due to Nonstationary Gaussian Excitation", Report 28-90, Institute of Engineering Mechanics, University of Innsbruck (1990).

[14] Atalik, T.S., Utku, S.: "Stochastic Linearization of Multidegree of Freedom Nonlinear Systems", *Journal of Earthquake Engineering and Structual Dynamics*, 1976, Vol.4, pp.411-420.

[15] Chen, X.-W.: "The Response of Hysteretic MDOF-Systems Driven by an Evolutionary Process Evaluated by the Equivalent Linearization Technique", Dissertation, Institute of Engineering Mechanics, University of Innsbruck (1988).

[16] Hampl, N.C., Schuëller, G.I.: "Probability Densities of the Response of Non-Linear Structures under Stochastic Dynamic Excitation", *Probabilistic Engineering Mechanics*, Vol.4, No.1, March 1989, pp.2-9.

[17] Nataf, A.: "Determination des Distribution dont les Marges sont Donnees", Comptes Rendus de l'Academie des Sciences, Paris, 1962, 225, pp.42-43.

[18] Liu, P-L., Der Kiureghian, A.:"Multivariate Distribution Models with Prescribed Marginals and Covariances", *Probabilistic Engineering Mechanics*, 1986, Vol.1, No.2, pp.105-112.

[19] Pradlwarter, H.J., Schuëller, G.I. 1989. Consideration of Non-Gaussian Hysteretic Properties of MDOF-Systems by Use of Equivalent Linearization. Proc. of ICOSSAR'89, the 5th Intern. Conf.on Structural Safety and Reliability, San Francisco, Aug. 7-11.1989. A.H-S. Ang, M. Shinozuka, G.I. Schuëller (eds.), Vol.II:1333-1340.New York:ASCE.

[20] Pradlwarter, H.J.: "Non-Gaussian Linearization - an Efficient Tool to Analyse Nonlinear MDOF-Systems", to appear: *Nuclear Engineering and Design*, 128, pp. 175-192, 1991.

Distribution of the Response of Hysteretic Oscillators with Wide-Band Random Excitation

J. B. Roberts and A. H. Sadeghi
University of Sussex, School of Engineering, Falmer, Brighton
BN1 9QT, U.K.

Abstract

It is shown that, under certain conditions, the normal method of stochastic averaging can be applied to predict the probability distribution of the response of a wide class of hysteretic oscillator to wide-band random excitation. Some specific results are obtained for oscillators with hysteresis of the curvilinear, Bouc-Wen type. Good agreement is obtained between theoretical predictions and experimental results, derived from dynamic tests on a mass-cable system.

1. Introduction

In designing systems to withstand random environmental loading, it is often vital to fully account for the effect of non-linearity in their dynamic characteristics. Non-linearity can have a profound influence on the probability distribution of the response, and hence on the probability of failure. A linear analysis may, at best, lead to an excessively conservative design or, at worst, to a system liable to catastrophic failure. Hysteretic behaviour is an important kind of non-linearity, which occurs in many practical applications. Despite its apparently complex nature, involving non-linearity with respect to both damping and stiffness, of an hereditary character, it is often possible to model hysteresis in terms of a single, first-order non-linear differential equation [1]. This can be combined with the usual system equations to yield an extended set of equations. The problem of predicting the probability distribution of the response can then be tackled by a variety of well established analytical techniques, in the field of non-linear random vibrations, including statistical linearisation, moment closure and perturbation [1].
Nearly all the available theoretical methods give only limited information on the response, usually in terms of statistical moments. However, one method, usually referred to as "stochastic averaging" [2] allows one, in at least some cases, to derive very simple asymptotic approximations to the probability density function of the response. Earlier, this

N. Bellomo · F. Casciati (Eds.)
Nonlinear Stochastic Mechanics
IUTAM Symposium Turin

method has been applied to the case of an oscillator with bilinear hysteresis [3-6]. A generalisation of the stochastic averaging method, based on a consideration of the energy envelope of the response, has also been applied to oscillators with both bilinear [7] and curvilinear [8] types of hysteresis.
In this paper it is shown that the normal stochastic averaging method can be applied to a wide class of hysteretic oscillators under certain conditions, which are clearly identified. Some very simple analytical results for the oscillator response statistics are obtained, for oscillators with curvilinear hysteresis. To validate the theoretical approach, a comparison is given between some experimental estimates of the probability distribution of the response of a mass-cable system, under random excitation, and corresponding theoretical predictions.

2. Equations of motion

The equation of motion for an oscillator with viscous damping and a hysteretic restoring force can be written in the form

$$m\ddot{u} + c\dot{u} + \Phi(u,t) = f(t) \tag{1}$$

where m and c are, respectively, the mass and viscous damping coefficients, $\Phi(u,t)$ represents the hysteretic restoring term and f(t) is the external excitation. It is convenient to express Φ as a combination of a linear, non-hysteretic term and a hysteretic component; thus

$$\Phi = \alpha k u + (1-\alpha) k z_1 \tag{2}$$

Here the relative contribution of the non-hysteretic term is measured by the non-dimensional parameter α and k is a linear stiffness parameter.
The "purely hysteretic" term, z_1, can, in many cases, be modelled with good accuracy by a first-order differential equation of the form

$$\dot{z}_1 = G(\dot{u}, z_1) \tag{3}$$

where G is a non-linear function. Specific forms for G exist for bilinear hysteresis and curvilinear hysteresis of the Bouc-Wen type [1].
It is useful to non-dimensionalise the governing equations at this stage. If

$$y = u/u^*, \ \tau = \omega_o t, \text{ where } \omega_o = (k/m)^{1/2} \tag{4}$$

is the "linear natural frequency", and u* is a characteristic length scale associated with the response, then equations (1) and (2) may be rewritten as

$$\ddot{y} + 2\zeta\dot{y} + \alpha y + (1-\alpha)z = N(\tau), \ \dot{z} = g(\dot{y},z) \tag{5}$$

where

$$\zeta = \frac{c}{2\sqrt{km}}, \ N(\tau) = \frac{f(t)}{mu^*\omega_o^{3/2}}, \ z = \frac{z_1}{u^*} \tag{6}$$

In the following $N(\tau)$ is assumed to be a broad-band, stationary random process, with zero mean.

3. Application of stochastic averaging

Let $y = a\cos\Phi, \ \dot{y} = -a\omega_c\sin\Phi$, where $\Phi = \omega_c t + \phi$ (7)

and ω_c is a "characteristic frequency" of the response process y. Henceforth it will be assumed that the energy dissipated in a typical cycle of the response is low, compared with the total energy associated with that cycle, so that the response process is narrow-band. ω_c can then be associated with the frequency at which the response spectrum peaks.

On transforming from $(y, \dot{y})$ variables to (a, ϕ) variables, equation (5) can be re-expressed as the following pair of coupled first-order equations:

$$\omega_c\dot{a} = H\sin\Phi - N(\tau)\sin\Phi, \ \omega_c a\dot{\phi} = H\cos\Phi - N(\tau)\cos\Phi \tag{8}$$

where

$$H = -2\zeta a\omega_c\sin\Phi + (\alpha-\omega^2{}_c)a\cos\Phi + (1-\alpha)z \tag{9}$$

Since y is assumed to be narrow band, a and ϕ change little over one cycle of oscillation. The first terms on the right of equations (11) and (12), which relate to energy dissipation, can therefore be averaged over one cycle, on the basis that a and ϕ are constant. Hence one obtains the following partially averaged equations

$$\dot{a} = -a[\zeta+h(a)] - \frac{N(\tau)\sin\Phi}{\omega_c} \tag{10}$$

$$\dot{\phi} = \frac{1}{2}[\omega^2(a)-\omega_c^2] - \frac{N(\tau)\cos\Phi}{a\omega_c} \tag{11}$$

where

$$h(a) = \frac{(1-\alpha)}{2\pi a^2 \omega_c} \cdot A(a), \quad A(a) = \oint z dy \tag{12}$$

$$\omega^2(a) = \alpha + (1-\alpha)B(a) \tag{13}$$

and

$$B(a) = \frac{1}{\pi a} \int_0^{2\pi} z\cos\Phi d\Phi \tag{14}$$

Here $\omega(a)$ is an amplitude dependent equivalent natural frequency, in the Krylov-Bogoliubov sense, and $A(a)$ is the area enclosed by a hysteretic loop, of amplitude a.

To complete the averaging procedure, we will now require that ζ, $h(a)$ and $[\omega^2(a)-\omega_c^2]$ are all *small* and of the same order; i.e. of order $O(\epsilon)$, say, where ϵ is a small parameter (it is noted that we have already assumed that the energy dissipation, as measured by ζ and $h(a)$ is small). If this requirement is satisfied then, with rescaling to ensure that the response is of order ϵ^0, one finds that the excitation is of order $\epsilon^{1/2}$. The Stratonovitch-Khasminskii theorem [2] is then applicable, effectively enabling the second terms on the right-hand side of equations (10) and (11) to be averaged.

Referring to equations (12) and (13), and noting that typical hysteretic loops for z appear as in Fig. 1(a), while the loops for the corresponding restoring term

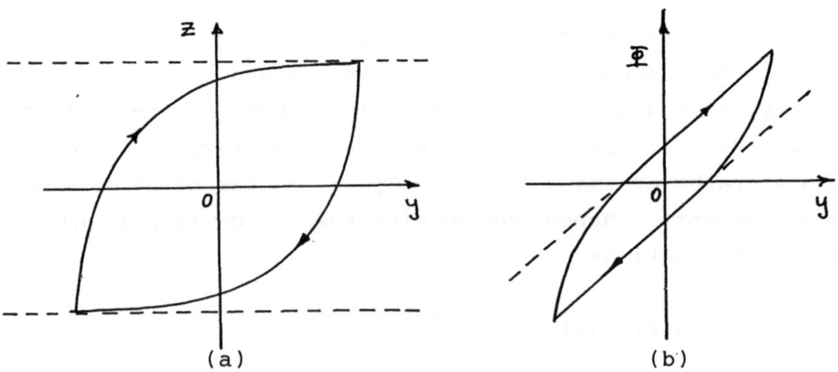

Fig. 1 Typical hysteretic loops (a) z (b) Φ

$$\phi = \alpha y + (1-\alpha)z \tag{15}$$

appear as in Fig. 1(b), one can conclude that h(a) and $[\omega^2(a)-\omega_C^2]$ will both be small if one or more of the following conditions are met:

1. $(1-\alpha)$ is of order ϵ, and ω_C is set to unity. This is the case where the secondary slope, α, of ϕ, is close to the primary slope of unity.
2. A(a) and [B(a)-1] are of order ϵ, and ω_C is set to unity. Generally the loop area will increase monotonically with increasing a. Thus, this condition implies a weak level of excitation, so that a is small. For a bilinear characteristic A(a) and [B(a)-1] are actually zero from $a<1$. For curvilinear loops we further require that $A(a)/a^2$ is small, for small a.
3. The excitation level is relatively high, so that much of the response is associated with the near horizontal portions of the z characteristic. For large amplitudes of response, relative to the "yield" amplitude, A(a) will grow linearly with a (see Section 5); hence $A(a)/a^2$ will decrease as a increases and h(a) will become small. In these circumstances it is natural to set the characteristic frequency, ω_C, to equal the secondary stiffness, $\alpha^{1/2}$ - i.e. $\omega_C^2 = \alpha$. Then, since B(a) decreases as a increases, for large a [1], it follows that $[\omega^2(a)-\omega_C^2]$ becomes small.

On applying the Stratonovitch-Khasminskii limit theorem, on the assumption that one of the above three conditions holds, one obtains a Markov approximation for the joint process (a,ϕ). The governing differential equations are

$$da = -a[\zeta+h(a)] + \frac{\pi S(\omega_c)dt}{2a\omega_c^2} dt - \frac{[\pi S(\omega_c)]^{1/2}}{\omega_c} dW_1(t) \tag{16}$$

and

$$d\phi = \frac{1}{2\omega_c}[\omega^2(a)-\omega_c^2]dt - \frac{[\pi S(\omega_c)]^{1/2}}{a\omega_c} dW_2(t) \tag{17}$$

where $S(\omega)$ is the power spectrum of y(t) and $W_1(t)$ and $W_2(t)$ are independent unit Wiener processes.

It follows from the above two equations that the amplitude process is uncoupled from the phase - i.e. is a one-dimensional Markov process, governed by the FPK equation

$$\frac{\partial p}{\partial \tau} = \frac{\partial}{\partial a}\left[(a[\zeta+h(a)] - \frac{\pi S(\omega_c)}{2a\omega_c^2})p\right] + \frac{\pi S(\omega_c)}{2\omega_c^2}.\frac{\partial^2 p}{\partial a^2} \qquad (18)$$

where $p(a|a_0;\tau)$ is the transition density function for a.

4. Stationary solution

The stationary density function

$$w(a) = \lim_{\tau\to\infty} p(a|a_0;\tau) \qquad (19)$$

may be obtained from equation (17) by setting $\partial p/\partial\tau = 0$, and p = w. The result is

$$w(a) = ca \exp\left\{-\frac{2\omega_c^2}{\pi S(\omega_c)}\int_0^a \Lambda(\xi)d\xi\right\} \qquad (20)$$

where

$$\Lambda(a) = a\,[\zeta+h(a)] \qquad (21)$$

and c is a normalisation constant. Moreover, by considering the FPK equation for the joint process (a,ϕ), derived from equations (16) and (17) one finds that the stationary joint density function for this process is given by

$$w(a,\phi) = \frac{w(a)}{2\pi} \qquad (22)$$

and hence the density function for the joint process $(y,\dot{y})$ is given by

$$w(y,\dot{y}) = \frac{w(a)}{2\pi\omega_c a} \qquad (23)$$

From this, various statistics for y and $\dot{y}$ are easily calculated: for example, the standard deviation of y, σ, is given by

$$\sigma^2 = \int_{-\infty}^{\infty}\int_{-\infty}^{\infty} y^2 w(y,\dot{y})dyd\dot{y} = \frac{1}{2}\int_0^{\infty} w(a)a^2 da \qquad (24)$$

Within the above theoretical framework, some simple analytical results have previously been given, for the case of a bilinear oscillator [3,4].

5. Oscillators with curvilinear hysteresis

Oscillators with curvilinear type hysteresis can be represented by the following Bouc-Wen model [1]:

$$\dot{z} = -\gamma|\dot{y}|z|z|^{n-1} - \beta\dot{y}|z|^{n} + A\dot{y} \tag{25}$$

where γ, β, Ak and n are "loop parameters", which govern the shape of the loop.

Referring to the loops sketched in Fig. 2, it is evident that, asymptotically, for large a (condition 3 in Section 3), A(a) will grow linearly with a; thus

$$A(a) \rightarrow \mu + \lambda a \tag{26}$$

where μ and λ are constants, dependent on the loop parameters. It is also evident from Fig. 2 that

$$\lambda = 4z_m^{*} \tag{27}$$

where z_m^* is the limiting value of the maximum value of z, z_m, as $a \rightarrow \infty$.

It is possible to derive analytical expressions for μ in certain special cases. For example, for n = 1 (see also [8])

$$A(a) = 2A\left\{\left[\frac{\ln(1-\chi_o)+\chi_o}{\lambda_d^2}\right] - \left[\frac{\ln(1-\theta_o)+\theta_o}{\lambda_a^2}\right]\right\} \tag{28}$$

where

$$\chi_o = \frac{\lambda_d z_m}{A}, \quad \theta_o = \frac{\lambda_u z_m}{A} \tag{29}$$

$$\lambda_u = \beta+\gamma, \quad \lambda_d = \beta-\gamma \tag{30}$$

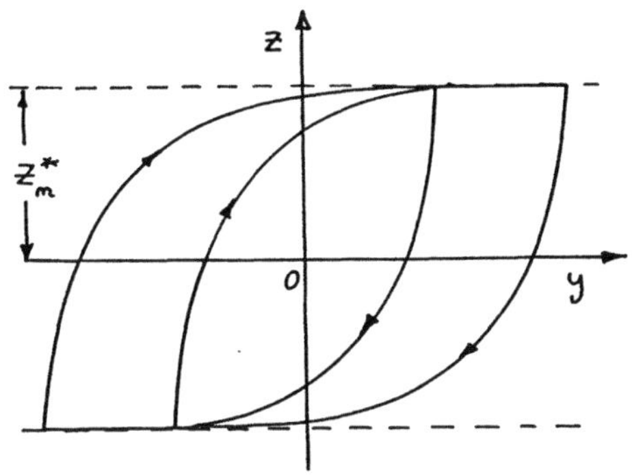

Fig. 2 Two hysteresis loops for z, for different amplitudes

From this result one finds that

$$\mu = \frac{2A}{\lambda_d} \ln\left[1-\frac{\lambda_d}{\lambda_u}\right]\left[\frac{1}{\lambda_d} + \frac{1}{\lambda_u}\right] - \frac{2A}{\lambda_u}\left[\frac{1}{\lambda_u} - \frac{1}{\lambda_d}\right] \tag{31}$$

and

$$z_m^* = A/\lambda_u \tag{32}$$

On substituting the asymptotic form for A(a), given by equation (26) into equation (20), the following simple asymptotic expression for the density of a is obtained

$$w(a) = ca^{(1+\rho)} \exp(-\Lambda a)$$

where

$$\rho = -\frac{\mu\omega_c(1-\alpha)}{\pi^2 S(\omega_c)} \tag{34}$$

$$\Lambda = \frac{\lambda\omega_c(1-\alpha)}{\pi^2 S(\omega_c)} \tag{35}$$

From equations (23) and (33), some simple analytical results can be obtained for the statistics of y and $\dot{y}$. For example, σ is given by

$$\sigma^2 = \frac{(3+\rho)(2+\rho)}{2\Lambda^2} \tag{36}$$

For high levels of excitation, $\rho \to 0$ and equation (36) reduces to

$$\sigma^2 = \frac{3}{\Lambda^2} \tag{37}$$

This agrees with the corresponding asymptotic result for the bilinear oscillator, found previously [3].

An expression for the density function for y, w(y), can also be found from equations (23) and (33). Thus,

$$w(y) = 2cy^{(1+\rho)}\omega_c \int_1^\infty \frac{v^{1+\rho}\exp(-\gamma v)}{\sqrt{v^2-1}}\, dv \tag{38}$$

where

$$\gamma = \Lambda y^2 \tag{39}$$

For large y the following asymptotic expansion can be used:

$$w(y) \rightarrow 2cy^{(1+\rho)}\omega_c \sqrt{\frac{\pi}{2y}}\, e^{-y}\left[1 + \frac{(1+\rho)}{2y} + \ldots.\right] \qquad (40)$$

Similarly, from equations (23) and (33) the following expressions for the upward crossing rate, ν^+, of a fixed level, b, may be obtained:

$$\nu^+ = cb^{\rho+2}\omega_c^{\,2} y^{-(2+\rho)}\Gamma(2+\rho, y) \qquad (41)$$

For large y, one finds the simple asymptote

$$\nu^+ \rightarrow \frac{cb^{\rho+2}\omega_c^{\,2}\exp(-y)}{y} \qquad (42)$$

6. Comparison with experimental data

Some comparisons between theoretical predictions and experimental results will now be presented.

The experimental results were obtained from dynamic tests on a simplified version of a Stockbridge damper, as shown in Fig. 3 (see also [9,10]). Here the stranded cable acts as the hysteretic restoring element, and the transverse vibration of the system can be modelled, to a good degree of approximation, as a single degree of freedom oscillator.

Fig.4(a) shows the hysteretic behaviour of the cable, obtained by applying a transverse fluctuating force to the tip of the cantilever, with the mass removed. This behaviour can be accurately modelled by the Bouc-Wen representation of equation (25), applying parametric identification techniques to estimate the loop parameters [9]. In the present case a very good fit can be obtained by presetting n = 1. Fig. 4(b) shows the theoretical hysteretic behaviour predicted from the Bouc-Wen model, using the identified parameters.

Since the loops are slim it is possible to apply stochastic averaging, in this case. In fact the experimental conditions relate to condition 3 in Section 3, since most of the response relates to the near horizontal portions of the corresponding z loops (see Fig. 1).

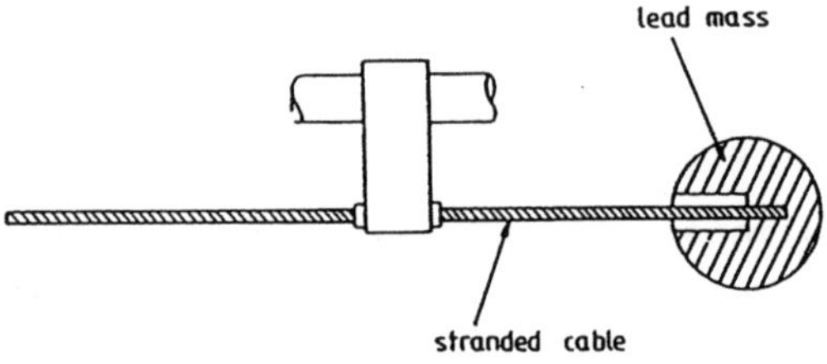

Fig. 3 Mass-cable system

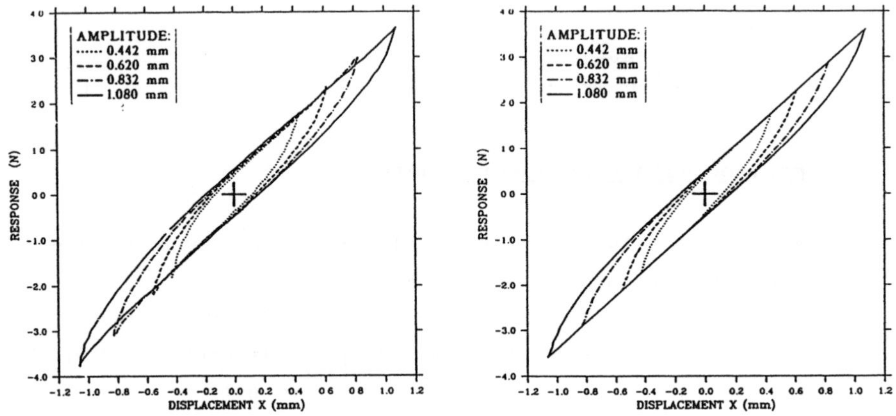

(a) (b)

Fig. 4 (a) Experimental hysteretic loops

(b) Reconstructed hysteretic loops using identified parameters

To obtain a comparison between theoretical and experimentally determined probability density functions, for the response, one long data set, consisting of a simultaneous record of applied transverse force and resulting transverse displacement were obtained (81920 samples, with a sampling interval of 1 ms). The force was a random signal with an approximately Gaussian distribution, and power mostly in the range 0 to 25 Hz. Fig. 5 shows a comparison between the power spectra of the force and displacement response, as estimated from this data. In the vicinity of the response peak, at about 5 Hz, the force spectrum is moderately smooth; and the bandwidth requirement imposed by the stochastic averaging method is satisfied; namely, that the bandwidth of the response is significantly smaller than the bandwidth of the excitation.

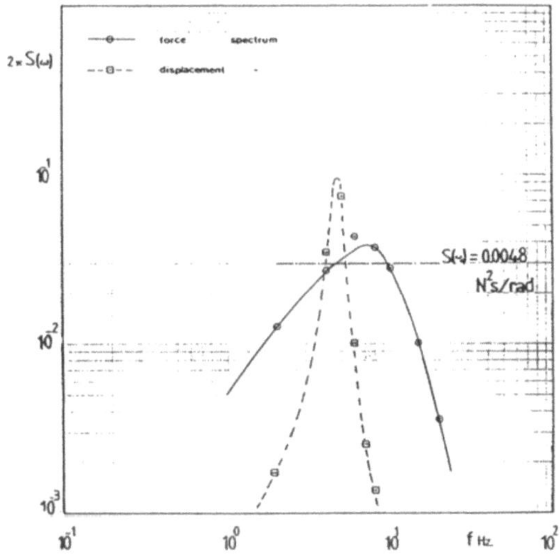

Fig. 5 Power spectra of the force excitation and displacement response processes

For the application of the stochastic averaging theory, it is necessary to know the spectral level of the response at the peak frequency, $S(\omega_c)$. From Fig. 5 one obtains $S(\omega_c) = 0.0048$ N^2s/rad. Using this value, and the estimated system parameters, it was found, from equations (33) and (34) that

$$\rho = 0.437 \text{ and } \Lambda = 3.58(mm^{-1}). \qquad [10]$$

Based on these values, and equations (33) and (38), theoretical density functions for a and y were computed. Comparisons between these results and the corresponding experimental density functions are shown in Figs. 6(a) and (b), respectively. In both cases it is seen that the stochastic averaging theory gives an excellent agreement with the experimental results, correctly predicting the effect of non-linearity on the distribution of the response, both quantitatively and qualitatively. For comparison in Fig. 6(a), the 'linear' Rayleigh density function is shown, based on the measured standard deviation of the response Similarly Fig. 6(b) shows the 'linear' Gaussian density function, based on the measured standard deviation. These comparisons clearly show that a linearisation will, in this case, lead to a serious under-prediction of the probability of reaching relatively large amplitude levels.

7. Conclusions

The main conclusions can be summarised as follows:

1. Under certain specific conditions the normal method of stochastic averaging can be applied to the case of an hysteretic oscillator under wide-band random excitation. These conditions imply that the response band-width is significantly smaller than the excitation band-width.
2. The stochastic averaging method gives the possibility of deriving very simple analytical results for the probability distribution of the response.
3. Specific theoretical results, obtained for a case of an oscilator with hysteresis of the Bouc-Wen type, agree well with some experimental results obtained from tests on a randomly excited mass-cable system.

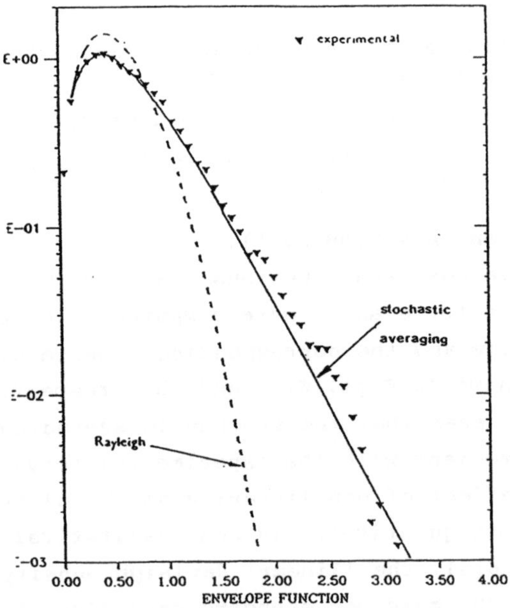

(a)

Fig. 6 (a) Probability density functions for the envelope process, a

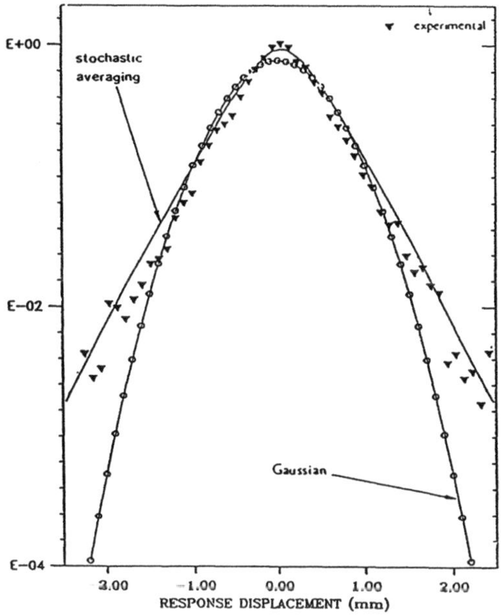

(b)

(b) Probability density functions for the displacement response process, y

References

1. J. B. Roberts and P. D. Spanos, "Random Vibration and Statistical Linearization", J. Wiley, Chichester, UK, 1990.

2. J. B. Roberts and P. D. Spanos, "Stochastic Averaging: An Approximate Method of Solving Random Vibration Problems", Int. J. Non-linear Mechanics, Vol. 21, 1986, pp. 111-134.

3. J. B. Roberts, "The Response of an Oscillator with Bilinear Hysteresis to Stationary Random Excitation", J. App. Mechs., ASME, Vol. 45, 1978, pp. 923-928.

4. J. B. Roberts, "The Yielding Behaviour of a Randomly Excited Elasto-Plastic Structure", J. Sound Vib., Vol. 72, 1980, pp.71-86.

5. J. B. Roberts and A. H. Sadeghi, "Nonstationary Response of Oscillators with Bilinear Hysteresis to Random Excitation", in 'Stochastic Structural Dynamics', published by Elsevier, Amsterdam (eds. S. T. Ariaratnam et al), 1988, pp. 197-218.

6. M. Yar and J. K. Hammond, "On Modelling and Response of Bilinear Hysteretic Systems", J. Eng. Mechs., ASCE, Vol. 113, 1987, pp. 3-10.

7. W. Q. Zhu and Y. Lei, "Stochastic Averaging of the Energy Envelope of Bilinear Hysteretic Systems", Proc. IUTAM Symp. on Nonlienar Stochastic Dynamic Engineering Systems, Innsbruck, Austria, June 1987, published by Springer-Verlag, Berlin, 1988 (eds. F. Ziegler and G. I. Schueller), pp. 361-380.

8. J. B. Roberts, "Application of Averaging Methods to Randomly Excited Hysteretic Systems", Proc. IUTAM Symp. on Nonlinear Stochastic Dynamic Engineering Systems, Innsbruck, Austria, June 1987, published by Springer-Verlag, Berlin, 1988 (eds. F. Ziegler and G. I. Schueller), pp. 361-380.

9. J. B. Roberts, "Sequential Parametric Identification and Response of Hysteretic Oscillators with Random Excitation", Structural Safety, Vol. 8, 1990, pp. 45-68.

10. A. H. Sadeghi, "Identification and Stochastic Response of Hysteretic Systems", DPhil Thesis, University of Sussex, UK, 1991.

First Passage and Wave Density Analysis by Means of the Computer Package CROSSREG

Igor Rychlik and Georg Lindgren

Dept of Mathematical Statistics, Lund Institute of Technology
Box 118, S-221 00, Lund, Sweden

1 Passage times

First passage times and intercrossing times for critical, constant or timedependent, levels are of central interest in reliability and many other engineering and statistical applications of stochastic processes. For Gaussian processes, very effective numerical algorithms can be constructed by means of Slepian models and regression approximations. A Slepian model is a process that describes the process behaviour conditioned on a level crossing. By conditioning on suitably chosen random variables one can obtain quite accurate integral expressions of low dimension, for the passage time density, simply by neglecting a small residual. The regression approximation has been systematically developed by Rychlik (1987). For an account of Slepian models and regression approximations, see Lindgren & Rychlik (1991). CROSSREG is a package of FORTRAN subroutines, which perform intelligent transformation and numerical integration in regression aprroximations in order to produce high accuracy approximations of the density of first passage and intercrossing times; see Rychlik & Lindgren (1990).

Suppose $x(s), s \geq 0$, is a stochastic process, given in the form

$$x(s) = -a(s) + Rb(s) + \Delta(s), \tag{1}$$

where R is a standard Rayleigh distributed random variable, and $\Delta(s)$, $\Delta(0) = \Delta'(0) = 0$, is a zero-mean continuously differentiable Gaussian process, independent of R.

Example 1: Let T_u be the distance between an upcrossing of the level u and the following u-downcrossing, for a Gaussian process $z(s)$ with covariance function $r_z(\tau)$, say. In the sequel, we shall use Gaussian low frequency white noise to illustrate the results, i.e. the covariance function

$$r_z(\tau) = \frac{\sin t\sqrt{3}}{t\sqrt{3}}. \tag{2}$$

N. Bellomo · F. Casciati (Eds.)
Nonlinear Stochastic Mechanics
IUTAM Symposium Turin

Using the theory of Slepian models, the density of T_u is the same as the first passage density T of the process $x(s)$ given by

$$x(s) = ur_z(s) - u - Rr_z^{'}(s) + \Delta(s),$$

which is of the form (1) with $a(s) = -ur_z(s) + u$ and $b(s) = -r_z^{'}(s)$. □

Example 2: Let T, H be the wavelength and amplitude of a Gaussian process $z(s)$, i.e. the horizontal and vertical distances between a local maximum of $z(s)$ and the following local maximum. Here, we assume that $z(s)$ is twice continuously differentiable, and such that $Var(z(0)) = Var(z'(0)) = 1$, which is a matter of proper scaling. The empirical distribution of T, H can be studied using the theory of Slepian models. Let $x(s)$ be a Slepian model process for z' after zero downcrossing,

$$x(s) = Rr''(s) + \Delta_0(s) = Rb_0(s) + \Delta_0(s).$$

From the theory of Slepian models it follows that $T = T_0(x)$, where T_0 is the time for the first zero upcrossing of x. In order to study the distribution of H we have to introduce a second process $y(s)$, which is a Slepian model for $z(0) - z(s)$ after a zero downcrossing of the derivative z' at zero, i.e.

$$y(s) = R\frac{Cov(z(0) - z(s), z''(0))}{Var(z''(0))^{\frac{1}{2}}} + \Delta_1(s) = -a_1(s) + Rb_1(s) + \Delta_1(s).$$

Then $(T, H) = (T_0(x), y(T))$. □

The present version of the program CROSSREG is directed to applications related to wave and crossing problems. In many cases wavelength or excursionlength can be written as the first upcrossing time of a certain Slepian model process, which has a downcrossing of a fixed level at time zero. Then the Rayleigh variable R is the derivative of the process at the downcrossing, see Example 1 and 2. However, by a minor modifications, CROSSREG can be adapted to problems with a general density of R. Such an extension was used in a nonstationary wave problem, see Rychlik (1991).

2 ONEREG

The basic unit in CROSSREG is ONEREG, which contains routines for evaluation of the regression approximation for the density of the first upcrossing zero T of the process $x(s)$ defined by (1). We shall evaluate the density of T at a fixed point $t > 0$. Taking $Y = \Delta(t)/Var(\Delta(t))^{1/2}$, $x(s)$ can be written in the following form, depending on t,

$$x(s) = -a(s) + Rb(s) + Yc(s) + \Delta_1(s), \tag{3}$$

where $\Delta_1(s)$ is another nonstationary Gaussian process. Note that the function $c(s)$ as well as the random variables Y and $\Delta_1(s)$ also depend on t, but in the following we always suppress this dependence in the notation.

If $b(t) \neq 0$, one then has the following exact density of T,

$$\begin{aligned} f_T(t) &= \int E\Big[I\Big(t; -a(\cdot) + p(t,y)b(\cdot) + yc(\cdot) + \Delta_1(\cdot)\Big) \cdot \\ &\quad \cdot\Big(-a'(t) + p(t,y)b'(t) + yc'(t) + \Delta_1'(t)\Big)\Big] \frac{p(t,y)^+}{|b(t)|} e^{-\frac{1}{2}p(t,y)^2} \frac{1}{\sqrt{2\pi}} e^{-\frac{1}{2}y^2}\, dy \\ &= \int E(y) \frac{p(t,y)^+}{|b(t)|} e^{-\frac{1}{2}p(t,y)^2} \frac{1}{\sqrt{2\pi}} e^{-\frac{1}{2}y^2} dy, \end{aligned} \tag{4}$$

where $p(t,y) = (a(t) - yc(t))/b(t)$, and the integral is extended over all y such that $p(t,y) > 0$. Further, $I(t;x)$ is an indicator function, defined for any function $x(\cdot)$ by

$$I(t;x) = \begin{cases} 1 & \text{if } x(s) < 0 \text{ for all } s, 0 < s < t, \\ 0 & \text{otherwise.} \end{cases}$$

In (4) we have also introduced the function

$$\begin{aligned} E(y) &= E\Big[I\Big(t; -a(\cdot) + p(t,y)b(\cdot) + yc(\cdot) + \Delta_1(\cdot)\Big) \cdot \\ &\quad \cdot\Big(-a'(t) + p(t,y)b'(y) + yc'(t) + \Delta_1'(t)\Big)\Big]. \end{aligned} \tag{5}$$

The explicit evaluation of the expectation in (5) is intractable, since the indicator I depends on the whole sample path of y in the interval $(0,t)$. The programs in CROSSREG use the regression method to compute accurate approximations of $E(y)$. The main idea of the regression method is to find a vector of random variables $\mathbf{Y}$, defined on the process Δ_1, containing the most relevant information about the indicator I. Then the residual process $\Delta_1(s)$ is approximated by its regression curve $E[\Delta_1(s)|\mathbf{Y}]$. The subprogram RIND selects $\mathbf{Y}$ in a recursive way and allows a random number N of elements in the vector $\mathbf{Y} = (Y_1, ..., Y_N)$. The regression approximation of order n is obtained by truncation of N to $n-1$, ($N \leq n-1$). Then the expectation in (5) becomes at most an n-dimensional integral, which can be evaluated numerically. For simplicity, in the following we use only a deterministic number of variables N.

The simplest (first order) approximation is obtained by replacing the Δ_1-process by 0, i.e. taking $E(y) \approx E_0(y) = -a'(t) + p(t,y) \cdot b'(t) + yc'(t)$, for any t such that $-a(s) + p(t,y)b(s) + yc(s) < 0$ in the whole interval $0 < s < t$.

Higher order approximations are obtained by replacing $\Delta_1(s)$ by a conditional expectation on n suitable regressors, $Y_1, \ldots, Y_n$, i.e.

$$\begin{aligned} E(y) \approx E_n(y) &= E\Big[I\Big(t; -a(\cdot) + p(t,y)b(\cdot) + yc(\cdot) + E[\Delta_1(\cdot)|Y_1, \ldots, Y_n]\Big) \cdot \\ &\quad \cdot\Big(-a'(t) + p(t,y)b'(t) + yc'(t) + E[\Delta_1'(t)|Y_1, \ldots, Y_n]\Big)\Big]. \end{aligned}$$

The n'th order regression approximation of the density of T can then be written as

$$f_{T_n}(t) = \int E_{n-1}(y) \frac{p(t,y)^+}{|b(t)|} e^{-\frac{1}{2}p(t,y)^2} \frac{1}{\sqrt{2\pi}} e^{-\frac{1}{2}y^2}\, dy. \tag{6}$$

To obtain best possible numerical accuracy in the approximation of the T-distribution, one can renormalize the problem and expand the integrals near a point (r, y) such that $-a(t) + rb(t) + yc(t) = 0$. CROSSREG contains the numerical algorithms needed to perform these transformations and integrations, which are made by means of Legendre-Gaussian quadrature. It further contains an algorithm RIND for approximation of the function $E(y)$.

Example 1 (contd.): We have used ONEREG to approximate the excursion length density for zero-mean Gaussian low frequency white noise with covariance function (2). Figure 1 shows approximate densities of different orders for the length of an excursion over levels $u = 0, 1, 2, 3$.

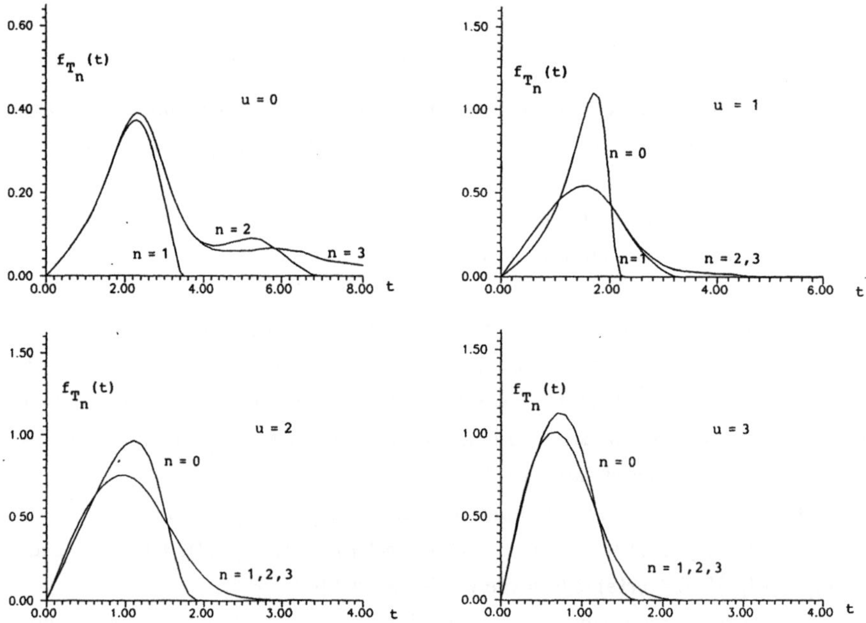

Figure 1: Regression approximations (6) to excursion length density for low frequency Gaussian noise after upcrossing of a level u.

As seen, for an excursion over the zero level, it is necessary to include a total of 3 regressors ($n = 3$ in (6)), in order to catch the bimodal distribution after a zero crossing. For higher levels even the zero order approximation gives reasonable results, and in fact for $u = 2, 3$ the curves for $n = 1, 2, 3$ coincide. □

Example 1 (contd.): In many applications the studied processes have a finite memory, and then the functions $b(s)$ and $c(s)$ in (3) tend to zero, while the residual process $\Delta_1(s)$ converges in distribution to a stationary Gaussian process, as s goes to infinity. The functions describe the stochastic structure of the process, while $a(s)$ in (3) is related to the type of crossing or waveform problem under consideration; see Example 1. For positive and large functions a, e.g. u in Example 1 being negative, the waiting time T can be very long, and hence for large t-values, the regression approximation method is inconvenient since it requires large values of n, i.e. high dimensional integrals to reach satisfactory accuracy. In such cases the program uses asymptotic convergence of the levelcrossings of the Gaussian processes to a Poisson point processes to approximate the tail of the T-density. □

3 TWOREG

CROSSREG also contains a unit, TWOREG, which calculates a marked first passage density of the process $x(t)$, defined by (1), and a companion process $y(t)$ of similar type, acting as a "mark" on the x-process at the crossing time. An example of a marked first passage problem is the joint wavelength and amplitude distribution in Gaussian noise described in Example 2.

The program computes the regression approximation of the joint density of T and $y(T)$, at a fixed point (t, h), where T is the first upcrossing zero of the process $x(s)$; $(T = t,\ y(T) = h)$. One assumes that the processes x, y are given in the form

$$x(s) = -a_0(s) + Rb_0(s) + \Delta_0(s), \quad y(s) = -a_1(s) + Rb_1(s) + \Delta_1(s), \tag{7}$$

where R is the same Rayleigh variable as in (1), independent of the $\Delta_i(s)$–processes. Further, the processes Δ_0, Δ_1 are zero-mean continuously differentiable jointly Gaussian processes, and $\Delta_0(0) = \Delta_0'(0) = 0$. For a fixed $t > 0$, taking $X_1 = \Delta_1(t)/Var(\Delta_1(t))^{1/2}$, $x(s)$ can be written as

$$x(s) = -a_0(s) + Rb_0(s) + X_1c_0(s) + \Delta_0^*(s), \tag{8}$$

where $\Delta_0^*(s)$ is another nonstationary Gaussian process. Further, since $(T, H) = (t, h)$ implies $y(t) = h$, the random variables R and X_1 satisfy a linear equation $h = -a_1(t) + Rb_1(t) + X_1c_1(t)$ which we write as

$$0 = -a_1 + Rb_1 + X_1c_1, \tag{9}$$

with $a_1 = a_1(t) + h$. As before, let $Y = \Delta_0^*(t)/Var(\Delta_0^*(t))^{1/2}$, then $x(s)$ is decomposed as follows

$$x(s) = -a(s) + Rb(s) + X_1 c(s) + Yd(s) + \Delta_1(s),$$

(the indices 0 and * are omitted to simplify the notation), and Δ_1 is a new residual process.

If the vectors $\mathbf{g_0} = (b(t), c(t))$ and $\mathbf{g_1} = (b_1, c_1)$ are linearly independent, i.e. if $\det|G(t)| > 0$, where $G(t)^T = (\mathbf{g_0}^T, \mathbf{g_1}^T)$, one has the following exact density for (T, H),

$$\begin{aligned}
f_{T,H}(t,h) &= \\
&= \int E\Big[I\Big(t; -a(\cdot) + p_1(t,h,y)b(\cdot) + p_2(t,h,y)c(\cdot) + yd(\cdot) + \Delta_1(\cdot)\Big)\cdot \\
&\quad \cdot\Big(-a'(t) + p_1(t,h,y)b'(t) + p_2(t,h,y)c'(t) + yd'(t) + \Delta_1{}'(t)\Big)\Big]\cdot \\
&\quad \cdot \frac{p_1(t,h,y)^+}{2\pi|\det G(t)|} e^{-\frac{1}{2}||\mathbf{p}(t,h,y)||^2} e^{-\frac{1}{2}y^2}\, dy, \\
&= \int E(y) \frac{p_1(t,h,y)^+}{2\pi|\det G(t)|} e^{-\frac{1}{2}||\mathbf{p}(t,h,y)||^2} e^{-\frac{1}{2}y^2}\, dy,
\end{aligned} \tag{10}$$

where $(p_1(t,h,y), p_2(t,h,y))^T = G(t)^{-1} \cdot (a(t) - yd(t), a_1)^T$, and the integral is over y such that $p_1(t,h,y) > 0$.

As in (5) the $E(y)$-function can be written in the general form

$$E(y) = E[I(t; m(\cdot; y) + \Delta_1(\cdot)) \cdot (\frac{d}{dt} m(t; y) + \Delta_1'(t))],$$

where, for a fixed y, $m(\cdot; y)$, is an explicitly defined function and Δ_1 is a zeromean Gaussian residual. The regression approximation of $E(y)$ in (10) is computed using the subprogram RIND described above. To obtain best possible numerical accuracy in the approximation of the (T, H)-density, one renormalizes the integral (10) by expanding the integral near a point (r, x, y) such that

$$-a(t) + rb(t) + xc(t) + yd(t) = 0, \quad -a_1 + rb_1 + xc_1 = 0.$$

Example 2 (contd.): We have used TWOREG to approximate the wavelength (T) and amplitude (H) density for Gaussian low frequency white noise with covariance function (2), i.e. the horizontal and vertical distances between a local maximum and the following local minimum.

Figure 2 shows the very accurate results for approximations of order $n = 1, 2, 3$. The zero order approximation is truncated at the first local minimum for $r_x(t)$ and is therefore not very accurate. □

Example 2 (contd.): The algorithm TWOREG can be easily extended to approximate the density of $T, \mathbf{x}(T)$, where $\mathbf{x}$ is a vector valued Gaussian process. A suitable program is in preparation. □

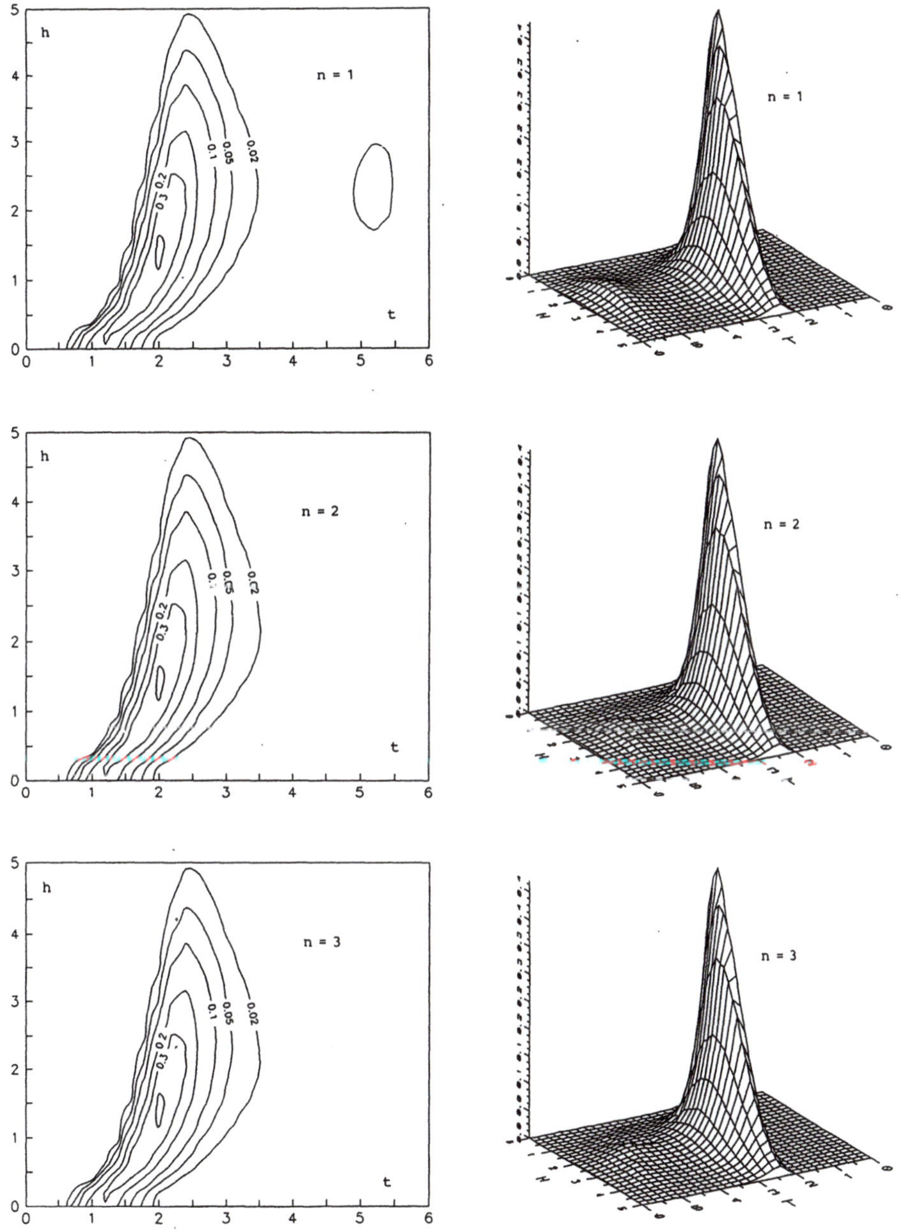

Figure 2: Regression approximations to wavelength and amplitude density for low frequency Gaussian noise.

4 FORTRAN routines

In Sections 1 and 2 we have presented algorithms for approximation of the first zero upcrossing time density and the marked crossing time density for a differentiable stochastic process of a form given by (1). The algorithms are implemented in a collection of FORTRAN subroutines, with two main routines called ONEREG and TWOREG.

The routine ONEREG gives the density of the first zero upcrossing T by the process $x(s) = -a(s) + Rb(s) + \Delta(s)$, see (1). The routine TWOREG gives the joint density of T and the value at T of a supplementary process $y(s)$, see (7).

Before calling ONEREG and TWOREG, one has to call the subroutine INITINTEG which initiates all constants used in the routines, as well as nodes and weights for numerical integrations:

```
CALL INITINTEG(EPS0,IAC,NIT).
```

All input parameters should be written on files:

ACCUR.DAT	-	constants
Z.DAT	-	nodes
H.DAT	-	weights

The nodes and weights in Z.DAT and H.DAT have been taken from Abramowitz & Stegun (1968). The input file ACCUR.DAT contains parameters for output and constants used in stopping criteria which control the numerical accuracy. Here is an example of an input file ACCUR.DAT:

OUTPUT.DAT	Name of output file
1	NIT
0	IAC
0.001	EPS - constant ϵ discussed in [18], Section 1.5
0.0001	EPS0 - constant ϵ_0 in (21) in [18]
4.5	C - constant c in (25) in [18]
20.	PMAX - threshold used to avoid numerical underflow in the exp-function
12	NNW-number of diff quadr. used in subroutine GAUSS1
2	quadr. with 2 nodes
3	quadr. with 3 nodes
⋮	⋮
16	quadr. with 16 nodes
20	quadr. with 20 nodes

The first zero upcrossing time density of the process (1) at the point $t > 0$ is computing by calling ONEREG:

```
CALL ONEREG(F,R,A,B,DAN,DBN,VDER,EPS0,IAC,N,NIT),
```

where N > 2 is a number of grid points T, say, in the interval $[0, t]$ (T(1)=0,... ,T(N)=t). The matrix R, with dimension N × N × 2, contains the covariances between $\Delta(T), \Delta(T)$ and $\Delta(T), \Delta'(T)$. The vectors A, B with dimension N contain values of functions a, b in (1) at the grid points T, respectively. The derivatives $a'(t), b'(t)$ are stored in DAN, DBN, respectively, and the variance of $\Delta'(t)$ is given in VDER. Finally EPS0, IAC and NIT are accuracy constants, see [18] for more detailed discussion.

Similarly, by calling the routine TWOREG, one can compute the joint density of the first zero upcrossing time of x at $t > 0$ and the value at t of a supplementary process y, see (7):

```
CALL TWOREG(F,R,A,B,C,XN,YN,DXN,VDER,EPS0,IAC,N,NIT).
```

Here F, EPS0, IAC, N, NIT are the same parameters as in ONEREG, R, VDER are defined as in ONEREG with the process Δ replaced by Δ_0^*, see (8), A, B, C are the values of functions a_0, b_0, c_0 in (8) at the grid T. Finally, XN = $(a_0(t), b_0(t), c_0(t))$, YN = (a_1, b_1, c_1), see (9) and DXN = $(a_0'(t), b_0'(t), c_0'(t))$.

5 EXTENSIONS

The decomposition of the x-process in (1) in a form of an explicit functional defined on the random variables R, X, and the residual Δ, is the key to derivation of formulas (4) and (10) for the density of T and (T, H), respectively. In the case of Gaussian processes the decomposition is linear in R, X and Δ. Analogues of (4) and (10) can be derived for more complicated nonlinear decompositions of the x-process, for example, memoryless nonlinear transformations of vectorvalued Gaussian processes. This approach was used to derive regression approximations for χ^2-processes, e.g. envelopes of Gaussian processes; see the description of CHIREG in [18].

The programs ONEREG and TWOREG have been used to analyse properties of a certain simple class of nonlinear mechanical systems with stochastic Gaussian inputs. Generally, one assumes that there exists a Gaussian process z, say, and events A_i, $i = 0, ..., n$, such that during the visit of the process z to the region defined by the event A_i, the nonlinear system can be approximated by a linear system. The exit times from the regions A_i describe the evolution of the nonlinearities. The regression method is then used to approximate the marked exit times. The modified ONEREG and TWOREG programs have been used to analyse properties of switching oscillators, a "jump and bump" problem, see ([18], [5]), and for reliability analysis of a degenerating system of fibers, a "Daniels system", see [17] for a detailed presentation.

References

[1] Abrahams, J. (1986): A survey of recent progress on level crossing problems for random processes. In *Communications and Networks, A survey of recent advances.* Eds.: Blake, I.F. & H.V. Poor. Springer-Verlag, New York, 6-25.

[2] Abramowitz, M., Stegun, I.A. (1968): *Handbook of Mathematical Functions.* National Bureau of Standards, Washington DC.

[3] Ditlevsen, O. (1985): Survey on applications of Slepian model processes in structural reliability. *Proc. ICOSSAR '85*, Eds.: I. Konishi, A. H.-S. Ang, M. Shinozuka, IASSAR Japan, Vol. 1, 241-250.

[4] Ditlevsen, O., Lindgren, G. (1988): Empty envelope excursions in stationary Gaussian processes. *J. Sound and Vibration*, 122, 571-587.

[5] Jogréus, C. (1991): *Methods for the analysis of switching stochastic systems.* Univ. of Lund, Dissertation.

[6] Leadbetter, M.R., Lindgren, G. and Rootzén, H. (1983): *Extremes and Related Properties of Random Sequences and Processes.* Springer-Verlag, New York.

[7] Lindgren, G. (1983): On the shape and duration of FM-clicks. *IEEE Trans. Inform. Theory*, IT-29, 536-543.

[8] Lindgren, G., Rootzén, H. (1987): Extreme values: theory and technical applications. *Scand. J. Statist.* 14, 241-279.

[9] Lindgren, G., Rychlik, I. (1991): Slepian models and regression approximations in crossing and extreme value theory. *Intern. Statist. Rev.*, 59, 195-225.

[10] Rychlik, I. (1987): Regression approximations of wavelength and amplitude distributions. *Adv. Appl. Prob.* 19, 396-430.

[11] Rychlik, I. (1988): Rain flow cycle distribution for ergodic load processes. *SIAM J. Appl. Math.* 48, 662-679.

[12] Rychlik, I. (1989a): On the distribution of random waves and cycles. *Springer Lecture Notes in Statistics, No 51, Extreme Value Theory*, 1989, 100-113.

[13] Rychlik, I. (1989): The two barriers problem for continuously differentiable processes. *Univ. of North Carolina, Dept. of Statistics Res. Rep.* No 277, 1-30. To appear in *Adv. Appl. Prob.* 1992.

[14] Rychlik, I. (1990a): New bounds for the first passage, wave-length and amplitude densities. *Stochastic Process. Appl.* 34, 313-339.

[15] Rychlik, I. (1990b): Confidence bands for linear regressions. *Univ. Lund Stat. Res. Rep.*, No 1990:4, 1-15. To appear in *Commun. in Statist.*, 1992.

[16] Rychlik, I. (1991): Rainflow cycles in random loads. *Univ. Lund Stat. Res. Rep.*, No 1991:5, 1-10. To appear.

[17] Rychlik, I., Gregoriu, M. (1991): Reliability of Daniels systems with equal load sharing rule subject to stationary Gaussian dynamic loads. *Univ. Lund Stat. Res. Rep.*, No 1990:20, 1-26. To appear in *Probabilistic Engineering Mechanics*, 1992.

[18] Rychlik, I., Lindgren, G. (1990): CROSSREG – a computer package for first passage and wave density analysis. *Univ. Lund Stat. Res. Rep.*, No 1990:2, 1-28.

Stochastic Stabilization by Multiplicative Noise

M. Shinozuka and K.Y. Billah
Program in Structures & Mechanics
Department of Civil Engineering & Operations Research
Princeton University

SUMMARY

The effect of parametric colored noise on the stabilization of a nonlinear bistable system is presented. Theoretical study suggests that when dealing with stability issues the replacement of correlated noise by white noise may not be always appropriate. Moreover, the interpretation of stochastic differential equation may not be as simple as an option between the Ito and Stratonovich interpretation. A particular study of such a stability problem is presented in this paper and corroboration to some theoretical predictions by numerical studies are included. These numerical verifications using the spectral density method are the first of its kind as no conclusive results to date have been obtained by other simulation methods. The broad conclusion obtained in this study are similar in nature to another study carried out for stochastic stability studies of suspension bridges.

N. Bellomo · F. Casciati (Eds.)
Nonlinear Stochastic Mechanics
IUTAM Symposium Turin

Studies involving parametric noise are analytically more involved than that of forced ones. The situation becomes much more complicated if stability and colored noise issues are considered together.

To understand the complexity of such physical process let us consider the following equation

$$\frac{1}{\gamma}\ddot{x} + \dot{x} = d\,x - b\,x^3 \tag{1}$$

where x is macroscopic variable, $1/\gamma$ = m (mass)d is control parameter and b is positive constant provided for nonlinear stabilization (when d> 0). The threshold of instability of the state x=0 is d=0. Here the parameter can fluctuate as $[d(t) = d_0 + y(t)]$ where d_0 is a constant, y(t) is the noisewith with a bandwidth $\Delta\omega$ (correlation time τ_c), assumed to be Gaussian and given by the equation

$$\dot{y}(t) = -\Delta\omega\, y + \Delta\omega \sqrt{Q}\,\xi(t) \tag{2a}$$

($\langle\xi(t)\rangle = 0$, and $\langle\xi(t)\,\xi(o)\rangle = \delta(t)$), with power spectral density given by

$$\langle y(t)\, y(0)\rangle = \frac{Q}{2}\Delta\omega\, e^{-\Delta\omega\,|t|} \tag{2b}$$

Now, if

if $\frac{1}{\gamma} \to \infty$ i.e., $m \to 0$

Eq.1 reduces to the following

$$\dot{x} = [d - b\,x^3] + y(t)\;x \tag{3}$$

Let us concentrate on the effect of the noise y(t) (in Eq. 3) on the deterministic instability condition $d_0=0$. If d is fixed at d_0 the above system has two time scales namely $\tau_s = 1/\gamma$ and $\tau_l = |1/d_0|$. For sufficiently close to the threshold value $d_0=0$, the realtion $\tau_l >> \tau_s$ is satisfied i.e., the relaxtion time of the system is large compared to the correlation time. Under this circumstance the noise is usually taken as white i.e.,

$$\langle y(t)\, y(t')\rangle = Q\delta(t-t') \tag{4}$$

The problem defined by Eq. 1 and 2 thus reduces to Eq. 3 and 4 for small mass of the system for $d=d_0+y(t)$, where d_0 is close to zero.

In Eq. 3 the noise is parametric; the interpretation of the equation is thus necessary i.e., whether it is an Ito or Stratonovich stochastic differential equation. In engineering, the Stratonovich interpretation is usually considered since it usually believd to correspond to physical noise. If we make this assumption then Eq. 3 and 4 for $m\to 0$, i.e., (mass small) and $\tau_c \to 0$ ("whitish" noise) the Stratonovich interpretation holds.

However some circumspection is required in the above assumption. Theoretical analysis[1,2] (through adibatic elimination) indicates that the actual interpretation of the stochastic differential equation is not so simple. The theoretical results may be summarized under the following:

- $\lambda = \frac{\tau_s}{\tau_c} = \frac{1}{\gamma\tau_c} \to 0$; small mass and not totally white noise. This is the Stratonovich interpretation and no stabilization is introduced by the noise. It is to be noted that this is not the usually interpreted Stratonovich case with $m \to 0$, and $\tau_c \to 0$ as noted above.

• $\lambda = \frac{\tau_s}{\tau_c} = \frac{1}{\gamma\tau_c} \rightarrow \infty$; large mass and "whitish" noise. This is the Itò interpretation and stabilization is observed and the shifts are related to the system and noise parameters.

•For intermediate values of λ neither Ito nor Stratonovich interpretation hold; some other interpretations are required.

The above results are the summary of the theoretical calculation done by Graham and Schenzle.[1] In general, Graham and Schenzle showed that the $\langle x^2(t)\rangle$ values for the above two limiting cases and for intermediate values of λ are given by

$$\lim_{t\to\infty} \langle x^2(t)\rangle = 1/b \begin{cases} d_0 - \frac{1}{2}\frac{\lambda Q}{1+\lambda}, & d_0 > \frac{1}{2}\frac{\lambda Q}{1+\lambda} \\ 0, & d_0 \le \frac{1}{2}\frac{\lambda Q}{1+\lambda} \end{cases} \tag{5}$$

Eq. 5 is the core of the theoretical study for the verification of which numerical studies are important. In the above equation if $\lambda \rightarrow 0$ noise does not cause any shift in deterministic stability. If however, $\lambda \rightarrow \infty$ then shift is $(d_0 - Q/2)$.

These two exteme cases may be obtained from the following one-dimensional. stochastic differential equation

$$\dot{x} = [d - b x^3] + x\mu(t) \tag{6a}$$

with

$$\langle \mu(t)\, \mu'(t')\rangle = Q\delta(t-t') \tag{6b}$$

For the $\lambda \rightarrow 0$ case the interpretation is Stratonovich and $\lambda \rightarrow \infty$ the interpretation is Ito.(Note that $\gamma \rightarrow 0$ is not necessarily Stratonovich interpretation)

To corroborate the above study, analog simulation[3-7] studies have been undertaken to date which concluded that there are substantial limitations of the theory; these are shown in Fig. 1. Subsequently, to commensurate with those numerical studies, a variant theoretical interpretation was also made.[4] On the other hand, it is also discussed in the literature that the inherent presence of weak additive stochastic forces in analog simulation measurements may have prevented the substantiation of the theory.[8] In this communication digital simulation results are presented which provide a variant perspective for the theoretical study.

In the present work numerical simulation results are presented in which noise was generated from the corresponding power spectral density[10] (Eq. 3) which provide certain advantages over other methods when used in conjunction with a physical system.[11] The results presented are for different λ values and the verification or otherwise relates to Eq. 5.

Fig. 2 presents the results for large λ values, the limiting case of which represents the Ito interpretation. Fig. 3 is for small λ values the corresponding limiting case of which is the Stratonovich interpretation. The figures show that both interpretation along with "intermediate" interpretations are needed as predicted by Graham and Schenzle. It is important to note that for high Q, the $<x^2>$ values go to zero for all presented cases. On the other hand, in analog simulation studies[4-7] the initial part of the curve (for increasing Q values) matched quite well but the curve did not tend towards zero for larger values of Q but increased to higher values (in a nonlinear fashion) and even exceeded the threshold value (here 0.3) for the no-noise case. Since the analog simulation observations were interpreted as accurate a different theory was proposed in Ref. 4 to rectify the perceived discrepancy of Ref. 1. However, in subsequent publications[8-9] it has been commented that there might be an inherent difficulty with the analog simulation method namely that there could be error due to the inherent presence of inertia in the system.

In the present undertaking the numerical calculation conform to the theory of Ref. 1 in contrast to the analog simulation studies. The good matchings shown in Figs. 2 & 3 for different λ values may be attributed largely to the proper consideration of the ratios $\lambda 1 = \tau_s/\tau_l$ and $\lambda 2 = \tau_c/\tau_l$ in the numerical computations which are integral part of Graham and Schenzles's theory. For Eq. 5 to match with numerical results two conditions namely, $\lambda 1 = \tau_s/\tau_l \rightarrow 0$ and $\lambda 2 = \tau_c/\tau_l \rightarrow 0$ have to be satisfied.

In numerical simulation, what ratio constitutes such extremes values of $\lambda 1$ and $\lambda 2$ can only be known by gradually taking smaller ratios until stable and nonchanging results are obtained. For Figs. 2 & 3 the ratios $\lambda 1$, $\lambda 2$ used were in the range of 0.001 to 0.005. To shed more light on the importance of the limiting values of $\lambda 1$ and $\lambda 2$ further calculations are presented for $\lambda=1$. Fig. 4 shows the results for values of the ratios ranging from 0.001 to 0.05 whereas Fig. 5 is for higher ratios namely, 0.2 and 0.5. These show that the time scale of inertia (τ_s) and noise (τ_c) with respect to slow time scale of instability (τ_l) play a pivotal role in the deviation or otherwise of the numerical calculation. When higher ratios (Fig. 5) are used the curves tend to move away from the zero mean square response value. It is the authors' belief that the present work which involved many parameters (γ, d_0, b, τ_s, τ_l, τ_c, τ_s/τ_l, τ_c/τ_l, τ_s/τ_l) was possible since the noise was generated independent of the system equation and from its PSD.

It is pointed out that both the Ito ($\lambda \rightarrow \infty$) and the Stratonovich interpretation ($\lambda \rightarrow 0$) were obtained near the limit in the presented numerical simulations using only one numerical integration algorithm: the Runga-Kutta method used commonly for deterministic system. This is in contrast to views expressed in the literature that for digital simulation to achieve that end twe algorithms are needed.[9] Finally it is noted that the broad conclusion obtained in this study are similar in nature to another study carried out for stochastic stability studies of suspension bridges.[12]

[1]R. Graham & A. Schenzle, Phy. Rev. A26, 1676 (1982).

[2]C. W. Gardiner, Phy. Rev. A29, 2814 (1984).

[3]C. Festa, L. Fronzoni, P. Grigolnini, & F. Marchesoni, Phy. Lett. A102, 95 (1984).

[4]S. Faetti, C. Festa, L. Fronzoni, P. Grigolnini, & P. Martano, Phy. Rev. A30, 3252 (1984).

[5]S. Faetti, C. Festa, L. Fronzoni, P. Grigolnini, Advan. Chem. Phys. 62, 445 (1985).

[6]S. Faetti, L. Fronzoni, & P. Grigolnini, Phy. Rev. A32, 1150 (1985).

[7]P. Grigolnini in *Noise in Nonlinear Dynamical Systems*, Edited by F. Moss and P. V. E. McClintock (Cambridge University Press, Cambridge, England, 1989), Vol. 1.

[8]R. Manella, S. Faetti, P. Grigolnini, P. V. E. McClintock and F. E. Moss, J. Phy. A19, L699 (1986).

[9]P. V. E. McClintock & F. Moss in *Nonlinear Dynamical Systems*, Edited by F. Moss and P. V. E. McClintock (Cambridge University Press, Cambridge, England, 1989), Vol. 3.

[10]M. Shinozuka and G. Deodatis, Applied Mechanics Review, 44, 191 (1991).

[11]K.Y.R. Billah & M. Shinozuka, Phy. Rev. A42, 7492(1989).

[12]K.Y.R. Billah & M. Shinizuka, submitted to *J. Applied Mechanics*.

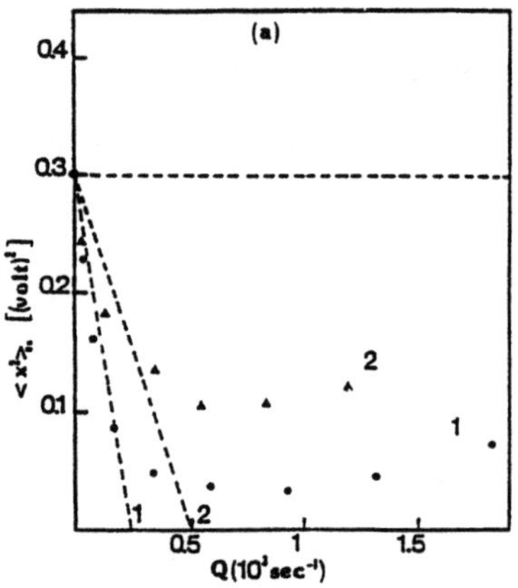

Fig.1a

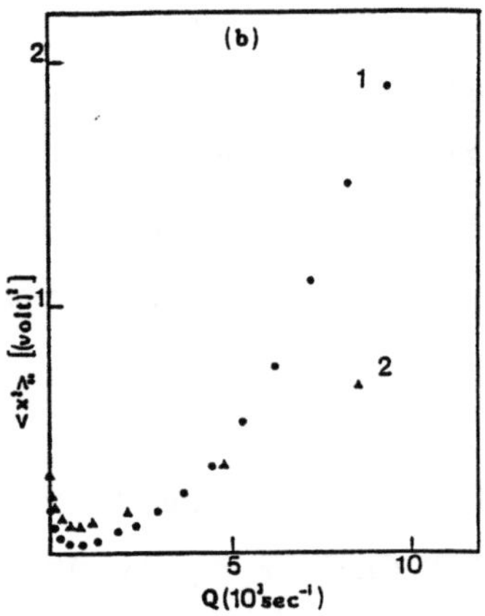

Fig. 1b

(———): Theory

(Δ): λ=0.71; (•): λ=10.

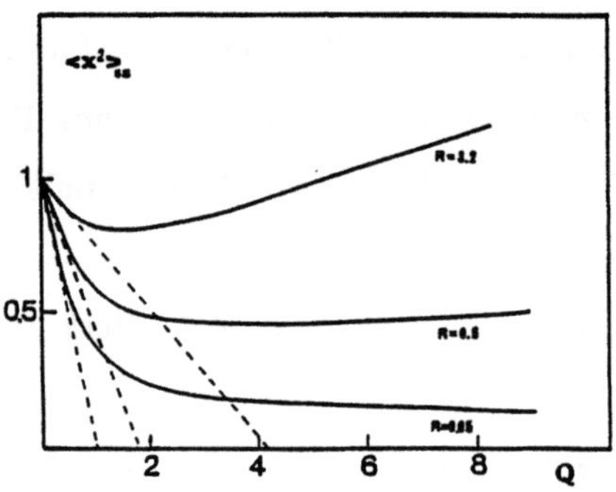

Fig. 1c: (———): Theory; Solid: Simulation.

Fig. 1. Analog simulation to date.

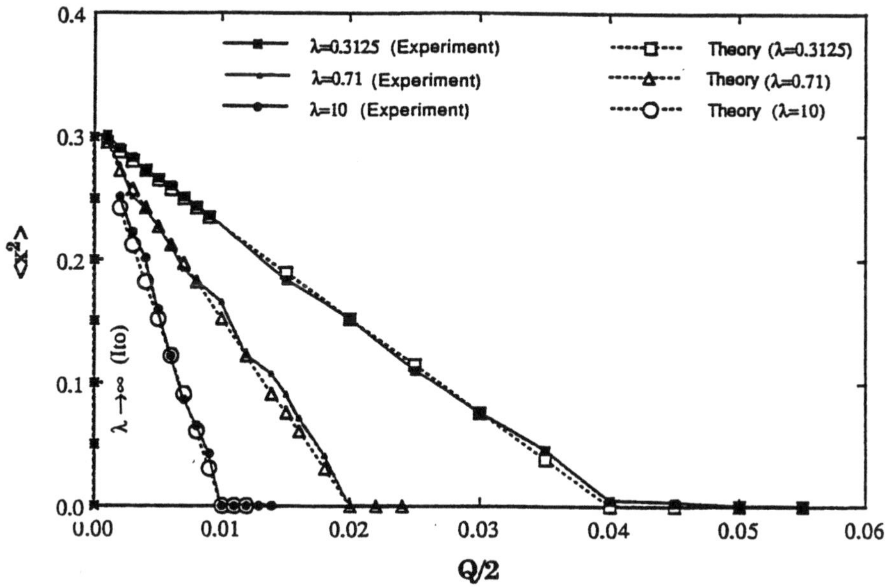

Fig. 2. Numerical simulation vs. theory for large λ values; Solid :Experiment, Dotted:Theory.

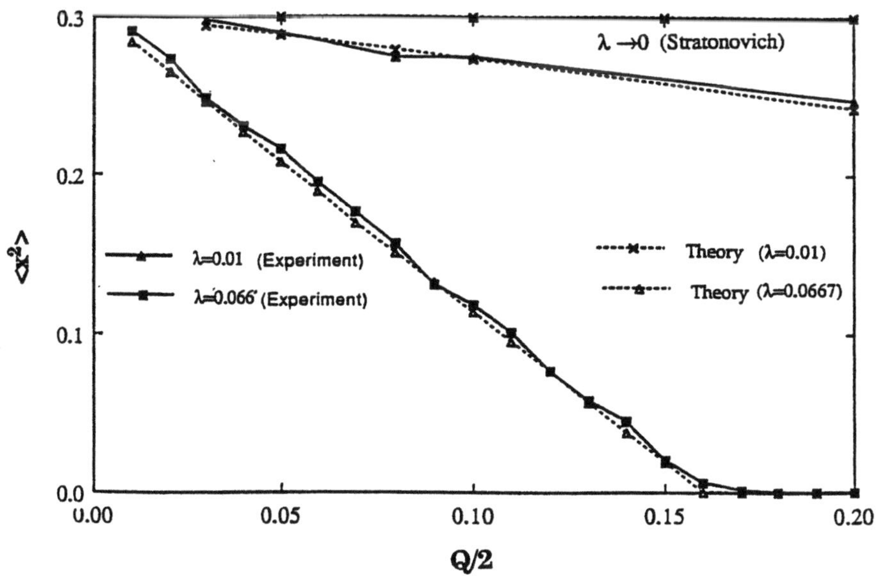

Fig. 3. Numerical simulation vs. theory for small λ values; Solid:Experiment, Dotted: Theory.

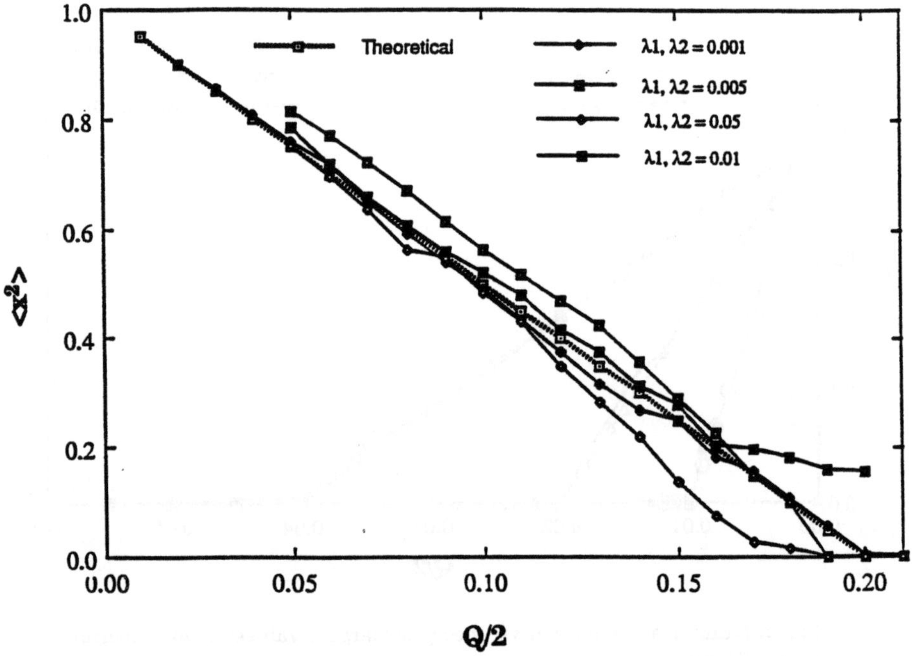

Fig. 4. Numerical simulation vs. theory for small values of λ1 and λ2; λ = 1.

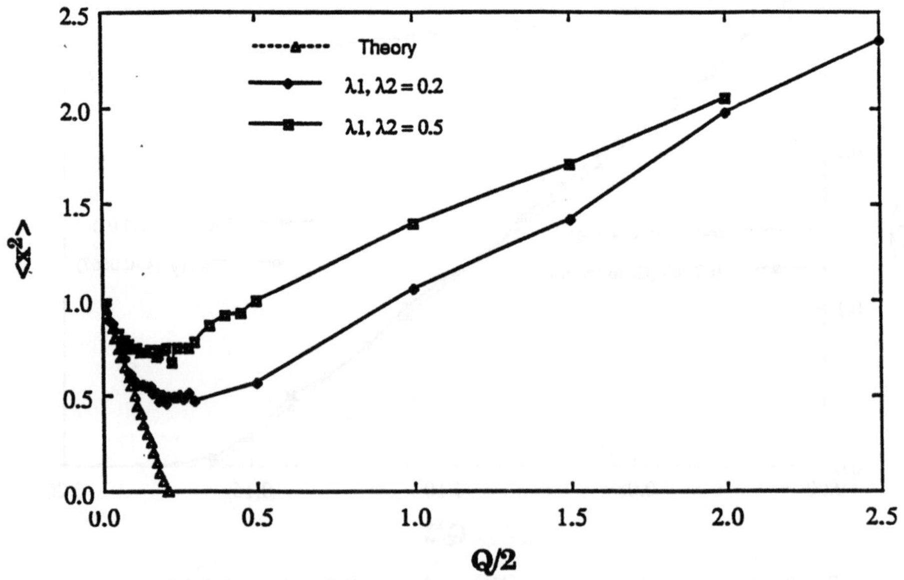

Fig. 5. Numerical simulation vs. theory for large values of λ1 and λ2; λ = 1.

Random Seismic Response of Equipment on Hysteretic Structures

M. P. Singh and G.O. Maldonado
Department of Engineering Science and Mechanics
Virginia Polytechnic Institute and State University
Blacksburg, Virginia 24061 U.S.A.

Summary

The paper presents a method for calculating the seismic response of equipment supported on hysteretically behaving inelastic structures subjected to random ground motions. The hysteretic behavior is modeled by the well–known Bouc–Wen model. Stochastic linearization approach is used to deal with the nonlinear problem. The linearized system of equations are solved by a generalized modal analysis approach to obtain the stochastic seismic response. The stochastic approach is extended to propose a response spectrum approach for generating floor response spectra directly from the input ground response spectra. Numerical results showing the effect of hysteretic nonlinearity on the floor motion response are presented.

Introduction

The research on the seismic response of equipment or secondary systems supported on linear primary structures has been of continued interest for some time now, whereas the investigations about the behavior of the secondary systems supported on inelastic structures have started only recently. As the structures are likely to go in the inelastic range and respond hysteretically under a design level earthquake, the study of the seismic response of secondary systems supported on inelastic structures is of significant practical importance.

The inelastic behavior causes an increase in the displacement response but a reduction in the force and acceleration responses in the primary structures. Several studies have been done to evaluate this inelastic effect. However, only a few studies have been done to examine the effect of the material nonlinearity and hysteretic behavior of primary structures on the response of supported secondary systems. The work of Lin and Mahin [1] and Sewell et al [2] are quite illuminating and noteworthy.

In this paper, we present a stochastic method to evaluate the effect of hysteretic nonlinearity of a primary structure on its floor response spectra for ground motions defined in stochastic form such as by a power spectral density function. The hysteretic

N. Bellomo · F. Casciati (Eds.)
Nonlinear Stochastic Mechanics
IUTAM Symposium Turin

behavior is modeled by the well–known Bouc–Wen model [3–6]. The hysteretic nonlinearity is handled through the stochastic equivalent linearization approach. A generalized modal analysis approach is used to solve the system equations for calculating the primary and secondary responses. The use of this modal analysis approach has also allowed us to develop a response spectrum approach for generating seismic floor response spectra. The approach is used to evaluate the effect of hysteretic nonlinearity on the response of single degree of freedom secondary systems and also to generate floor response spectra for multi degree of freedom primary structure for seismic inputs defined in terms of ground response spectra.

Analytical Formulation

Equations of Motion: To examine the effect of hysteretic nonlinearity on equipment response, we consider, a multi–degree–of–freedom shear structure shown in Figure 1. The restoring force offered by i^{th} story is modeled by a nonlinear hysteretic element with following characteristics, as proposed by Wen [4]:

$$g(u_i,v_i) = \alpha_i k_i u_i + (1-\alpha_i)k_i v_i \tag{1}$$

where u_i is the deformation of the element; k_i is the initial stiffness of the hysteretic model; α_i = proportionate contribution of the linear elastic element to the restoring force; and v_i is the auxiliary displacement variable of the hysteretic element. The hysteretic behavior is modeled by the popular Bouc–Wen [3,4] model

$$\dot{v}_i = \dot{u}_i - \gamma_i v_i\,|\dot{u}_i|\,|v_i|^{\eta_i-1} - \beta_i\,\dot{u}_i|v_i|^{\eta_i} \tag{2}$$

with parameters γ_i, β_i and η_i. By choosing different values of these parameters, hysteretic models with wide variety of shapes and characteristics can be obtained [5,7].

With the restoring force of equation (1), one can write the equation of motion of ith mass as follows:

$$m_i\ddot{x}_i - c_i\dot{x}_{i-1} + (c_i + c_{i+1})\,\dot{x}_i - c_{i+1}\,\dot{x}_{i+1} + g(u_i,v_i) - g(u_{i+1},v_{i+1})$$
$$= -m_i\ddot{x}_g \;; \qquad i = 1\ldots n \tag{3}$$

where x_i is the displacement of the mass m_i with an overdot representing the time derivative; c_i = discrete damping coefficients of the ith story damper; and $\ddot{x}_g(t)$ = ground acceleration. In the stochastic equivalent linearization approach, we replace equation (2) by the following equation

$$\dot{v}_i = a_i\dot{u}_i + b_i v_i \;; \qquad i = 1\ldots n \tag{4}$$

and obtain the linearization coefficients by minimization of the mean square error. Under the restrictive assumption of Gaussian response, these linearization coefficients can be obtained as [8],

$$a_i = E\left[\frac{\partial g}{\partial \dot{u}_i}\right] \; ; \quad b_i = E\left[\frac{\partial g}{\partial v_i}\right] \; ; \quad i = 1 \ldots n \tag{5}$$

Since the linearization coefficients depend upon the statistics of the response, the approach to solve the equations of motion must be iterative.

Substituting for $u_i = x_{i+1} - x_i$ and combining equations (1), (3) and (4), we obtain the following state vector equation:

$$[G_1]\{\dot{y}\} + [G_2]\{y\} = -\{F(t)\}\ddot{x}_g(t) \tag{6}$$

where

$$\left.\begin{aligned} &[G_1] = \begin{bmatrix} M & 0 & 0 \\ 0 & I & 0 \\ 0 & 0 & I \end{bmatrix} ; \quad [G_2] = \begin{bmatrix} M & K & H \\ -I & 0 & 0 \\ -a & 0 & -b \end{bmatrix} ; \quad \{F\} = \begin{Bmatrix} M\,r \\ 0 \\ 0 \end{Bmatrix} \\ &\{Y\}^T = \left[\{\dot{x}\}^T \; , \; \{x\}^T \; , \; \{v\}^T\right] \end{aligned}\right\} \tag{7}$$

in which $[M]$ = diagonal mass matrix containing m_i; $[c]$ = damping matrix defined in terms of damping coefficients c_i; $[K]$ = stiffness matrix defined in terms of $\alpha_i k_i$; $[H]$ = matrix containing hysteretic element coefficients; $[a]$ and $[b]$ are the matrices defined in terms of the linearization coefficients a_i and b_i and $\{r\}$ is the vector of ground motion influence coefficient. All the submatrices are of size $n \times n$. $[G_1]$ and $[G_2]$ are of size $N \times N$ where $N = 3n$.

Modal Analysis: Since the matrices in equation (6) are not symmetric, one needs to solve the following two adjoint eigenvalue problems [9]

$$[G_2]\{\phi_j\} = \lambda_j[G_1]\{\phi_j\} \; ; \quad [G_2]^T\{\psi_j\} = \lambda_j[G_1]^T\{\psi_j\} \tag{8}$$

where λ_j is the jth eigenvalue and $\{\phi_j\}$ and $\{\psi_j\}$ are the corresponding right and left eigenvectors, respectively. It is relevant to note that for n hysteretic elements, there will be exactly n number of real eigenvalues and eigenvectors. The remaining eigenvalues and eigenvectors will occur in the pairs of complex and conjugates. To decouple equation (6), we use the transformation $\{Y\} = [\Phi]\{z\}$ (where $[\Phi]$ is the matrix of eigenvectors $\{\phi_j\}$, and $\{z\}$ = the vector of principal coordinates) and premultiply by $\{\psi_j\}^T$ to obtain:

$$\dot{z}_j + \lambda_j z_j = -\{\psi_j\}^T\{F(t)\}\, \ddot{x}_g(t) \, , \quad j = 1 \ldots N \tag{9}$$

where $[\psi_j]$ and $\{\phi_k\}$ are biorthonormalized as $\{\psi_j\}^T[G_1]\{\phi_k\} = \delta_{jk}$. For zero–start initial conditions, the solution of equation (9) can be written as:

$$z_j(t) = -\gamma_j \int_0^t e^{-\lambda_j(t-\tau)} \ddot{x}_g(\tau)d\tau \tag{10}$$

where $\gamma_j = \{\psi_j\}^T\{F(t)\}$. This provides the displacement response of, say, m^th^ floor as:

$$x_m = \sum_{j=1}^{N} \phi_{mj} z_j(t) \tag{11}$$

For the analysis of a supported secondary system, the absolute acceleration of the supporting floor is of main interest. It can be shown that the vector of absolute acceleration $\{\ddot{x}_a\}$ can be written as:

$$\{\ddot{x}_a\} = -\sum_{j=1}^{N} \{\bar{\phi}_j\}\lambda_j z_j \tag{12}$$

where $\{\bar{\phi}_j\}$ is the first part of the eigenvector $\{\phi_j\}$ which is associated with the relative velocity response.

Floor Response Spectrum: A floor response spectrum is the maximum response of a single degree of freedom oscillator of certain frequency and damping ratio subjected to the motion of the floor at its base. For an oscillator placed on m^th^ floor this response is obtained from the following equation of motion

$$\ddot{x}_o + 2\beta_o\omega_o\dot{x}_o + \omega_o\dot{x}_o + \omega_o^2 x_o = -\ddot{x}_{am} = \sum_{j=1}^{N} \bar{\phi}_{m,j}\lambda_j z_j \tag{13}$$

where x_o is the displacement response of the oscillator; $\ddot{x}_{am}$ = absolute acceleration of the floor; and ω_o and β_o, respectively, are the frequency and damping ratio of the oscillator. To obtain the maximum value of the oscillator response, we first obtain the mean square response value. For assumed stationarity of the ground motion and the responses of the primary structure and supported oscillator, it can be shown that the mean square value of the absolute acceleration response of the oscillator is given by:

$$E[\ddot{x}_{ao}^2] = \sum_j^N \sum_k^N q_j q_k \int_{-\infty}^{\infty} \frac{(\omega_o^4 + 4\beta_o^2\omega_o^2\omega^2)}{(\lambda_j + i\omega)(\lambda_k - i\omega)} |H_o|^2 \, \Phi_g(\omega)d\omega \tag{14}$$

where $\Phi_g(\omega)$ is the spectral density function of the ground acceleration; $q_j = (\bar{\phi}_{mj}\lambda_j\gamma_j)$ and H_o is the transfer function defined as

$$H_o = 1/\{\omega_o^2 - \omega^2 + 2i\beta_o\omega_o\omega\} \tag{15}$$

For a given ground motion spectral density function, equation (14) can be used to obtain the mean square value of the oscillator acceleration response. To obtain the maximum response, the root mean square value must be multiplied by an appropriate peak factor. An approximate value for the peak factor can be calculated by using one of the several methods available in the literature on the subject. To calculate the oscillator response in terms of the input ground response spectra, however, we must express equation (14) in terms of the modal frequencies and damping ratios of the supporting structure. For this, we separate the terms associated with the real and imaginary eigenvalues and express a complex eigenvalue as:

$$\lambda_j = \beta_j\omega_j + i\omega_j\sqrt{1-\beta_j^2} \tag{16}$$

where ω_j and β_j can be identified as the modal frequency and damping ratio, respectively. We further separate terms with j = k and j ≠ k, and the use partial fractions to write the mean square value of the oscillator acceleration as:

$$E[\ddot{x}_{ao}^2] = R_1 + R_2 + R_3 \tag{17}$$

where

$$\left.\begin{aligned}
R_1 &= \sum_j^n \bar{q}_j^2\, I_{rj} + \sum_{j\neq k}^n\sum^n \bar{q}_j\bar{q}_k\{A\ I_{rj} + B\ I_{rk}\} \\
R_2 &= \sum_j^n \left[\bar{q}_j\{A'(I_{rj} + I_{sk}(B', C'))\} + \bar{q}_k\{A''I_{rk} + I_{sj}(B'',C'')\}\right] \\
R_3 &= \sum_j^n I_{sj}(c_j^2, a_j^2) + \sum_{j\neq k}^n\sum^n \{I_{sj}(\bar{A},\bar{B}) + I_{sk}(\bar{C},\bar{D})\}
\end{aligned}\right\} \tag{18}$$

and

$$\left.\begin{aligned}
I_{rj} &= A_1 I_{1j} + B_1 I_{20} + C_1 I_{30} \\
I_{sj}(B', C') &= A_2 I_{2j} + B_2 I_{3j} + C_2 I_{20} + D_2 I_{30}
\end{aligned}\right\} \tag{19}$$

where (A,B), (A′,B′,C′), (A″,B″,C″), ($\bar{A}$,$\bar{B}$,$\bar{C}$), (A_1,B_1,C_1) and (A_2,B_2,C_2,D_2) are the coefficients of partial fractions; $\bar{q}_j$ is same as q_j but associated with the real eigenvalue; and

$$c_j = \omega_j\left[a_j\beta_j - b_j\sqrt{1-\beta_j^2}\right] \tag{20}$$

in which a_j and b_j are the real and imaginary parts of the complex quantity q_j. I_{1j}, I_{2j} and I_{3j} are frequency integrals defined as

$$\left.\begin{aligned}
I_{ij} &= \int_{-\infty}^{\infty} \Phi_g(\omega)\ \frac{1}{(\nu_j^2+\omega_j^2)}\ d\omega\ ;\ I_{2j} = \int_{-\infty}^{\infty} \Phi_g(\omega)|H_j|^2 d\omega \\
I_{3j} &= \int_{-\infty}^{\infty} \Phi_g(\omega)\omega^2 |H_j|^2 d\omega
\end{aligned}\right\} \tag{21}$$

where H_j is the frequency response function similar to H_o defined in equation (17) but for parameters β_j and ω_j. It is noted that I_{2j} and I_{3j} are the mean square values of the relative displacement and relative velocity responses of an oscillator of parameters β_j

and ω_j, excited by ground motion whose spectral density function is $\Phi_g(\omega)$. I_{1j} is mean square value of the velocity response of a massless oscillator the motion of which is defined by the following equation:

$$\dot{v} + \nu_j v = \ddot{x}_g(t) \tag{22}$$

The definition of the secondary oscillator response in equation (17) in terms of the frequency integrals I_{1j}, I_{2j} and I_{3j} facilitates the use of ground motion spectra as these integrals can be defined in terms of ground spectra as:

$$I_{2j} = \left[R_{pj}/\omega_j^2\, F_{pj}\right]^2; \quad I_{3j} = \left[R_{rj}/F_{rj}\right]^2 \tag{23}$$

where R_{pj} and R_{rj}, respectively, are the commonly used pseudo acceleration and relative velocity response spectrum values for the oscillator parameters of ω_j and β_j. F_{pj} and F_{ij} are the peak factor values associated with pseudo acceleration and relative velocity responses. Although not commonly available, input spectrum similar to the pseudo acceleration and relative velocity ground spectra can also be developed for the velocity response of a massless oscillator. Figure 2 shows one such spectrum obtained for an ensemble of 75 synthetically generated ground acceleration time history.

By substitution of the frequency integrals I_1, I_2 and I_3 defined in terms of input spectra in equation (17) and multiplying the resulting quantity by the square of the peak factor for the $\ddot{x}_{ao}$ response, we will obtain the floor response spectrum value for the oscillator parameters (ω_o, β_o). It is often reasonable to assume that the peak factor for $\ddot{x}_{ao}$ response and the peak factors introduced in equation (23) are all equal, and thus they cancel out from the final expression for the floor response spectrum value. Thus equation (17) is also an expression to define the floor response spectrum value, if the frequency integrals in equations (18) and (19) are simply replaced by their respective response spectrum values. The numerical results for floor response spectra obtained by this approach are shown in the next section.

Numerical Results

For the structure shown in Figure 1, the absolute acceleration floor response spectra are obtained for various floors. The floor mass and story stiffness for this structure are: $m_1 = 3$, $m_2 = 2$, $m_3 = 2$ and $m_4 = 1$, all in kips–sec^2/in; $k_1 = 3200$, $k_2 = 2400$, $k_3 = 1600$ and $k_4 = 800$, all in kips/in. units. The structural damping matrix was constructed such that the damping ratio in each mode was .05. The post–yield stiffness parameter, α was chosen to be 0.05 for each story, which represents a nearly elasto–plastic system. The other parameters of the hysteretic relationship are $\beta = \gamma =$

$.5/y_e^{\eta}$ where y_e is the yield level in inches. The exponent η was chosen to be equal to 3 and 9 to provide a rather sharp knee near the yield in the hysteresis loop. The yield level parameter has been assumed to be the same for different stiffness elements. However, the results have been obtained for different yield levels to examine its effect on the floor response spectrum value.

The seismic inputs defined in terms of a Kanai–Tajimi type of spectral density function as well as the pseudo–acceleration and relative velocity ground response spectra have been used. The velocity spectrum for the massless oscillator shown in Figure 2 is also used.

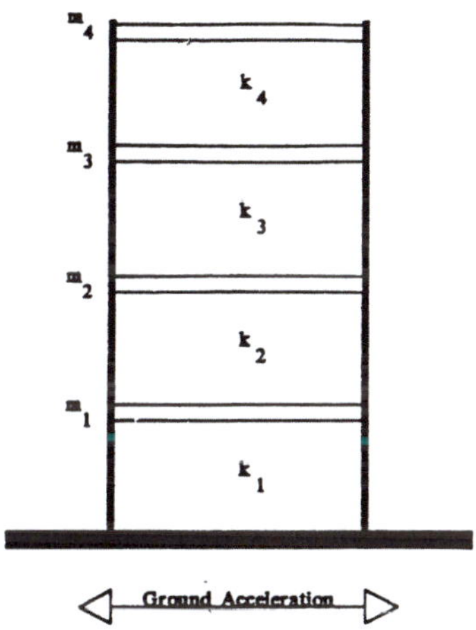

FIG. 1: Shear building

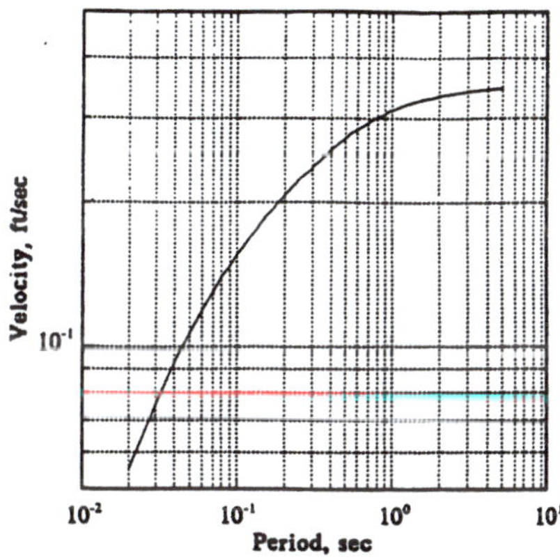

FIG. 2: Massless oscillator spectrum

Figure 3 shows the maximum accelerations of various floors for increasing levels of maximum ground acceleration. The story yield level for this figure was fixed at 0.4". Decreasing slope of the curves for various stories indicates the softening of structure due to yielding at higher intensities of the ground motion. Figure 4 shows again the same response, but now for a fixed level of ground motion intensity (0.4g) and increasing value of yield levels. The response increases asymptotically to its elastic response level as the yield level is increased. Thus, yielding in the supporting structure is seen to reduce the maximum acceleration which a supported secondary system will experience.

To show the frequency content of a floor acceleration, in Figure 5 we plot the root mean square value of the floor response spectra for floor 2 obtained for the spectral density function input. Similarly Figure 6 shows the floor response spectra obtained for floor 1 but for response spectra input. The elastic spectra is seen to provide the highest peaks always. For the lower floors, however, the valleys in the elastic spectra

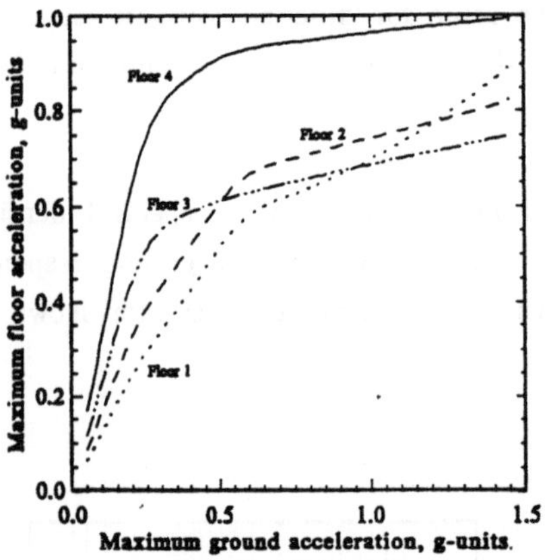

FIG. 3: Maximum floor acceleration for increasing levels of ground acceleration

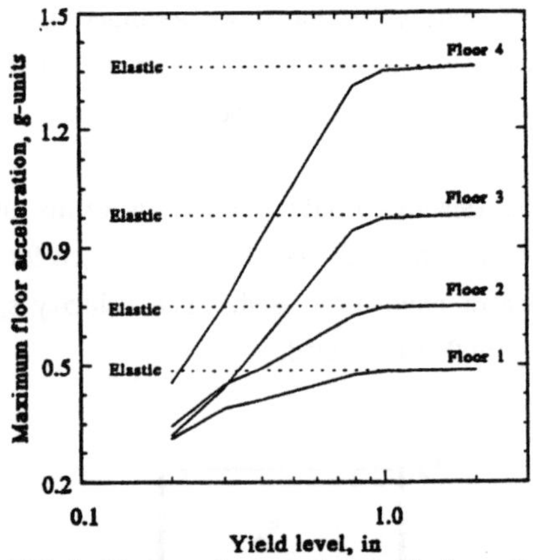

FIG. 4: Maximum floor acceleration for increasing yield levels

are also the lowest. Thus, the secondary systems at some frequencies may even experience a higher level of response when the supporting structure yields. But in general, the highest peaks in the floor response spectra are reduced by yielding. The floor spectra for decreasing yielding level are seen to flatten out, showing the effect of increased damping provided by hysteresis cycles.

In Figures 7 and 8 we plot the ratio of the inelastic to elastic floor response spectra for floors 1 and 4 for various yield levels. It is seen that the inelastic floor response

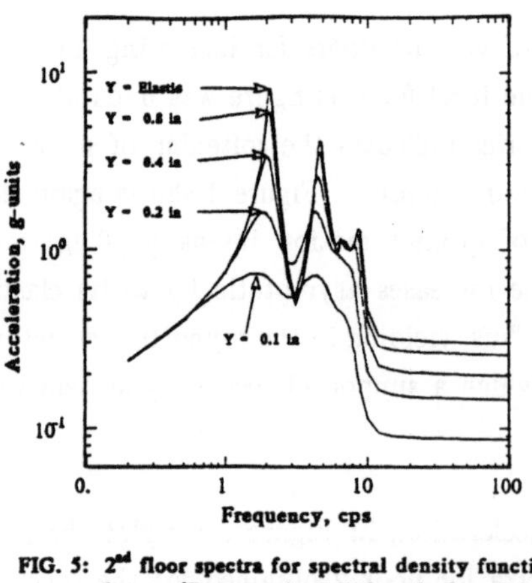

FIG. 5: 2nd floor spectra for spectral density function input. (Damp. ratio=0.005, n=3, G_{max}=0.4 g)

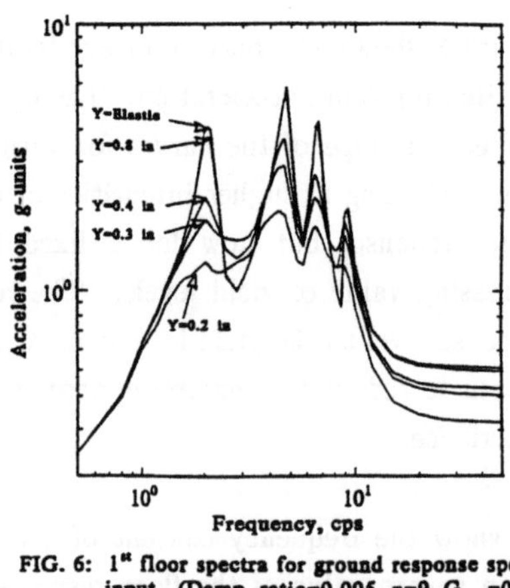

FIG. 6: 1st floor spectra for ground response spectra input. (Damp. ratio=0.005, n=9, G_{max}=0.4 g)

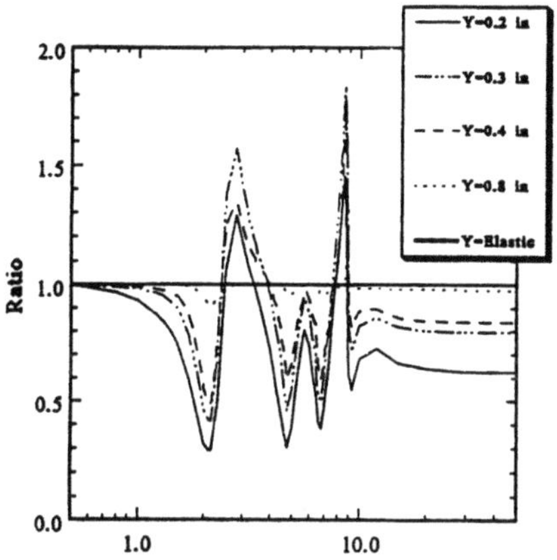

FIG. 7: Floor spectrum ratio for floor 1 for various yield levels. (Damp. ratio=0.005, n=9, G_{max}=0.4 g)

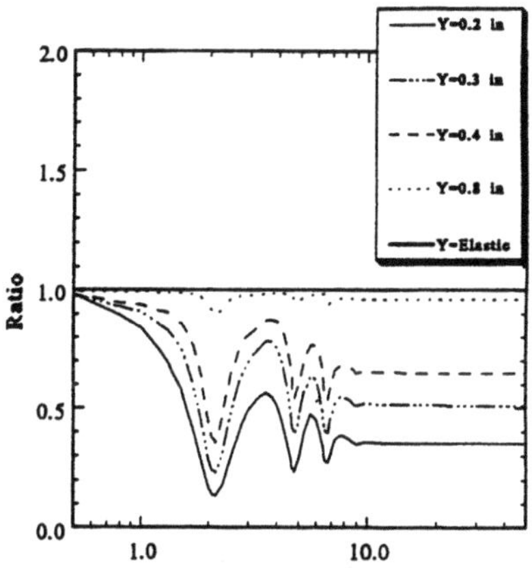

FIG. 8: Floor spectra ratio for floor 4 for various yield levels. (Damp. ratio=0.005, n=9, G_{max}=0.4 g)

spectrum values are not always less than elastic response spectrum values as the ratio greater than 1 are also seen for some frequencies. Somewhat similar observations have also been made in Reference 2 where several multi–degree of freedom structures were analyzed for recorded ground motions.

Concluding Remarks

A response spectrum method is presented to obtain floor response spectra for seismic design of secondary systems supported on nonlinear hysteretic primary structure. Stochastic linearization is used to deal with the hysteretic nonlinearity. The equations of motion are solved by a generalized modal analysis approach to develop the response spectrum approach. The effect of hysteretic nonlinearity on the secondary system response is shown to be significant. It reduces the acceleration of a floor as well as the floor response spectrum peaks. Because of the energy dissipation in the hysteresis loop, the floor acceleration peaks are flattened and the valleys are raised. Thus, some secondary systems may even experience higher response due to yielding effect.

Acknowledgements

This research work is supported by the National Science Foundation through Grant No. CEE–8412830 with Dr. S. C. Liu as its Program Director. This support is gratefully acknowledged. The opinions, findings and conclusions or recommendations expressed in the paper are those of the writers and do not necessarily reflect the views of the National Science Foundation.

References

1. Lin, J.; Mahin, S.A.: Seismic response of light subsystems in inelastic structures, J. of Structural Engineering, ASCE, Vol. 111, No. 2, Feb. (1985), 400–417

2 Sewell, R. T.; Cornell, C. A.; Toro G. R.; McGuire R. K.: A study of factors influencing floor response spectra in nonlinear multi–degree–of–freedom structures, Report No. 82, Dept. of Civil Eng., Stanford University, Nov.(1986)

3. Bouc, R.: Forced vibration of mechanical systems with hysteresis. Abstract, Proc. 4th Conf. on Nonlinear Oscillations, Prague, Czechoslovakia, 1967.

4. Wen, Y. K.: Equivalent linearization of hysteretic systems under random excitation. J. of Appl. Mechanics, ASME, 47 (1980), 150–154.

5. Baber, T. T.; Wen, Y. K.: Stochastic equivalent linearization for hysteretic, degrading, multistory structures. Dept. of Civil Engr., Univ. of Ill., SRS No. 471, Dec. 1979.

6. Casciati, F.; Faravelli, L.: Method of nonlinear stochastic dynamics for the assessment of structural fragility. Nuclear Engineering & Deisgn, (90) 1985.

7. Maldonado, G. O.; Singh, M. P.: Stochastic response of single degree of freedom hysteretic oscillators. Report No. VPI–E–87–5, Virginia Polytechnic Institute and State University, College of Engr., April 1987.

8. Atalik, T. S.; Utku, S.: Stochastic linearization of multidegree of freedom nonlinear systems. J. Earthq. Engr. Str. Dyn., 4(1976), 411–420.

9. Meirovitch, L.: Computational Methods in Structural Dynamics. Sijthoff & Noordhoff Int. Publishers, The Netherlands, 1980.

Analysis of Stochastic Systems Via Maximum Entropy Principle

K. SOBCZYK
J. TRĘBICKI

Institute of Fundamental Technological Research
Polish Academy of Science
Świętokrzyska 21, 00-049 Warsaw

Summary

The principle of maximum entropy introduced first in statistical physics and successfully applied in other fields (statistics, reliability theory, simulation etc.) has recently been developed to analyze the systems governed by stochastic differential equations (cf. [2], [3]). In this paper we extend the maximum entropy methodology to stochastic vibratory nonlinear systems.

Starting from general formulation of the system dynamics and assuming that an information about the response (solution) is available in the form of the equations for moments, the probability distribution of the stationary response is derived as a result of maximization of the entropy functional. Along with the basic scheme of the method the comparisons of the results with prediction of other known procedures is given.

1. Introduction

In stochastic analysis of dynamically systems the exact probability distribution of the response can be found only in the simplest cases. Most often, all that we are able to obtain (from the governing stochastic equations) are the moments of the response, or the equations for moments.

An important and challenging problem which occurs can be stated as: how can the partial information contained in given moments, or equations for moments be used for the best characterization of the probability distribution of the response? Among possible approaches to this problem, the principle of maximum entropy (introduced first in statistical physics by Jaynes [1]) seems to be especially attractive and effective (cf. Sobczyk [2], and Sobczyk, Trębicki [3]).

In general, the principle of maximum entropy states that, of all the distributions that satisfy the constraints, one should choose the one with the largest entropy. Since the entropy H(X) of random variable X characterizes the uncertainty or randomness associated with possible values of X, maximizing the entropy leads to a distribution with the largest informational content (within the given constraints).

There are many reasons for the usage of the maximum entropy principle to the prediction of the "most rational" approximate probability distribution of the response of stochastic dynamical systems on the basic of available incomplete information. For example, all of the best known distributions in statistics are maximum entropy distributions given appropriate simple moment

N. Bellomo · F. Casciati (Eds.)
Nonlinear Stochastic Mechanics
IUTAM Symposium Turin

constraints (even the Cauchy distribution of random variable X is a maximum entropy distribution over all distributions satisfying $E[\ln(1+X^2)]=\beta$ where β is given, and E[.] denotes a mean value). The principle of maximum entropy has been successfully applied in a variety of fields including statistical mechanics and thermodynamics (cf. de Beauregard, Tribus [4]), statistics (cf. Kullback [5]), reliability estimation (cf. Tribus [6], Spencer, Bergman [7]), system simulation (cf. Chan [8]), spectrum estimation (cf. Ulrych, Bishop [9]). In our recent paper [3] we showed that the maximum entropy principle can also serve as effective tool in the analysis of a wide class of stochastic dynamical systems

In this paper we extend the maximum entropy methodology to stochastic vibratory nonlinear systems, along with the comparison of the results with predictions of other known procedures.

2. General Idea

Let the system of interest be described by the following stochastic equation for the vector process $\mathbf{Y}(t)=[Y_1(t),\dots,Y_n(t)]$

$$\frac{d\mathbf{Y}(t)}{dt} = \mathbf{F}\left[\mathbf{Y}(t),\mathbf{X}(t,\gamma)\right] \tag{2.1}$$

where $\mathbf{X}(t,\gamma)$ is a given stochastic process; $\gamma\in\Gamma$, $(\Gamma,\mathcal{F},P)$ is the basic probability space. Assume that system (2.1) has a stationary solution. Most often the available information about $\mathbf{Y}(t)$ is given in the form

$$E\left[G_{r_1 r_2 \dots r_n}\left(Y_1,Y_2,\dots,Y_n\right)\right] = \beta_{r_1 r_2 \dots r_n} \tag{2.2}$$

$$M \geq r_1 + \dots + r_n \neq 0 \;;\qquad r_i=0,1,2,\dots,M \;;\qquad i=1,2,\dots,n$$

where $G_{r_1 r_2 \dots r_n}(\mathbf{Y})$ are functions specified in particular situations, and $\beta_{r_1 r_2 \dots r_n}$ are given numbers, M is a specified number - the highest order of correlation moment $E\left[Y_1^{r_1} Y_2^{r_2} \dots Y_n^{r_n}\right]$.

Restrictions (2.2) can also be represented in the form of a sequence of equations

$$E\left[G_k\left(Y_1,Y_2,\dots,Y_n\right)\right] = \beta_k \;,\qquad k=1,2,\dots,N \tag{2.3}$$

where N is the number of moment equations assumed as constraints.

The appropriate (approximate) probability distribution $p(\mathbf{y})$ of the stationary solution of (2.1) is that one for which the entropy functional reaches its maximum, i.e.

$$H = - \int p(\mathbf{y}) \ln \left[p(\mathbf{y}) \right] d\mathbf{y} = \max \tag{2.4}$$

under the restrictions (2.2) and the normalization condition

$$\int p(\mathbf{y}) d\mathbf{y} - 1 = 0 \tag{2.5}$$

Restrictions (2.2) include both:

(i) the equations for moments (from infinite hierarchy of equations); in this case the all $\beta_{r_1 r_2 \dots r_n} = 0$.

(ii) prescribed moments of the response; in this case

$$E\left[G_{r_1 r_2 \dots r_n} \left(Y_1, Y_2, \dots, Y_n \right) \right] = E\left[Y_1^{r_1} Y_2^{r_2} \dots\dots Y_n^{r_n} \right] = \beta_{r_1 r_2 \dots r_n} \tag{2.6}$$

The method of Lagrange multipliers leads to the following general expression for the maximum entropy probability density

$$p(\mathbf{y}) = C \exp\left\{ - \sum_{r_1 + \dots + r_n = 1}^{M} \lambda_{r_1 \dots r_n} G_{r_1 \dots r_n} \left(y_1, y_2, \dots, y_n \right) \right\} \tag{2.7}$$

where C is the normalizing constant, and $\lambda_{r_1 \dots r_n}$ are the Lagrange multipliers which should be determined from the moment equations. In special case where the prescribed moments form restrictions, that is (2.6) holds, then maximum entropy distribution is

$$p(\mathbf{y}) = C \exp\left\{ - \sum_{r_1 + \dots + r_n = 1}^{M} \lambda_{r_1 \dots r_n} y_1^{r_1} y_2^{r_2} \dots\dots y_n^{r_n} \right\} \tag{2.8}$$

In the one dimensional case (n=1) formulae (2.7) and (2.8) take the form respectively

$$p(y) = C \exp\left\{ \sum_{k=1}^{M} \lambda_k G_k \left(y \right) \right\} \tag{2.9}$$

$$p(y) = C \exp\left\{ - \sum_{k=1}^{M} \lambda_k y^k \right\} \tag{2.10}$$

In two dimensional case (n=2)

$$p(y_1,y_2) = C \exp\left\{ - \sum_{i+j=1}^{M} \lambda_{ij} G_{ij} \left[y_1,y_2 \right] \right\} \tag{2.11}$$

When the moment constraints are taken in the form (2.3) then the maximum entropy distribution (2.11) can be represented as:

$$p(y_1,y_2) = C \exp\left\{ - \sum_{k=1}^{N} \lambda_k G_k \left[y_1,y_2 \right] \right\} \tag{2.12}$$

It can be shown that if the maximum entropy distribution exists for given constraints, then this distribution is unique (that is, there is only one set of Lagrange multipliers that satisfies the moment constraints).

In order to determine the Lagrange multipliers we substitute to the appropriate expression for the probability density into the specified constraints. This leads to the system of algebraic nonlinear equations for the Lagrange multipliers. The standard numerical procedures (e.g. the Newton method) constitute, in general, an effective tool for obtaining solution.
In two-dimensional case a total number of moment equations (including the normalization condition) is equal to:

$$\sum_{k=0}^{M} \left[M + 1 - k \right] = \frac{(M+1)(M+2)}{2} \tag{2.13}$$

The corresponding system of nonlinear algebraic equations for the Lagrange multipliers (when (2.12) is used) takes the form

$$\int_{-\infty}^{+\infty}\int_{-\infty}^{+\infty} G_k \left[y_1,y_2 \right] p(y_1,y_2)dy_1 dy_2 = 0 \tag{2.14}$$

3. Specific Problems

In order to show usefulness of the maximum entropy method in the analysis of stochastic systems we shall consider few specific problems.

3.1 First-order nonlinear system (cf. [3].)

Among the problems considered in paper [3] the following dynamic system was discussed

$$\frac{dY}{dt} + a_1 Y + a_2 Y^2 + a_3 Y^3 \ln\frac{|Y|}{\beta} = D\xi(t,\gamma) \tag{3.1}$$

where $a_3>0$, $\beta>0$ and $\xi(t,\gamma)$ is a Gaussian white noise. The exact stationary probability density of the solution is

$$f_{st}(y) = C \exp\left\{ -\frac{y^2}{D^2}\left[a_1 + \frac{2}{3}a_2 y + \frac{1}{2}a_3 y^2\left(\ln\frac{|y|}{\beta} - \frac{1}{4}\right)\right]\right\} \tag{3.2}$$

where C is a normalizing constant. Making use of the above formula, the exact moments of the stationary solution can be evaluated. Figure 1 visualizes the comparison of the exact density (3.2) with the approximate maximum entropy densities when two, four and six first moments are known. The values of parameters are: $a_1=0.5$, $a_2=-0.25$, $a_3=5.2$, $\beta=1.0$, $D=1.0$. It is clear that the case of N=2 (two first moments given) provides the Gaussian distribution, which constitutes not acceptable approximation. However, the curves for N=4 and N=6 show very satisfactory approximation of the true distribution.

Let us assume now that our available information about the stationary solution is given by the first equations for moments (from the infinite hierarchy of equations). The first N equations of the hierarchy are:

$$E\left[a(Y)Y^{k-1} + \frac{1}{2}(k-1)b(Y)Y^{k-1}\right] = 0\ , \qquad k=1,2,\ldots,N \tag{3.3}$$

and

$$\begin{aligned} a(Y) &= -\left[a_1 Y + a_2 Y^2 + a_3 Y^3 \ln\frac{|Y|}{\beta}\right] \\ b(Y) &= D^2 \end{aligned} \tag{3.4}$$

According to the general scheme of the method the maximum entropy distribution has the form (2.9) where

$$G_k(y) = \frac{1}{2}(k-1)D^2 y^{k-2} - a_1 y^k - a_2 y^{k+1} - a_3 y^{k+2}\ln\frac{|y|}{\beta} \tag{3.5}$$

$$k=1,2,3,\ldots$$

and $\beta_k=0$. The usage of Newton method for resulting system of algebraic equations allows us to evaluate the Lagrange multipliers. Figure 2 shows a comparison of the exact distribution and the maximum entropy distributions when two equations (N=2) and four moment equations (N=4) are given. It is seen that the approximation based on four moment equations provides very satisfactory agreement with the exact distribution. It should also be noticed that the case N=2 gives much better approximation than the case N=2 at Fig.1 when exact moments are given.

3.2 Nonlinear oscillator with random excitation

A wide class of nonlinear vibratory systems can be represented by the following stochastic equation

$$\rho\frac{d^2Y}{dt^2} + \beta\frac{dY}{dt} + g(Y) = \xi(t,\gamma) \tag{3.6}$$

where g(Y) is a specified nonlinear function of the response and $\xi(t,\gamma)$ is a Gaussian white noise with intensity 2D. The corresponding system of first order equations is ($Y_1=Y$, $Y_2=\dot{Y}$)

$$\begin{aligned} \frac{dY_1}{dt} &= Y_2 \\ \frac{dY_2}{dt} &= -\frac{1}{\rho}\left[\beta Y_2 + g(Y_1) - \xi(t,\gamma)\right] \end{aligned} \tag{3.7}$$

The equations for moments (constituting the restrictions for the maximum entropy principle) have the form

$$\begin{aligned} \frac{d}{dt}E\left[Y_1^i Y_2^j\right] &= E\left\{iY_1^{i-1}Y_2^{j+1} - \frac{j}{\rho}\left[\beta Y_2 + g(Y_1)\right]Y_1^i Y_2^{j-1} + \frac{D}{\rho^2}\, j(j-1)Y_1^i Y_2^{j-2}\right\} \\ &\equiv E\left[G_{ij}(Y_1,Y_2)\right] \end{aligned} \tag{3.8}$$

where $i,j=0,1,2,\dots,M$, $M\geq i+j\neq 0$, and M is the highest degree of given moments.

In the stationary case $\frac{d}{dt}E\left[Y_1^i Y_2^j\right]=0$, and the moment equations are

$$E\left[G_{ij}(Y_1,Y_2)\right] = 0 \qquad i,j=0,1,2,\dots,M, \quad M\geq i+j\neq 0 \tag{3.9}$$

We are looking for the probability distribution of stationary response of (3.7) which maximizes the entropy

$$H = -\int_{-\infty}^{+\infty}\int_{-\infty}^{+\infty} p(y_1,y_2)\ln\left[p(y_1,y_2)\right]dy_1\,dy_2 \tag{3.10}$$

under restrictions (3.9) and normalization condition (2.5). In the case where $0\neq i+j\leq M=2$ the equations for stationary moments are:

$$\begin{aligned} &i = 1,\ j = 0: && E[Y_2] = 0 \\ &i = 0,\ j = 1: && E[g(Y_1)] = 0 \\ &i = 2,\ j = 0: && E[Y_1Y_2] = 0 \\ &i = 1,\ j = 1: && E\left[Y_2^2 - \frac{1}{\rho}Y_1(\beta Y_2 + g(Y_1))\right] = 0 \\ &i = 0,\ j = 2: && E\left[\beta Y_2^2 + Y_2 g(Y_1) - \frac{D}{\rho}\right] = 0 \end{aligned} \tag{3.11}$$

Following the general scheme of the method (described in Sec.2) and performing appropriate analysis we obtain :

$$\lambda_{10} = \lambda_{01} = \lambda_{20} = 0, \qquad \lambda_{11} + \beta\lambda_{02} = \frac{\beta}{2D} \tag{3.12}$$

The maximum entropy probability density takes the form:

$$p(y_1, y_2) = p_1(y_1)\, p_2(y_2) = = C_1 \exp\left[-\lambda_{11} y_1 g(y_1)\right] \frac{1}{\sqrt{2\pi}\,\sigma_{Y_2}} \exp\left[-\frac{y_2^2}{2\sigma_{Y_2}^2}\right] \tag{3.13}$$

where $\sigma_{Y_2}^2 = \frac{D}{\beta\rho}$, C_1 is normalizing constant of density $p_1(y_1)$, and λ_{11} is determined from the algebraic equation:

$$C_1 \int_{-\infty}^{+\infty} \left[y_1 g(y_1) - \frac{D}{\beta} \right] \exp\left[-\lambda_{11} y_1 g(y_1)\right] dy_1 = 0 \tag{3.14}$$

To show the effectiveness of the maximum entropy method, the case

$$g(Y_1) = Y_1 + \varepsilon Y_1^3, \qquad \begin{matrix} \rho=1 \\ \beta=1 \\ D=1 \end{matrix}, \quad \varepsilon>0 \tag{3.15}$$

is considered in detail. In this case the exact stationary probability density is following

$$f_{st}(y_1) = C \exp\left\{-\left[\frac{y_1^2}{2} + \frac{\varepsilon y_1^4}{4}\right]\right\} \tag{3.16}$$

the constant C is determined by the normalization condition

$$C = \frac{(\varepsilon)^{1/2} \exp(-\varepsilon^{-1})}{K_{\frac{1}{4}}(\varepsilon^{-1})} \tag{3.17}$$

where K_ν is the Bessel function of the second order with imaginary argument. Figures 3-5 show the results of comparisons with the exact probability density and with approximations obtained by statistical linearization and Gram-Charlier expansions and maximum entropy principle. At Figures 3-4 the probability densities are compared while the Fig.5 shows the departure of the approximate densities from the exact one measured according to the integral distance

$$\Delta_1 = \left\{ \int \left[f(y_1) - p(y_1) \right]^2 \right\}^{1/2} \tag{3.18}$$

where $f(y_1)$ is the exact density and $p(y_1)$ - its approximation.

3.3 Nonlinear oscillator under random parametric and external excitations

Another case which has been chosen to show the effectiveness of the maximum entropy principle is a system considered previously by Dimentberg (cf. [10]) and Wu and Lin (cf. [11]). The governing equation of the system is

$$\ddot{Y} + 2\alpha\dot{Y}\left[1+\xi_1(t,\gamma) \right] + \beta_1\dot{Y}\left[Y^2 + \dot{Y}^2/\Omega^2 \right] + \Omega^2 Y\left[1 + \xi_2(t,\gamma) \right] = \xi_3(t,\gamma) \tag{3.19}$$

where α, β and Ω are given constants, and ξ_1, ξ_2, ξ_3 are mutually independent Gaussian white noises with spectral densities S_1, S_2, S_3, respectively.

Dimentberg has found the explicit exact stationary solution of equation (3.19) in the case when $\Omega^2 S_2 = 4\alpha^2 S_1$; this solution has the form

$$p(y_1, y_2) = \frac{C_3 \exp[-\beta(y_1^2 + y_2^2/\Omega^2)]}{\left[y_1^2 + y_2^2/\Omega^2 + \zeta \right]^{\delta-\beta\zeta}} \tag{3.20}$$

where $\zeta = S_3/S_2\Omega^4$, $\delta = 2\alpha/S_2\Omega^2 + 0.5$, $\beta = \beta_1/S_2\Omega^2$ and C_3 is the normalization constant; Wu and Lin performed comparisons of the exact variance of the response with various closure approximations. Here we use these findings for comparisons with the solution obtained by maximum entropy method.

The moment equations given by Wu, Lin [11] when written down for a stationary case after appropriate rearrangements, can be represented as:

$$E\left[Y_1 \right] = E\left[Y_2 \right] = E\left[Y_1 Y_2 \right] = 0, \tag{3.21}$$

$$\Omega^2 E\left[Y_1 \right] - E\left[Y_2 \right] = 0, \tag{3.22^I}$$

$$4\alpha E\left[Y_2^2 \right] - 8\alpha^2 S_1 E\left[Y_2^2 \right] + 2\beta_1 E\left[Y_1^2 Y_2^2 \right] + \frac{2\beta_1}{\Omega^2} E\left[Y_2^4 \right] + \\ - \Omega^4 S_2 E\left[Y_1^2 \right] - S_3 = 0, \tag{3.22^{II}}$$

$$3E\left[Y_1^2 Y_2^2 \right] - \Omega^2 E\left[Y_1^4 \right] = 0, \qquad (3.22^{III})$$

$$3E\left[Y_1^2 Y_2^2 \right] - \Omega^2 E\left[Y_2^4 \right] = 0, \qquad (3.22^{IV})$$

$$4\alpha E\left[Y_1^2 Y_2^2 \right] - 8\alpha^2 S_1 E\left[Y_1^2 Y_2^2 \right] + 2\beta_1 E\left[Y_1^4 Y_2^2 \right] + \frac{2\beta_1}{\Omega^2} E\left[Y_1^2 Y_2^4 \right] +$$
$$- \Omega^4 S_2 E\left[Y_1^4 \right] - S_3 E\left[Y_1^2 \right] = 0, \qquad (3.22^{V})$$

$$8\alpha E\left[Y_2^4 \right] - 32\alpha^2 S_1 E\left[Y_2^4 \right] + 4\beta_1 E\left[Y_1^2 Y_2^4 \right] + \frac{4\beta_1}{\Omega^2} E\left[Y_2^6 \right] +$$
$$- 6\Omega^4 S_2 E\left[Y_1^2 Y_2^2 \right] - 6S_3 E\left[Y_2^2 \right] = 0, \qquad (3.22^{VI})$$

The above equations constitute the constraints in the maximum entropy principle.

According to the general idea the maximum entropy probability density has the form (2.12) where the functions $G_k(Y_1, Y_2)$ are defined by the left-hand sides of equations (3.21) and (3.22), but the Lagrange multipliers associated with constraints (3.21) turn out to be zero.

Numerical calculations performed for the case: α=0.5, Ω=1, β_1=1, $S_1 = S_2$=0.1 and for different values of S_3 are shown at Figures 6-9. Figures 6-7 show the comparison of the maximum entropy distributions with the exact one. Figure 8 visualizes the departure of approximate densities from the exact one according to the distance.

$$\Delta_2 = \left\{ \iint \left[f(y_1, y_2) - p(y_1, y_2) \right]^2 dy_1 dy_2 \right\}^{1/2} \qquad (3.23)$$

where $f(y_1, y_2)$ is the exact probability density (3.20) and $p(y_1, y_2)$ - its approximation. For a varying S_3, the exact stationary mean square and their approximation obtained by a cumulant-neglect scheme and maximum entropy principle are visualized at Fig. 9.

4. Conclusions

The analysis reported in this paper along with numerical calculations in specific cases indicates - similarly as results of paper [1] - that the maximum entropy principle can serve as an effective tool for determination of the approximate stationary probability density of the response of a wide class of non-linear systems with random parametric and external noises.

As far as oscillating systems are concerned (Problems 3.2 and 3.3) the comparison with the exact probability distribution shows, in general, very satisfactory agreement. Calculations and graphical results for the Duffing oscillators presented at Fig. 3-4 show, however essential differences in departure from the exact probability density depending on the magnitude of ε. For small ε (weak nonlinearity) maximum entropy method gives worse results than Gram-Charlier expansion of "6-th order" whereas for $\varepsilon=8.0$ (very strong nonlinearity) maximum entropy method (for the same moment constraints) is better than Gram-Charlier expansion of "6-th order". This indicates that maximum entropy method can be especially useful in the situations where most approximations give not satisfactory results (strong nonlinearities). This feature of the method can also be observed at Fig. 5 where departure of the maximum entropy distribution from the exact one is plotted as function of ε.

Figures 6-8 show the behavior of maximum entropy density for oscillator with parametric and external noises. Plots of a densities $p_1(y_1)$ and $p_2(y_2)$ which are obtained by maximum entropy principle, show generally very satisfactory agreement with the exact densities $f_1(y_1)$ and $f_2(y_2)$.

References

1. Jeynes E. T., Information theory and statistical mechanics, Physical Rev., 1957,106(4),620-630.

2. Sobczyk K., Stochastic Differential Equations with Applications to Physics and Engineering, Kluwer Acad. Publ., Dordrecht, 1991.

3. Sobczyk K., Trębicki J., Maximum entropy principle in stochastic dynamics, Probability Eng. Mech., 1990, vol.5, No.3.

4. de Beauregard O. C., Tribus M., Information theory and thermodynamics, Helvetica Physica Acta, vol.47 238-247, 1974.

5. Kullback S., Information Theory and Statistics, Chapman and Hall, N. York, 1959.

6. Tribus M., Rational Descriptions, Decision and Designs, N. York, Pergamon Press, 1969.

7. Spencer B. F., Bergman L.A., On the estimation of failure probability having prescribed statistical moments of first passage time, Probabilistic Eng. Mech., vol.1, No.3, 1986.

8. Chan M., System simulation and maximum entropy, Operations Research, vol.19, 1751-1753, 1971.

9. Ulrych T.J., Bishop T.N., Maximum entropy spectral analysis and autoregressive decomposition, Rev. Geophysics and Space Physics, vol.43, No.1, pp. 183-200, 1975.

10. Dimentberg M.F., An exact solution to a certain non-linear random vibration problem, Intern. J. Non-linear Mech. vol.17, 231-236, 1982.

11. Wu W.F., Lin Y.K., Cumulant-neglect closure for nonlinear oscillators under random parametric and external excitations, Int. J. Nonlinear Mech. vol.17, No.4, 349-362, 1984.

12. Hampl N. C., Schüeller G. I., Probability densities of the response of structures under stochastic dynamics excitation, Prob. Eng. Mech. 1989, Vol. 4, No.1.

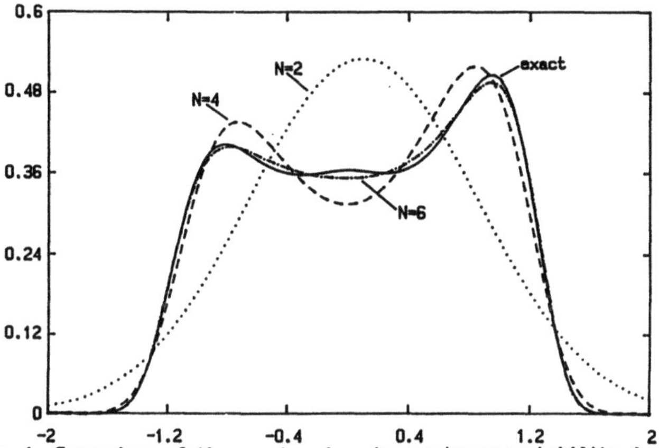

Fig. 1. Comparison of the exact and maximum entropy probability densities: N - number of given moments, (example 3.1)

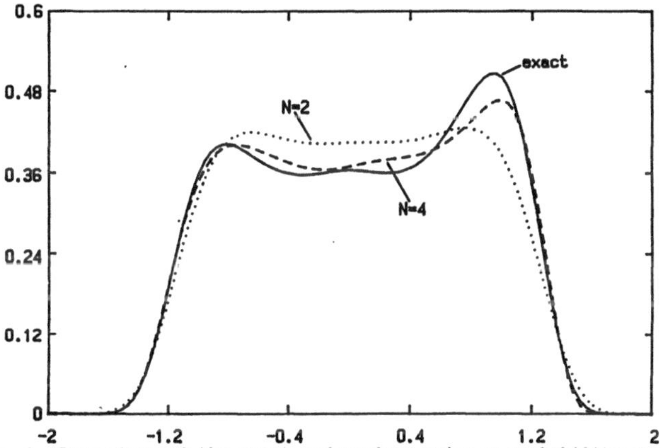

Fig. 2. Comparison of the exact and maximum entropy probability densities: N - number of moment equations, (example 3.1).

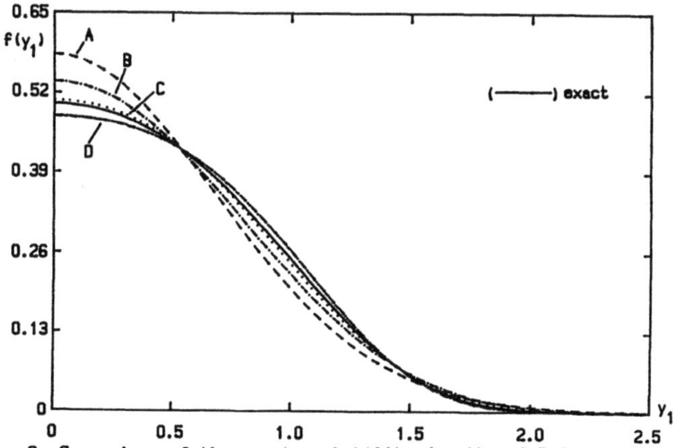

Fig. 3. Comparison of the exact probability density of Duffing oscillator with densities obtained by: A - statistical linearization, B - Gram-Charlier exp. 4-th order, C - Gram-Charlier exp. 6-th order, D - maximum entropy; ε=0.8, (example 3.2)

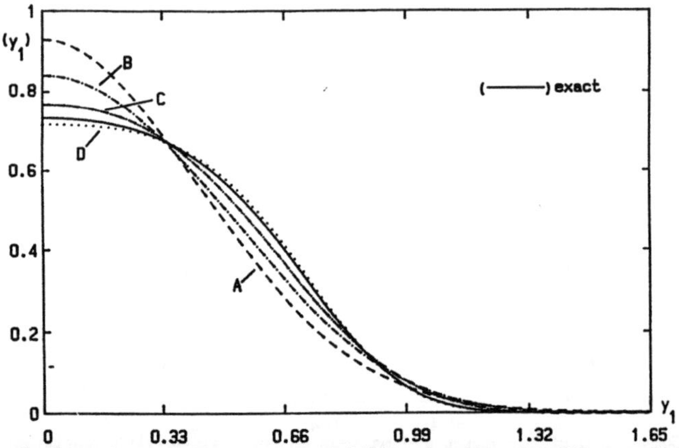

Fig. 4. Comparison of the exact probability density of Duffing oscillator with densities obtained by: A - statistical linearization, B - Gram-Charlier exp. 4-th order, C - Gram-Charlier exp. 6-th order, D - maximum entropy; ε=8.0, (example 3.2)

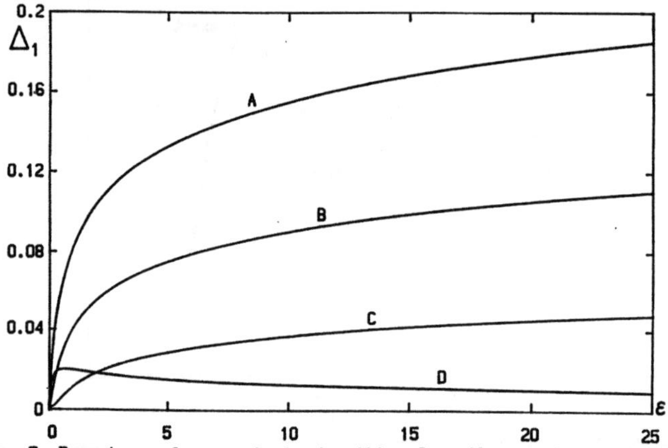

Fig. 5. Departure of approximate densities from the exact one: A, B, C, D, - have the same meaning as at Fig. 3-4, (example 3.2)

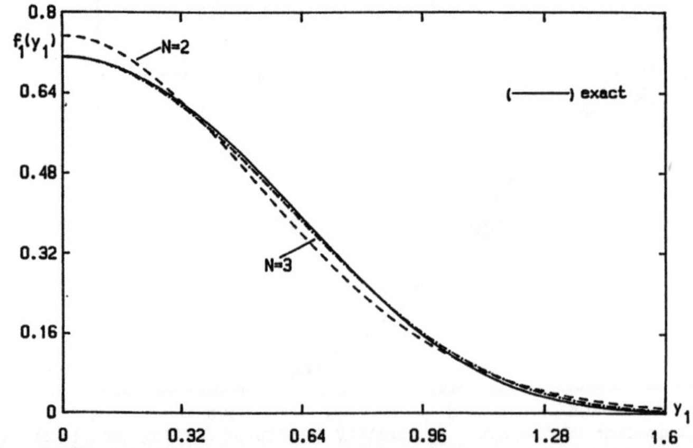

Fig. 6. Exact probability density of the displacement and its maximum entropy approximations; two moment equations (N=2) and three moment equations (N=3) are taken from (3.22); S_3=1.0, (example 3.3)

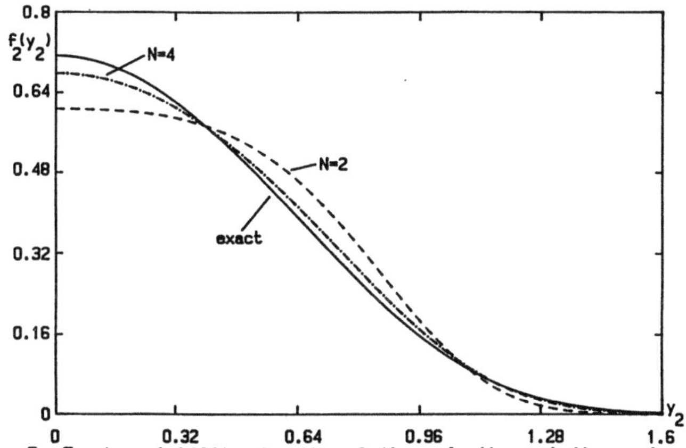

Fig. 7. Exact probability density of the velocity and its maximum entropy approximations; two moment equations (N=2) and three moment equations (N=3) are taken from (3.22); S_3=1.0, (example 3.3)

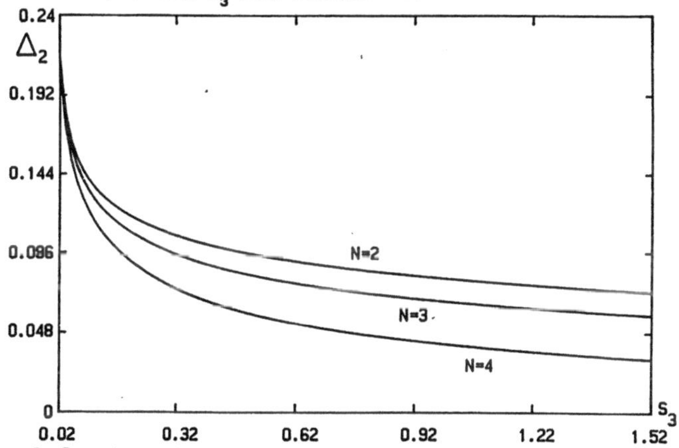

Fig. 8. Departure of maximum entropy approximations from the exact probability density: N - number of moments equations which are taken from equations (3.22).(example 3.3)

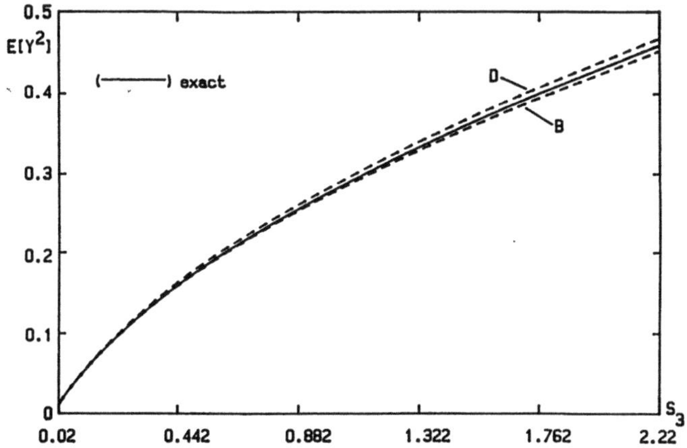

Fig. 9. Stationary mean square of displacement and velocity: B - cumulants beyond the 4-th order neglected, D - maximum entropy principle when four moments equations from equations (3.22) are taken, (example 3.3)

System Identification of Nonlinear Dynamic Structures Based on Stochastic Time Series Model Fitting

K. Suzuki and K. Kawanobe

Department of Mechanical Engineering
Tokyo Metropolitan University
1-1 Minami-osawa Hachioji-shi Tokyo 192-03, Japan

Summary

A new technique for dynamic system identification by use of a stochastic time series models such as the AR model or the ARMA model, is proposed and investigated. After showing the advantage of the methods over former major identification methods based on FFT, this paper focuses on a double variate AR model method which can be developed to identify nonlinear dynamic systems. Here the dynamic characteristics are evaluated in the form of equivalently linearized system parameters. As a typical example, a simple nonlinear system having a bi-linear stiffness characteristic is adopted. Through numerical simulation for the model, the validity and applicability of the method to such a nonlinear system is examined.

Introduction

Together with the development of the modal analysis technique, which has been applied to the dynamic design of mechanical structure systems, system identification and modeling methodologies have played an important role in the field of dynamics. Many studies dealing with system identification methods have been proposed from various technical viewpoints. If the problem is restricted to an identification of system parameters, such as natural frequency, damping and modal shape of motion for dynamic structure systems, identification procedures can be classified into the following two categories; frequency domain methods and time domain methods. A major method, known as the curve fitting technique, has been investigated under the assumption that a frequency response function or an impulse response function can be evaluated in advance by some adequate shaking test and data processing. The method we have proposed as "an AR identification method" has a different approach from curve fitting techniques. The AR method depends on AR (autoregressive) model or the ARMA (autoregressive and moving average) model fitting, without utilizing frequency response functions or impulse response functions [1]. According to the AR method, the system parameters can be obtained in the time domain through a stochastic time series. At this stage, Burg's algorithm is introduced in order to improve models such as the AR or ARMA models. The main advantages of the AR method over former major structural identification methods, which are mainly based on the FFT curve fitting technique, are as follows:

N. Bellomo · F. Casciati (Eds.)
Nonlinear Stochastic Mechanics
IUTAM Symposium Turin

(1) The method is effective for a structure which has random excitation under normal operating conditions and is not suitable for applying shaking tests such as hammering, or sinusoidal excitation.

(2) The identification can be carried out even when the total volume of observed data in the time domain is not plentiful, due to unavoidable practical restrictions.

Considering the merits mentioned above, this particular paper presents a new identification procedure in which observed time histories of the random input excitation and the associated dynamic system response are simulated with a double variate AR model. This method is basically categorized as a time domain method and the procedure is highly developed from the viewpoint of computational convenience by using an order update algorithm. After examining the validity of the method when applied to the system parameter identification of a linear elastic structure system, the study develops the method for nonlinear system identification. A simple single-degree-of-freedom lumped mass model having a bi-linear stiffness characteristic is adopted. Linearized parameters for the natural frequencies and damping ratios are evaluated by the proposed identification procedure. By calculating stochastic properties relating to the identified system parameters, the applicability of the method to such a nonlinear system is investigated.

Double-Variate AR Model Method

The system identification method proposed here, based on a double variate AR model (DVARM), is briefly explained. This method is allows a simultaneous modeling of both system input excitation process and system output response process. At this stage, any type of stationary random process can be adopted as the input process (broad band or narrow band process). The double variate autoregressive (AR) model can be written as

$$\boldsymbol{x}(j) = \sum_{k=1}^{m} \boldsymbol{A}_{mk}\boldsymbol{x}(j-k) + \boldsymbol{f}_m(j) \tag{1}$$

where

$$\boldsymbol{x}(j) = \begin{Bmatrix} u(j) \\ y(j) \end{Bmatrix}, \quad \boldsymbol{A}_{mk} = \begin{bmatrix} a_{mk}^{(uu)} & a_{mk}^{(uy)} \\ a_{mk}^{(yu)} & a_{mk}^{(yy)} \end{bmatrix} \quad \text{and} \quad \boldsymbol{f}_m(j) = \begin{Bmatrix} f_m^{(u)}(j) \\ f_m^{(y)}(j) \end{Bmatrix}$$

In this equation, $\boldsymbol{x}(j)$ is a vector containing the input process $u(j)$ and the associated output process $y(j)$, $\boldsymbol{A}_{mk}$ is a matrix of AR coefficients $a_{mk}^{(pq)}$ and $f_m^{(u)}(j)$ and $f_m^{(y)}(j)$ denote prediction errors related to the input error process $u(j)$ and the output error process $y(j)$, respectively. In order to estimate the AR coefficients $a_{mk}^{(pq)}(j)$, Burg's algorithm is utilized, whereby a formerly proposed procedure based on a single variate AR (SVARM) model has been improved [2]. Figure 1 illustrates schematically the relations in equation (1); it is seen that the prediction errors $f_m^{(u)}(j)$ and $f_m^{(y)}(j)$ correspond exactly to the input $u(j)$ and the output $y(j)$, respectively. Equation (1) can be reformulated by a z-transformation as

$$\boldsymbol{x}(z) = -\left[\sum_{k=1}^{m} \boldsymbol{A}_{mk} z^{-k}\right]^{-1} \boldsymbol{f}_m(z) \tag{2}$$

In this equation, it is assumed that $\boldsymbol{A}_{m0} = -\boldsymbol{I}$ ($\boldsymbol{I}$; a 2×2 identity matrix). The covariance matrix $\boldsymbol{V}_m$ relating to the prediction errors $f_m^{(u)}(j)$ and $f_m^{(y)}(j)$ is found by assuming that they have Gaussian white noise characteristics.

$$\boldsymbol{V}_m = \mathrm{E}[\boldsymbol{f}_m(j)\boldsymbol{f}_m^T(j)] = \boldsymbol{f}_m(z)\boldsymbol{f}_m^T(z^{-1}) \tag{3}$$

The pulse transfer function, which provides a relation between the system input and output, can be explicitly described as

$$G_{uy}(z) = \frac{\displaystyle\sum_{k=0}^{m}\sum_{r=0}^{m}\{-a_{mk}^{(yu)} \quad a_{mk}^{(uu)}\}\,\boldsymbol{V}_m\,\{a_{mk}^{(yy)} \quad -a_{mk}^{(uy)}\}^T z^{r-k}}{\displaystyle\sum_{k=0}^{m}\sum_{r=0}^{m}\{a_{mk}^{(yy)} \quad -a_{mk}^{(uy)}\}\,\boldsymbol{V}_m\,\{a_{mk}^{(yy)} \quad -a_{mk}^{(uy)}\}^T z^{r-k}} \tag{4}$$

in which the inverse matrix in equation (2) has been expanded. In equation (2), it is necessary to determine the AR parameters such as $a_{mk}^{(pq)}$, $\boldsymbol{V}_m$ and an optimum order of m by referring to the FPE criteria or to the AIC [3].

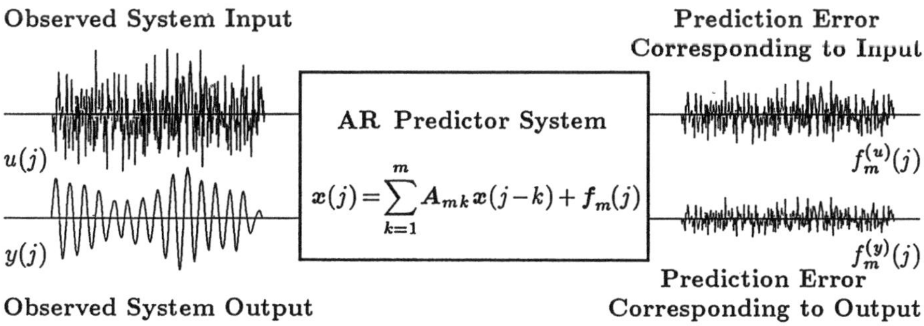

Figure 1. Double variate AR model

It is possible to determine the optimum system parameters by considering the fact that the AR model order m should be twice the mode number n for the system, within the frequency range under consideration. However, for practical convenience, the optimum value of m must be chosen to be selected larger than $2n$. Actually, if we introduce the FPE criterion, the optimum order of m is usually found to be larger than $2n$. At this stage, Burg's algorithm is introduced in order to improve the accuracy of model fitting.

Based upon the concept of modal analysis, it is assumed that the transfer function (compliance) $H(s)$ for a multi-degree-of-freedom system can be expressed by a superposition of transfer functions for the required number of single-degree-of-freedom systems.

$$H(s) = \sum_{j=1}^{n} \frac{\phi_j}{s^2 + 2\zeta_j \omega_j s + \omega_j^2} \tag{5}$$

The identification procedure for general multi-degree-of-freedom system can be performed depending on the relationship between equation (4) and (5). That is, by taking account the fact that the characteristic roots (poles) obtained from equation (4) in the z-domain should exactly correspond to those from equation (5) in the s-domain, we can identify the required natural frequency ω_j and damping ratio ζ_j of the system directly from the optimally determined AR paramenters. The actual identification procedure is summarized below:

(1) The coefficients of the AR model are calculated by the Levinson-Wiggins-Robinson (LWR) algorithm using observed input and response time histories [4].

(2) By solving the characteristic equation of equation (4), the roots (poles) are obtained in the z-domain.

(3) Using a time shift operator $z = \exp(s\tau)$, these roots are transformed into the s-plane, as shown in Figure 2.

(4) The system parameters are calculated by solving the equation

$$z_j = \exp\left\{-\omega_j \tau(\zeta_j - i\sqrt{1-\zeta^2})\right\} \tag{6}$$

where the j-th natural frequency ω_j and the j-th damping ratio ζ_j are determined as the j-th modal system parameters corresponding to the j-th characteristic root z_j from equation (4). It should be noted that, in the DVARM method proposed here, averaging is exclusively performed in the time domain, while in the FFT-based curve fitting method, averaging is mainly done in the frequency domain. That is, this method has the merit that the system parameters such as ω_j and ζ_j can be directly obtained from the optimally determined DVARM without any calculation of the frequency transfer function or the

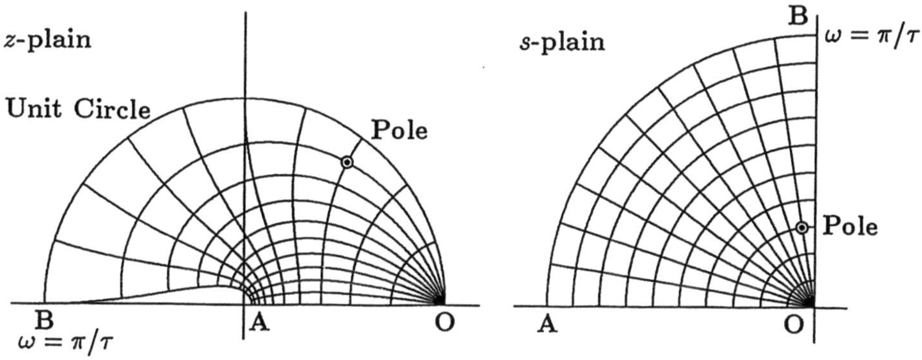

Figure 2. Mapping Relation between z-domain and s-domain

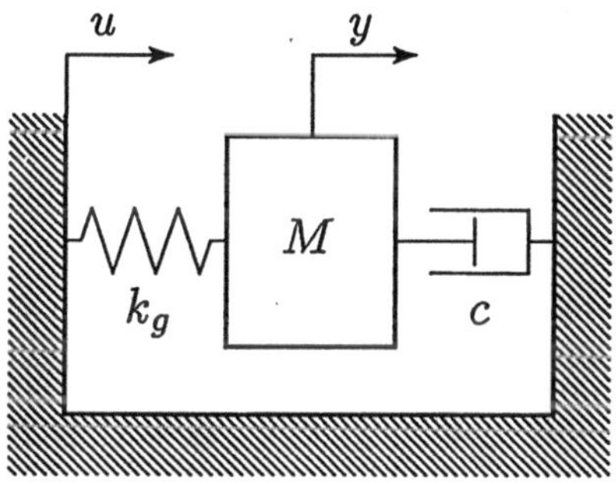

Figure 3. Single degree of freedom system

impulse response function. Moreover, high speed calculation algorithms such as the LWR algorithm and the EFFT (extended FFT) algorithm are introduced in order to reduce computer time [5].

Numerical Simulation for Linear System

In order to examine the applicability of the identification method, the system parameters are estimated for a simple linear system having a single-degree-of-freedom, as shown in Figure 3. In this example, Burg's algorithm is utilized for the determination of the AR coefficients and the order of model m is selected by referring to the minimized FPE (final prediction error). Random time histories are simulated by a superposition of a large number of the sinusoidal waves having different frequencies and phases. Taking the histories as the input excitation (the sampling time step being 0.01 [sec]), the response process of the mass acceleration is calculated by using the 4th order Runge-Kutta algorithm.

First, we try to identify damping ratio for the model. It is well known that damping estimation, from vibration test data and also simulation data, is quite troublesome for many actual cases. The natural frequency ω_n of the system is set as one tenth of the Nyquist frequency and the damping values, prescribed in advance and distributed within a fairly wide range, from 0.0001 to 0.5, are identified by the DVARM method. In this case, the system order is determined as $m = 6$, irrespective of the damping value. Figure 4 demonstrates the relation between the given damping values and those identified; it is noted that some errors appear when the damping is very small. This suggests a limitation of the method for very lightly damped systems. However, in almost all cases, the identification provides quite good accuracy. Even when the damping becomes very low, higher accuracy can be obtained by increasing the amount of data processed.

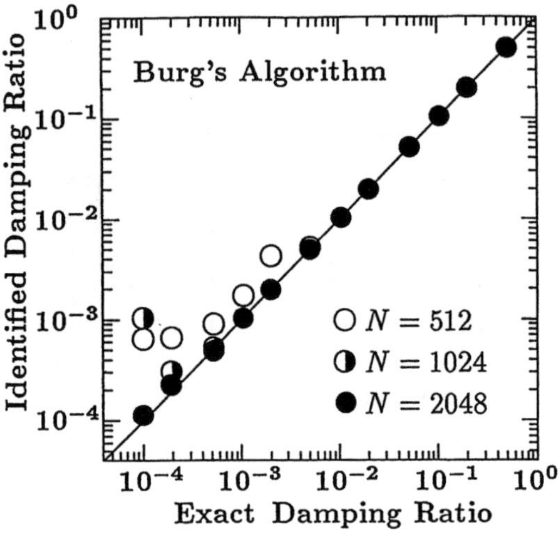

Figure 4. Comparison damping ratios and exact value

Further investigation was carried out in order to compare results by the DVARM method with those obtained by other methods, based on curve fitting (FFT) and on the formerly proposed single variate AR model (SVARM) method. Here, the simple model in Figure 3 is taken again as an example. The system parameters, given in advance, are as follows,

$$\begin{cases} \omega_n/2\pi = 5.0[\mathrm{Hz}] \\ \zeta_n = 1, 2, 3, 5, 7, 10[\%] \end{cases}$$

Typical results for the identified system parameters are shown in Table 1.

From this, it can be easily seen that the DVARM method gives the most reliable results, among the three methods, for almost all cases. Particularly, even when number of data

Table 1. Identified parameters for SDOF system
(Comparison of the effect of data number)

Number of Data	Exact Value		Identified Value					
			FFT†		SVARM†		DVARM†	
	f_n[Hz]	ζ_n[%]	$\hat{f}_n$[Hz]	$\hat{\zeta}_n$[%]	$\hat{f}_n$[Hz]	$\hat{\zeta}_n$[%]	$\hat{f}_n$[Hz]	$\hat{\zeta}_n$[%]
2048	5.00	1.00	5.00	1.18	5.00	1.07	5.00	1.00
		2.00	5.00	2.23	5.00	1.97	5.00	2.00
		3.00	5.00	3.21	5.00	2.89	5.00	3.00
		5.00	5.00	5.20	5.00	4.79	5.00	5.00
		7.00	5.00	7.17	5.00	6.73	5.00	7.00
		10.00	5.00	10.08	5.00	9.80	5.00	10.00
1024	5.00	1.00	5.01	0.83	4.97	0.82	5.00	1.00
		2.00	5.00	1.87	4.97	1.59	5.00	2.00
		3.00	4.99	2.92	4.98	2.41	5.00	3.00
		5.00	4.97	5.04	5.00	4.21	5.00	5.00
		7.00	4.96	7.26	5.02	6.20	5.00	7.00
		10.00	4.95	10.51	5.06	9.48	5.00	10.00
512	5.00	1.00	5.00	6.48	4.87	1.82	4.96	1.04
		2.00	5.00	7.54	4.84	2.85	4.95	2.08
		3.00	5.00	8.53	4.82	3.97	4.94	3.11
		5.00	5.02	10.37	4.79	6.34	4.94	5.15
		7.00	5.04	12.17	4.78	8.91	4.94	7.16
		10.00	5.08	14.72	4.76	13.16	4.95	10.17

†FFT; Curve fitting method based on FFT
SVARM; Single variate AR model method ($m = 6$)
DVARM; Double variate AR model method ($m = 6$)

points is reduced to $N = 512$ or 1024, the proposed DVARM method, provides a fairly accurate identification of the system parameters f_n and ζ_n. In contrast, in such cases, the identified values for the damping are considerably distorted by the FFT method and the SVARM method. It can be expected from these, and other investigations not presented here, that the DVARM method gives a stable and robust identification of the system parameters for linear dynamic structural systems.

Application to Nonlinear Structure Systems

Encouraged by the stability and accuracy of the DVARM method for linear systems mentioned above, we tried to apply the method to nonlinear structural systems. As a typical example of a nonlinear structural model, a single-degree-of-freedom system having a bi-linear spring characteristic, as shown in Figure 5, is adopted.

This model is frequently introduced for the practical analysis of dynamic systems, such as pipings having a clearance with the supporting device, mechanical equipments which are

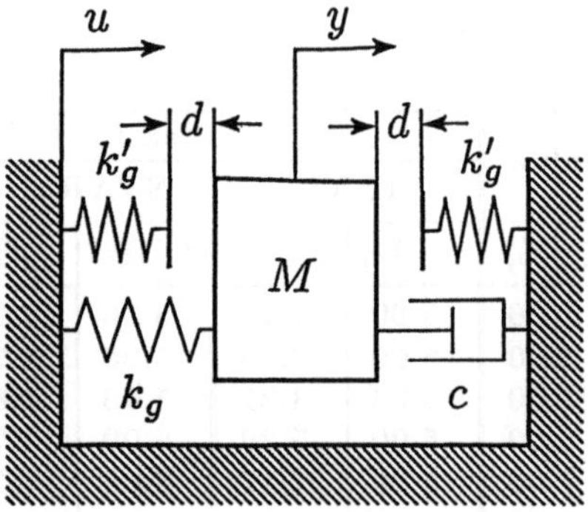

Figure 5. Single degree of freedom bi-linear system (impact model)

liable to impact disturbance with their foundation and so on. The equation of motion for the model can be written as

$$M\ddot{x} + c\dot{x} + k(x)x = M\ddot{u} \tag{7}$$

$$k(x) = \begin{cases} k_g & \text{when } |x| < d \\ k_g + k_g' & \text{when } |x| \geq d \end{cases} \tag{8}$$

Here x is the relative displacement $(y - u)$. The system obviously behaves nonlinearly. However, if the system is vibrating entirely in the region $|x| < d$ or in the region $|x| \geq d$ respectively, it has a linear elastic response. The natural frequencies and damping ratios for these two cases are given by

$$f_n = \frac{1}{2\pi}\sqrt{\frac{k_g}{M}}, \zeta_n = \frac{c}{2\sqrt{Mk_g}} \text{ for } |x| < d, \qquad f_n' = \gamma f_n, \zeta_n' = \frac{\zeta_n}{\gamma} \text{ for } |x| \geq d \tag{9}$$

where $\gamma = \sqrt{(k_g + k_g')/k_g}$. Except for these two extreme cases (in other words, when the system is excited by random forces which have a stochastically varying amplitude), the system response must be treated as nonlinear. The objective of the study here is to identify this nonlinear behaviour and to describe the characteristics by the use of linearized dynamic system parameters.

Figure 6 showns some examples of numerically computed responses (relative velocity) of the model for several "gap" sizes, d. In this example $f_n = 15$ [Hz], $\zeta_n = 1$ [%], $f_n' = 3$ [Hz] and $\zeta_n' = 5$ [%] are given in advance. The middle three response time histories clearly show nonlinear behaviour; in particular, in the case where $d = 2.0$ [mm], some "spikes" appear at the instant of impact arising between the mass and the supporting device. The DVARM

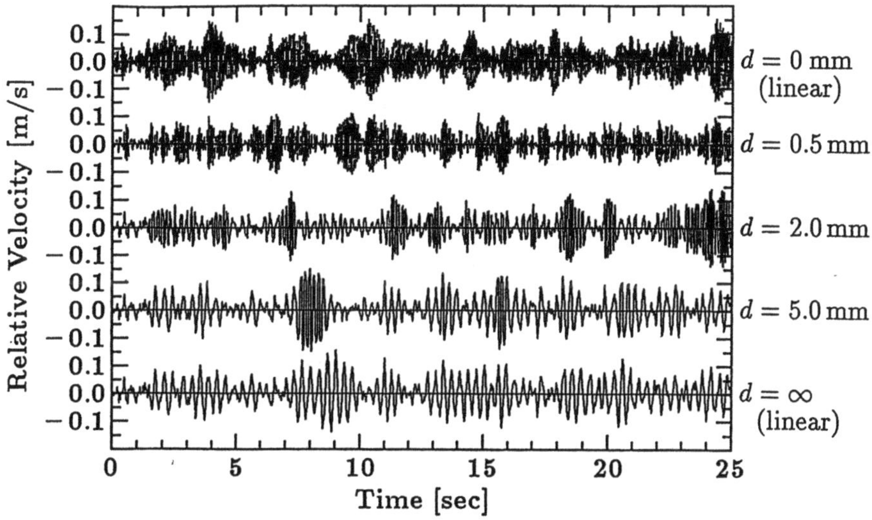

Figure 6. Examples of computed relative velocity responses for the "Gap" model

identification method is applied to estimate the natural frequencies and damping, taking models with various gap sizes d. Computation for each model is repeated 100 times in order to get stochastic properties by taking the input process as a sample process among a population of numerically generated random process data.

Therefore, the mean value and standard deviation over 100 samples can be calculated for the individual "gap" system. Table 2 gives the obtained stochastic properties for the linearized model, so obtained. The shift in the natural frequency and damping, due to changing gap size can be recognized. Especially for the damping estimation, a significant scatter in the estimated values appears around 1 [mm] through to 8 [mm]. Such a scattering is very clearly seen in the distribution of the poles (roots) in the s-domain, as shown in Figure 7. Among these cases, the roots of the system having $d = 5$ [mm] gap are most scattered. Therefore, it can be understood that for such a "strongly nonlinear model", identification based on the DVARM method, using an equivalent linear model becomes impossible.

Conclusions

(1) A new system identification method for dynamic structure systems is proposed and some advantages over former methods are presented.

(2) After examining the validity of the method, as applied to linear elastic systems, the applicability to nonlinear systems is investigated, taking a "bi-linear gap system" as an

Table 2. Stochastic properties as for linealized system parameters of bi-linear "Gap" model

Gap Size	Natural Frequency [Hz]			Damping Ratio [%]		
d [mm]	$\mu\{f_n\}^\dagger$	$\sigma\{f_n\}^\dagger$	C.O.V†	$\mu\{\zeta_n\}^\dagger$	$\sigma\{\zeta_n\}^\dagger$	C.O.V†
0.0	15.00	0.00	0.00	1.00	0.00	0.00
0.1	14.99	0.01	0.00	1.00	0.01	0.01
0.2	14.92	0.01	0.00	1.12	0.04	0.03
0.5	14.41	0.12	0.01	2.28	0.44	0.19
1.0	13.12	0.46	0.04	7.12	2.27	0.32
2.0	9.55	0.79	0.08	18.38	4.69	0.26
4.0	6.13	0.61	0.10	34.48	6.99	0.20
5.0	5.32	0.51	0.10	53.24	10.42	0.27
6.0	4.40	0.59	0.14	44.00	13.45	0.35
7.0	3.90	0.80	0.21	24.90	18.18	0.73
8.0	3.39	0.62	0.18	15.88	17.91	1.13
∞	3.00	0.00	0.00	5.00	0.00	0.00

$^\dagger\mu\{\}$; mean value over 100 sample process
$\sigma\{\}$; standard deviation
C.O.V; coefficient of variation

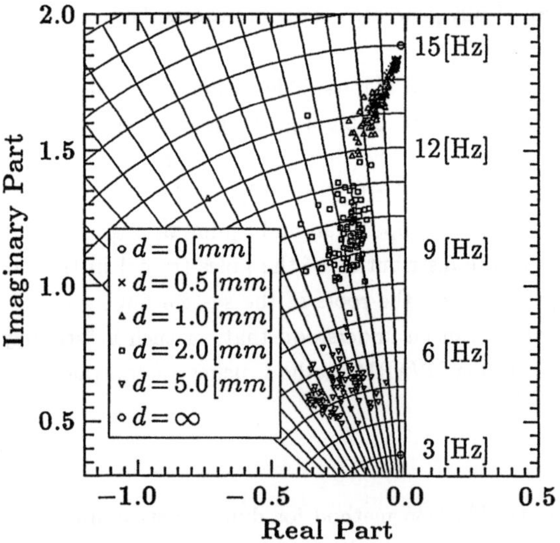

Figure 7. Distribution of identified parameters in the s-plane

example. The stochastic characteristics of the identified parameters for the linearized system were evaluated and examined.

Acknowledgements

The authors cordially thank Professor S. Nakagiri of University of Tokyo, who kindly suggested that they present this work to the present symposium. They also wish to express their appreciation to Mr. S. Inomata, who studied part of this work when a former graduate student in their laboratory.

References

[1] Suzuki, K. and Kawanobe, K.: Dynamic System Identification Based on Two- Dimensional Autoregressive Model Fitting. *JSME Transactions* (in Japanese), Vol. 53, C (1987), 550-559.

[2] Suzuki, K.: Identification of Dynamic Characteristics of a Structural System with the Aid of Auto-Regressive Model Fitting *ASME PVP Conference, Bound Vol.* PVP-Vol. 113 (1986), 15-22.

[3] Akaike, H.: Autoregressive Model Fitting for Control. *Ann. Inst. Statist. Math.*, Vol. 23 (1971), 163-180.

[4] Morf, M. et al.: Recursive Multichannel Maximum Entropy Spectral Estimation. *IEEE Transaction on Geoscience Electronics*, Vol. GE-16 (1978), 85-94.

[5] Gan, R.G. et al.: An Extended FFT Algorithm for ARMA Spectral Estimation. *IEEE Transaction on Acoustics, Speech, and Signal Processing* Vol. ASSP-32, 168-170.

Nonlinear Stochastic Mechanics in Structural Safety and Reliability Monitoring of Gas Lifelines

S.A.Timashev,
Sci. & Engineering Center for Reliability and Resource of Large Machine Systems, USSR Acad. of Sci., Ural Dept., Sverdlovsk, USSR

Abstract

The interaction between the gas industry and its demands and the fundamental science and its possibilities relative to the safety and reliability of the gas lifelines (GLL) is presented. The whole procedure of structural safety and reliability monitoring (SSARM) with respect to GLL is described and the position of nonlinear stochastic dynamics, as the kernel of this system, is defined.
A sophisticated generalization of the correlation theory as applied to distributed physically and stochastically nonlinear systems is developed to solve a bunch of new problems and to precisely meet the demands of the industry.
The possible transition from correlation theory results to posing and solving GLL reliability problems is indicated. The terminal goals of gas lifeline SSARM as an expert system for the decision making persons are briefly described.

Introduction

Gas lifelines (GLL) consist of compressor stations and main pipe lines. USSR GLL are approximately 35 000 km long. Most of them were constructed 15...20...25 years ago and are aging rather quickly. The most efficient way to pull down the performance expenses for these GLL is to organize for them and to introduce the structural safety and reliability monitoring system (SSARMS) proposed and developed by the author of this paper [1]. In brief, the SSARMS for the linear component of GGL boils down to following.
The GLL is considered as a large-scale distributed mechanical system (as a sequence of thin wall cylindrical shells or beams with geometrical and mechanical imperfections) that continuously evolves in time due to change of environment, to the action of external forces and to the degradation of pipeline material properties. From the point of view of responsibility the GLL may be objectively enough divided into separated specific parts (straight and curved in horizontal or vertical plane sections of GLL in unhabited regions, compensators, loopings, intersections of GLL with waterways, railways and highways, as well as GLL parts located in the vicinity of settlements and important railways or highways).
Considering GLL design and performance they may also be divided into sections, for which the following parameters may be considered as homogeneous (and quasi-stationary):

- properties of pipe material, welds and isolation;
- viscoelastic properties of the soil surrounding the pipe;
- corrosion properties of the pipe environment;
- properties of the load vector;
- geometry of the pipe cross - sections;
- geometry of the pipe longitudinal axis in 3 D space.

It is important to note that during performance (in fact, during the whole life cycle) of the GLL the pipe sizes, the boundary conditions, the mathematical operators (differential equations in partial derivatives that

N. Bellomo · F. Casciati (Eds.)
Nonlinear Stochastic Mechanics
IUTAM Symposium Turin

govern the output of the system), as well as the material properties of the pipe itself are subject to great changes.
The proposed classification is performed using the usual information, which is collected by linealoges, "intelligent pigs" of all types, by soil pressiometry, local extensometry and by spotting all buckled sections of the pipe using the helicopter or sputnik technique.
Having such a classification for all sections of the GLL pipe and the numerical values of aforementioned parameters, it is possible to solve the relevant stress-strain problem for every section of GLL using nonlinear stochastic dynamics.
Further, using theory of reliability it is possible to obtain fairly precise on-line assessments of readiness, maintainability, reliability and residual lifetime of every part of GLL pipe and pass this information on the working station monitor of the decision making person (DMP).
So the GLL SSARM concept consists of:

- collecting a data-base on each and every GLL section and constantly updating it;
- constructing a software package allowing identification and diagnostics of the GLL pipe condition, on-line assessments of GLL residual lifetime, reliability;
- giving recommendations for the DMP on how to use the GLL at every moment of time.

Definition of the problem

It is evident that the kernel of such a system lies in non-linear stochastic mechanics as applied to solutions of relevant problems of stress-strain analysis for all GLL sections.
Although it is obvious that a full group of solutions needed to assess the whole GLL's reliability is large, we proceed with only one specific but very important show case, because all others may be derived from it by one or another evident generalization.
Consider part of a GLL as a beam of cylindrical form consisting of 3 sections (Fig.1). The flank sections of the GLL are supposed to be of semiinfinite length and to be lying on a dense foundation with high bearing capacity (gravel, clampy sand, low and medium humidity clay, frost soil, rock) and subgrade reaction modulus (SRM) value in the range of 5...20...100(1500).
The central section is considered to be of finite length and lying on a weak foundation with low bearing capacity (peat, bog, marsh, wet clay, quick-sand) and with SRM value in the range of 0.1...0.5. The GLL pipe is generally loaded by an uniformly distributed normal load q and internal pressure p. Both these loads are random functions of time and the GLL coordinate, but lower are considered to be constant. The main stochastic component here is the SRM, considered as a homogeneous random function of GLL coordinate. On-site research carried out in 3 West Siberia regions (Yamburg, Surgut, Nizhnevartovsk) on 14 GLL sections showed that 3 of them had plastic deformations.
We arrived in a natural way at the necessity to obtain stochastic properties of the stress-strain state of a 3-section GLL part along its length taking into account the stochastic properties of SRM value and with respect to possible plastic deformations in the cross-section of the GLL. Evidently it is most expedient to use here the correlation theory.
Beginning from 1965, when V.V. Bolotin and D.N. Sobolev simultaneously proposed techniques to solve correlation problems as applied to beams lying on elastic foundation with homogeneous properties of SRM as a random function, this technique was used and developed as applied to various kinds of foundation problems (see for instance [2,3]). But multisectional beams in the sense of SRM change, that may deform in the plastic range under combination of loads, as yet were not considered. This illustrates that precise meeting the

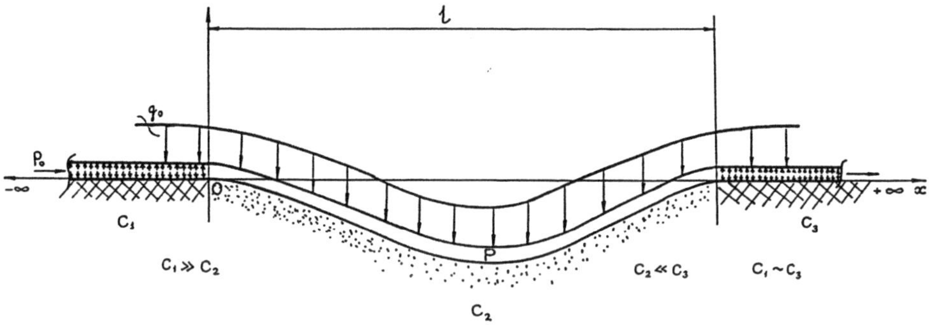

Figure 1 - A typical part of GLL as a 3-section beam of cylindrical form

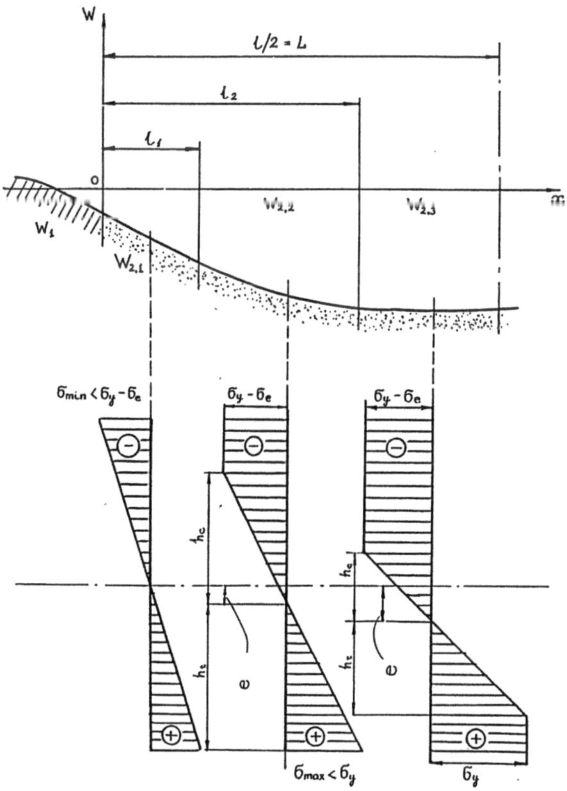

Figure 2 - Three freedom deformation zones (elastic, one-side yield, two-sides yield)

demands of an industry often leads to posing and solving new problems in stochastic dynamics.

Correlation problems for semiinfinite sections

Consider first deflections of the left-hand semiinfinite part of the pipeline. Assuming that

$$C_1(x) = C_{01} + \tilde{C}_1(x) \quad , \quad w_1(x) = w_{01}(x) + \tilde{w}_1(x)$$

where $< C_1(x) > = C_{01}$, $<w_1(x)> = w_{01}(x)$, the stochastic differential equation of beam deflection has the form

$$EI \frac{d^4 w}{dx^4} + [C_{01} + \tilde{C}_1(x)]\, w_1(x) = q_0 \, ,$$

where EI is the beam rigidity; q_0 is the normal load intensity consisting of beam's own mass and of the filling soil. This equation is stochastically nonlinear, since it contains a product of random functions $\tilde{C}_1(x)\, \tilde{w}_1(x)$.
Using explicitly the perturbation method $C_1(x) = C_{01} + \mu\, C_1(x)$, $w_1(x) = w_{01}(x) + \mu w_1(x) + \ldots$ we arrive at the following equation

$$EI \frac{d^4 w_0}{dx^4} + C_{01}\, w_0 = q_0 \, ,$$

$$EI \frac{d^4 \tilde{w}}{dx^4} + C_{01}\, \tilde{w} = - \tilde{C}_1 w_0 \qquad (1)$$

....

For the mean deflection w_{01} we have

$$w_{01}(x) = \frac{q_0}{C_{01}} [1 + e^{k_0 x}\, C_1 \cos k_0\, x + C_2 \sin k_0\, x)] ,$$

where $k_{01} = C_{01}/EI$. C_1 and C_2 should be determined from the conditions of conjugation with the middle section at the origin (Fig.1).
For the cross correlation function we have

$$\frac{d^4 K_{wc1}}{d\, x^4} + 4\, k_{01}\, K_{wc1}\, (x, x') = - \frac{w_{01}(x)}{EI} K_{c1}\, (x,x') . \qquad (3)$$

From on-site experimental investigations it is known that

$$K_{c1}(x,x') = \sigma^2{}_1\, e^{-\alpha_1 |x'-x|} = \sigma^2{}_1\, e^{-\alpha_1 |\xi|} \, , \qquad (4)$$

where $\sigma^2{}_1$ is the variance, α_1 is the bandwidth coefficient of the SRM as a random function.
If we presume that

$$K_{wc1}(x,x') = K_1(x)\, \sigma^2{}_1\, e^{-\alpha_1 |\xi|} \, , \qquad (5)$$

where $K_1(x)\sigma^2{}_1$ is the correlation between w_1 and c_1 in coinciding points x, then eq.(3) transforms to a fourth order ODE

$$K^{IV} + 4k_{01}K_1 = - w_{01}(x)/EI. \qquad (6)$$

If $w_{01}(x)$ is found, then

$$K_1(x) = K(x) + K^*(x)$$

where

$$K(x) = e^{k_{01}x} (B_1 \cos k_0 x + B_2 \sin k_0 x) \qquad (7)$$

is the general solution of the homogeneous form of eq.(6), and

$$K^*(x) = \frac{q_0}{c_{01}^2} + \frac{q_0 k_{01}}{c_{01}^2} x e^{k_{01}x} [(C_1 + C_2) \cos k_{01} x + + (C_1 + C_2) \sin k_{01} x] \qquad (8)$$

is the partial solution of the nonhomogeneous part.
Equations (7) and (8) are written as boundary effect solutions, that satisfy the boundary condition at $x \to -\infty$. The same technique holds for the autocorrelation functions. The solution for it includes two new variant coefficients (D_1 and D_2) which should be also derived. It is important here to point out that if the fluctuations $\tilde{C}_1(x)$ and therefore $\tilde{w}_1(x)$ are not small, then the equations for $w_{01}(x)$ and $K_{wC1}(x,x')$ as was shown by B.P. Makarov [3], become interconnected and must be solved simultaneously.

Correlation problems for a finite elasto-plastic section

Proceed to deformation analysis of the GLL central part. Using the same reasoning as before, we obtain for the elastic case

$$w_2(x) = \frac{q_0}{C_2} [1 + e^{k_2 x} (C_1 \cos k_2 x + C_2 \sin k_2 x) + $$

$$+ e^{k_2 x} (C_3 \cos k_2 x + C_4 \sin k_2 x)], \qquad (9)$$

where $w_2(x) = \langle w_2(x)\rangle$, $4k_2^4 = C_2/EI$. Here, again the cross correlation function may be expressed in the following form

$$K_{wC}(x,x') = K_2(x)\, \sigma_2^2\, e^{-\alpha_2 |\xi|},$$

where σ_2^2 and α_2 are the parameters of the SRM' correlation function.
The equations for the deflection function $w_2(x)$ accounting for elasto-plastic deformations are derived by a two-fold differentiation of the bending moment with respect to x:

$$M'' = q_0 - C_2 w_2 ,$$

so the main problem here is to evaluate the elasto-plastic moment for different cases. For the case when p = 0 (stage of GLL construction) the plastic deformations in the pipe cross section are symmetrical. Therefore we have

$$Bw_2^{IV}\left(\frac{1}{A} \arcsin \frac{A}{w''_2} - \frac{\sqrt{(w_2''^2-A^2)}}{w_2''^2}\right) + 2B(w_2''')^2\left(\frac{\sqrt{(w''^2-A^2)}}{w_2''^3} - \frac{1}{w_2'\sqrt{(w_2''^2- A^2)}}\right) + C_2 w_2 = q_0 \qquad (10)$$

(Here each prime denotes a differentiation with respect to x), where $A = 2 \sigma_y/dE$, $B = d^2 \delta \sigma_y/2$, σ_y is the yield stress, d is the overall pipe diameter, δ is the pipe wall thickness.

If p is present (stage of pipe performance), then the stress state in the pipeline wall becomes two dimensional: $\sigma=\sigma(\sigma_x,\sigma_0)$, where σ_x are the bending stresses and $\sigma = pR/\delta$, $(R=d/2)$ are the circumferential ones (due to pressure p). In this case the yield state of the tube wall material is defined according to Saint Venant criteria $\sigma_{equiv} = \sigma_1 - \sigma_3$ where σ_1 and σ_3 are the principal stresses.

In the tension zone of the pipe wall cross section we have

$$\{\sigma_1=\sigma_x;\ \sigma_2=\sigma\quad;\ \sigma_3=0\}\ :\ \sigma_{equiv} = \sigma_x = \sigma_y \tag{11}$$

In the compression zone

$$\{\sigma_1=\sigma\Theta;\ \sigma_2=0;\ \sigma_3=-\sigma\}\quad :\ \sigma_{equiv} = \sigma_\Theta + \sigma_x = \sigma_y \tag{12}$$

From eqs.(11),(12) it follows that yield appears first in the compression zone (this phenomenon should be exploited when choosing places for installation of strain gauges), so the stress distribution along the pipe cross section is asymmetrical relatively to the pipeline axis and in general, 3 different deformation zones can appear along the central section of GLL (Fig.2): purely elastic cross sections $(0\ l_1)$, cross sections with one-side plasticity $(l_2 - l_1)$ and with two-side plasticity $(L - l_3)$.

In order to calculate the moment of internal forces we first calculate the neutral line position and then determine the before mentioned moments for the elasto-plastic zones with one-side yield $(M_{2,2})$ and two-sides yield $(M_{2,3})$ respectively

$$M_{2,2} = \phi_1\ (R,\delta,h_c,e,\sigma_y,\sigma_\Theta)\ ;\quad M_{2,3} = \phi_2\ (R,\delta,h_c,h_t,e,\sigma_y,\sigma_\Theta)$$

where e is the eccentricity of the neutral line expressed through the pipeline curvature $x=w_2''$; h_c and h_t are the dimensions of the elastic kernel zone that are in compression and tension, respectively.

The expressions for M_1 and M_2 are not written down explicitly because they are cumbersome.

The derived equations are essentially non-linear and allow neither exact solution nor sufficiently well-grounded approximate analytical solution. Therefore a numerical method should be used. To solve the stochastic boundary problem the shooting method was used based on reduction of the boundary problem to the Cauchy problem and on integrating it according to the Runge-Kutta (R-K) method.

If the plasto-elastic stresses are symmetrical over the pipe cross-section (p=0) and q_o=const, the solutions for mean deflections $w_{02}(x)$ and correlation functions $K_{w2C2}(x,x')$ and $K_{w2}(x,x')$ will be symmetrical across the middle section span. In this case there will be 4 specific parts along the length of the pipe in consideration: a semi-infinite and finite length section in purely elastic state and 2 finite length sections in the elasto-plastic state. The boundary between the elastic and plasto elastic zones in found in a natural way

$$\sigma_{max}\big|_{x = l_i} = \sigma_y\quad ,\qquad i = 1,2$$

The first stage of calculation consists in deriving the joint solution for $w_{01}(x)$, $w_{02}(x)$ along the GLL. To achieve this select arbitrarily C_1 and C_2 from a set $\{C_1,C_2\}$ that includes non-elastic solutions. With C_1, C_2 given, solution of eq.(2) determines 4 initial conditions at x = 0:

$$w_2 = w_1;\ w'_2 = w'_1;\ w''_2 = w''_1;\ w'''_2 = w'''_1$$

Using this the R-K integration method may be developed. Thus found elastic solution holds true up to some section $x=L_1$, where $\sigma_{max} = \sigma_y$ for the first time. In this cross-section 4 initial values for further integration are determined

$$x = l_1:\ w_3 = w_2;\ w'_3 = w'_2;\ w''_3 = w''_2;\ w'''_3 = w'''_2,$$

so we switch over to one of the elasto-plastic deflection equations and again follow the R-K method.
At each step of integration in the elasto-plastic region the sign of w'_2 and w''_2 is checked. According to symmetry conditions should be

$$w'\Big|_{x = 1/2} = w'''\Big|_{x = 1/2} = 0$$

But due to arbitrariness of chosed C_1 and C_2, w'_2 and w'''_2 do not vanish simultaneously. So generally, we get two values. Repeat this calculation several times with fixed C_1 and various C_2 (or vice versa) and obtain a set of curves

$$\bar{l} = \bar{l}(C_1|C_2);\ \bar{\bar{l}} = \bar{\bar{l}}(C_1|C_2)$$

Inter- or extrapolation of these curves allows to obtain a unique couple $\{C_1,C_2\}$, at which $\bar{l} = \bar{\bar{l}} = l(\bar{C}_1,\bar{C}_2)$ and to obtain such solutions for a whole range of lengths l that happen to be identified on a given GLL.
Now we procede with deriving the correlation equations for the elasto-plastic zone. Expression for the elasto-plastic bending moment (when p=0) is

$$M = F(\kappa) = B[\frac{\kappa}{A} \arcsin \frac{A}{\kappa} - \frac{A}{\kappa} \sqrt{1-(A/\kappa)^2}] \tag{13}$$

After twofold differentiation of M with respect to x we arrive at

$$w_2^{IV} \frac{\partial F}{\partial \kappa} + (w''')^2 \frac{\partial^2 F}{\partial \kappa^2} + C_2(x)\ w_2(x) = q_0 \tag{14}$$

If we present the full deflection $w_2(x)$ and the SRM $C_2(x)$ as in the elastic case in the form: mean values plus small fluctuations, then the fluctuation equation for eq.(14) will be

$$w_2^{VI}\ b_1(w_{02})+2\ w'''_2\ b_2(w_{02})+w''_2\ b_3(w_{02})+C_{02}w_2=w_{02}C(x) \tag{15}$$

Coefficients $b_j(w_{02})$ are expressed through the derivatives of $F(w_{02}) = F(w_2)\Big|_{w_2=w_{02}}$ in the following manner:

$$b_1 = (\partial F/\partial \kappa)\Big|_{\kappa=w''_{02}};\quad b_2 = w'''(\partial^2 F/\partial \kappa^2)\Big|_{\kappa=w''_{02}};$$

$$b_3 = w^{IV}_{02}(\partial^2 F/\partial \kappa^2)\Big|_{\kappa=w''_{02}} + w'''_{02}(\partial^3 F/\partial \kappa^3)\Big|_{\kappa=w''_{02}}$$

Using eq.(15) it is easy to obtain the equation for the crosscorrelation function

$$\frac{\partial^4 K_{w1C2}}{\partial x^4} b_1(w_{02}) + 2 \frac{\partial^3 K_{w2C2}}{\partial x^3} b_2(w_{02}) + \frac{\partial^2 K_{w2C2}}{\partial x^2} b_3(w_{01}) +$$

$$C_{02}K_{w2C2}(x,x') = - w_{02} K_{C2}(x,x') = - w_{02} \sigma^2{}_2 e^{-\alpha_2|\xi|} \tag{16}$$

If, again it is possible to write

$$K_{w2C2}(x,x') = K_2(x) \sigma^2{}_2 e^{-\alpha_2|\xi|},$$

then

$$K_2^{IV}b_1(w_{02}) + K_2'''a_1(w_{02}) + K_2''a_2(w_{02}) + K_2'a_3(w_{02}) + + K_2a_4(w_{02}) = w_{02}(x) \tag{17}$$

Using eq.(15), we also get the equation for the autocorrelation function

$$\frac{\partial^4 K_{w2}}{\partial x'^4} b_1 + 2 \frac{\partial^3 K_{w2}}{\partial x'^3} b_2 + \frac{\partial^2 K_{w2}}{\partial x'^2} b_3 + C_{02}K_{w2}(x,x') =$$

$$= - w_{02}(x') K_{w2C2}(x,x') = - w_{02}(x')\sigma_2^2 K_2(x) e^{-\alpha_2|\xi|} \tag{18}$$

If, again, we have

$$K_{w2}(x,x') = \bar{K}_2(x')\sigma_2^2 K_2(x)e^{-\alpha_2|\xi|},$$

then we arrive at

$$\bar{K}_2^{IV}b_1 + \bar{K}_2'''\bar{a}_1 + \bar{K}_2''\bar{a}_2 + \bar{K}_2'\bar{a}_3 + \bar{K}_2\bar{a}_4 = -w_{02}(x') \tag{19}$$

where $\bar{a}_j$, j=1,...,4 are conjugated with coefficients a_j of eq.(17). Thus, the relevant correlation equations are reduced to fourth order ODE but with variable coefficients. So they also may be integrated by the R-K method.

Numerical Results

The special feature of the derived equations is that their coefficients depend on the found above mean value of $w_2(x)$. Further, for numerical analysis of eqs.(17) and (19) an explicit Newton Rapson type procedure was used. This procedure permits to perform successive approximation aimed at finding the initial constants of integration (B_1, B_2 and D_1, D_2) for a given length l of GLL middle section. It is done using the values of the Jacobian from functions that characterize the discrepancies of symmetry for K_{w2} and K_{w2c2}. The results of computer calculations for a pipe with d = 1420 mm, δ = 32 mm made out of a high grade steel with σ_y = 450MPa showed the following.
The elasto-plastic deformation zone appears at the line that divides the sections with high and low SRM (Fig.3). Here the negative moments are maximal (for C_2 = 5MPa). When C_2=1MPa two elasto-plastic zones appear: on the boundary and in the central part of the GLL middle section. The relative mean curvature modulus of the GLL in elasto-plastic zones are more than unity (Fig.4), the variances of GLL curvature quickly grow from the origin (Fig.5) and have extremes near the boundary and in the central zone of the middle section.
Summarized data on maximal and minimal bending moments as functions of the GLL middle section length and its SRM (no matter where they are located) is given on Fig.6. According to the figure, the extreme moments themselves have maxima and minima that are dependent on the middle section length and the SRM value. Therefore, these specific lengths are most dangerous. They become less and

tend to the origin with the rise of SRM value. The moments (negative and positive) may be in the elasto-plastic range for big values of l.

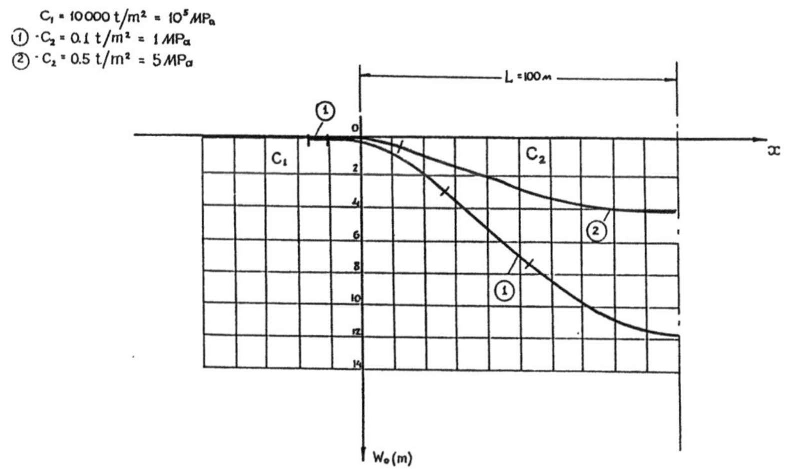

Fig. 3 Location of elasto-plastic zones along the GLL for two values of C_2

Reliability assessment

The results obtained allow to proceed to GLL reliability evaluation at the stage of its construction. We use the following reliability criterion

$$h_{pl} = d - h_e \leq h^* \tag{20}$$

where h_e is the size of the pipe cross section elastic zone. It may be explicitly derived as

$$h_e = A/\kappa = A/w'' \tag{21}$$

But from correlation approximation we know that curvature $\kappa(x)$ is normally distributed, therefore

$$p(\kappa) = \frac{1}{\sqrt{2\pi}\ \sigma} \exp\left[-(\ -\langle\kappa\rangle)^2/2\sigma^2\right] \tag{22}$$

where

$$\langle\kappa\rangle = w_{01}''(w_{02}'')$$
$$\sigma^2 = \frac{\partial^4 K_{w1(2)}(x,x)}{\partial x^2\ \partial x'^2}\bigg|_{x'=x} \tag{23}$$

This means that the reliability criterion (20) may be rewritten as follows

$$\kappa \leq \kappa_* = A/(h - h_*) \tag{24}$$

and therefore reliability

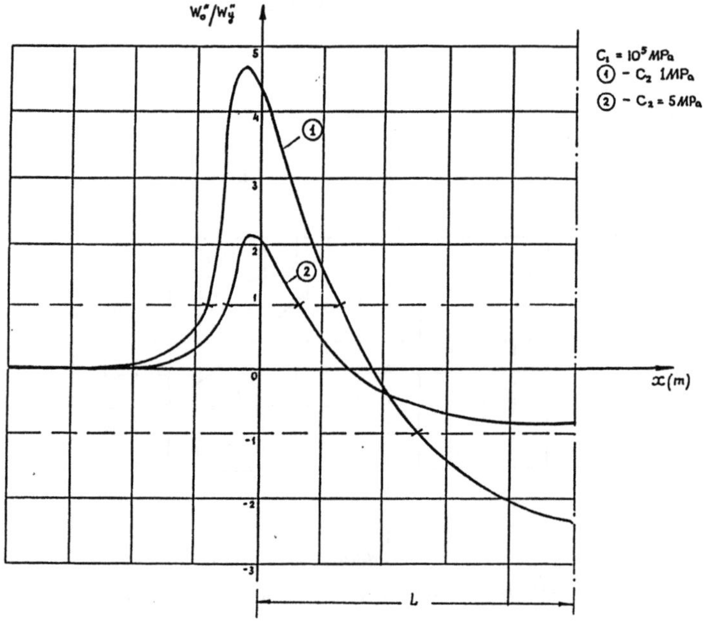

Figure 4 - The change of relative mean curvature modulus along the GLL (in elasto-plastic zones they are more than utility)

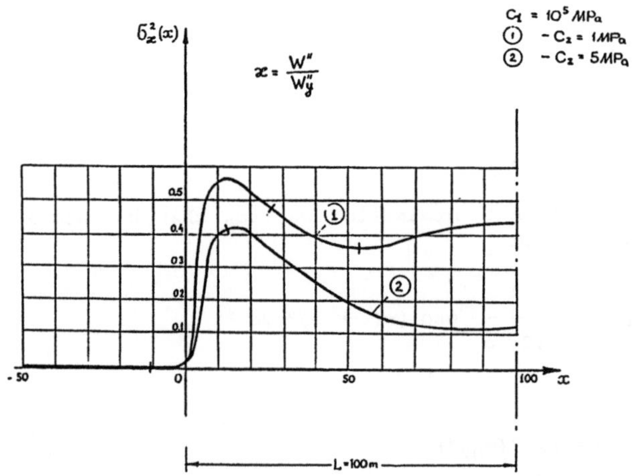

Figure 5 - The change of the relative variance of GLL curvature along its axis X

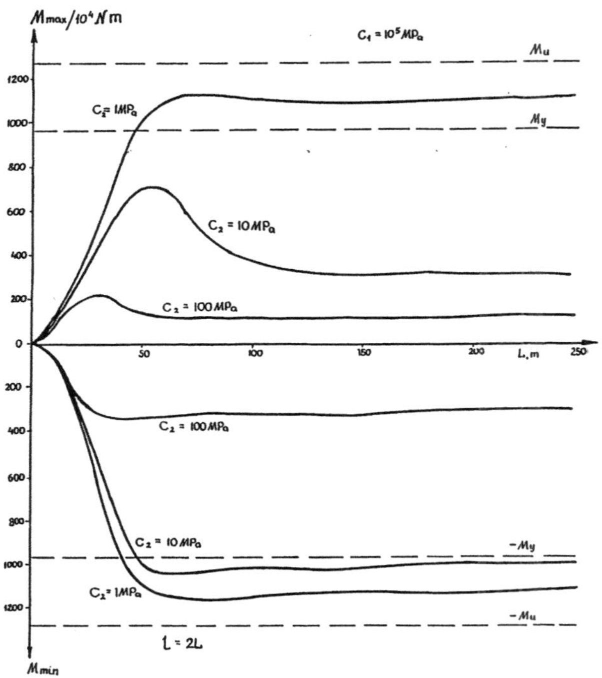

Figure 6 - Summarized data on maximal and minimal bending moments and functions of the GLL middle section length and its SRM

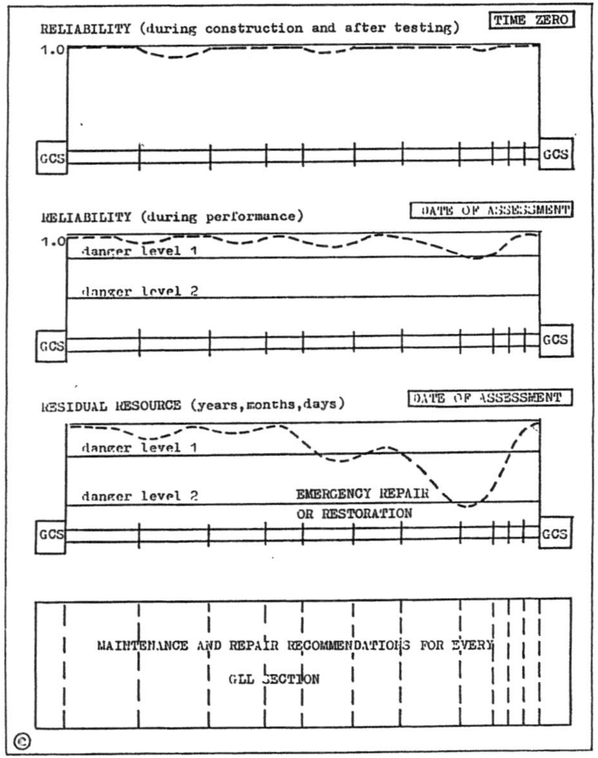

Figure 7 - Results of GLL's real time reliability and residual resource assessment as presented on the decision making person's (DMP) working station monitor (vertical dashes denote the boundaries between GLL sections)

$$R(x) = \int_{-\infty}^{A/(h-h_*)} \frac{1}{(\sqrt{2\pi})\,\sigma_K(x)} \exp\,[-K(x) + - \langle K_x\rangle^2/2\sigma_{K(x)}{}^2]\; dK(x) \tag{25}$$

This result may be produced on the decision making person DMP working station monitor (Fig.7), so that reliability of a GLL between two adjacent gas compressor stations (GCS) would be visualized. After solving more generalized problems (when $c(x,t)$ approximates a viscoelastic media and $q(t)$ and $p(t,x)$ are random functions of time and of coordinate, GLL's reliability during performance and it residual lifetime may be calculated and also visualized on the DMP's monitor. Further, maintenance and repair recommendations may be worked out for every GLL section (Fig.7). Using the results of optimal control of random functions in the two level policy [4], it is possible to obtain the values of the warning danger level (1) and of the prefailure level (2), that calls for emergency repair or restoration. As a whole the said above comprises a part of a SSARM expert system that is now under construction.

Conclusions

1. Precise meeting of the ever growing demands of industry inevitably leads to discovery of new not evident problems of nonlinear stochastic mechanics.
2. As a result of such precise meeting the industry demands a bunch of new stochastic dynamics problems was posed and solved. They envelope problems of distributed one-dimension finite and infinite stochastically and material nonlinear systems behaviour under the influence of a normal load, internal pressure and of a sequence of SRM that are considered as pieces of homogeneous random functions of system coordinate.
3. The results obtained in the present paper may be used for parametrical identification of environment (subgrade) parameters, when the conditions of performance suddenly change due to a meteorological or seismic action. They also give some limits on what to use as diagnostic parameters when monitoring the GLL condition.
4. It seems that for GLL most expedient is to use the longitudinal axis curvature and the local linear deformations as current diagnostic parameters.
5. The proposed solutions allow to reduce the GLL (a distributed system) to a multi DOF system, where "multi" is reasonable.
6. There are some especially dangerous combinations of C_1, C_2 and l for the GLL performance, reliability and residual lifetime, that should be accounted for.
7. Solution of a full group of design and performance schemes for a given GLL allows to give on-line assessment of its reliability and residual resource.
8. SSARM expert system for GLL is now under construction.

Reference

[1] S.A.TIMASHEV. Structural Safety and Reliability Monitoring Systems, Center for Applied Stochastic Research, Florida Atlantic University, Boca Raton, FLA 1990, 63 pp.

[2] V.V.BOLOTIN. Probabilistic and statistical methods in structural mechanics, 2nd Edition, Moscow, Stroiizdat, 1982, 324 pp. (in Russian)

[3] B.P.MAKAROV, D.N.SOBOLEV, B.V.KOCHETKOV. Design of Structures Foundations Lying on Viscoelastic Randomly Nonhomogeneous Subgrade. Moscow, Stroiizdat, 1987, pp.256 (in Russian)

[4] S.A.TIMASHEV. Reliability of Large Mechanical Systems, Moscow, Nauka, 1982; 182 pp., in Russian SEAG, Pavia, 1984, pp.155

Simulation and Analysis of Mechanical Systems with Parameter Fluctuation

Walter V. Wedig
Institut für Technische Mechanik, Kaiserstraße 12
Universität Karlsruhe, D - 7500 Karlsruhe 1, BRD

Summary

Mechanical systems with parameter fluctuations lead to bifurcation problems. With increasing fluctuation intensity the equilibrium of the system becomes unstable and bifurcates into non-trivial stationary solutions. To illustrate this effect we consider the typical example of a simply supported beam under axial loading. Its governing partial differential equation of motion can be reduced to a non-linear ordinary one. For the special case of white noise fluctuations we investigate Lyapunov exponents and bifurcation points of both solution forms.

1. Bifurcation problems of mechanical systems

As a representative example of bifurcations in mechanical systems, we consider a uniform beam under axial loading $P(t)$. According to figure 1, the beam has the length l, the bending stiffness EI, the axial stiffness EA, the mass μ per unit length and the external viscous damping β. Following Weidenhammer (1969), the transverse motion $w(x,t)$ of a beam section is described by a boundary value problem, given as follows:

$$EIw_{xxxx} + \beta w_t + \mu w_{tt} - \tfrac{1}{l}EA[u(t) + \tfrac{1}{2}\int_0^l w_{xx}^2 dx]w_{xx} = 0, \tag{1}$$

$$w(0,t) = w(l,t) = w_{xx}(0,t) = w_{xx}(l,t) = 0, \qquad 0 \le x \le l. \tag{2}$$

Herein, sub t and sub x denote partial derivations of $w(x,t)$ with respect to the time t and to the length coordinate x, respectively. The governing equation of motion (1) is a non-linear integro-differential equation which provides Kirchhoff's assumption that longitudinal waves $u(x,t)$ are negligible. Correspondingly, $u(t)$ represents only the end deflection of the beam determined by the axial force $P(t)$.

As illustrated in figure 1, we restrict our interest to the special case of simple hinged supports that the displacements $w(x,t)$ and moments $EIw_{xx}(x,t)$ at both ends of the beam possess the homogeneous boundary conditions (2) valid for each time $t \ge 0$. Obviously, these conditions are satisfied by the first mode $w(x,t) = T(t)\sin \pi x/l$ which is associated to the first natural frequency ω_1 of the unloaded and undamped beam.

$$w(x,t) = T(t)\sin \pi x/l, \qquad \omega_1^2 = \pi^4 EI/(\mu l^4), \tag{3}$$

$$\ddot{T}(t) + 2D\omega_1\dot{T}(t) + \omega_1^2[1 + \frac{Al}{I\pi^2}u(t) + \gamma T^2(t)]T(t) = 0. \tag{4}$$

Insertion of the first mode reduces the partial differential equation (1) to an ordinary differential equation which is noted in (4). Herein, the parameter γ characterizes the cubic restoring of the beam and D denotes a dimensionless damping value. Both parameters are calculabe by inspection.

N. Bellomo · F. Casciati (Eds.)
Nonlinear Stochastic Mechanics
IUTAM Symposium Turin

The equation (4) possesses two different solutions: the trivial solution $T(t) \equiv 0$ and the bifurcated one $T^2(t) > 0$. The stability of both solution forms is decided by the top Lyapunov exponents λ_0 and λ_S, respectively. Herein, the index 0 denotes the top Lyapunov exponent of the trivial solution and correspondingly, the index S belongs to the stationary bifurcation. Negative Lyapunov exponents λ_0 determine an exponentially decreasing of perturbed solutions to the equilibrium. In the present paper, we restrict our interest to λ_0.

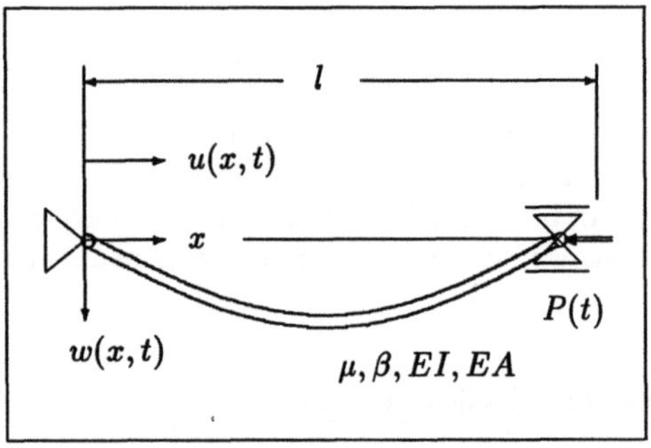

Figure 1: Beam under axial loading $P(t)$

2. Simulation of stochastic bifurcation problems

As already mentioned, we are interested in the special case that the parameter fluctuation $Alu(t)/(\pi^2 I)$ can be modelled by stationary white noise $\sigma\dot{W}_t$ where σ is the intensity and $\dot{W}_t$ is a normalized process with zero mean and the delta-correlation $E(\dot{W}_t\dot{W}_s) = \delta(t-s)$. Consequently, $T(t)$ is a stochastic process to be noted by X_t. Therewith, the bifurcation equation (4) reads as follows:

$$\ddot{X}_t + 2D\omega_1\dot{X}_t + \omega_1^2\left[1 + \sigma\dot{W}_t + \gamma X_t^2\right]X_t = 0. \tag{5}$$

Following the concept of Khasminskii (1967), we introduce polar coordinates via the amplitude process A_t and the phase process Ψ_t

$$X_t = A_t\cos\Psi_t, \qquad A_t = \sqrt{X_t^2 + (\dot{X}_t/\omega_1)^2}, \tag{6}$$
$$\dot{X}_t = \omega_1 A_t\sin\Psi_t, \qquad \Psi_t = \arctan\dot{X}_t/(\omega_1 X_t) \tag{7}$$

and transform the second order equation (5) into a first order system of stochastic Itô differential equations for the increments of A_t and Ψ_t.

$$\begin{aligned} dA_t = &-2D\omega_1 A_t\sin^2\Psi_t dt + \tfrac{1}{2}\omega_1^2\sigma^2 A_t\cos^4\Psi_t dt \\ &-\omega_1\gamma A_t^3\sin\Psi_t\cos^3\Psi_t dt - \omega_1\sigma A_t\sin\Psi_t\cos\Psi_t dW_t, \end{aligned} \tag{8}$$
$$\begin{aligned} d\Psi_t = &-\omega_1(1 + 2D\sin\Psi_t\cos\Psi_t)dt - \omega_1\sigma\cos^2\Psi_t dW_t \\ &-\omega_1\gamma A_t^2\cos^4\Psi_t dt - \omega_1^2\sigma^2\sin\Psi_t\cos^3\Psi_t dt. \end{aligned} \tag{9}$$

Herein, W_t is the normalized Wiener process. It is normally distributed with zero mean and uncorrelated increments. The $(\omega_1\sigma)^2$- terms in both equations result from the application of Itô's calculus. More details of this calculus are given e.g. by Arnold (1974). Finally, we introduce the natural logarithm $L_t = \log A_t$ of the amplitude process and calculate its increment.

$$d \log A_t = -2D\omega_1 \sin^2 \Psi_t dt - \omega_1 \gamma A_t^2 \sin \Psi_t \cos^3 \Psi_t dt$$
$$-\tfrac{1}{2}\omega_1^2 \sigma^2 \cos^2 \Psi_t (\sin^2 \Psi_t - \cos^2 \Psi_t) dt - \omega_1 \sigma \sin \Psi_t \cos \Psi_t dW_t. \qquad (10)$$

This form is needed for the determination of Lyapunov exponents derived in the last section of the paper.

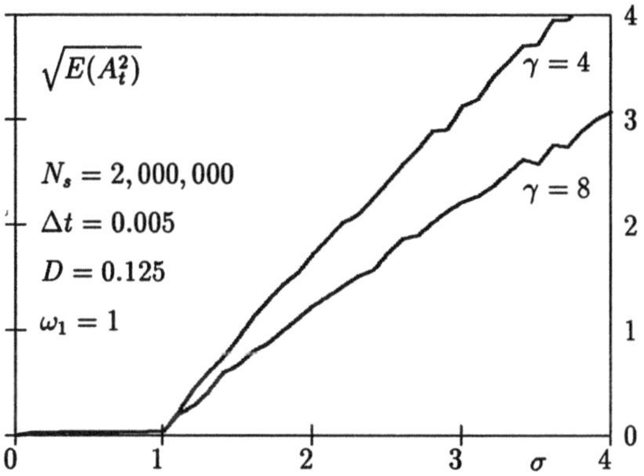

Figure 2: Simulated bifurcation diagram

Before going into more details of the stability analysis we perform simple numerical experiments to show some basic effects of the bifurcation problem of interest. The realized simulation routine is based on the forward Euler scheme applied to the phase equation (9) and to the log-amplitude equation (10).

$$\Psi_{n+1} = \Psi_n - \omega_1 \Delta t (1 + 2D \sin \Psi_n \cos \Psi_n) - \omega_1 \sigma \sqrt{\Delta_t} R_n \cos^2 \Psi_n$$
$$-\omega_1 \gamma \Delta t \exp(2L_n) \cos^4 \Psi_n - \omega_1^2 \sigma^2 \Delta t \sin \Psi_n \cos^3 \Psi_n, \qquad (11)$$
$$L_{n+1} = L_n - 2D\omega_1 \Delta t \sin^2 \Psi_n - \tfrac{1}{2}\omega_1^2 \sigma^2 \Delta t \cos^2 \Psi_n (\sin^2 \Psi_n - \cos^2 \Psi_n)$$
$$-\omega_1 \sigma \sqrt{\Delta t} R_n \sin \Psi_n \cos \Psi_n - \omega_1 \gamma \Delta t \exp(2L_n) \sin \Psi_n \cos^3 \Psi_n. \qquad (12)$$

Herein, Δt denotes the time step. The Ψ_n and L_n are time-discrete versions of the phase process Ψ_t and of the log-amplitude process L_t. R_n is a sequence of uncorrelated numbers which are normally distributed with zero mean and $E(R_n^2) = 1$. This corresponds to the time-continuous versions $E(dW_t) = 0$ and $E(dW_t^2) = dt$. Figure 2 shows numerical results obtained for the system data $\omega_1 = 1$, $D = 0.125$ and for the two cubic restoring parameters $\gamma = 4$ and $\gamma = 8$. The time step selected was $\Delta t = 0.005$ applied for $N_s = 2,000,000$ sample points. The recursion formulas (11) and (12) produced a stationary sequence of L_n-values from which we calculated the time-discrete amplitudes A_n. Squaring them and summing up we computed the time average and plotted the associated root mean square against the intensity parameter σ. The plots show clearly that there is a critical intensity parameter or bifurcation point σ_c where the trivial solution $A_t \equiv 0$ bifurcates into a stationary solution $A_t^2 > 0$. In particular, we recognize that the bifurcation point is not influenced by the non-linear parameter γ.

3. Invariant measure of the stationary phase

To investigate the stability of the trivial solution $A_t \equiv 0$, the amplitude equation (8) and the phase equation (9) are linearized with respect to $A_t << 1$. This leads to an uncoupled phase equation of the following form:

$$d\Psi_t = -\omega_1[1 + (2D + \omega_1\sigma^2 \cos^2 \Psi_t) \sin \Psi_t \cos \Psi_t]dt - \omega_1\sigma \cos^2 \Psi_t dW_t. \qquad (13)$$

By means of the forward difference scheme applied in (11), the phase process is simulated for $\omega_1 = 1, D = 0.125$ and for the two different excitation intensities $\sigma = 1$ and $\sigma = 3$. In figure 3, we show histograms of the estimated phase densities $p(\psi)$ obtained for a step size $\Delta t = 0.005$ and $N_s = 1,000,000$ sample points. From this it follows that the phase process is stationary. It rotates clockwise mainly driven by the inhomogeneous drift term $-\omega_1 dt$. For the special angles $\psi = \pi/2 \pm k\pi$ $(k = 0, 1, 2, ..)$, the diffusion term in (13) is vanishing so that the estimations of $p(\psi)$ converge rapidly in these singular points. Since the equation (13) is doubly periodic with respect to ψ, the phase density $p(\psi)$ needs only to be evaluated in the limited angle range $-\pi/2 \leq \psi \leq +\pi/2$.

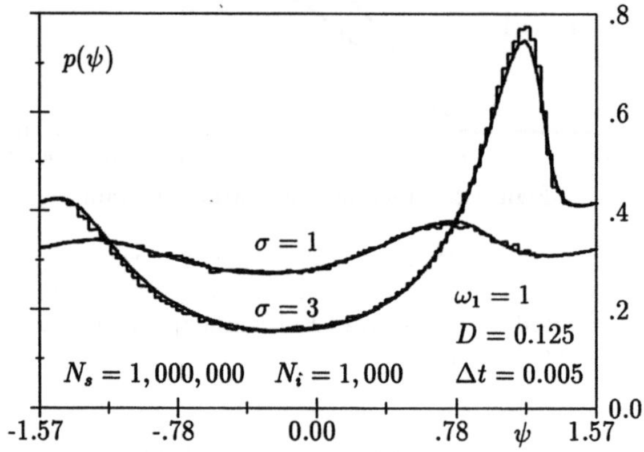

Figure 3: Stationary density of the invariant measure

The Itô equation (13) defines a Markov process which is equivalently described by the associated Fokker-Planck equation. In the stationary case, this diffusion equation can be integrated with respect to the angle variable ψ leading to a linear first order differential equation for the phase density $p(\psi)$.

$$\tfrac{1}{2}\omega_1\sigma^2 \cos^4 \psi\, p'(\psi) + [1 + (2D - \omega_1\sigma^2 \cos^2 \psi) \sin \psi \cos \psi]p(\psi) = C, \qquad (14)$$

$$-\pi/2 \leq \psi \leq +\pi/2, \qquad p(\pm\pi/2) = C, \qquad \int_{-\pi/2}^{+\pi/2} p(\psi)d\psi = 1. \qquad (15)$$

Herein, C is a constant of integration to be determined by the normalization condition, noted in (15). Derivatives with respect to ψ are denoted by primes. Obviously, the Fokker-Planck equation (14) possesses two singularities at the boundaries $\psi = \pm\pi/2$. To select a regular physically meaningful solution $p(\psi)$, we can apply a singular perturbation method (see Wedig, 1988) for expansions around the two singular points. Inside the regular angle range, we can map the linear solution obtained by classical integration routines.

According to Wedig (1989), the method, mentioned above, can be simplified considerably by means of a backward difference scheme which regularize the numerical integration in (14). It leads to the following recurrence formula:

$$p'(\psi) = (p_n - p_{n-1})/\Delta\psi, \quad (n = 1, 2, \ldots N), \qquad \Delta\psi = \pi/N, \tag{16}$$

$$p_n = \frac{2\Delta\psi C + \omega_1\sigma^2 \cos^4\psi_n p_{n-1}}{2\Delta\psi\,[1 + (2D - \omega_1\sigma^2\cos^2\psi_n)\sin\psi_n\cos\psi_n] + \omega_1\sigma^2\cos^4\psi_n}. \tag{17}$$

The recursive computation in (17) begins with $\psi_0 = -\pi/2$ and $p_0 = p(-\pi/2) = C$. It ends at the upper angle with the same density value $p_N = p(+\pi/2) = C$. The recursion is numerically stable for sufficiently small step sizes $\Delta\psi = \pi/N$. Figure 3 shows a comparison of the density histograms simulated by (13) for $N_s = 1{,}000{,}000$ sample points with the smooth lines, calculated from (17) for the interval number $N_i = 1{,}000$.

4. Lyapunov exponents of the trivial solution

The stationary phase density $p(\psi)$ yields an invariant measure of the linearized system equation (5). It determines the exponentially growth behaviour of the associated state vector $\vec{X}_t$. This growth can be investigated by determining the top Lyapunov exponent λ_0 which is to be calculated from the linearized amplitude equation (8).

$$\begin{aligned} d\log A_t = &-\omega_1\sigma\sin\Psi_t\cos\Psi_t dW_t - \omega_1[2D\sin^2\Psi_t \\ &+\tfrac{1}{2}\omega_1\sigma^2\cos^2\Psi_t(\sin^2\Psi_t - \cos^2\Psi_t)]dt. \end{aligned} \tag{18}$$

Invoking the multiplicative ergodic theorem of Oseledec (1968) the top Lyapunov exponent λ_0 is obtained as

$$\lambda_0 = \lim_{t\to\infty} \tfrac{1}{t}\log \parallel \vec{X}_t \parallel / \parallel \vec{X}_0 \parallel = \lim_{t\to\infty} \tfrac{1}{t}\log(A_t/A_0), \tag{19}$$

where $\parallel \cdot \parallel$ denotes a suitable norm of the system vector $\vec{X}_t$. It is suitable to take the Euclidean norm, already introduced by the amplitude process A_t. The A_0 and $\parallel \vec{X}_0 \parallel$ represent associated initial conditions at the time $t_0 = 0$.

The linear amplitude equation (18) can easily be integrated leading to the solution

$$\begin{aligned} \log(A_t/A_0) = -\omega_1\sigma\int_0^t \sin\Psi_\tau\cos\Psi_\tau dW_\tau - \omega_1\int_0^t [2D\sin^2\Psi_\tau \\ +\tfrac{1}{2}\omega_1\sigma^2\cos^2\Psi_\tau(\sin^2\Psi_\tau - \cos^2\Psi_\tau)]d\tau. \end{aligned} \tag{20}$$

Inserted into the multiplicative ergodic theorem (19), the integrated form (20) produces time averages like

$$\lim_{t\to\infty}\tfrac{1}{t}\int_0^t(\cdot)d\tau = E(\cdot) = \int_{-\pi/2}^{+\pi/2}(\cdot)p(\psi)d\psi, \qquad \lim_{t\to\infty}\tfrac{1}{t}\int_0^t(\cdot)dW_\tau = 0. \tag{21}$$

The time average of the diffusion term dW_τ is vanishing because of the non-anticipating property of Itô integrals. The time average of the drift term $d\tau$ can be replaced by corresponding expected values $E(\cdot)$ calculable by the density $p(\psi)$ of the invariant measure of the phase process. Consequently, the top Lyapunov exponent λ_0 is given by

$$\lambda_0 = -2D\omega_1 E(\sin^2\Psi_t) - \tfrac{1}{2}\omega_1^2\sigma^2 E[\cos^2\Psi_t(\sin^2\Psi_t - \cos^2\Psi_t)]. \tag{22}$$

Herein, the expected values are completely determined by the stationary density distribution (17) of the phase process. They are calculated by corresponding integrations over the angle range $-\pi/2 \leq \psi \leq +\pi/2$.

Figure 4 shows an evaluation of the top Lyapunov exponent λ_0 in the parameter range $0 \leq D \leq 2$ and $0 \leq \sigma \leq 4$ for the natural frequency $\omega_1 = 1$. The calculations are performed by means of the recurrence formula (17) applying Simpson's rule for the integrals of the expectation values in (22). The obtained Lyapunov exponents are given in form of contour lines with the level interval $\Delta\lambda = 0.1$. The line starting with the parameter values $\sigma = D = 0$ denotes the zero level or stability boundary of the linear system. The Lyapunov exponents are negative below this line, and they increase with increasing intensities of the parametric excitation. Right from the stability boundary, the Lyapunov exponents decrease with increasing damping ratios D up to a certain optimal line where they increase again. For the special deterministic case $\sigma = 0$, the Lyapunov exponents, calculated above, coincide with the eigenvalues of the unperturbed system. Further controls are possible in the stability boundary, already derived by Kozin and Mitchel (1974) via singularity methods. The contour lines of figure 4 can also be compared with simulated results of Kliemann and Arnold (1981) for parametric excitations by coloured noise.

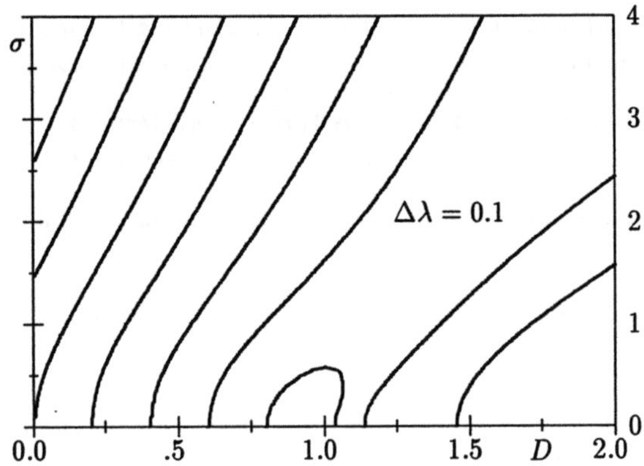

Figure 4: Contour lines of Lyapunov exponents

References

1. Weidenhammer, F. 1969. Biegeschwingungen des Stabes unter axial pulsierender Zufallslast. VDI-Berichte Nr. 135: 101-107.

2. Khasminskii, R.Z. 1967. Necessary and sufficient conditions for asymptotic stability of linear stochastic systems. Theor. Prob. and Appls. 12: 144-147.

3. Arnold, L. 1974. Stochastic Differential Equations. New York: Wiley.

4. Wedig, W. 1988. Pitchfork and Hopf bifurcations in stochastic systems - effective methods to calculate Lyapunov exponents. To appear in: Effective Stochastic Analysis (ed. by P. Krée, W. Wedig), Heidelberg: Springer.

5. Wedig, W. 1989. Vom Chaos zur Ordnung. Gamm-Mitteilungen ISSN 0936-7195, Heft 2: 3-31.

6. Oseledec, V.I. 1968. A multiplicative ergodic theorem, Lyapunov characteristic numbers for dynamical systems. Trans. Moscow Math. Soc. 19: 197-231.

7. Kozin, F. & Mitchel, R.R. 1974. Sample stability of second order linear differential equations with wide band noise coefficients. SIAM J. Appl. Math., 17: 571-605.

8. Arnold, L. & Kliemann, W. 1981. Qualitative theory of stochastic systems. In: Probabilistic Analysis and Related Topics (ed. by A.T. Bharucha-Reid). Vol. 3, New York: Academic Press.

Optimal Aseismic Hybrid Control of Nonlinear and Hysteretic Structures using Velocity and Acceleration Feedbacks

J. N. Yang* and Z. Li* and S. C. Liu**

*Department of Civil Engineering, University of California, Irvine, CA 92717
**National Science Foundation, Washington, D.C. 20550

Summary

For practical implementations of aseismic hybrid control systems, it may be easier to measure the acceleration response rather than the displacement response of the building. In this paper, a version of the instantaneous optimal control algorithm for nonlinear or hysteretic structural systems is presented. The control algorithm utilizes the acceleration response rather than the displacement response for the determination of the control vector. The optimal algorithm is applied to an aseismic hybrid control system consisting of a lead-core rubber-bearing isolation system connected to an actuator. It is demonstrated numerically that the performance of the present algorithm may be better than the previous algorithm utilizing the displacement response of the structure for the determination of the control vector.

Introduction

It has been demonstrated recently that a combined use of active and passive control systems, referred to as the hybrid control system, is more effective, beneficial and practical, in some cases, for reducing the building response under strong earthquakes [e.g., 1-7]. The idea of hybrid control systems is to utilize the advantages of both the passive and active control systems to extend the range of applicability of control systems to protect the integrity of the structure. In particular, under extreme environments, such as strong earthquakes, hybrid control systems are superior. However, the use of hybrid control systems involves active control of nonlinear or hysteretic structural systems, since passive control systems, such as base isolation systems, usually behave nonlinearly or inelastically under dynamic loads. Instantaneous optimal control algorithms for nonlinear or hysteretic structures were presented in Refs. 5-9. These control algorithms require the measurements of the displacement and velocity responses for the determination of the control vector.

Recent laboratory experiments indicated possible difficulties involved in measuring the displacement response of the building subjected to earthquake ground motions. This is because both the ground and the building are moving during earthquakes, so that there is no absolute reference for the determination of the displacement response. Further, the displacement response obtained by numerically integrating the measured velocity response for linear structures is not satisfactory due to serious error accumulations. On the other hand, the acceleration response can easily be measured by installing acceleration sensors. Consequently, it is desirable to develop control algorithms utilizing acceleration sensors rather than displacement sensors.

N. Bellomo · F. Casciati (Eds.)
Nonlinear Stochastic Mechanics
IUTAM Symposium Turin

In this paper, an alternate version of the instantaneous optimal control algorithm for nonlinear or hysteretic structures, utilizing the acceleration response rather than the displacement response, is proposed following the approach presented by Yang, et al [10] for linear structures. The proposed control algorithm is applied to an important type of aseismic hybrid control system, i.e., a lead-core rubber-bearing isolation system connected to an actuator. The behavior of the base isolation system is highly inelastic. Numerical examples are worked out to demonstrate the performance of the present control algorithm.

Formulation

Nonlinear Hysteretric Model for Inelastic Systems

Various hysteretic models for the restoring force of an inelastic system have been developed in recent years [e.g. 11-14]. In this paper, the differential equation model presented by Wen, et al [11-14] will be used for both buildings and passive protective systems. Let $x_i(t)$ be the relative displacement between the ith floor and the i-1th floor, i.e., the deformation of the ith story unit. The stiffness restoring force, $F_{si}(t)$, of the ith story unit is given by

$$F_{s_i}(t) = \alpha_i k_i x_i + (1 - \alpha_i) k_i D_{y_i} v_i \tag{1}$$

in which k_i=elastic stiffness of the ith story unit, α_i= ratio of post-yielding to pre-yielding stiffness, D_{yi}=yield deformation=constant, and v_i is a nondimensional variable introduced to describe the hysteretic component of the deformation, with $|v_i| \leq 1$, where

$$D_{y_i} \dot{v}_i = A_i \dot{x}_i - \beta_i |\dot{x}_i| \, |v_i|^{n_i - 1} v_i - \gamma_i \dot{x}_i |v_i|^{n_i} \tag{2}$$

In Eq. (2), parameters A_i, β_i and γ_i govern the scale and general shape of the hysteresis loop, whereas the smoothness of the force-deformation curve is determined by the parameter n_i. The hysteretic model given by Eqs. (1) and (2) provides an explicit mathematical expression with many parameters flexible enough to reflect hysteretic behaviors of different base isolation systems, such as rubber bearings and sliding systems, which are the main concern of this study.

Equations of Motion

For simplicity, consider a base-isolated one-dimensional inelastic building implemented by a hybrid control system as shown in Fig. 1. The building system is idealized by an n-degree-of-freedom system and subjected to a one-dimensional earthquake ground acceleration $\ddot{X}_0(t)$. The matrix equation of motion of the entire building system can be expressed as follows

$$\underline{M}\ddot{\underline{X}}(t) + \underline{C}\dot{\underline{X}}(t) + \underline{K}_e\underline{X}(t) + \underline{K}_I\underline{V}(t) = \underline{H}\underline{U}(t) + \underline{F}\ddot{X}_0(t) \tag{3}$$

in which a linear viscous damping model has been assumed, $\underline{X}(t)=[x_1,x_2,\ldots,x_n]'$ = an n vector denoting the deformation of each story unit, and $\underline{V}(t)=[v_1,v_2,\ldots,v_n]'$ = an n vector denoting the hysteretic variable of each story unit. For the notations above, an under bar denotes a vector or a matrix and a

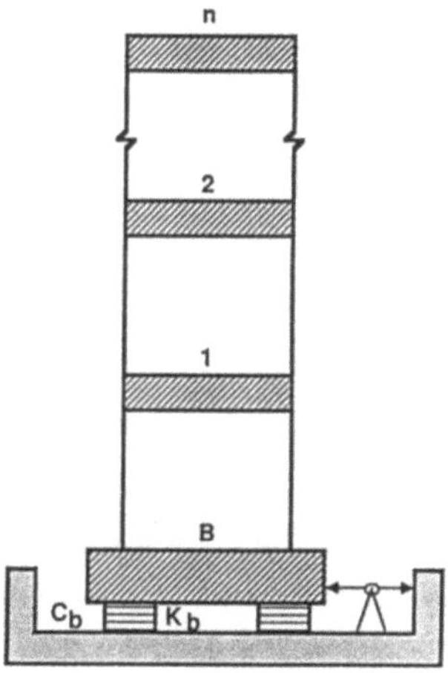

Fig.1 : Structural Model of a Building with Base Isolation System and Actuator

prime indicates the transpose of a vector or a matrix. In Eq. (3), $\underline{M}$ is a (nxn) mass matrix with the i-jth element $M(i,j)=m_i$ for $i=1,2,\ldots,n$ and $j=1,2,\ldots,i$ and $M(i,j)=0$ for $j>i$, where m_i is the mass of the ith floor; $\underline{C}$ is an (nxn) band-limited damping matrix with all elements equal to zero except $C(i,i)=c_i$ for $i=1,2,\ldots,n$ and $C(i,i+1)=-c_{i+1}$ for $i=1,2,\ldots,n-1$, where c_i is the damping coefficient of the ith story unit. $\underline{K}_e$ and $\underline{K}_I$ are (nxn) band-limited elastic stiffness matrix and hysteretic stiffness matrix, respectively [5-7]. All elements of $\underline{K}_e$ and $\underline{K}_I$ are zero except $K_e(i,i)=\alpha_i k_i$, $K_I(i,i)=(1-\alpha_i)k_i D_{yi}$ for $i=1,2,\ldots,n$ and $K_e(i,i+1)=-\alpha_{i+1}k_{i+1}$, $K_I(i,i+1)=-(1-\alpha_{i+1})k_{i+1}\,D_{yi+1}$ for $i=1,2,\ldots,n-1$. $\underline{H}$ is an (nxr) matrix denoting the location of r controllers and $\underline{F}$ is an n vector denoting the influence of the earthquake ground acceleration.

From Eq. (2), the vector $\underline{V}$ appearing in Eq. (3) is given by

$$\dot{\underline{V}} = \underline{f}(\dot{\underline{X}},\underline{V}) \tag{4}$$

in which the ith element of $\dot{\underline{V}}$ or $\underline{f}(\dot{\underline{X}}, \underline{V})$, denoted by $\dot{v}_i$ or $f_i(\dot{x}_i, v_i)$, is given by

$$\dot{v}_i = f_i(\dot{x}_i,v_i) = D_{yi}^{-1}\left[A_i\dot{x}_i - \beta_i\,|\dot{x}_i|\,|v_i|^{n_i-1}\,v_i - \gamma_i\dot{x}_i\,|v_i|^{n_i}\right] \tag{5}$$

By introducing a 3n state vector $\underline{Z}(t)$ and 3n vectors $\underline{B}$ and $\underline{W}_1$

$$\underline{Z}(t) = \begin{bmatrix}\underline{X}\\ \underline{V}\\ \dot{\underline{X}}\end{bmatrix} \;;\; \underline{B} = \begin{bmatrix}\underline{0}\\ \underline{0}\\ \underline{M}^{-1}\underline{H}\end{bmatrix} \;;\; \underline{W}_1 = \begin{bmatrix}\underline{0}\\ \underline{0}\\ \underline{M}^{-1}\underline{F}\end{bmatrix} \tag{6}$$

the second-order nonlinear matrix equation of motion, Eq. (3), can be converted into a first order matrix equation as follows:

$$\underline{\dot{Z}}(t) = \underline{g}[\underline{Z}(t)] + \underline{B}\underline{U}(t) + \underline{W}_1\ddot{X}_0(t) \tag{7}$$

in which g[$\underline{Z}$(t)] is a 3n vector consisting of nonlinear functions of components of $\underline{Z}$(t),

$$\underline{g}[\underline{Z}(t)] = \begin{bmatrix} \underline{\dot{X}} \\ - - - - - - - - - - - \\ \underline{f}(\underline{\dot{X}},\underline{V}) \\ - - - - - - - - - - - \\ -\underline{M}^{-1}\left(\underline{C}\underline{\dot{X}} + \underline{K}_e\underline{X} + \underline{K}_I\underline{V}\right) \end{bmatrix} \tag{8}$$

<u>Instantaneous Optimal Control with Acceleration and Velocity Feedbacks</u>

The time dependent performance index proposed by Yang, et al [10] for linear structures is used as follows

$$J^*(t) = \underline{\dot{Z}}'(t)\underline{Q}^*\underline{\dot{Z}}(t) + \underline{U}'(t)\underline{R}\underline{U}(t) \tag{9}$$

in which $\underline{Q}^*$ is a (3nx3n) positive semi-definite weighting matrix and R is a (rxr) positive definite weighting matrix. Although the displacement response vector $\underline{X}$(t) does not appear in the performance index J^*(t), Eq. (9), it is expected that a minimization of J^*(t) will also reduce $\underline{X}$(t).

To solve the general nonlinear equation of motion given by Eq. (7), the state vector $\underline{Z}$(t) is expressed by a backward finite difference form, i.e., $\underline{Z}(t) = \underline{Z}(t-\Delta t) + \Delta t\underline{\dot{Z}}(t-\Delta t)$. Then, Eq. (7) can be written as

$$\underline{\dot{Z}}(t) = \underline{g}[\underline{Z}(t-\Delta t) + \Delta t\underline{\dot{Z}}(t-\Delta t)] + \underline{B}\,\underline{U}(t) + \underline{W}_1\ddot{X}_0(t) \tag{10}$$

with zero initial conditions, i.e., $\underline{\dot{Z}}(0) = \underline{Z}(0) = 0$.

The Hamiltonian H^*(t) is obtained from Eqs. (9) and (10) as

$$\begin{aligned} H^*(t) = \underline{\dot{Z}}'(t)\underline{Q}^*\underline{\dot{Z}}(t) + \underline{U}'(t)\underline{R}\underline{U}(t) + \underline{\lambda}'\{\underline{\dot{Z}}(t) - \underline{g}[\underline{Z}(t-\Delta t) + \Delta t\underline{\dot{Z}}(t-\Delta t)] \\ - \underline{B}\,\underline{U}(t) - \underline{W}_1\ddot{X}_0(t)\} \end{aligned} \tag{11}$$

The necessary conditions for minimizing the time dependent objective function, J^*(t), subjected to the constraint of the equation of motion, Eq. (10), are as follows

$$\frac{\partial H^*(t)}{\partial \underline{\dot{Z}}} = 0 \quad ; \quad \frac{\partial H^*(t)}{\partial \underline{U}} = 0 \quad ; \quad \frac{\partial H^*(t)}{\partial \underline{\lambda}} = 0 \tag{12}$$

The optimal closed-loop control vector $\underline{U}$(t) is obtained by substituting Eq. (11) into Eq. (12); with the result

$$\underline{U}(t) = -\underline{R}^{-1}\underline{B}'\underline{Q}^*\underline{\dot{Z}}(t) \tag{13}$$

Substituting Eq. (13) into Eq. (7), one obtains the nonlinear matrix equation of motion with the optimal control vector given by Eq. (13)

$$\dot{Z}(t) = \left(I + BR^{-1}B'Q^{*}\right)^{-1} g[Z(t)] + \left(I + BR^{-1}B'Q^{*}\right)^{-1} W_1 \ddot{X}_o(t) \tag{14}$$

The response state vector $\underline{Z}(t)$ can be simulated numerically from Eq. (14) using the Fourth-Order Runge-Kutta method described in Refs. 5-7.

Estimation of Hysteretic Component Vector V

The optimal control vector $\underline{U}(t)$ given by Eq. (13) is regulated by the 3n response state vector $\dot{\underline{Z}}(t)=[\dot{\underline{X}}'(t), \dot{\underline{V}}'(t), \ddot{\underline{X}}'(t)]'$. The velocity and acceleration response vectors, $\dot{\underline{X}}(t)$ and $\ddot{\underline{X}}(t)$, are available from measurements using velocity and acceleration sensors. However, the derivative of the hysteretic component vector $\dot{\underline{V}}(t)=[\dot{v}_1(t),\dot{v}_2(t),\ldots,\dot{v}_n(t)]'$ is not measurable and hence it should be estimated. An approach to estimate $\dot{v}_i(t)$ for $i=1,2,\ldots,n$ is by use of a reduced order nonlinear state observer. In the present case, the observer can be constructed easily, since $\dot{v}_i(t)$ is only a function of the interstory velocity $\dot{x}_i(t)$ given by Eq. (5). As a result, the estimate of the hysteretic component $\dot{v}_i(t)$, denoted by $\dot{v}_i^*(t)$, follows from Eq. (5) as

$$\dot{v}_i^*(t) = f_i(\dot{x}_i, v_i^*) = D_{yi}^{-1}\left[A_i\dot{x}_i - \beta_i \mid \dot{x}_i \mid \mid v_i^* \mid^{n_i-1} v_i^* - \gamma_i \dot{x}_i \mid v_i^* \mid^{n_i} \right] \tag{15}$$

in which the argument t for $\dot{x}_i(t)$ and $v_i^*(t)$ has been omitted for simplicity and $\dot{x}_i(t)$ is obtained from measurements. Given the measured $\dot{x}_i(t)$, Eq. (15) can be integrated on-line to obtain the estimates $v_i^*(t)$ and $\dot{v}_i^*(t)$ for $v_i(t)$ and $\dot{v}_i(t)$, respectively, without any difficulty. It should be noted that the performance of the control algorithm is quite satisfactory without a feedback of the derivative of the hysteretic component vector $\dot{V}(t)$. Consequently, the estimation of $\dot{\underline{V}}(t)$ is usually not necessary as will be demonstrated in the numerical example.

Numerical Examples

An eight-story building that exhibits bilinear elasto-plastic behavior is considered. The stiffness of each story unit is designed such that yielding occurs simultaneously for each story unit. The properties of the building are as follows: (i) the mass of each floor is identical with $m_i = m = 345.6$ metric tons; (ii) the preyielding stiffnesses of the eight-story units are k_{i1} ($i=1.2,\ldots,8$) $= 3.4\text{x}10^5$, $3.26\text{x}10^5$, $2.85\text{x}10^5$, $2.69\text{x}10^5$, $2.43\text{x}10^5$, $2.07\text{x}10^5$, $1.69\text{x}10^5$ and $1.37\text{x}10^5$ kN/m, respectively, and the postyielding stiffnesses are $k_{i2} = 0.1\, k_{i1}$ for $i=1,2,\ldots,8$, i.e., $\alpha_i=0.1$ and $k_i=K_{i1}$; and (iii) the viscous damping coefficients for each story unit are $c_i = 490$, 467, 410, 386, 348, 298, 243 and 196 kN.sec/m, respectively. The damping coefficients given above result in a classically damped building with a damping ratio of 0.38% for the first vibrational mode. The natural frequencies of the unyielded building are 5.24, 14.0, 22.55, 30.22, 36.89, 43.06, 49.54 and 55.96 rad./sec. The yielding level for each story unit varies with respect to the stiffness; with the results, $D_{yi} = 2.4$, 2.3, 2.2, 2.1, 2.0, 1.9, 1.7, and 1.5 cm. The bilinear elasto-plastic behavior can be described by the hysteretic model, Eqs. (1) and (2), with $A_i=1.0$, $\beta_i=0.5$, $n_i=95$ and $\gamma_i=0.5$ for $i=1.2,\ldots,8$. A simulated earthquake with a maximum ground acceleration of 0.3g as shown in Fig. 2 is used as the input excitation.

With the eight-story bilinear elasto-plastic building and the earthquake ground acceleration shown in Fig. 2, time histories of all the response quantities have been computed. Within 30 seconds of the earthquake episode, the maximum interstory deformation, x_{mi}, and the maximum absolute acceleration of each floor, a_i, are shown in Table 1. The time history of the first story deformation, $x_1(t)$, is presented in Fig. 3(a). As observed from Table 1 and Fig. 3(a), the deformation of the unprotected building is excessive and that yielding takes place in each story unit.

To reduce the structural response, a lead-core rubber-bearing isolation system is implemented as shown in Fig. 1. The restoring force of the lead-core rubber-bearing system is modeled by Eq. (1)

$$F_{sb} = \alpha_b k_b x_b + (1-\alpha_b) k_b D_{yb} \nu_b \tag{16}$$

in which ν_b is given by Eq. (2) with i=b. The mass of the base isolation system is m_b=450 metric tons and the viscous damping coefficient is assumed to be linear with c_b=26.17 kN.sec/m. The restoring force of the base isolation system given by Eq. (16) is not bilinear elasto-plastic and the parameter values are given as follows: k_b=18050 kN/m, α_b=0.6 and D_{yb}=4cm. The parameters governing the scale and general shape of the hysteresis loop of the isolation system, Eq. (2), were assumed to be A_b=1.0, β_b=0.5, n_b=3 and γ_b=0.5. The hysteretic characteristics of such a base isolation system, i.e., x_b versus ν_b, is shown in Fig. 4. For the building with the base isolation system alone, the first natural frequency of the preyielding structure is 2.21 rad/sec. The damping ratio for the first vibrational mode is 0.15%. It is observed that the fundamental frequency is reduced significantly by the implementation of the base isolation system. The response vector $\underline{X}(t)$ is given by $\underline{X}=[x_b,x_1,\ldots,x_8]'$.

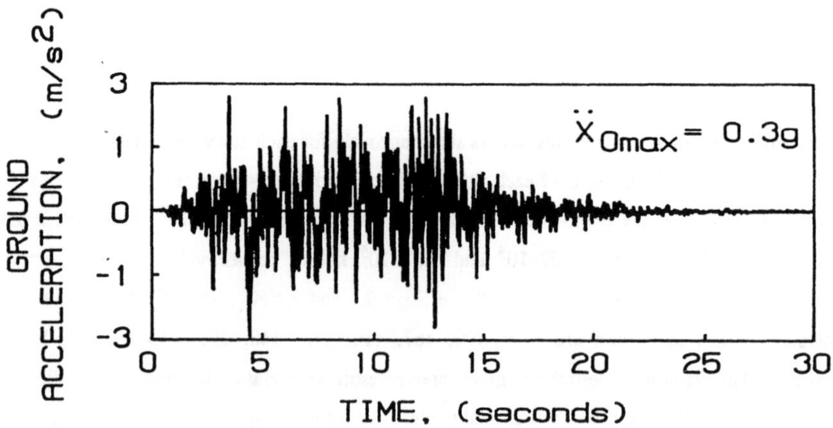

Fig.2 : A Simulated Earthquake Ground Acceleration

Table 1 : Maximum Response Quantities of Building : BIS = Base Isolation System
AF^1 = Actuator with $\underline{Z}$ feedback; AF^2 = Actuator with $\dot{\underline{Z}}$ feedback

FLOOR NO	Yield Displ	without control		with BIS		BIS & AF^1 U_m=808.6kN		BIS & AF^2 U_m=807kN	
	D_{yi} (cm)	x_{mi} (cm)	a_i (cm/s^2)	x_{mi} (cm)	a_i (cm/s^2)	x_{mi} (cm)	a_i (cm/s^2)	x_{mi} (cm)	a_i (cm/s^2)
B	-	-	-	42.70	158	26.11	117	25.77	109
1	2.4	3.05	486	1.31	155	0.86	118	0.82	111
2	2.3	2.80	456	1.25	152	0.82	123	0.78	114
3	2.2	3.14	571	1.28	157	0.83	122	0.79	116
4	2.1	2.95	529	1.17	164	0.74	116	0.70	113
5	2.0	2.70	558	1.07	177	0.66	125	0.62	116
6	1.9	2.91	720	0.97	185	0.58	117	0.55	113
7	1.7	2.90	580	0.82	198	0.48	131	0.46	124
8	1.5	2.08	617	0.52	206	0.33	129	0.30	120

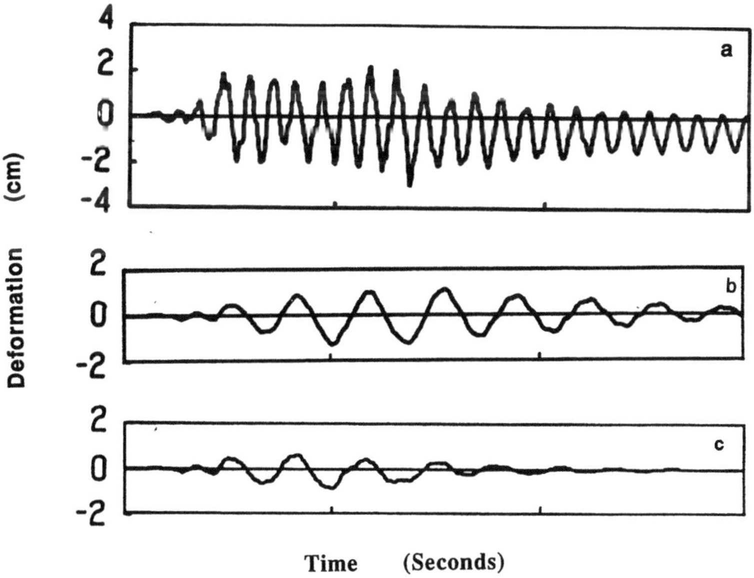

Fig. 3 : Deformation of First-Story Unit : (a) Without Control; (b) Base Isolation System alone (c) Base Isolation System and Actuator with $\dot{\underline{Z}}$ feedback

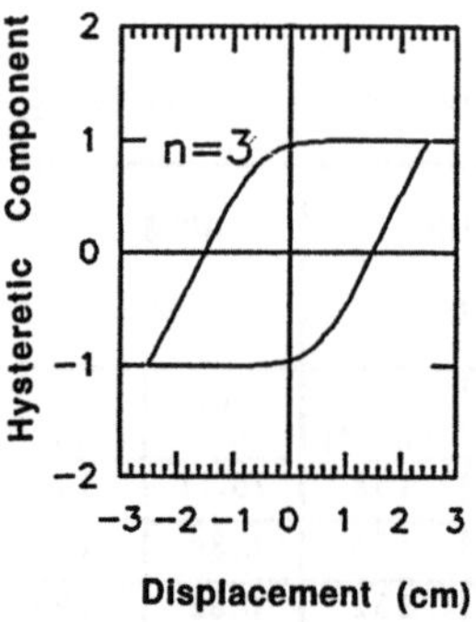

Fig. 4 : Hysteretic Characteristics of Base Isolation System

Time histories of all the response quantities were computed and presented in Ref. 7. The time history, $x_1(t)$, of the deformation of the first story unit is shown in Fig. 3(b) and that of the deformation of the base isolation system, $x_b(t)$, is presented in Fig. 5(a). The maximum response quantities of the building in 30 seconds of the earthquake episode are shown in the columns of Table 1 designated as "With BIS". As observed from Table 1 and Fig. 3(b), the interstory deformation and the floor acceleration are drastically reduced. However, the deformation of the base isolation system shown in row B of Table 1 may be excessive.

To protect the safety and integrity of the base isolation system, an actuator is connected to the base isolation system as shown in Fig. 1. With the actuator applying active control force $\underline{U}(t)$ to the base isolation system, the structural response depends on the weighting matrices $\underline{R}$ and $\underline{Q}^*$. For this example, the weighting matrix $\underline{R}$ consists of only one element, denoted by R_0, whereas the dimension of the $\underline{Q}^*$ matrix is (27x27). The $\underline{Q}^*$ matrix is partitioned as follows

$$\underline{Q}^* = \alpha \left[\begin{array}{ccc} \underline{0} & \underline{0} & \underline{0} \\ \hline \underline{Q}^*_{21} & \underline{Q}^*_{22} & \underline{Q}^*_{23} \end{array} \right] \tag{17}$$

in which $\underline{0}$ is a (18x9) matrix with all elements equal to zero, and $\underline{Q}^*_{21}$, $\underline{Q}^*_{22}$ and $\underline{Q}^*_{23}$ are (9x9) matrices.

Because of the fact that there is only one actuator installed on the basement floor as shown in Fig. 1, the control vector $\underline{U}(t)$ depends only on the difference between the first and the second rows of $\underline{Q}^*_{21}$, $\underline{Q}^*_{22}$ and $\underline{Q}^*_{23}$ matrices, i.e., $\underline{Q}^*_{21}(1,j) - \underline{Q}^*_{21}(2,j)$, $\underline{Q}^*_{22}(1,j) - \underline{Q}^*_{22}(2,j)$ and $\underline{Q}^*_{23}(1,j) - \underline{Q}^*_{23}(2,j)$. As a result, all elements of the second rows of $\underline{Q}^*_{21}$, $\underline{Q}^*_{22}$ and $\underline{Q}^*_{23}$ matrices are set to be zero for simplicity.

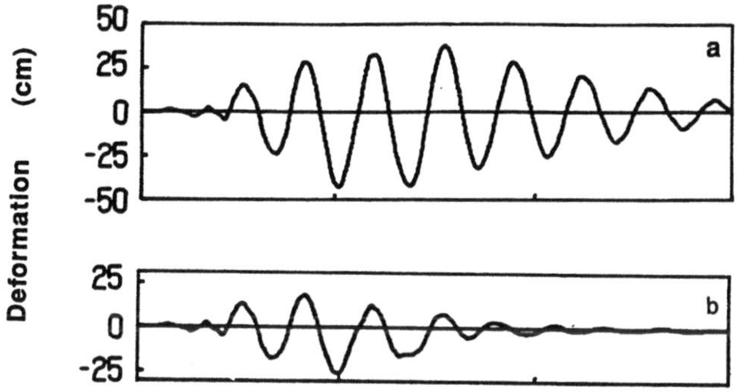

Fig.5 : Deformation of Base Isolation System : (a) Rubber Bearing Isolation System; (b) Base Isolation System and Actuator with $\dot{\underline{Z}}$ feedback

Further, all other elements which have no effect on the control vector U(t) are also set to be zero for convenience. It is mentioned that matrices $\underline{Q}^*_{21}$, $\underline{Q}^*_{22}$ and $\underline{Q}^*_{23}$ are associated with the feedbacks of $\dot{\underline{X}}(t)$, $\dot{\underline{V}}(t)$ and $\ddot{\underline{X}}(t)$, respectively. For instance, setting all the elements of $\underline{Q}^*_{21}$ equal to zero, the optimal control vector $\underline{U}(t)$ given by Eq. (13) is no longer a function of $\dot{\underline{X}}(t)$. If all the elements of $\underline{Q}^*_{22}$ are zero, there is no feedback of $\dot{\underline{V}}(t)$ and hence it is not necessary to estimate $\dot{\underline{V}}(t)$. In this paper, all elements of the $\underline{Q}^*_{22}$ matrix are set to be zero and hence there is no need to estimate $\dot{\underline{V}}(t)$ for the determination of the control vector $\underline{U}(t)$.

With a base isolation system, the building usually behaves like a rigid body and its deformation is relatively small compared to that of the base isolation system. As a result, the performance of the control system will not be affected adversely if sensors are not installed on the building as demonstrated in Refs. 5 and 7 thorugh extensive numerical simulation results. Since the installation of sensors is quite expensive, only a velocity sensor and an acceleration sensor will be installed on the base isolation system. Consequently, all the elements in $\underline{Q}^*_{21}$ and $\underline{Q}^*_{23}$ matrices are zero except $Q^*_{21}(1,1)$ and $Q^*_{23}(1,1)$. For illustrative purpose, the following values are assigned to these two elements: $Q^*_{21}(1,1)=35$ and $Q^*_{23}(1,1)=0.5$.

Time histories of all the response quantities have been computed using Eq. (14) for $\alpha=18{,}750$ and $R_0=1$. The time history, $x_1(t)$, of the deformation of the first story unit is presented in Fig. 3(c) and that of the deformation of the base isolation system, $x_b(t)$, is shown in Fig. 5(b). Within 30 seconds of the earthquake episode, the following quantities are presented in the columns of Table 1, designated as "BIS&AF^2": (i) the maximum interstory deformations, x_{mi}, (ii) the maximum floor accelerations a_i, and (iii) the maximum active control force U_m. Also presented in the columns designated as "BIS&AF^1" of Table 1 for comparison are the corresponding results using $\underline{Z}(t)$ feedback taken from Refs. 5 and 7.

It is observed from Table 1 that the performance of the optimal algorithm using velocity and acceleration feedbacks is slightly better than that using the displacement and velocity feedbacks. Extensive numerical results indicate that the response quantities of both the base isolation system and the building can further be reduced if the active control force is increased. Likewise, for the base-isolated building, it is not necessary to install sensors on the building.

Conclusions

A version of the instantaneous optimal control algorithm for nonlinear or hysteretic structural systems is presented. The optimal algorithm utilizes the acceleration and velocity responses of the building as feedbacks, thus eliminating the possible difficulty involved in the measurement of the displacement response of the building subjected to strong earthquakes. The optimal algorithm has been applied to an aseismic hybrid control system consisting of a rubber-bearing isolation system and an actuator.

The performance of the present control algorithm has been investigated, evaluated and compared numerically with that of the previous algorithm [Refs. 5 and 7]. It is shown numerically that the performance of the present optimal algorithm is slightly better than that of the previous algorithms for the type of aseismic hybrid control system investigated herein. It is further demonstrated that (i) the aseismic hybrid control system investigated is very effective in reducing the building response under strong earthquakes, and (ii) for aseismic hybrid control systems consisting of a base isolation system, it is not necessary to install sensors on the building. Finally, the assignment of the weighting matrix $\underline{Q}^*$ is based on the stability criterion. Due to space limitation, a systematic approach for the assignment of the weighting matrix $\underline{Q}^*$ will be presented elsewhere [15].

Acknowledgements

This research is partially supported by the National Science Foundation Grant No. NSF-BCS-91-96089 and National Center for Earthquake Engineering Research Grant No. NCEER-90-2202.

References

1. Kelly, J.M., Leitmann, G. and Soldators, A.G. (1987), "Robust Control of Base-Isolated Structures Under Earthquake Excitation", Journal of Optimization Theory and Applications, Vol. 53, No. 2, pp. 159-180.

2. Reinhorn, A.M., Soong, T.T. and Yen, C.Y. (1987), "Base Isolated Structures with Active Control", Seismic Engineering, Recent Advances in Design, Analysis, Testing and Qualification Methods, ASME, PVP-Vol. 127, pp. 413-419.

3. Yang, J.N, Danielians, A. and Liu, S.C. (1990) "Aseismic Hybrid Control Systems For Building Structures Under Strong Earthquakes", Intelligent Structures, edited by K.P. Chong, S.C. Liu and J.C. Li, Elsevier Applied Science, pp. 179-195.

4. Yang, J.N., Danielians, A., and Liu, S.C. (1991), "Aseismic Hybrid Control Systems for Building Structures", Journal of Engeneering Mechanics, ASCE, Vol. 117, No. 4, April, pp. 836-853.

5. Yang, J.N., Li, Z., Danielians, A., and Liu, S.C., "Instantaneous Optimal Control of Nonlinear Hybrid Structural Systems," paper to appear in Proc. of Eighth VPI&SU Symposium on Dynamics and Control of Large Structures, May 6-8, 1991, Blacksburg, VA.

6. Yang, J.N., Li, Z., Danielians, A., and Liu, S.C. (1991), "Optimal Hybrid Control of Seismic-Excited Nonlinear and Inelastic Structures", paper to appear in Proc. US-Japan-Italy Workshop/Seminar on Intelligent Systems, June 27-29, Peruga, Italy.

7. Yang, J.N., Li, Z. and Danielians, A. (1991), "Hybrid Control of Seismic-Excited Nonlinear and Inelastic Structural Systems", Technical Report, NCEER-91-0020, SUNY, Buffalo.

8. Yang, J.N., Akbarpour, A., and Ghaemmaghami, P. (1987) "New Optimal Control Algorithms for Structural Control", Journal of Engineering Mechanics, ASCE, Vol. 113, No. 9, pp. 1369-1386.

9. Yang, J.N., Long, F.X., and Wong, D.(1988) "Optimal Control of Nonlinear Structures", Journal of Applied Mechanics, ASME, Vol. 55, December, pp. 931-938.

10. Yang, J.N., Li, Z., and Liu, S.C. (1991) "Instantaneous Optimal Control With Acceleration and Velocity Feedback", Journal of Probabilistic Engineering Mechanics, Vol. 23, No. 5, September.

11. Baber, T.T., and Wen, Y.K.(1981),"Random Vibration of Hysteretic Degrading Systems", Journal of Engineering Mechanics, ASCE, Vol. 107, No. EM6, pp. 1069-1987.

12. Li, Z., Katukura, H., and Izumi, M. (1991), "Synthesis and Extension of One-Dimensional Nonlinear Hysteretic Models", Journal of Engineering Mechanics, ASCE, Vol 117, No. 1, pp. 100-109.

13. Park, Y.J., Wen, Y.K., and Ang, A.H-S. (1986), "Two Dimensional Random Vibration of Hysteretic Structures", Journal of Earthquake Engineering and Structural Dynamics, Vol. 14, pp. 543-557.

14. Wen, Y.K. (1989), "Methods of Random Vibration for Inelastic Structures", Journal of Applied Mechanical Reviews, ASME, Vol. 42, No. 2, pp. 39-52.

15. Yang, J.N., Li, Z. (1991), "Active and Hybrid Control of Civil Engineering Structures", paper to be presented at the US-Korea-Japan Trilateral Seminar on Frontier R&D for Constructed Facilities to be held on Oct. 21-24 in Hawaii.

Stochastic Excited Hamiltonian Systems

W. Q. ZHU, G. Q. CAI and Y. K. LIN

Center for Applied Stochastics Research
Florida Atlantic University
Boca Raton, FL 33431, USA

Summary

The method of generalized stationary potential is applied to obtain exact probability densities for multi-degree-of-freedom nonlinear Hamiltonian systems excited by Gaussian white noises. The excitations can be additive, or multiplicative, or both, and governing equations are more general than those previously reported in the literature. Further extension is made to a still more general class, which include the Hamiltonian systems as special cases.

Introduction

Numerous publications on randomly excited nonlinear dynamic systems have appeared in the literature. Some mathematically exact solutions have been obtained when the excitations can be idealized as Gaussian white noises, so that the system response, represented as a vector in a state space, is a Markov vector. In this case, the probability density of the system response, conditional on a given initial state, is governed by a parabolic partial differential equation, called the Fokker-Planck equation, which can be derived from the equation of motion. Still, full exact solutions for the Fokker-Planck equations are known only for very special first order nonlinear systems.

For second and higher order nonlinear systems, solutions have been obtained only for limited cases and only for the reduced Fokker-Planck equations, namely, Fokker-Planck equations without the time-derivative term. The unknown in a reduced Fokker-Planck equation is the unconditional probability density of the response vector when it reaches statistical stationarity after the system is exposed to stationary excitations for a sufficiently long time. When the excitations are additive, appearing only as inhomogeneous terms on the right hand sides of the equations of motion, exact solutions have been found for a number of nonlinear systems, which can be grouped in a general class as shown by Caughey and Ma [3].

When multiplicative excitations are also present, solution to a reduced Fokker-Planck equation becomes more difficult. Dimentberg [4] found the exact solution for a specific system under

N. Bellomo · F. Casciati (Eds.)
Nonlinear Stochastic Mechanics
IUTAM Symposium Turin

both additive and multiplicative excitations. A more general case was considered by Yong and Lin [11] where a systematic procedure was given to obtain the exact solutions under an assumption called detailed balance. Further extension was made by Lin and Cai [6] and Cai and Lin [1] to systems belonging to the class of generalized stationary potential, which may or may not be in detailed balance. Additional exact solutions were obtained by Zhu [12], Soize [9], and Zhu, Cai and Lin [13]. In the last two references, stochastic systems of many degrees of freedom were formulated in the Hamiltonian form, and conditions were obtained for systems having exact solutions.

In this paper, attention is again focused on nonlinear multi-degree-of-freedom Hamiltonian systems subject to both additive and multiplicative Gaussian white noise excitations. These Hamiltonian systems are more general than those considered by soize [9], and are solvable using the procedure developed by Cai and Lin [1]. The procedure is further extended to still more general systems which include the Hamiltonian systems [9,13] as special cases. It is shown that the present results are reducible to earlier exact solutions given by Caughey and Ma [2,3], Soize [9], Zhu [12] and Zhu,Cai and Lin [13].

Generalized Stationary Potential

Consider a general dynamic system governed by the following equations of motion

$$\frac{d}{dt}X_i = \psi_i(X) + \theta_{ij}(X)\ W_j(t) \tag{1}$$

where X_i are components of the system response vector X, function ψ_i and θ_{ij} are generally nonlinear, and $W_j(t)$ are Gaussian white noises in the sense of Stratonovich, with

$$E[\,W_i(t)\ W_j(t+\tau)\,] = 2\pi K_{ij}\,\delta(\tau) \tag{2}$$

It is known that the system response vector X is a Markov vector and its stationary probability density is governed by the reduced Fokker-Planck equation

$$\frac{\partial}{\partial x_i}G_i = 0 \tag{3}$$

$$G_i = A_i(x)\,p_s(x) - \frac{1}{2}\frac{\partial}{\partial x_j}[B_{ij}(x)\,p_s(x)] \tag{4}$$

where $p_s(x)$ is the stationary probability density of X, G_i is the probability flow in the ith direction, and A_i and B_{ij} are the drift and diffusion coefficients obtainable using the well-known rules of Wong and Zakai [10]:

$$A_i(x) = \psi_i(x) + \pi K_{ls}\theta_{rs}\frac{\partial}{\partial x_r}\theta_{il}(x) \tag{5-1}$$

$$B_{ij}(x) = 2\pi K_{ls}\theta_{il}\theta_{js} \tag{5-2}$$

In Eqs.(3) and (4), a lower case symbol represents the possible value of the corresponding upper case random quantity. This convention will be followed throughout the paper. Equation (3) is solved subject to the following boundary conditions:

$$G_i = 0 \qquad \textit{at the boundaries} \tag{6}$$

We split both the drift and diffusion coefficients as follows [1]:

$$A_i = A_i^{(1)} + A_i^{(2)} \tag{7-1}$$

$$B_{ij} = B_{ij}^{(i)} + B_{ji}^{(j)} \tag{7-2}$$

Eq.(7-2) retains the symmetry condition for diffusion coefficients, $B_{ij}=B_{ji}$. Substituting Eqs.(7-1) and (7-2) into Eqs.(3) and (4), one obtains

$$\frac{\partial}{\partial x_i}[A_i^{(1)}p_s - \frac{\partial}{\partial x_j}B_{ij}^{(i)}p_s] + \frac{\partial}{\partial x_i}[A_i^{(2)}p_s] = 0 \tag{8}$$

Let $A_i^{(1)}$ be associated with the vanishing probability flow, and $A_i^{(2)}$ associated with the circulatory probability flow, namely,

$$\frac{\partial}{\partial x_i}[A_i^{(2)}(x)p_s(x)] = 0 \tag{9-1}$$

$$A_i^{(1)}(x)p_s(x) = \frac{\partial}{\partial x_j}[B_{ij}^{(i)}(x)p_s(x)] \tag{9-2}$$

where the superscript (i) does not imply a summation. Eqs.(9-1) and (9-2) are sufficient conditions for Eq.(8). The system is said to belong to the class of generalized stationary potential, if a consistent function $p_s(x)$ can be found to satisfy Eqs.(9-1) and (9-2). Much success has been achieved in finding exact solutions for second order nonlinear stochastic systems based on this solution technique [1,6]. Thus, the present paper is focused on higher order systems; in particular, higher order Hamiltonian systems.

Stochastically Excited Hamiltonian Systems

Consider an n-degree-of-freedom stochastically excited Hamiltonian system governed by the following equations of motion:

$$\dot{Q}_j = \frac{\partial H}{\partial P_j} ; \quad j = 1, 2, ..., n \tag{10-1}$$

$$\dot{P}_j = -\frac{\partial H}{\partial Q_j} - h(H)\, c_{jk}(Q,P) \frac{\partial H}{\partial P_k} + g(H) f_{jl}(Q,P) W_l(t) \tag{10-2}$$

$$j, k = 1, 2, ..., n ; \quad l = 1, 2, ..., m$$

where Q_j and P_j are generalized displacements and generalized momenta, respectively, $H=H(Q,P)$ is a Hamiltonian with continuous first order derivatives, $h(H)$ and $g(H)$ are twice-differentiable functions, $c_{jk}(Q,P)$ are differentiable functions, $f_{jl}(Q,P)$ are twice-differentiable functions, and $W_l(t)$ are Gaussian white noises in the sense of Stratonovich, with correlation functions given by Eq.(2).

The system governed by Eqs.(10-1) and (10-2) is generally nonlinear and includes both additive and multiplicative excitations. It is a generalization of the one considered by Soize [9] by introducing the variable coefficients $c_{jk}(Q,P)$ and $f_{jl}(Q,P)$ and allowing the random excitations to be correlated.

Letting $Q_j=X_j$ and $P_j=X_{j+n}$, Eqs.(10-1) and (10-2) are set in the same form as Eq.(1). Then the drift and diffusion coefficients for the Hamiltonian system can be obtained from Eqs.(5-1) and (5-2) as follows:

$$A_j = \frac{\partial H}{\partial p_j}, \quad A_{j+n} = -\frac{\partial H}{\partial q_j} - h\, c_{jk} \frac{\partial H}{\partial p_k} + \pi g K_{ls} f_{ks} \frac{\partial}{\partial p_k}(g f_{jl})$$

$$B_{j,k} = B_{j,k+n} = B_{j+n,k} = 0 , \quad B_{j+n,k+n} = g^2 b_{jk}$$

$$j, k = 1, 2, ..., n ; \; l, s = 1, 2, ..., m$$

where $b_{jk} = 2\pi K_{ls} f_{jl} f_{ks}$. By splitting these coefficients into

$$A_j^{(1)} = 0 , \quad A_j^{(2)} = \frac{\partial H}{\partial p_j}$$

$$A_{j+n}^{(1)} = -h\,c_{jk}\frac{\partial H}{\partial p_k} + \pi g K_{ls} f_{ks}\frac{\partial}{\partial p_k}(g f_{jl})\ , \qquad A_{j+n}^{(2)} = -\frac{\partial H}{\partial q_j}$$

$$B_{j,k}^{(j)} = B_{j,k+n}^{(j)} = B_{j+n,k}^{(j+n)} = 0\ , \qquad B_{j+n,k+n}^{(j+n)} = g^2 b_{jk}^{(j)}$$

$$j,k = 1,2,\ldots\ldots,n;\ \ l,s = 1,2,\ldots\ldots,m$$

where $b_{jk}^{(j)} + b_{kj}^{(k)} = b_{jk}$, and substituting them into Eqs.(9-1) and (9-2), one obtains

$$-\frac{\partial}{\partial q_j}(\frac{\partial H}{\partial p_j}p_s) + \frac{\partial}{\partial p_j}(\frac{\partial H}{\partial q_j}p_s) = 0 \tag{11-1}$$

$$[h\,c_{jk}\frac{\partial H}{\partial p_k} - \pi g K_{ls} f_{ks}\frac{\partial}{\partial p_k}(g f_{jl})]p_s + \frac{\partial}{\partial p_k}[g^2 b_{jk}^{(j)} p_s] = 0 \tag{11-2}$$

The general solution for Eq.(11-1) may be written as

$$p_s(q,p) = C\,exp[-\phi(H)] \tag{12}$$

where C is a normalization constant, and ϕ is an arbitrary function of H. Substituting Eq.(12) into Eq.(11-2) leads to

$$g^2 b_{jk}^{(j)}\frac{d\phi}{dH}\frac{\partial H}{\partial p_k} = h\,c_{jk}\frac{\partial H}{\partial p_k} - \pi g K_{ls} f_{ks}\frac{\partial}{\partial p_k}(g f_{jl}) + \frac{\partial}{\partial p_k}[g^2 b_{jk}^{(j)}] \tag{13}$$

These are n equations for ϕ. To find a consistent function ϕ satisfying all these equations, certain restrictions must be imposed on the system parameters, as discussed below.

I. Consider a simple case where g is a constant, c_{jk} and f_{jl} are functions of $\boldsymbol{q}$ only. In this case, Eq.(13) is reduced to

$$g^2 b_{jk}^{(j)}\frac{d\phi}{dH}\frac{\partial H}{\partial p_k} = h\,c_{jk}\frac{\partial H}{\partial p_k} \tag{14}$$

which has a solution

$$\frac{d\phi}{dH} = \frac{\eta\,h(H)}{g^2} \tag{15}$$

under the condition $c_{jk} = \eta\,b_{jk}^{(j)}$, or equivalently, $c_{jk} + c_{kj} = \eta\,b_{jk}$, where η is an arbitrary constant. The stationary probability density is then

$$p_s(q,p) = C\,exp[-\frac{\eta}{g^2}\int_0^H h(u)\,du] \tag{16}$$

II. Let g be a function of H, whereas c_{jk} and f_{jl} remain functions of $\boldsymbol{q}$. Then Eq.(13) takes the form of

$$\frac{d\phi}{dH} = \frac{c_{jk}\dfrac{\partial H}{\partial p_k}}{b_{jk}^{(j)}\dfrac{\partial H}{\partial p_k}}\frac{h}{g^2} - \frac{b_{jk}\dfrac{\partial H}{\partial p_k}}{b_{jk}^{(j)}\dfrac{\partial H}{\partial p_k}}\frac{1}{2g}\frac{dg}{dH} + \frac{2}{g}\frac{dg}{dH} \tag{17}$$

In order for the right hand side to be just a function of H, we must have $c_{jk}=c_{kj}=\eta b_{jk}/2$. Under this restriction, we obtain

$$p_s(q,p) = C\,g^{-1}(H)\,exp[-\eta\int_0^H \frac{h(u)}{g^2(u)}du] \tag{18}$$

III. If $W_l(t)$ in Eq.(10-2) are Gaussian white noises in the sense of Ito, then the Wong-Zakai correction terms $\pi g K_{ls} f_{ks}\dfrac{\partial}{\partial p_k}(g f_{jl})$ are no longer required in the above analysis. Furthermore, if g is a function of H, and if c_{jk} and f_{jl} are functions of $\boldsymbol{q}$ only, then a similar derivation leads to the following stationary probability density

$$p_s(q,p) = C\,g^{-2}(H)\,exp[-\eta\int_0^H \frac{h(u)}{g^2(u)}du] \tag{19}$$

under the condition $c_{jk}+c_{kj}=\eta b_{jk}$. The result obtained by Soize [9] is a special case of Eq.19 with constant coefficients c_{jk} and f_{jl} and $\eta=2$.

It is of interest to note that an identical stationary probability density is obtained for all the systems with an identical pair of h and g, but with different c_{jk} and f_{jl} satisfying certain conditions given above. Therefore, they belong to the same family of stochastically equivalent systems [7].

Extension

The above procedure can be further extended to a more general class of systems, which includes the Hamiltonian systems as special cases. Consider a system governed by

$$\dot{Q}_j = D(Q)\frac{\partial H}{\partial P_j}\,; \qquad j = 1, 2, \ldots\ldots, n \tag{20-1}$$

$$\dot{P}_j = -D(Q)\frac{\partial H}{\partial Q_j} - h(H)\,c_{jk}(Q,P)\frac{\partial H}{\partial P_k} + g(H)\,f_{jl}(Q,P)\,W_l(t) \tag{20-2}$$

$$j,k = 1,2,......,n; \quad l = 1,2,......,m$$

where $D(Q)$ is an arbitrary function. The Hamiltonian system described by Eqs.(10-1) and (10-2) is a special case with $D(Q)$=1. We obtain the following sufficient conditions for the above system to possess an exact stationary probability density $p_s^*(q,p)$:

$$-\frac{\partial}{\partial q_j}(D\frac{\partial H}{\partial p_j}p_s^*) + \frac{\partial}{\partial p_j}(D\frac{\partial H}{\partial q_j}p_s^*) = 0 \tag{21-1}$$

$$[h\,c_{jk}\frac{\partial H}{\partial p_k} - \pi g K_{ls} f_{ks}\frac{\partial}{\partial p_k}(g f_{jl})]p_s^* + \frac{\partial}{\partial p_k}[g^2 b_{jk}^{(j)} p_s^*] = 0 \tag{21-2}$$

By comparing Eqs.(21-1) and (21-2) with Eqs.(11-1) and (11-2), it can be seen that

$$p_s^*(q,p) = p_s(q,p)/D(q) \tag{22}$$

Thus, for the generalized Hamiltonian system (20-1) and (20-2),

$$p_s^*(q,p) = C\frac{1}{D(q)}exp[-\frac{\eta}{g^2}\int_0^H h(u)du] \tag{23}$$

if g is a constant, and c_{jk} and f_{jl} depend only on q,

$$p_s^*(q,p) = C\frac{1}{D(q)}g^{-1}(H)\,exp[-\eta\int_0^H \frac{h(u)}{g^2(u)}du] \tag{24}$$

if g is a function of H, and c_{jk} and f_{jl} depend only on q, and

$$p_s^*(q,p) = C\frac{1}{D(q)}g^{-2}(H)\,exp[-\eta\int_0^H \frac{h(u)}{g^2(u)}du] \tag{25}$$

if g is a function of H, c_{jk} and f_{jl} depend only on q, and W_l(t) in (20-2) are Gaussian white noises in the sense of Ito.

Examples

I. Consider a multi-degree-of-freedom oscillatory system

$$\ddot{X}_j + h(H)c_{jk}(X,\dot{X})\dot{X}_k + \frac{\partial U(X)}{\partial X_j} = g(H)f_{jl}(X,\dot{X})\,W_l(t) \tag{26}$$

with a Hamiltonian $H = \frac{1}{2}\dot{X}_j\dot{X}_j + U(X)$. Let $Q_j = X_j$, $P_j = \dot{X}_j$. The results (16), (18) and (19) for the three different cases are applicable. The system investigated by Scheurkogel and Elishakoff [8] is a special case of Eq.(26) with two degrees of freedom and with constant h, g, c_{jk} and f_{jl}.

II. Let $Q_j = X_j$, $P_j = X_{n+j}$, and

$$H = \sum_j \int_0^{X_{n+j}} r_j(u)\,du + U(X_1,X_2,......) = \sum_j \int_0^{P_j} r_j(u)\,du + U(Q) \tag{27}$$

Then the Hamiltonian system (10-1) and (10-2) is reduced to the following

$$\dot{X}_j = r_j(X_{n+j}) \tag{28-1}$$

$$\dot{X}_{n+j} = -h(H)c_{jk}r_k(X_{n+k}) - \frac{\partial}{\partial X_j}U(X_1,X_2,......) + g(H)f_{jl}W_l(t) \tag{28-2}$$

These equations represent a generalization of the class of problems considered by Caughey and Ma [2]. The generalization allows for certain types of parametric excitations. The special case without damping coupling has been considered previously by Zhu [12].

III. Consider a dynamic system with varying inertia coupling

$$\frac{d}{dt}[m_{jk}(X)\dot{X}_k] + h(H)c_{jk}(X,\dot{X})\dot{X}_k + \frac{\partial H}{\partial X_j} = g(H)f_{jl}(X,\dot{X})\,W_l(t) \tag{29}$$

where m_{jk} are functions of X. Equations of the form of (29) occur in certain aeroelastic systems [5]. In this case, the generalized displacements and momenta are $Q_j = X_j$ and $P_j = m_{jk}\dot{X}_k$, respectively, and the Hamiltonian is

$$H = \frac{1}{2}(m^{-1})_{jk}m_{jl}m_{ks}\dot{X}_l\dot{X}_s + U(X) \tag{30}$$

Then the exact stationary probability density of x and $\dot{x}$ is given by

$$p_s(x,\dot{x}) = |J|\,p_s(q,p) \tag{31}$$

where J is a Jacobian defined by

$$J = \begin{vmatrix} \frac{\partial q_1}{\partial x_1} & \frac{\partial q_1}{\partial \dot{x}_1} & \frac{\partial q_1}{\partial x_2} & \frac{\partial q_1}{\partial \dot{x}_2} & \cdots\cdots \\ \frac{\partial p_1}{\partial x_1} & \frac{\partial p_1}{\partial \dot{x}_1} & \frac{\partial p_1}{\partial x_2} & \frac{\partial p_1}{\partial \dot{x}_2} & \cdots\cdots \\ \frac{\partial q_2}{\partial x_1} & \frac{\partial q_2}{\partial \dot{x}_1} & \frac{\partial q_2}{\partial x_2} & \frac{\partial q_2}{\partial \dot{x}_2} & \cdots\cdots \\ \frac{\partial p_2}{\partial x_1} & \frac{\partial p_2}{\partial \dot{x}_1} & \frac{\partial p_2}{\partial x_2} & \frac{\partial p_2}{\partial \dot{x}_2} & \cdots\cdots \\ \cdots & \cdots & \cdots & \cdots & \cdots\cdots \end{vmatrix}, \tag{32}$$

and $p_s(q,p)$ is selected from Eqs.(16), (18) or (19) for three different cases. It can be shown that J is equal to the determinant of matrix $\boldsymbol{m}$. The special case of the system (29) with constant c_{jk} and f_{jl} was investigated by Soize [9].

IV. The system

$$\ddot{X}_j + \frac{1}{D(X)} h(H) c_{jk}(X,\dot{X})\dot{X}_k + D(X)\frac{\partial H}{\partial X_j} = g(H) f_{jl}(X,\dot{X})\, W_l(t) \tag{33}$$

where $H = \frac{1}{2D(X)}\dot{X}_j\dot{X}_j + U(X)$ is equivalent to the generalized Hamiltonian system (20-1) and (20-2) with $Q_j = X_j$, $P_j = \dot{X}_j$. Thus, its stationary probability density is given by Eq.(23), (24) or (25) corresponding to each of the three cases. The results obtained previously by Caughey and Ma [3], and Zhu [12] were special cases.

Concluding Remarks

The key to obtain an exact probability density for a stochastically excited Hamiltonian system is the ability to identify a circulatory probability flow which depends only on the Hamiltonian itself. This implies that the solution may be expressed as a function of Hamiltonian alone, which is valid only under certain conditions. Further extension to a more general class, however, permits that the probability density depends also on the generalized displacements.

Acknowledgement

This paper is prepared under grant AFOSR-91-0073 from the Air Force Office of Scientific Research, Air Force Systems Command, USAF, monitored by Dr. Spencer Wu. The U.S. Government is authorized to reproduce and distribute reprints for governmental purposes, notwithstanding any copyright notation thereon.

References

1. Cai, G. Q. and Lin Y. K., "On Exact Stationary Solutions of Equivalent Non-Linear Stochastic Systems," *International Journal of Non-Linear Mechanics,* Vol. 23, No. 4, 1988, pp. 315-325.

2. Caughey, T. K. and Ma, F., "The Steady-State Response of a Class of Dynamic Systems to Stochastic Excitations," *Journal of Applied Mechanics,* Vol. 49, 1982, pp. 629-632.

3. Caughey, T. K. and Ma, F., "The Exact Steady-State Solution of a Class of Nonlinear Stochastic Systems," *International Journal of Non-Linear Mechanics,* Vol. 17, No. 3, 1983, pp. 137-142.

4. Dimentberg, M. F., "An Exact Solution to a Certain Nonlinear Random Vibration Problem," *International Journal of Non-Linear Mechanics,* Vol. 17, 1982, pp. 231-236.

5. Ibrahim, R. A., *Parametric Random Vibration,* Research Studies Press, Letchworth, Hertfordshire, England, 1986.

6. Lin, Y. K. and Cai, G. Q., "Exact Stationary-Response Solution for Second-Order Nonlinear Systems Under Parametric and External White-Noise Excitations: Part II," *Journal of Applied Mechanics, ASME,* Vol. 55, 1988, pp. 702-705.

7. Lin, Y. K. and Cai, G. Q., "Equivalent Stochstic Systems," *Journal of Applied Mechanics, ASME,* Vol. 55, No. 4, 1988, pp. 918-922.

8. Scheurkogel, A. and Elishakoff, I., "Non-Linear Random Vibration of a two-degree-0f-Freedom System," *Non-Linear Stochastic Dynamic Engineering Systems,* Edited by Ziegler, F. and Scheller, G. I., Spring-Verlag, Berlin, 1988, pp. 285-299.

9. Soize, C., "Steady-State Solution of Fokker-Planck Equation in High Dimension," *Probabilistic Engineering Mechanics,* Vol. 3, No. 4, 1988, pp.196-206.

10. Wong, E. and Zakai, M., "On the Relation between Ordinary and Stochastic Equations," *International Journal of Engineering Science,* Vol. 3, 1965, pp. 213-229.

11. Yong, Y., and Lin, Y. K. "Exact Stationary-Response Solution for Second-Order Nonlinear Systems Under Parametric and External White-Noise Excitations," *Journal of Applied Mechanics, ASME,* Vol. 54, 1987, pp. 414-418.

12. Zhu, W. Q., "The Exact Stationary Response Solution of Several Classes of Nonlinear Systems to White Noise Parametric and/or External Excitations," *Applied Mathematics and Mechanics,* Vol. 11, No. 2, 1990, pp. 165-175.

13. Zhu, W. Q., Cai, G. Q. and Lin, Y. K., "On Exact Stationary Solutions of Stochastically Perturbed Hamiltonian Systems;" *Probabilistic Engineering Mechanics,* Vol. 5, No. 2, 1990, pp. 84-87.

MIX
Papier aus verantwortungsvollen Quellen
Paper from responsible sources
FSC® C105338

If you have any concerns about our products,
you can contact us on
ProductSafety@springernature.com

In case Publisher is established outside the EU,
the EU authorized representative is:
Springer Nature Customer Service Center GmbH
Europaplatz 3, 69115 Heidelberg, Germany

Printed by Libri Plureos GmbH
in Hamburg, Germany